Two-Dimensional Nanostructures for Energy-Related Applications

Two-Dimensional Nanostructures for Energy-Related Applications

Editor

Kuan Yew Cheong

School of Materials & Mineral Resources Engineering
Engineering Campus, Universiti Sains Malaysia
Pulau Pinang
Malaysia

CRC Press
Taylor & Francis Group
Boca Raton London New York

CRC Press is an imprint of the
Taylor & Francis Group, an **informa** business
A SCIENCE PUBLISHERS BOOK

Cover illustration reproduced by kind courtesy of Dr. M. A. Fraga, Dr. G. Leal, Dr. M. Massi and Dr. V. J. Trava-Airoldi (authors of Chapter 7)

CRC Press
Taylor & Francis Group
6000 Broken Sound Parkway NW, Suite 300
Boca Raton, FL 33487-2742

Printed on acid-free paper
Version Date: 20160923

International Standard Book Number-13: 978-1-4987-3293-2 (Hardback)

Library of Congress Cataloging-in-Publication Data

Names: Cheong, Kuan Yew, editor.
Title: Two-dimensional nanostructures for energy-related applications / editor, Kuan Yew Cheong, School of Materials & Mineral Resources Engineering, Engineering Campus, Universiti Sains Malaysia, Pulau Pinang, Malaysia.
Description: Boca Raton, FL : CRC Press, [2016] | "A science publishers book." | Includes bibliographical references.
Identifiers: LCCN 2016028127| ISBN 9781498732932 (hardback) | ISBN 9781498732949 (e-book)
Subjects: LCSH: Nanostructured materials. | Energy conversion.
Classification: LCC TA418.9.N35 T88 2016 | DDC 621.31/24--dc23
LC record available at https://lccn.loc.gov/2016028127

Visit the Taylor & Francis Web site at
http://www.taylorandfrancis.com

and the CRC Press Web site at
http://www.crcpress.com

Preface

Energy crisis is one of the major challenges of this century. In order to resolve this issue, numerous strategies have been proposed and implemented. Through the understanding of fundamental nanoscience and application of nanomaterials, innovative nanotechnology and engineering solutions has been developed to address this crisis. The main focus of this edited book is on the latest advances of two-dimensional nanostructures for energy and its related applications. This is part of the collective effort to gather and compile knowledge and know-how of synthesis, characterization and utilization of nanomaterial with one of the dimensions below 100-nm thick (two-dimensional nanomaterials) for applications in energy saving and conversion type of devices; namely light emitting diode (LED), photovoltaic cell, photoelectrochemical cell, fuel cell, thermoelectric and photo-thermoelectrochemical cells, etc.

A total of thirteen (13) chapters have been authored by well-known experts in their respective field with the presentation channeling towards applied physics point of view. This edited book serves as a useful source of reference for postgraduates, researchers and engineers who intend to work in this area. Chapter 1 is devoted to layered materials, such as graphene and MoS_2, as active materials of supercapacitors and as oxygen reduction reaction (ORR) catalysts of fuel cell. Enhanced characteristics of the supercapacitors and activities of the electrocatalytic ORR with the incorporation and utilization of nanostructured materials have been systematically discussed. The subsequent two chapters focus on layered materials such as Bi_2Te_3, metal sulfide and indium selenide, as well as thin film of organic materials namely tetracyanoquinodimethane complexes for thermoelectric and photo-thermoelectrochemical applications. Potential application of graphene as electrodes for LEDs, as electron transport layers for dye-sensitized solar cells and as field emitters for field emission devices has been elaborated in Chapter 4 and III-nitrides epitaxial layer on Si substrate for low-cost and high-efficient LED has been presented in Chapter 5. The next three chapters are dedicated to materials for photovoltaic application, which covers development of chemical vapor deposited nanodiamond thin film, metal-containing diamond-like carbon and ZnO thin film as part of the functional materials for this application.

In Chapters 9 and 10, nanoporous α-Fe_2O_3 thin film and TiO_2 thin film, respectively, used as photoelectrodes in photochemical and photovoltaic cells have been presented. Chapter 11 focuses on low-temperature synthesis techniques to produce two-dimensional ZnO nano-architectures with multi-functional properties.

A review on the technique and strategy to produce high quality and reliable native oxide thin film on SiC substrate has been given in Chapter 12 and last but not least, atomic structures, energetic and electronic structures of various Ge thin films on Si(001) substrates based on first-principles total-energy calculations have been critically reviewed and presented in Chapter 13.

In summary, this valuable edited book summarizes the latest advances and development in the area of two-dimensional nanostructures for energy related applications. The success of this book is attributed to the hard work of all contributed authors of each chapter who were generous enough to share their knowledge and willing to dedicate their time to put their thoughts in a well written format. My heartful thanks to the contributions given by my postgraduate students, in particular Z. X. Lim, for their help in the initial stage of this project. Finally, my deepest gratitude goes to my lovely wife, two wonderful kids and parents for their patience, understanding and support given to me.

Kuan Yew Cheong
Universiti Sains Malaysia, Malaysia
17 December 2015

Contents

1

Supercapacitors and Oxygen Reduction Reaction Catalysts Based on Graphene, MoS_2 and Related Layered Materials

K. Gopalakrishnan[1, 2, a] and C. N. R. Rao[1, 2, *]

ABSTRACT

Graphene, MoS_2 and similar layered materials are emerging with a variety of potential applications in energy devices. In particular, they perform well as catalysts in supercapacitors as well as in the Oxygen Reduction Reaction (ORR) of fuel cells. This chapter presents a brief survey of supercapacitors and ORR catalysts based on graphene, MoS_2 and related materials.

Keywords: Graphene, MoS_2, supercapacitors, oxygen reduction reaction.

1. INTRODUCTION

Two-dimensional (2D) materials such as graphene and molybdenum disulfide (MoS_2) are materials with many potential applications. Their use in supercapacitors and in the oxygen reduction reaction (ORR) arise due to high surface area, unique crystal structure as well as electrical and related properties (Wang et al. 2014a, Rao et al. 2014a, Hu et al. 2015, Raccichini et al. 2015, Rao et al. 2015). Thus, graphene has been investigated extensively for electrochemical energy storage applications (Rao et al. 2015). Recent studies have shown that MoS_2 can also be utilized for electrochemical applications in supercapacitors and ORR (Rao et al. 2015). Doped graphenes are found to show enhanced capacitance relative to that of pristine graphene (Rao et al. 2014a). MoS_2, is a graphene analogue composed

[1] Chemistry and Physics of Materials Unit, New Chemistry Unit, International Centre for Materials Science, CSIR Centre of Excellence in Chemistry and Sheik Saqr Laboratory.

[2] Jawaharlal Nehru Centre for Advanced Scientific Research, Jakkur, P.O., Bangalore-560064, India.

[a] E-mail: gopalakrishnan@jncasr.ac.in

* Corresponding author: cnrrao@jncasr.ac.in

of a layered structure like graphite with the layers stacked together by weak van der Waals interactions (Rao et al. 2013, Rao et al. 2014b). This weak interaction enables exfoliation to form single-layer nanosheets. Graphene, nitrogen-doped graphene and MoS_2 exhibit excellent electrocatalytic activity in oxygen reduction reactions comparable to platinum deposited carbon (Rao et al. 2015). Nitrogen in nitrogen-doped graphene is an n-type dopant and can give electrons to the π orbitals of carbon, which in turn can provide an electron to the π^* orbital of oxygen and thereby enhance the catalytic activity (Wang et al. 2012a). In the case of MoS_2, the exposed Mo edges are found to be responsible for the high ORR catalytic activity (Wang et al. 2013a). In this chapter, we discuss supercapacitor characteristics and the electrocatalytic ORR activity of 2D materials of graphene and MoS_2.

2. SUPERCAPACITORS

2.1 Characteristics of Supercapacitors

Capacitive properties of electrodes are characterized by the specific capacitance, energy and power density, cyclability and charge-discharge processes. In order to evaluate the capacitive performances Cycling Voltammetry (CV), galvanostatic Charging-Discharging (CD) and Electrochemical Impedance Spectroscopy (EIS) are utilized. The measurements are performed by a two- or three-electrode cell configuration (Stoller and Ruoff 2010).

Specific Capacitance (*C*) is calculated using the following formula from CV,

$$C = \frac{1}{mVv}\int_{V_-}^{V_+} I(V)\,dv \tag{1}$$

where, *m* is the mass of the active material, $V\ (= V_+ - V_-)$ is the potential window between the positive and negative electrode, *v* (V/s) is the scan rate.

Specific capacitance is calculated from galvanostatic charge-discharge curves using the formula,

$$C = \frac{i\,dt}{m\,dv} \tag{2}$$

where, *i* is the discharge current and *dt/dv* is the slope of the discharge curve.

Energy densities (*E*) and Power densities (*P*) are calculated from the formulae,

$$E = \frac{1}{2}CV^2 \tag{3}$$

$$P = \frac{C}{t} \tag{4}$$

where, *C* is the specific capacitance, *V* the operational potential window and *t* the discharge time.

2.2 Graphene-Based Supercapacitors

Graphene, the one-atom-thick planar sheet of sp^2-bonded carbon atoms, has many novel properties including high current density, ballistic transport, chemical inertness, high thermal conductivity, optical transmittance and super hydrophobicity (Rao et al. 2009, Novoselov et al. 2012). Graphene was first obtained by micromechanical cleavage from high-quality graphene crystallites (Novoselov et al. 2005). Several methods are now available for the synthesis of graphene. Thus, graphene has been synthesized in various ways on different substrates (Rao et al. 2009). Besides mechanical cleavage of graphite (Novoselov et al. 2004), the other important methods employed to produce graphene samples are epitaxial growth on an insulator surface (such as Silicon Carbide (SiC)) and Chemical Vapor Deposition (CVD) on the surfaces of single crystals of metals (Rao et al. 2009). Arc-discharge is a good method to synthesize few-layer graphene (two to four layers) in gram quantities (Subrahmanyam et al. 2009). There are many reviews on the various synthesis of single and few-layer graphene (Rao et al. 2009, Choi et al. 2010, Novoselov et al. 2012, Rao et al. 2012).

The performances of graphene-based supercapacitor were first explored by (Vivekchand et al. 2008). The graphene samples were prepared by Exfoliation of Graphitic oxide (EG) or transformation of nanodiamond (DG). The Brunauer-Emmett-Teller (BET) surface areas of EG and DG were 925 and 520 m^2/g, respectively. EG showed the highest specific capacitance of 117 F/g when compared to DG (35 F/g) in 1 M H_2SO_4 aqueous electrolyte. In an ionic liquid electrolyte EG and DG showed specific capacitance values of 75 F/g and 40 F/g respectively. The values of the maximum energy density stored using these capacitors were 31.9 and 17.0 Wh/kg for EG and DG respectively. Stoller et al. (2008) have used chemically modified graphene with a surface area of 705 m^2/g to obtain specific capacitances of 135 and 99 F/g in aqueous (5.5 KOH) and organic ($TEABF_4$) electrolytes, respectively. Theoretically, the specific surface area of single-layer graphene can reach up to 2675 m^2/g and can show a specific capacitance of 550 F/g (Xia et al. 2009). Microwave exfoliated graphite oxide gives a specific capacitance as high as 191 F/g^{-1} in KOH with the surface area of 463 m^2/g (Zhu et al. 2010).

Peng et al. (2015a) have reported a scalable method to make Laser-Induced Graphene (LIG) for fabricating microsupercapacitors (MSCs) using laser induction of commercial polyimide (PI) sheets. Two flexible solid-state supercapacitors (SCs) have been described as LIG-SCs and LIG-MSCs. These devices show a real capacitance of >9 mF/cm^2 at a discharge current density of 0.02 mA/cm^2 which is more than twice that achieved when using aqueous electrolytes. Figures 1.1a and b show stacked solid-state LIG-SCs assembled in series and parallel, respectively. Figures 1.1c and d show charge-discharge curves of 3-stack solid-state series and parallel LIG-SC, respectively. The stacked series LIG-SC has 2*x* higher working voltage window, whereas the stacked parallel LIG-SC shows a 2*x* longer discharge time when operated at the same current density, resulting in a 2*x* higher capacitance compared to a single LIG-SC. In both configurations, the charge-discharge curves have triangular shapes with a minimum voltage drop indicating least internal and contact resistances.

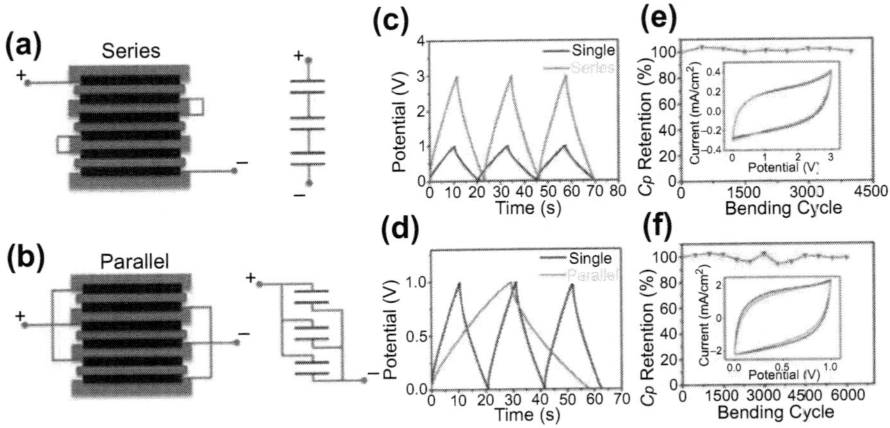

Fig. 1.1 Electrochemical performances of stacked LIG-SCs in series and parallel circuits: (a) Illustration of a stacked series LIG-SC and its corresponding circuit diagram. (b) Illustration of a stacked parallel LIG-SC and its corresponding circuit diagram. (c) Galvanostatic CC curves comparing a single LIG-SC to a stacked series LIG-SC at a current density of 0.5 mA/cm^2. (d) Galvanostatic CC curves comparing a single LIG-SC to a stacked parallel LIG-SC at a current density of 0.5 mA/cm^2. (e) Cyclability testing of a flexible stacked series LIG-SC at a current density of 0.5 mA/cm^2. Inset shows the initial CV curves (black) and the 4000th CV curve (red) at a scan rate of 0.1 V/s. (f) Cyclability testing of a flexible, stacked parallel LIG-SC at a current density of 1.0 mA/cm^2. Inset shows the initial CV curves (black) and the 6000th CV curve (red) at a scan rate of 0.1 V/s. (Reproduced with permission from Ref. (Peng et al. 2015a)).

Figures 1.1e and f show that the capacitance of the stacked LIG-SC circuits are nearly 100% of their initial value even after being subjected to several thousand bending cycles at a bending radius of 17 mm. Additionally, the CV curves at different bending cycles are nearly overlapped (see Fig. 1.1e and f shown as insets), indicating well-maintained flexibility. Yoo et al. (2011) have reported ultrathin supercapacitors of pristine graphene and multilayer Reduced Graphene Oxide (RGO) with specific capacitances of 80 $\mu F/cm^2$ and 394 $\mu F/cm^2$, respectively. Laser written RGO patterns on hydrated Graphene Oxide (GO) films exhibit a capacitance of 0.51 mF/cm^2 with volumetric capacitance of 3.1 F/cm^3. The energy density is 4.3×10^{-4} Wh/cm^3, with a power density of 1.7 W/cm^3. El-Kady et al. (2012) have used a simple solid-state method to attain an aerial capacitance of 3.67 mF/cm^2 in aqueous 1.0 M H_3PO_4 and 4.04 mF/cm^2 in 1.0 M H_2SO_4 at 1 A/g. Wu et al. (2013) have described graphene-based in-plane inter-digital micro-supercapacitors on random substrates. An aerial capacitance of 78.9 $\mu F/cm^2$ and a volumetric capacitance of 17.5 F/cm^3 were obtained at 10 mV/s. The power density and energy density are 495 W/cm^3 and 2.5 mWh/cm^3 respectively, with excellent cycling stability.

Liu et al. (2015) have fabricated high performance supercapacitors by using all carbon electrodes, such as carbon nanotubes and graphene with volume energy in the order of 10^{-3} Wh/cm^3 and power densities in the range of 10 W/cm^3 in PVA-acid electrolyte. A low temperature vacuum assisted synthesis of graphene reduced oxide have a mesoporous surface with a pore size around 3.77 nm exhibited a specific capacitance of 284.5 F/g and an energy density of 131 Wh/kg (Yang et al. 2015). Bamboo-like graphene supercapacitors made by direct laser writing along a graphene oxide fiber reaches a capacitance of 14.3 mF/cm^2 (Liang et al. 2015). Importantly, large-scale integration of the in-fiber supercapacitors is possible and hundreds of device units can be fabricated within minutes using this technique. These supercapacitors exhibit excellent stability as well as repeatable electrochemical and mechanical performance even though they are large-scale integrated. Wang et al. (2013b) have employed the sugar-blowing technique to grow a 3D self-supported graphene product with a high specific surface area of 1005 m^2/g. The capacitance in H_2SO_4 solution was 250 F/g at 1 A/g which and decreased to 130 F/g at a high current of 100 A/g with a power density of 893 kW/g at 100 A/g.

A self-assembled graphene organogel supercapacitor is reported to exhibit a specific capacitance of 140 F/g at 1 A/g with an energy density of 43.5 Wh/kg (Sun et al. 2011). Graphene synthesized by reduction of CO_2 in magnesium and calcium metal flames shows a specific capacitance of 220 and 180 F/g at a current density of 0.1 A/g in 6 M KOH electrolyte (Zhang et al. 2014). Graphene hydrogels synthesized using a simple gelation process with a surface area of 514.9 m^2/g show an aerial capacitance of 33.8 mF/cm^2 at 1 mA/cm^2 (Maiti et al. 2014). An activated carbon aerogel containing graphene with a surface area of 1384 m^2/g has been employed as supercapacitor electrodes with a specific capacitance of 300 F/g which is better than activated carbon (189 F/g) (Lee et al. 2014). Electrolyte dependent supercapacitor measurements have been performed by (Liu et al. 2013). The performance and the specific capacitance varies in the order of $P_{4,4,4,4}$ BF_4/acetonitrile (~78 F/g) < BMIM BF_4/acetonitrile (~75 F/g) = BPy BF/acetonitrile (~73 F/g) < Et_4N BF_4/acetonitrile (~70 F/g) at a current density of 1 A/g. This increase in performance is understood in terms of the relative ionic sizes, which leads to different conductivities and charge densities at the electrode-electrolyte interlayer.

2.2.1 Doped Graphene Supercapacitors

Doped graphenes show enhanced supercapacitor performance compared to pristine graphene because doping affects the electronic structure in a desirable fashion. A change in the electronic structure of graphene increases the binding of electrolyte ions in solution (Rao et al. 2014a, Rao et al. 2015, Wang et al. 2012a). Nitrogen-doped graphene hydrogels show a specific capacitance of 308 F/g at 3 A/g (Guo et al. 2013). It has been recently found that increase in the nitrogen content of graphene increases the specific capacitance (Gopalakrishnan et al. 2013a). With a nitrogen content varying in the 2 to 8 wt.% range, the specific capacitance was 126 F/g at 10 mV/s in an aqueous electrolyte. Nitrogen-doped graphene oxide prepared under microwave irradiation using urea as a nitrogen source with a maximum nitrogen content of 18 wt.% showed maximum specific capacitance of around 465 F/g at 5 mV/s in an aqueous electrolyte (Gopalakrishnan et al. 2013b). We show typical cyclic voltammogram in Fig. 1.2a and Nyquist plots in Fig. 1.2b. The cyclic stability was excellent with capacitance retention of 97% after 1000 cycles at 0.5 A/g (see Fig. 1.2c. We show the Ragone plot in Fig. 1.2d. The energy density is 44.4 Wh/kg at a current density of 0.3 A/g and while the power density is 852 W/kg. Hydrothermally synthesized nitrogen-doped graphene shows a specific capacitance of 326 F/g at 0.2 A/g in an aqueous electrolyte with superior cyclic stability and nearly 99% coulombic efficiency (Sun et al. 2012). The maximum energy density obtained is 35.05 Wh/kg at a power density of 175 W/kg. Lei et al. (2012) obtained a specific capacitance of 255 F/g at 0.5 A/g in an aqueous electrolyte (6 M KOH).

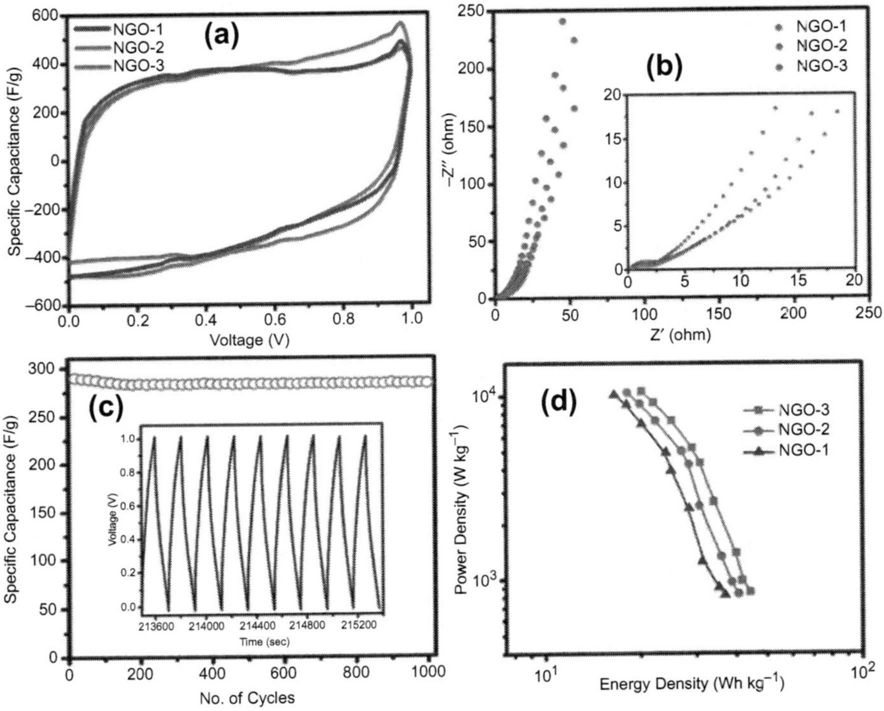

Fig. 1.2 (a) Cyclic voltammograms of NGOs at a scan rate of 20 mV/s. (b) Nyquist curves for NGO electrodes. (c) Specific capacitance versus the cycle number of NGO-3 measured at a current density of 0.5 Ag^{-1} within an operational window of 0.0-1 V (the inset shows the charge-discharge curves of the last few cycles for NGO-3). (d) Ragone plots of NGO based supercapacitors. (Reproduced with permission from Ref. (Gopalakrishnan et al. 2013b)).

Crumpled nitrogen-doped graphene-like nanosheets with a high specific surface area (1169 m^2/g) and large pore volume (2.58 cm^3/g), prepared from a macroporous resin via simultaneous urea gasification expansion and $CaCl_2$ activation, show a specific capacitance of 294 F/g at 0.5 A/g (Peng et al. 2015b). The highest energy density obtained is 17.0 Wh/kg at a power density of 225 W/kg and remains at 8.3 Wh/kg at a power density of 4540 W/kg. Nitrogen-doped graphene produced from dried distillers grains with solubles exhibits a high capacitance ~300 F/g and a relatively low inner resistance of 0.1 Ω (Jin et al. 2015). Graphene sheets with pyridinic and pyrollic nitrogen have been synthesized by functionalizing graphene oxide with different oxygen groups on its surface (Xie et al. 2015). Electrochemical test showed that specific capacitance reaches up to 217 F/g at a discharge current density 1 A/g and stable cycling performance up to 88.8% after 500 cycles. Nitrogen-doped graphene hydrogel prepared at low temperatures shows a specific capacitance of 217.8 F/g at 1 A/g. Even at 200 A/g, the electrode has a specific capacitance of 189.8 F/g with a retention rate of 95.8% of its initial capacitance after 1000 cycles (Xin et al. 2015).

Boron-doped Graphene (BG) made by the fried ice method gives a specific capacitance of 281 F/g in an aqueous electrolyte medium (2M H_2SO_4) (Zuo et al. 2013). Solid-state supercapacitors based on three-dimensional nitrogen- and boron co-doped monolithic graphene aerogels show a specific capacitance of ~62 F/g, an enhanced energy density of ~8.65 Wh/kg and a power density of ~1600 W/kg (Zhu et al. 2011). BG prepared by a simple pyrolysis process using GO and boric acid shows a specific capacitance of 172 F/g at 0.5 A/g and maintains 96.5% of the initial capacity after a continuous cycling of 5000 times (Niu et al. 2013).

High surface-area borocarbonitrides ($B_xC_yN_z$), prepared by the urea route have been examined for supercapacitor applications (Gopalakrishnan et al. 2013a). $BC_{4.5}N$ shows a specific capacitance of 178 F/g in aqueous electrolyte. In an ionic liquid, the specific capacitance value of $BC_{4.5}N$ is 240 F/g. In an aqueous medium, $BC_{4.5}N$ exhibits supercapacitor characteristics superior to those of nitrogen-doped graphene or reduced graphene oxide. Recently, (Kumar et al. 2015) have covalently functionalized different compositions of boron nitride with graphene studied their supercapacitor performance. Cyclic Voltammograms (CV) of carboxylated graphene, $BN_{0.25}G_{0.75}$, $BN_{0.5}G_{0.5}$ and $BN_{0.75}G_{0.25}$ are measured at different scan rates (2-100 mV/s) at a voltage window of -0.2-0.6 V. The CV curves measured at 40 mV/s are shown in Fig. 1.3a. The CV curves show a rectangular feature even at higher scan rates which indicates that these materials are good charge storage supercapacitor electrodes. A maximum capacitance of 217 F/g is found at a scan rate of 2 mV/s in the case of $BN_{0.25}G_{0.75}$. The specific capacitance values of carboxylated graphene, $BN_{0.5}G_{0.5}$ and $BN_{0.75}G_{0.25}$ are 89, 180, 92 F/g, respectively, at 2 mV/s. The galvanostatic charge–discharge curves of the BNG samples were measured in a voltage window from -0.2 to 0.6 V at different current densities. Figure 1.3b shows the charge–discharge curves of carboxylated graphene and $BN_{1-x}G_x$ composites measured at 1 A/g. The discharge time of $BN_{0.25}G_{0.75}$ was longer when compared to carboxylated graphene and other two composites. The charge–discharge curves look symmetric for the $BN_{1-x}G_x$ composites resembling those of ideal capacitors. The specific capacitance values decrease with increase in the current density as shown in Fig. 1.3c. The specific capacitance of $BN_{0.25}G_{0.75}$, $BN_{0.5}G_{0.5}$ and $BN_{0.75}G_{0.25}$ is 238, 204 and 116 F/g, respectively, at 0.3 A/g, while carboxylated graphene alone exhibits a value of 87 F/g. The high specific capacitance of $BN_{1-x}G_x$ is attributed to the enhanced surface area along with the presence of slit-like microporous channels created in the composite due to covalent cross-linking. $BN_{0.25}G_{0.75}$ shows excellent capacitance retention of 98% as shown in Fig. 1.3d.

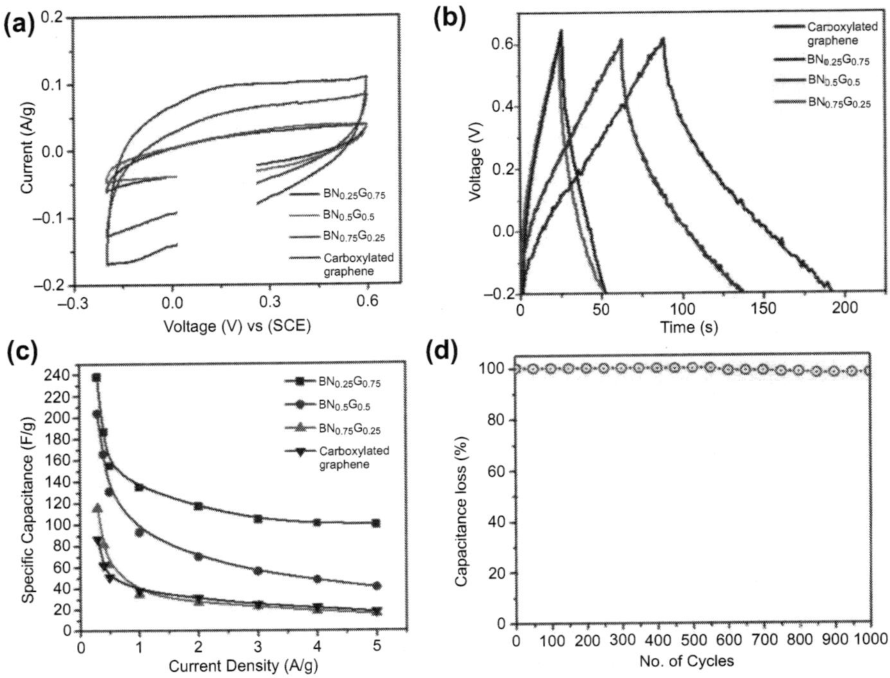

Fig. 1.3 (a) Cyclic voltammograms of carboxylated graphene and $BN_{1-x}G_x$ electrodes at a scan rate of 40 mV/s. (b) Galvanostatic charge-discharge curves for carboxylated graphene and $BN_{1-x}G_x$ electrodes (at 1 A/g). (c) Specific capacitance as a function of discharge current. (d) % capacitance loss versus the cycle number of $BN_{0.25}G_{0.75}$ measured at a current density of 1 A/g within an operational window of -0.2-0.6 V. (Reproduced with permission from Ref. (Kumar et al. 2015)).

2.2.2 Supercapacitors Based on Graphene-Conducting Polymer Composites

Conducting π-electrons polymers are excellent electrode materials for supercapacitors because of their high specific capacitance, high conductivity and ease of preparation. Polyaniline (PANI) possesses high conductivity and chemical stability. Its electrochemical behavior is complex due to the presence of several oxidation states of PANI including the fully reduced leucoemeraldine form, the intermediate emeraldine form and the fully oxidized pernigraniline form. Chemically modified graphene and polyaniline nanofiber composites have been prepared by the *in situ* polymerization of aniline monomer in the presence of graphene oxide under acid conditions and the highest specific capacitance achieved was 480 F/g at a current density of 0.1 A/g. PANI sandwiched graphene layers prepared by mixed dispersion of chemically converted graphene and polyaniline nanofibers show high conductivity with a capacitance of 210 F/g at 0.3 A/g with energy density of ~9.2 Wh/Kg (Wu et al. 2010). Doping of PANI with graphene also shows good electrochemical performance with capacitance of 531 F/g (Wang et al. 2009a).

Lai et al. (2012) have investigated the effect of graphene surface chemistry on the electrochemical performance of graphene/polyaniline composites as supercapacitor electrodes. Graphene Oxide (GO), chemically reduced GO (RGO), nitrogen-doped RGO (N-RGO) and amine-modified RGO (NH_2-RGO) were loaded with about 9 wt.% of PANI. The NH_2-RGO/PANI composite exhibited highest capacitance (420 F/g) and good cyclability. Wang et al. (2009b) developed a free-standing and flexible graphene/polyaniline composite by an *in situ* anodic electropolymerization. This composite electrode showed a good tensile strength of 12.6 MPa and values of 233 F/g and 135 F/cm^3 for gravimetric and volumetric capacitances, respectively.

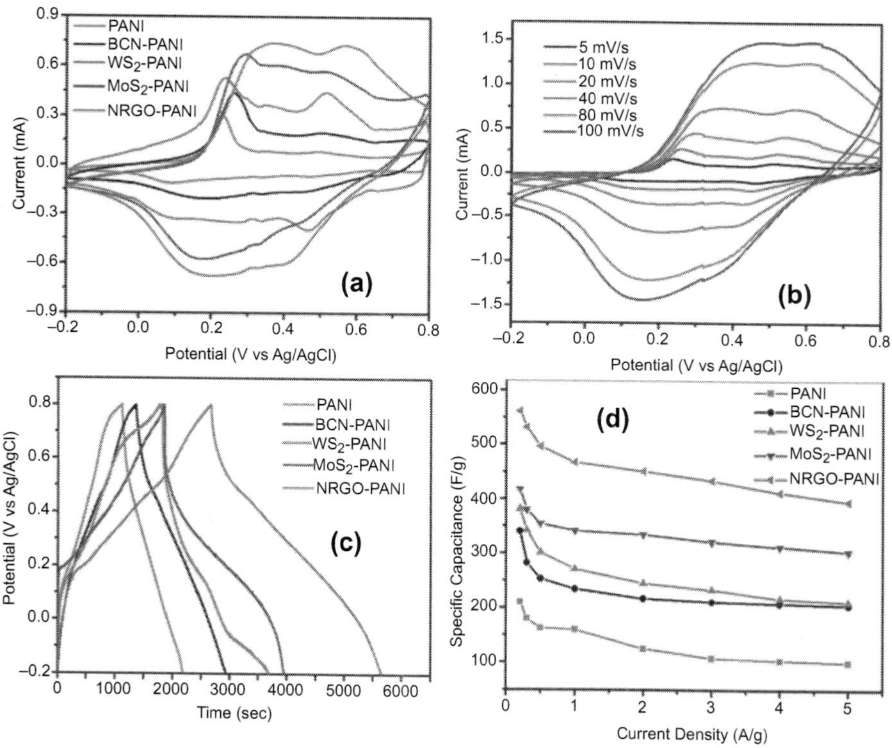

Fig. 1.4 (a) Cyclic voltammograms of PANI and 1:1 nanocomposites with NRGO, MoS_2, WS_2 and BCN at 40 mV/s. (b) Cyclic voltammograms of NRGO-PANI at different scan rates. (c) Galvanostatic charge discharge curves of PANI (1:1) and the 1:1 nanocomposites with NRGO, MoS_2, WS_2 and BCN at a current density of 0.2 A/g. (d) Specific capacitance of PANI and 1:1 nanocomposites with NRGO, MoS_2, WS_2 and BCN at different current densities. (Reproduced with permission from Ref. (Gopalakrishnan et al. 2015)).

Electrical properties of PANI nanocomposites with different 2D materials such as NRGO, MoS_2, WS_2 and BCN have been investigated (Gopalakrishnan et al. 2015). Cyclic voltammograms of PANI, NRGO, MoS_2, WS_2, BCN and its composites are measured at different scan rates (5-100 mV/s) and the CV curves measured at

40 mV/s are shown in Fig. 1.4a. Cyclic voltammograms show two pairs of redox peaks corresponding to reversible charge-discharge behavior and pseudocapacitance characteristic of PANI. The pair of peaks Q_1/Q_2 are attributed to the redox transition of PANI between leucoemeraldine and meraldine and the pair of peaks Q_3/Q_4 are attributed to the emeraldine–pernigraniline transformation. The increase in the integrated area of the CV loop indicates much higher capacitance for the composites than the pure PANI. NRGO-PANI composite shows the highest peak current of all the composites. The cyclic voltammograms of NRGO-PANI measured at different scan rates shown in Fig. 1.4b, clearly reveal that the peak current heightens with the increase in scan rates. The galvanostatic charge-discharge curves of PANI, NRGO, MoS_2, WS_2, BCN and its composites were measured in a potential window of -0.2-0.8 V vs Ag/AgCl (2 M KCl). Figure 1.4c shows charge-discharge curves measured at a current of 0.2 A/g. The discharging curve of the NRGO-PANI shows two voltage stages in the ranges of 0.8 to 0.4 V and 0.4 to -2.0 V, respectively. The former stage with a relatively short discharging duration is ascribed to Electrical Double Layer (EDL) capacitance while the latter stage with a much longer discharging duration is associated with the combination of EDL and Faradaic capacitances of NRGO and PANI. The other composites also show a similar behavior. Figure 1.4d shows the variation of specific capacitance at different current densities. The specific capacitance values decreases slightly as the current density increases, generally stabilizing above 1 A/g. The capacitance values of the PANI, NRGO, MoS_2, WS_2, $BC_{1.5}N$ are 210, 110, 60, 55 and 85 respectively. The highest capacitance is found with NRGO-PANI, the value being 561 F/g at 0.2 A/g.

2.3 MoS_2-Based Supercapacitors

MoS_2 is an important material for electrochemical devices. MoS_2 consists of a metal Mo layer sandwiched between two S layers, with these triple layers stacked together to form the layered structure (Rao et al. 2015). Such 2D electron-electron correlations among Mo atoms enhance the supercapacitance characteristics of MoS_2. Three-dimensional (3D) flower-like MoS_2 nanostructures prepared by a one-pot hydrothermal method show a maximum capacitance of 218 F/g at a scan rate of 5 mV/s and of about 217.6 F/g at a discharge current density of 0.1 A/g in aqueous electrolyte (1 M KCl) with a capacitance retention of about 76% after 1000 cycles (Pandey et al. 2015). The specific capacitance of this 3D-MoS_2 electrode in an organic electrolyte ($BMIMPF_6$) is higher than that of the MoS_2 electrode in aqueous electrolyte (~600 F/g at 5 mV/s). This is because ionic hopping in the organic medium improves the formation of electrical double layer and thus increases the capacitance. The energy density is calculated to be 60 and 166 W h/kg for the MoS_2 electrode in aqueous and organic media respectively. Flexible supercapacitors based on MoS_2 hierarchical nanospheres showed excellent electrochemical performance such as high capacitance 368 F/g at a scan rate of 5 mV/s and high power density of 128 W/kg at energy density of 5.42 Wh/kg (Javed et al. 2015). The supercapacitor inherits good characteristics such as lightweight, low cost, portability, high flexibility and long term cycling stability by retaining 96.5% after 5000 cycles at constant discharge current of 0.8 mA. Few-layered MoS_2 obtained by liquid phase exfoliation of bulk MoS_2 powder in 1-dodecyl-2-pyrrolidinone exhibited capacitances ~2 mF/cm^2 whereas for bulk MoS_2 the capacitance was ~0.5 mF/cm^2 in KOH electrolyte (Winchester et al. 2014). The charge storage significantly varies in organic solvents such as 1 M tetraethylammonium tetrafluoroborate in propylene carbonate (Et_4NBF_4 in PC) and ionic liquids [1-butyl-3-methylimidazolium hexafluorophosphate (BMIM-PF_6)]. The specific capacitance values obtained using Et_4NBF_4 in PC and BMIM-PF_6 were ~2.25 and ~2.4 mF/cm^2, respectively. Flower-like MoS_2 nanospheres synthesized through hydrothermal route with mean diameters of 300 nm shows supercapacitive behavior in 1 M KCl The specific capacitance of the flower-like MoS_2 is found to be 122 F/g at 1 A/g or 114 F/g at 2 mV/s (Zhou et al. 2014). Hydrothermally synthesized of MoS_2 nanosheets show a specific capacitance of 129.2 F/g at 1.0 A/g and maintained 73.8 F/g when the current density increased by 10 times (10 A/g) (Huang et al. 2014).

Acerce et al. (2015) have reported nanosheets of MoS_2 synthesized through chemical exfoliation to contain a high concentration of the metallic 1T phase. These chemically exfoliated MoS_2 nanosheets exhibited extraordinary efficiency and showed capacitance values ranging from ~400 to ~700 F/cm^3 in various aqueous electrolytes (0.5 M H_2SO_4, Li_2SO_4, Na_2SO_4 and K_2SO_4 together with KCl and KBr) in a three-electrode configuration. The reason for this is the high electrical conductivity (10-100 S/cm) of the 1T phase MoS_2 which is considerably higher than that of reduced graphene oxide electrodes (~100 S/cm). Figure 1.5a shows cyclic voltammograms (CV) of 1 T-MoS_2 in H_2SO_4, Li_2SO_4, Na_2SO_4 and K_2SO_4 for potentials ranging from -0.15 V to 0.85 V versus NHE. The rectangular nature of the CV curves indicate the capacitive behavior of the electrodes. The specific capacitance increases by a factor of nearly 20 in the 1T phase electrodes compared to the 2H phased electrodes. Figure 1.5b shows CV curves obtained from K_2SO_4 and KCl electrolytes to reveal whether the cations or anions intercalate into the electrodes. It is seen that despite the large difference in anion radii, the CVs for both electrolytes are nearly the same, suggesting that it is the cation that is intercalating. Figure 1.5c shows CV curves measured at different scan rates ranging from 5 mV/s to 1000 mV/s. Specific capacitance versus scan rate plot obtained from different electrolytes is shown in Fig. 1.5d. The volumetric capacitance ranges from 400 to 650 F/cm^3 at scan rates of 20 mV/s. Galvonastatic charge-discharge curves measured at different current densities in Na_2SO_4 electrolyte are shown in Fig. 1.5e. The curves show triangular capacitive behavior. This electrode shows good capacitive retention after performing more than 5,000 charge/discharge cycles at a current rate of 2 A/g (Fig. 1.5e). MoS_2 film-based micro-supercapacitors made from of MoS_2 nanosheets on Si/SiO_2 chip and subsequent laser patterning exhibited excellent electrochemical performance (Cao et al. 2013). The areal capacitance of hydrothermal synthesized MoS_2 based micro-supercapacitor is 8 mF/cm^2, much higher than that liquid exfoliated MoS_2 (~3.1 mF/cm^2) or MoS_2 powder (~1.8 mF/cm^2).

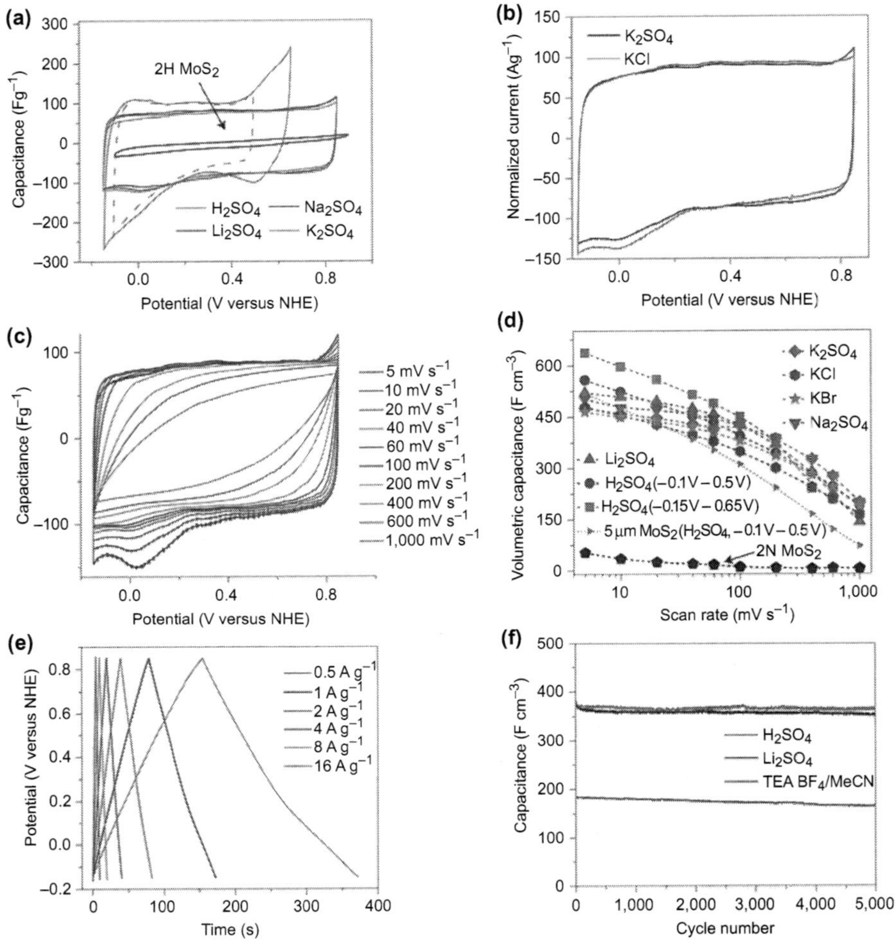

Fig. 1.5 (a) CVs of 1T phase MoS_2 nanosheet paper in 0.5 M sulfate-based electrolyte solutions at scan rates of 20 mV/s. (b) Comparison of the CV curves of 1T MoS_2 in 0.5 M K_2SO_4 and 1 M KCl. The absence of a noticeable difference between the two electrolytes supports the fact that only cations are being stored. (c) CVs of 1T phase MoS_2 electrodes in 0.5 M Na_2SO_4 from 5 mV/s to 1,000 mV/s. The capacitance remains virtually constant up to scan rate of 100 mV/s. (d) Evolution of the volumetric capacitance of the 1T phase MoS_2 electrodes with scan rate for different electrolytes and 1-µm- and 5-µm-thick films. The concentration of the cations in the electrolyte solutions was fixed at 1 M. (e) Galvanostatic cycles from 0.5 A/g-16 A/g in K_2SO_4. (f). Capacitance retention after 5,000 cycles in 0.5 M Li_2SO_4, H_2SO_4 and 1 M TEA BF_4 in acetonitrile. (Reproduced with permission from Ref. (Acerce et al. 2015)).

2.3.1 MoS_2-Graphene Supercapacitors

MoS_2-Graphene composites made by simple mixing of MoS_2 and graphene show a specific capacitance of ~11 cmF/cm^2 (weight ratio of 1:3) whereas the pristine MoS_2

showed 1.8 mF/cm^2 at 5 mV/s. The capacitance decreases up to 80% at 1000 mV/s for both MoS_2 as well as the composites is due to the pseudocapacitive nature of MoS_2. The large increase in capacitance of MoS_2-graphene is due to the difference in flake dimensions and surface energy. MoS_2-graphene composites form disordered heterostructures and prevent restacking, effectively increasing the specific surface area and the charge storing ability (Bissett et al. 2015). MoS_2@S-RGO microsupercapacitor consisting of sulfonated reduced graphene oxide (S-RGO) and MoS_2 nanoflowers exhibit high specific capacitance of 6.56 mF/cm^2, energy density (0.58 mWh/cm^3) and power density (13.4 mW/cm^3) in KOH/PVA gel electrolyte (Xiao et al. 2015). In addition, the printed microsupercapacitor loses only 9% of the maximum capacity after 1000 cycles, indicating excellent stability. MoS_2-RGO/MWCNT fiber-based supercapacitors with various amounts of MoS_2 were tested and the specific capacitance reached up to a value of 4.3 F/cm^3 at 6.3 wt% of MoS_2, which is significantly improved as compared to the devices made those from MoS_2/MWCNT or RGO/MWCNT in PVA-H_2SO_4 gel electrolyte (Sun et al. 2015). This asymmetric supercapacitor can be operated in a wide potential window of 1.4 V with high Coulombic efficiency, good rate and cycling stability and improved energy density due to electrochemical activity of MoS_2, electrical conductivity of well-aligned MWCNT fiber and the incorporation of RGO. Layered MoS_2–graphene composite synthesized from l-cysteine showed a maximum specific capacitance of 243 F/g at a current density of 1 A/g (Huang et al. 2013). The energy density and power density are 73.5 Wh/kg and 19.8 kW/kg, respectively. The enhancement in specific capacitance may be due to the interconnected conductive network which promotes the efficient ion transport and controls the aggregation of active material during charge–discharge process. A study of MoS_2 and Reduced Graphene Oxide (RGO) composites of different compositions has shown that the MoS_2/RGO (1:2) composite give the best results as shown in Fig. 1.6a (Gopalakrishnan et al. 2015a). The CV curves for the MoS_2/RGO nanocomposites show quasi-rectangular CV curves with small humps due to the redox process. The quasi-rectangular CV curves indicate excellent charge storage capability of the nanocomposites, which resemble those of an ideal supercapacitor, showing a maximum capacitance of 416 F/g with MoS_2/RGO (1:2) at a scan rate of 5 mV/s. It is smaller on either side of this composition, as in the 2:1 and 1:3 composites. When the scan rate heightens, the current increases and the capacitance decreases as shown in Figs. 1.6b and c show the charge-discharge curves measured at a current of 1 A/g. The discharge time of MoS_2/RGO (1:2) is significantly longer than the other composites and the curves look nearly symmetrical, indicating remarkable charge storing ability of this composite. The calculated specific capacitance of this composite is 249 F/g when measured at 0.3 A/g. The specific capacitance of MoS_2, RGO and MoS_2/RGO composites obtained at different current densities are shown in Fig. 1.6d. The synergistic effect is calculated to be 118% in the 1:2 composite. The electronic conductivity of RGO and surface properties of both MoS_2 and RGO in the composites seem to be responsible for the good specific capacitance.

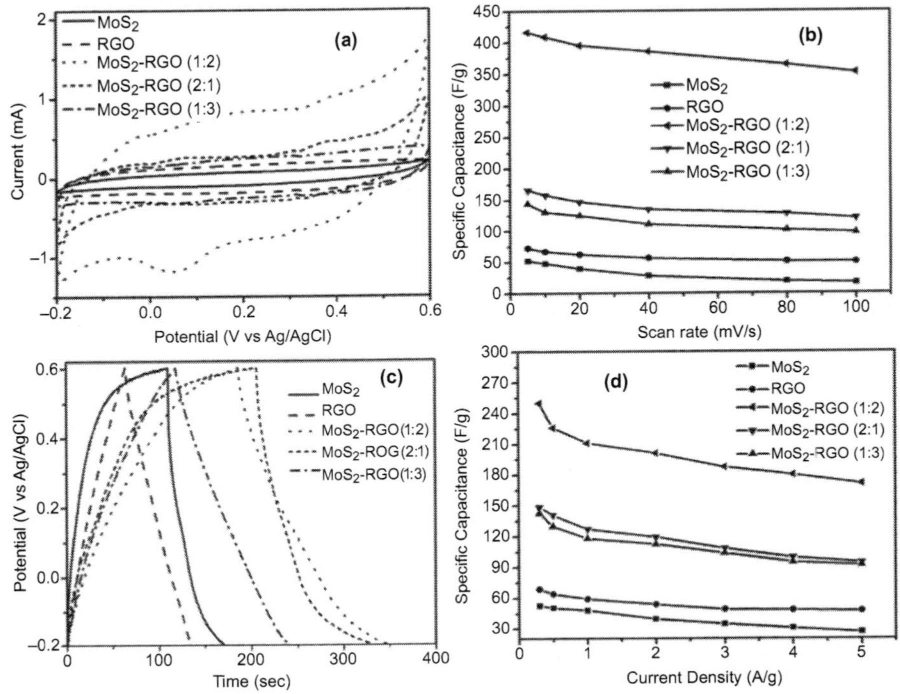

Fig. 1.6 (a) Cyclic voltammograms of MoS_2, RGO and MoS_2-RGO nanocomposites at 100 mV/s. (b) Specific capacitance of MoS_2, RGO and MoS_2-RGO nanocomposites at different scan rates. (c) Galvanostatic charge discharge curves of MoS_2, RGO and MoS_2-RGO nanocomposites at a current density of 1 A/g. (d) Specific capacitance of MoS_2, RGO and MoS_2-RGO nanocomposites at different current densities. (Reproduced with permission from Ref. (Gopalakrishnan et al. 2015a)).

2.3.2 MoS_2-Polyaniline Composites

Unique architectures composed of large surface area MoS_2 and conductive PANI are expected to render large electrode/electrolyte interaction and good ion diffusion when it is used as an electrode material for energy storage devices. Thus, MoS_2-Polyaniline electrodes consisting of MoS_2 thin nanosheets and PANI nanoarrays are fabricated via a large-scale approach showed superior capacitance of 853 F/g with at 1 A/g with a retention up to 91% after 4000 cycles and a high energy density of 106 Wh/kg at a power density of 106 kW/kg (Zhu et al. 2015). MoS_2/PANI composites with PANI in different compositions showed excellent supercapacitor performance (Gopalakrishnan et al. 2015). The 1:1 and 1:6 MoS_2/PANI composites show capacitance values of 417 and 567 F/g, respectively, with cyclic stability that is better than PANI alone. MoS_2 and polypyrrole nanocomposites prepared by *in situ* oxidative polymerization showed a specific capacitance of ~700 F/g at a scan rate of 10 mV/s.

Ni_3S_4-MoS_2 nanospheres are tested as supercapacitor electrodes, which exhibit high specific capacitance of 1440.9 F/g at 2 A/g and a good capacitance retention of 90.7% after 3000 cycles at 10 A/g (Zhang et al. 2015). Flexible all-solid-state supercapacitor constructed by combining the PVA/KOH gel electrolyte with MoS_2, $Ni(OH)_2$ and $MoS_2@Ni(OH)_2$ nanocomposite has shown average specific capacitances of 228 F/g, 344 F/g and 431 F/g, respectively, at a scan rate of 5 mV/s (Hao et al. 2014). These supercapacitors can be operated at high scan rates and have excellent long-life cycling stability, retaining 94.2% of the initial capacitance after 9000 cycles.

3. OXYGEN REDUCTION REACTION

3.1 Nitrogen-Doped Graphene

Nitrogen-doped Graphene (NG) is a promising electrocatalyst for the oxygen reduction reaction (ORR). Introduction of nitrogen in the carbon network enhances the ORR activity due to increase in the electron density of states near the Fermi level. Nitrogen doped porous graphene foams made by template method exhibits excellent ORR activity and long term stability in alkaline and acidic solutions. Its ORR activity is even better than that of the Pt-based catalyst in alkaline medium (Zhou et al. 2015). A two-step calcination strategy of α-hydroxy acids as carbon source and melamine as nitrogen source gives N-doped graphene with N content of 8.11 at.% shows excellent ORR activity. The better performance is attributed to its high pyridinic nitrogen content (Liao et al. 2015). Nitrogen-doped graphene with dominance of the pyridinic-N synthesized by chemical vapor deposition growth of graphene and post-doping with a solid nitrogen precursor of graphitic C_3N_4 contains N content up to 6.5 at.% when post-doped at 800°C (Wu et al. 2015). The NG so obtained exhibits high activity, long-standing stability and outstanding crossover resistance for electrocatalysis of ORR in alkaline medium. High-quality pyridinic N-rich nitrogen-doped graphene nanoplatelets achieved via an acid-catalyzed dehydration reaction between GO and primary amine-containing small molecules. The nitrogen content are in the of 3.27-15.28 wt.% by simple variation of the molar feed ratio of reactants (Chang et al. 2015). This high pyridinic nitrogen content allows superior electrocatalytic activity for ORR in an alkaline electrolyte.

Qu et al. (2010) propose a 4-electron pathway in alkaline solutions for nitrogen doped graphene, which shows a higher current density and good amperometric response for ORR compared to the commercial platinum-carbon catalyst. Nitrogen plasma treated graphene exhibits much higher electrocatalytic activity toward oxygen reduction than graphene under alkaline conditions (Shao et al. 2010). The ORR overpotential is significantly decreased in N-graphene thereby increasing the electrocatalytic activity of graphene toward ORR with a lower initial electrocatalytic activity than Pt/C, but much higher durability than Pt/C. The enhancement in the ORR activity is attributed to the pyridinic and pyrrolicnitrogens which increase the reactivity of the neighboring carbon atoms. Selectively synthesized pyridinic- and pyrrolic-nitrogen-doped graphene shows excellent oxygen reduction reactivity (Ding et al. 2013). The electrocatalytic activity of heavily nitrogen doped graphene made by microwave method with nitrogen content of 14 to 18 wt.% has been reported recently (Gopalakrishnan and Rao 2015b). Linear-Sweep Voltammetry (LSV) of NGs at a rotation speed of 1600 rpm was measured in the O_2-saturated

0.1 M KOH solution on Rotating Disk Electrode (RDE) and shown in Fig. 1.7a. NG samples exhibit much higher current density compared to Pt/C or RGO representing an enhanced activity for ORR. The onset potential of RGO is around 0.8 V, slightly shifting to 0.9 V vs RHE nitrogen-doped samples. This positive shift indicates that the nitrogen in the graphene reduces the ORR overpotential, thereby enhancing the catalytic activity. One of the NG samples shows a much more positive onset potential closer to that of platinum. Furthermore, a set of LSV curves for ORR based on the NG-3 catalyst recorded from 400 to 2400 rpm was shown in Fig. 1.7b. It is also seen that the current density heightens with the increase in rotation speed, since the diffusion distance shortens at higher rotation rates. Koutecky-Levich plots (i^{-1} vs $\omega^{-1/2}$) were obtained from the polarization curves at a potential of 0.8 V vs RHE for all the catalysts as shown in Fig. 1.7c. The transferred electron number (n) per O_2 from the slopes of Koutecky-Levich plots was calculated to be ~4 (0.8 V vs RHE) for all the NG samples while in RGO it is only ~3.34. This is considered to be due to the high nitrogen content in graphene. The *n* and kinetic current density (i_k) values of RGO and NG are shown in Fig. 1.7d . The i_k values of RGO, NG-1, –2, –3 and Pt/C are –0.1, –0.5, –0.4, –0.8 and –1.2 mA/cm^2, respectively, at 0.8 V vs RHE. NG-3 shows higher i_k values than graphene samples but lower than that of Pt/C.

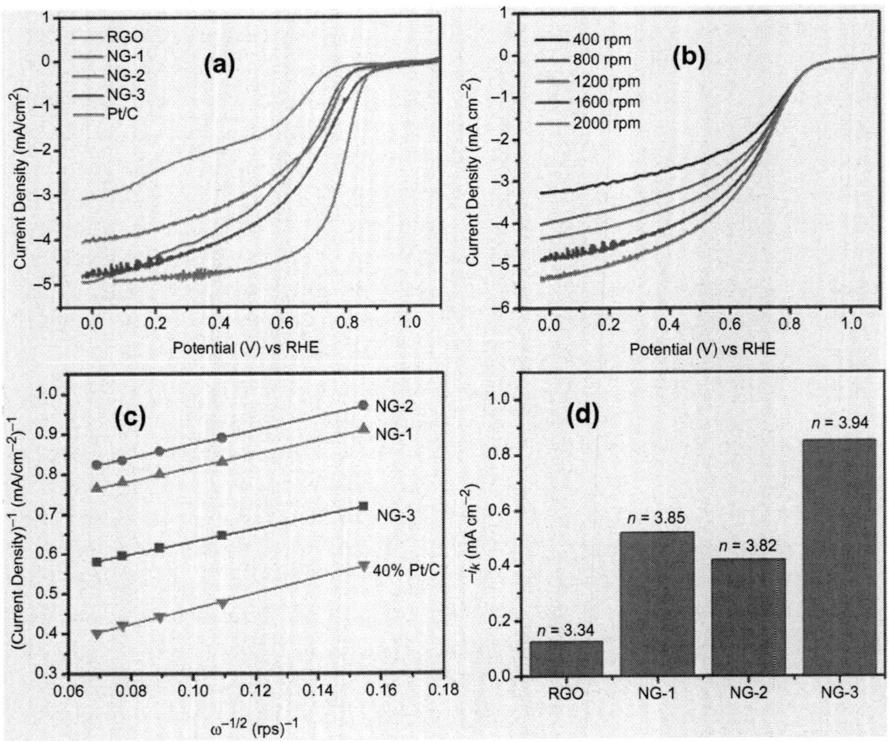

Fig. 1.7 (a) LSV curves of 40% Pt/Vul can X and NG samples in oxygen saturated 0.1 M KOH with a scan rate of 5 mV s^{-1} and a rotation rate of 1600 rpm. (b) RDE measurement of NG-3 in oxygen saturated 0.1 M KOH with a scan rate of 5 mV/s. (c) Koutecky-Levich plots obtained at 0.8 V vs RHE. (d) Variation of kinetic current density (i_k) at ~0.8 V vs RHE for RGO and NG samples. (Reproduced with permission from Ref. (Gopalakrishnan and Rao 2015b)).

Three-dimensional B, N-doped graphene foam performs as an excellent metal-free catalyst for ORR (Xue et al. 2013). BCN with tunable B/N co-doping levels obtained by thermal annealing GO in the presence of boric acid and ammonia performs as an excellent ORR electrocatalyst (Wang et al. 2012b). Composition-dependent ORR activity has been studied in few-layer borocarbonitrides (Moses et al. 2014). The *n* value was ~4 in all the compositions.

3.2 MoS_2

MoS_2 nanoparticles supported on gold nanoparticle films follow a four-electron pathway for the oxygen reduction reaction (ORR) in alkaline media with an onset potential of −0.10 V against the saturated calomel electrode (Wang et al. 2014b). These films exhibit superior stability and better electrocatalytic performance than commercial Pt/C. MoS_2 particles with different size distributions prepared by simple ultrasonication of bulk MoS_2 showed significantly improved catalytic activity toward the ORR. It is found that a decrease in particle size increases the catalytic activity (Wang et al. 2013a). Composites of MoS_2 with reduced graphene oxide show good ORR activity. The ORR onset peak potential of MoS_2/RGO is more positive than that of pristine RGO or pristine MoS_2. MoS_2/RGO composites (1:2 and 2:1) demonstrate better ORR activity than either RGO or MoS_2 alone. Such enhanced performance is attributed to a synergistic effect between RGO and MoS_2. The highest reduction potential and current density were obtained with the 1:2 MoS_2/RGO nanocomposites. The 1:2 MoS_2/RGO nanocomposite gives an electron transfer number closer to 4.

MoS_2/nitrogen doped graphene (NG) hybrid composites obtained through loading MoS_2 sheets onto NG through ultrasonication, exhibit improved electrocatalytic activity for ORR with dominant 4 electron pathway in alkaline solutions (Zhao et al. 2015). The synergistic effects along with exposed edges and increase conductivity make the MoS_2/NG composites highly competitive ORR catalysts. Phosphorus doped-ultrathin MoS_2 nanosheets synthesized via thermolytical process enhanced the ORR activity with four-electron selectivity and also seven-fold current increase in both onset and half-wave potentials than that of un-doped MoS_2 (Huang et al. 2015). The reason for this is that the less electronegative of phosphorus atoms in the plane of MoS_2 nanosheets are polarized by the surrounding sulfur atoms with larger electronegativity, endowing the phosphorus atoms with more positive charge than sulfur. The dissolved oxygen molecules in the electrolyte may prefer to be absorbed by the phosphorus-doped surface of the nanosheets rather than the pristine surface. The preferential absorption of oxygen molecules on P-MoS_2 nanosheets further accelerates the subsequent reduction processes. MoS_2 nanodots embedded 3-D porous frameworks of N-doped graphene show good activity towards oxygen reduction reaction in basic media (Du et al. 2015). This is because the MoS_2 nanodots have an abundance of exposed edge sites with good accessibility to oxygen molecules and also nitrogen-doped graphene played an important role in providing good conductivity, a large surface area and porous structure, which facilitate the electron and mass transfer.

4. CONCLUSIONS

The previous sections clearly bring out the useful properties of graphene, MoS_2 and related 2D materials, some of which may find useful applications. It is necessary to continue exploration of such 2D materials for possible applications in energy devices. By optimizing the compositions of these materials it should be possible to fabricate supercapacitors and fuel cells for practical use.

5. REFERENCES

Acerce, M., D. Voiry and M. Chhowalla. 2015. Metallic 1T phase MoS_2 nanosheets as supercapacitor electrode materials. Nat. Nano. 10(4): 313-318.

Bissett, M. A., I. A. Kinloch and R. A. W. Dryfe. 2015. Characterization of MoS_2–graphene composites for high-performance coin cell supercapacitors. ACS App. Mater. Interfaces 7(31): 17388-17398.

Cao, L., S. Yang, W. Gao, Z. Liu, Y. Gong, L. Ma, et al. 2013. Direct laser-patterned micro-supercapacitors from paintable MoS_2 films. Small 9(17): 2905-2910.

Chang, D. W., H.-J. Choi and J.-B. Baek. 2015. Wet-chemical nitrogen-doping of graphene nanoplatelets as electrocatalysts for the oxygen reduction reaction. J. Mater. Chem. A 3(14): 7659-7665.

Choi, W., I. Lahiri, R. Seelaboyina and Y. S. Kang. 2010. Synthesis of graphene and its applications: a review. Critical Rev. Solid State and Mater. Sci. 35(1): 52-71.

Ding, W., Z. Wei, S. Chen, X. Qi, T. Yang, J. Hu, et al. 2013. Space-confinement-induced synthesis of pyridinic- and pyrrolic-nitrogen-doped graphene for the catalysis of oxygen reduction. Ange. Chem. Inter. Ed. 52(45): 11755-11759.

Du, C., H. Huang, X. Feng, S. Wu and W. Song. 2015. Confining MoS_2 nanodots in 3D porous nitrogen-doped graphene with amendable ORR performance. J. Mater. Chem. A 3(14): 7616-7622.

El-Kady, M. F., V. Strong, S. Dubin and R. B. Kaner. 2012. Laser scribing of high-performance and flexible graphene-based electrochemical capacitors. Science 335: 1326-1330.

Gopalakrishnan, K., K. Moses, A. Govindaraj and C. N. R. Rao. 2013a. Supercapacitors based on nitrogen-doped reduced graphene oxide and borocarbonitrides. Solid State Commun. 175-176: 43-50.

Gopalakrishnan, K., A. Govindaraj and C. N. R. Rao. 2013b. Extraordinary supercapacitor performance of heavily nitrogenated graphene oxide obtained by microwave synthesis. J. Mater. Chem. A 1(26): 7563-7565.

Gopalakrishnan, K., K. Pramoda, U. Maitra, U. Mahima, M. A. Shah and C. N. R. Rao. 2015a. Performance of MoS_2-reduced graphene oxide nanocomposites in supercapacitors and in oxygen reduction reaction. Nanomaterials and Energy 4: 9-17.

Gopalakrishnan, K. and C. N. R. Rao. 2015b. Remarkable performance of heavily nitrogenated graphene in the oxygen reduction reaction of fuel cells in alkaline medium. Mater. Res. Express 2(9): 095503.

Gopalakrishnan, K., S. Sultan, A. Govindaraj and C. N. R. Rao. 2015. Supercapacitors based on composites of PANI with nanosheets of nitrogen-doped RGO, $BC_{1.5}N$, MoS_2 and WS_2. Nano Energy 12: 52-58.

Guo, H.-L., P. Su, X. Kang and S.-K. Ning. 2013. Synthesis and characterization of nitrogen-doped graphene hydrogels by hydrothermal route with urea as reducing-doping agents. J. Mater. Chem. A 1(6): 2248-2255.

Hao, C., F. Wen, J. Xiang, L. Wang, H. Hou, Z. Su, et al. 2014. Controlled incorporation of $Ni(OH)_2$ nanoplates into flowerlike MoS_2 nanosheets for flexible all-solid-state supercapacitors. Adv. Funct. Mater. 24(42): 6700-6707.

Hu, X., W. Zhang, X. Liu, Y. Mei and Y. Huang. 2015. Nanostructured Mo-based electrode materials for electrochemical energy storage. Chem. Soc. Rev. 44(8): 2376-2404.

Huang, H., X. Feng, C. Du and W. Song. 2015. High-quality phosphorus-doped MoS_2 ultrathin nanosheets with amenable ORR catalytic activity. Chem. Commun. 51(37): 7903-7906.

Huang, K.-J., L. Wang, Y.-J. Liu, Y.-M. Liu, H.-B. Wang, T. Gan, et al. 2013. Layered MoS_2–graphene composites for supercapacitor applications with enhanced capacitive performance. Inter. J. Hydro. Energy 38(32): 14027-14034.

Huang, K.-J., J.-Z. Zhang, G.-W. Shi and Y.-M. Liu. 2014. Hydrothermal synthesis of molybdenum disulfide nanosheets as supercapacitors electrode material. Electrochimica Acta 132: 397-403.

Huang, K.-J., L. Wang, J.-Z. Zhang and K. Xing. 2015. Synthesis of molybdenum disulfide/carbon aerogel composites for supercapacitors electrode material application. J. Electroanal. Chem. 752: 33-40.

Javed, M. S., S. Dai, M. Wang, D. Guo, L. Chen, X. Wang, et al. 2015. High performance solid state flexible supercapacitor based on molybdenum sulfide hierarchical nanospheres. J. Power Sources 285: 63-69.

Jin, H., X. Wang, Z. Gu, Q. Fan and B. Luo. 2015. A facile method for preparing nitrogen-doped graphene and its application in supercapacitors. J. Power Sources 273: 1156-1162.

Kumar, R., K. Gopalakrishnan, I. Ahmad and C. N. R. Rao. 2015. BN–Graphene composites generated by covalent cross-linking with organic linkers. Adv. Funct. Mater. Doi: 10.1002/adfm.201502166.

Lai, L., H. Yang, L. Wang, B. K. Teh, J. Zhong, H. Chou, et al. 2012. Preparation of supercapacitor electrodes through selection of graphene surface functionalities. ACS Nano 6(7): 5941-5951.

Lee, Y. J., G.-P. Kim, Y. Bang, J. Yi, J. G. Seo and I. K. Song. 2014. Activated carbon aerogel containing graphene as electrode material for supercapacitor. Mater. Res. Bull. 50: 240-245.

Lei, Z., L. Lu and X. S. Zhao. 2012. The electrocapacitive properties of graphene oxide reduced by urea. Energy & Environ. Sci. 5(4): 6391-6399.

Liang, Y., Z. Wang, J. Huang, H. Cheng, F. Zhao, Y. Hu, et al. 2015. Series of in-fiber graphene supercapacitors for flexible wearable devices. J. Mater. Chem. A 3(6): 2547-2551.

Liao, Y., Y. Gao, S. Zhu, J. Zheng, Z. Chen, C. Yin, et al. 2015. Facile fabrication of N-doped graphene as efficient electrocatalyst for oxygen reduction reaction. ACS Appl. Mater. Interfaces 7(35): 19619-19625.

Liu, J., F. Mirri, M. Notarianni, M. Pasquali and N. Motta. 2015. High performance all-carbon thin film supercapacitors. J. Power Sources 274: 823-830.

Liu, W.-w., X.-b. Yan, J.-w. Lang, J.-b. Pu and Q.-j. Xue. 2013. Supercapacitors based on graphene nanosheets using different non-aqueous electrolytes. New J. Chem. 37(7): 2186-2195.

Maiti, U. N., J. Lim, K. E. Lee, W. J. Lee and S. O. Kim. 2014. Three-dimensional shape engineered, interfacial gelation of reduced graphene oxide for high rate, large capacity supercapacitors. Adv. Mater. 26(4): 615-619.

Moses, K., V. Kiran, S. Sampath and C. N. R. Rao. 2014. Few-layer borocarbonitride nanosheets: platinum-free catalyst for the oxygen reduction reaction. Chem. Asian J. 9(3): 838-843.

Niu, L., Z. Li, W. Hong, J. Sun, Z. Wang, L. Ma, et al. 2013. Pyrolytic synthesis of boron-doped graphene and its application as electrode material for supercapacitors. Electrochimica Acta 108(0): 666-673.

Novoselov, K. S., A. K. Geim, S. V. Morozov, D. Jiang, Y. Zhang, S. V. Dubonos, et al. 2004. Electric field effect in atomically thin carbon films. Science 306(5696): 666-669.

Novoselov, K. S., A. K. Geim, S. V. Morozov, D. Jiang, M. I. Katsnelson, I. V. Grigorieva, et al. 2005. Two-dimensional gas of massless dirac fermions in graphene. Nature 438(7065): 197-200.

Novoselov, K. S., V. I. Falko, L. Colombo, P. R. Gellert, M. G. Schwab and K. Kim. 2012. A roadmap for graphene. Nature 490(7419): 192-200.

Pandey, K., P. Yadav and I. Mukhopadhyay. 2015. 3D flower-like MoS_2 nanostructures in aqueous and ionic liquid media. RSC Adv. 5(71): 57943-57949.

Peng, H., G. Ma, K. Sun, Z. Zhang, Q. Yang, F. Ran, et al. 2015b. A facile and rapid preparation of highly crumpled nitrogen-doped graphene-like nanosheets for high-performance supercapacitors. J. Mater. Chem. A 3(25): 13210-13214.

Peng, Z., J. Lin, R. Ye, E. L. G. Samuel and J. M. Tour. 2015a. Flexible and stackable laser-induced graphene supercapacitors. ACS App. Mater. Interfaces 7(5): 3414-3419.

Qu, L., Y. Liu, J.-B. Baek and L. Dai. 2010. Nitrogen-doped graphene as efficient metal-free electrocatalyst for oxygen reduction in fuel cells. ACS Nano 4(3): 1321-1326.

Raccichini, R., A. Varzi, S. Passerini and B. Scrosati. 2015. The role of graphene for electrochemical energy storage. Nat. Mater. 14(3): 271-279.

Rao, C. N. R., A. K. Sood, K. S. Subrahmanyam and A. Govindaraj. 2009. Graphene: the new two-dimensional nanomaterial. Angew. Chem. Inter. Ed. 48(42): 7752-7777.

Rao, C. N. R., U. Maitra and H. S. S. R. Matte. 2012. Synthesis, characterization and selected properties of graphene. *In*: Graphene: Synthesis, Properties and Phenomena. Wiley-VCH Verlag GmbH & Co. KGaA: 1-47.

Rao, C. N. R., H. S. S. Ramakrishna Matte and U. Maitra. 2013. Graphene analogues of inorganic layered materials. Angew. Chem. Inter. Ed. 52(50): 13162-13185.

Rao, C. N. R., K. Gopalakrishnan and A. Govindaraj. 2014a. Synthesis, properties and applications of graphene doped with boron, nitrogen and other elements. Nano Today 9(3): 324-343.

Rao, C. N. R., U. Maitra and U. V. Waghmare. 2014b. Extraordinary attributes of 2-dimensional MoS_2 nanosheets. Chem. Phys. Lett. 609: 172-183.

Rao, C. N. R., K. Gopalakrishnan and U. Maitra. 2015. Comparative study of potential applications of graphene, MoS_2 and other two-dimensional materials in energy devices, sensors and related areas. ACS App. Mater. Interfaces 7(15): 7809-7832.

Shao, Y., S. Zhang, M. H. Engelhard, G. Li, G. Shao, Y. Wang, et al. 2010. Nitrogen-doped graphene and its electrochemical applications. J. Mater. Chem. 20(35): 7491-7496.

Stoller, M. D., S. Park, Y. Zhu, J. An and R. S. Ruoff. 2008. Graphene-based ultracapacitors. Nano Lett. 8(10): 3498-3502.

Stoller, M. D. and R. S. Ruoff. 2010. Best practice methods for determining an electrode material's performance for ultracapacitors. Energy & Environ. Sci. 3(9): 1294-1301.

Subrahmanyam, K. S., L. S. Panchakarla, A. Govindaraj and C. N. R. Rao. 2009. Simple method of preparing graphene flakes by an arc-discharge method. J Phys. Chem. C 113(11): 4257-4259.

Sun, G., X. Zhang, R. Lin, J. Yang, H. Zhang and P. Chen. 2015. Hybrid fibers made of molybdenum disulfide, reduced graphene oxide and multi-walled carbon nanotubes for solid-state, flexible, asymmetric supercapacitors. Angew. Chem. Inter. Ed. 54(15): 4651-4656.

Sun, L., L. Wang, C. Tian, T. Tan, Y. Xie, K. Shi, et al. 2012. Nitrogen-doped graphene with high nitrogen level via a one-step hydrothermal reaction of graphene oxide with urea for superior capacitive energy storage. RSC Adv. 2(10): 4498-4506.

Sun, Y., Q. Wu and G. Shi. 2011. Supercapacitors based on self-assembled graphene organogel. Phys. Chem. Chem. Phys. 13(38): 17249-17254.

Vivekchand, S. R. C., C. Rout, K. S. Subrahmanyam, A. Govindaraj and C. N. R. Rao. 2008. Graphene-based electrochemical supercapacitors. J. Chem. Sci. 120(1): 9-13.

Wang, D.-W., F. Li, J. Zhao, W. Ren, Z.-G. Chen, J. Tan, et al. 2009b. Fabrication of graphene/polyaniline composite paper via *in situ* anodic electropolymerization for high-performance flexible electrode. ACS Nano 3(7): 1745-1752.

Wang, H., Q. Hao, X. Yang, L. Lu and X. Wang. 2009a. Graphene oxide doped polyaniline for supercapacitors. Electrochem. Commun. 11(6): 1158-1161.

Wang, H., T. Maiyalagan and X. Wang. 2012a. Review on recent progress in nitrogen-doped graphene: synthesis, characterization and its potential applications. ACS Catalysis 2(5): 781-794.

Wang, H., H. Feng and J. Li. 2014a. Graphene and graphene-like layered transition metal dichalcogenides in energy conversion and storage. Small 10(11): 2165-2181.

Wang, S., L. Zhang, Z. Xia, A. Roy, D. W. Chang, J.-B. Baek, et al. 2012b. BCN graphene as efficient metal-free electrocatalyst for the oxygen reduction reaction. Angew. Chem. Inter. Ed. 51(17): 4209-4212.

Wang, T., D. Gao, J. Zhuo, Z. Zhu, P. Papakonstantinou, Y. Li, et al. 2013a. Size-dependent enhancement of electrocatalytic oxygen-reduction and hydrogen-evolution performance of MoS_2 particles. Chem. Eur. J. 19(36): 11939-11948.

Wang, T., J. Zhuo, Y. Chen, K. Du, P. Papakonstantinou, Z. Zhu, et al. 2014b. Synergistic catalytic effect of MoS_2 nanoparticles supported on gold nanoparticle films for a highly efficient oxygen reduction reaction. Chem. Cat. Chem. 6(7): 1877-1881.

Wang, X., Y. Zhang, C. Zhi, X. Wang, D. Tang, Y. Xu, et al. 2013b. Three-dimensional strutted graphene grown by substrate-free sugar blowing for high-power-density supercapacitors. Nat. Commun. 4: 2905.

Winchester, A., S. Ghosh, S. Feng, A. L. Elias, T. Mallouk, M. Terrones, et al. 2014. Electrochemical characterization of liquid phase exfoliated two-dimensional layers of molybdenum disulfide. ACS App. Mater. Interfaces 6(3): 2125-2130.

Wu, J., L. Ma, R. M. Yadav, Y. Yang, X. Zhang, R. Vajtai, et al. 2015. Nitrogen-doped graphene with pyridinic dominance as a highly active and stable electrocatalyst for oxygen reduction. ACS App. Mater. Interfaces 7(27): 14763-14769.

Wu, Q., Y. Xu, Z. Yao, A. Liu and G. Shi. 2010. Supercapacitors based on flexible graphene/polyaniline nanofiber composite films. ACS Nano 4(4): 1963-1970.

Wu, Z. S., K. Parvez, X. Feng and K. Mullen. 2013. Graphene-based in-plane micro-supercapacitors with high power and energy densities. Nat. Commun. 4: 2487.

Xia, J., F. Chen, J. Li and N. Tao. 2009. Measurement of the quantum capacitance of graphene. Nat. Nano. 4(8): 505-509.

Xiao, Y., L. Huang, Q. Zhang, S. Xu, Q. Chen and W. Shi. 2015. Gravure printing of hybrid MoS_2@S-rGO interdigitated electrodes for flexible micro-supercapacitors. App. Phys. Lett. 107(1): 013906.

Xie, B., Y. Chen, M. Yu, X. Shen, H. Lei, T. Xie, et al. 2015. Carboxyl-assisted synthesis of nitrogen-doped graphene sheets for supercapacitor applications. Nanoscale Res. Lett. 10: 332.

Xin, H., D. He, Y. Wang, W. Zhao and X. Du. 2015. Low-temperature preparation of macroscopic nitrogen-doped graphene hydrogel for high-performance ultrafast supercapacitors. RSC Adv. 5(11): 8044-8049.

Xue, Y., D. Yu, L. Dai, R. Wang, D. Li, A. Roy, et al. 2013. Three-dimensional B, N-doped graphene foam as a metal-free catalyst for oxygen reduction reaction. Phys. Chem. Chem. Phys. 15(29): 12220-12226.

Yang, H., S. Kannappan, A. S. Pandian, J.-H. Jang, Y. S. Lee and W. Lu. 2015. Nanoporous graphene materials by low-temperature vacuum-assisted thermal process for electrochemical energy storage. J. Power Sources 284: 146-153.

Yoo, J. J., K. Balakrishnan, J. Huang, V. Meunier, B. G. Sumpter, A. Srivastava, et al. 2011. Ultrathin planar graphene supercapacitors. Nano Lett. 11(4): 1423-1427.

Zhang, J., T. Tian, Y. Chen, Y. Niu, J. Tang and L.-C. Qin. 2014. Synthesis of graphene from dry ice in flames and its application in supercapacitors. Chem. Phys. Lett. 591: 78-81.

Zhang, Y., W. Sun, X. Rui, B. Li, H. T. Tan, G. Guo, et al. 2015. One-pot synthesis of tunable crystalline Ni3S4@Amorphous MoS_2 core/shell nanospheres for high-performance supercapacitors. Small 11(30): 3694-3702.

Zhao, K., W. Gu, L. Zhao, C. Zhang, W. Peng and Y. Xian. 2015. MoS_2/Nitrogen-doped graphene as efficient electrocatalyst for oxygen reduction reaction. Electrochimica Acta 169: 142-149.

Zhou, X., B. Xu, Z. Lin, D. Shu and L. Ma. 2014. Hydrothermal synthesis of flower-like MoS_2 nanospheres for electrochemical supercapacitors. J. Nanosci. Nanotech. 14(9): 7250-7254.

Zhou, X., Z. Bai, M. Wu, J. Qiao and Z. Chen. 2015. 3-Dimensional porous N-doped graphene foam as a non-precious catalyst for the oxygen reduction reaction. J. Mater. Chem. A 3(7): 3343-3350.

Zhu, J., W. Sun, D. Yang, Y. Zhang, H. H. Hoon, H. Zhang, et al. 2015. Multifunctional architectures constructing of PANI nanoneedle arrays on MoS_2 thin nanosheets for high-energy supercapacitors. Small 11(33): 4123-4129.

Zhu, Y., S. Murali, M. D. Stoller, A. Velamakanni, R. D. Piner and R. S. Ruoff. 2010. Microwave assisted exfoliation and reduction of graphite oxide for ultracapacitors. Carbon 48(7): 2118-2122.

Zhu, Y., S. Murali, M. D. Stoller, K. J. Ganesh, W. Cai, P. J. Ferreira, et al. 2011. Carbon-based supercapacitors produced by activation of graphene. Science 332(6037): 1537-1541.

Zuo, Z., Z. Jiang and A. Manthiram. 2013. Porous B-doped graphene inspired by fried-ice for supercapacitors and metal-free catalysts. J. Mater. Chem. A 1(43): 13476-13483.

2

Layer-Structured Thermoelectric Materials: Fundamentals, Strategies and Progress

Zhi-Gang Chen[1,*] and Jin Zou[1,2,a]

ABSTRACT

Thermoelectric materials convert between heat and electricity without any emissions or vibrational parts, offering a sustainable solution to overcome the upcoming energy crisis. Compared with conventional bulk thermoelectric materials, layer-structured thermoelectric materials show unique advantage, such as super-low thermal conductivity and high power factor with enhanced thermoelectric performance, which has become a research focus in recent years. This chapter provides an overview on thermoelectric fundamentals and the strategies to enhance the thermoelectric performances. On the basis of fundamental analysis, several key strategies to engineer the thermoelectric parameters have been highlighted. Finally, the research progress of the state-of-art layer-structured thermoelectric materials has been summarized to understand and explain the underpinnings of the innovative breakthroughs in the last decade or so. We believe that recent achievements will augur the possibility for thermoelectric power generation and cooling and discuss several future directions which could lead to new exciting next generation of layer-structured thermoelectrics.

Keywords: Thermoelectric, layer-structured, fundamentals, strategies, engineer.

1. INTRODUCTION

The world faces numerous challenges relating to energy supply and consumption. Current average annual global power consumption is approximately 14 terawatt,

[1] Materials Engineering, The University of Queensland, ST Lucia, QLD 4072, Australia.

[2] Centre for Microscopy and Microanalysis, The University of Queensland, St Lucia, QLD 4072, Australia.

[a] E-mail: j.zou@uq.edu.au

* Corresponding author: z.chen1@uq.edu.au

with more than 80% consumption coming from fossil fuels, including oil, coal and natural gas (Chen et al. 2012). This consumption of unrenewable fossil fuel discharges greenhouse gas that result in global warming. All of these issues are driving the demand for obtaining more useful energy. For example, many new vehicle engines, transmissions and associated technologies are under development to increase transport vehicle fuel efficiency. However, these technologies are missing one important issue: much of the energy will still produce unusable heat in the vehicle exhaust or the cooling system. In fact, in internal combustion engines, approximately 40% of the fuel energy is wasted in exhaust gas, 30% is dissipated in the engine coolant, 5% is lost as radiation and friction and only 25% is useable for vehicle mobility and accessories (Bottner et al. 2006). Current waste energy assessments indicate that the waste energy, equivalent of 46 billion gallons of gasoline, is wasted annually from the exhaust pipes of ~200 million light-duty vehicles in the US alone (Bottner et al. 2006). In addition, far greater waste heat is generated in the end-to-end electrical production loses, which amounts to around two thirds of available energy in power-generating plants and manufacturing industries. Therefore, the potential of developing high-efficiency thermoelectric materials for waste heat recovery systems (such as thermoelectric generator) is huge.

Thermoelectric materials convert between heat and electricity without any emissions or vibrational parts, offering the opportunity to overcome the upcoming energy crisis. Thermoelectric technology plays a key role in our current challenge to develop alternative energy technologies to reduce our dependence on fossil fuels and to reduce greenhouse gas emissions. However, current available thermoelectric materials are not in common use, partly due to their low efficiency relative to mechanical cycles and engineer challenges related to using thermoelectrics for practical applications (Bell 2008). Therefore, the need of developing high-efficiency thermoelectric materials for waste heat recovery systems is urgent and will bring vast economic and environmental benefits.

Compared with conventional bulk thermoelectric materials, layer-structured thermoelectric materials show unique advantages, such as super-low thermal conductivity and high power factor with enhanced thermoelectric performance, which has become a research focus in recent years. In this chapter, we will provide an overview on thermoelectric fundamentals and the strategies to enhance the thermoelectric performances. On the basis of fundamental analysis, several strategies to engineer the thermoelectric parameters will be proposed. Finally, the research progress of the state-of-art layer-structured thermoelectric materials is highlighted.

2. FUNDAMENTALS FOR THERMOELECTRICS

The Seebeck effect (Seebeck coefficient, S) was first discovered in 1821 (Steele 1893). As illustrated in Fig. 2.1, electricity can be observed from heat gradient due to the fact that mobile electrons accumulate at the cold side to generate negative polar, while holes are left behind at the hot side to form positive polar. Eventually, an electric field is generated between the two ends to balance the thermal gradient effects on the electron movements (Steele 1893). Likewise, the Peltier effect (Peltier coefficient) (Yamanashi 1996), the reverse of the Seebeck effect, has been employed to produce thermoelectric refrigeration by applying electric current in the circuit to cause temperature difference on the junction of thermoelectric compounds. Based on the fact that S varies with temperature, a spatial gradient in temperature can result in a gradient in S. If a current is driven through this gradient, a continuous Peltier effect will occur. This was observed by William Thomson in 1851 and named as the Thomson effect (Sandoz-Rosado et al. 2013).

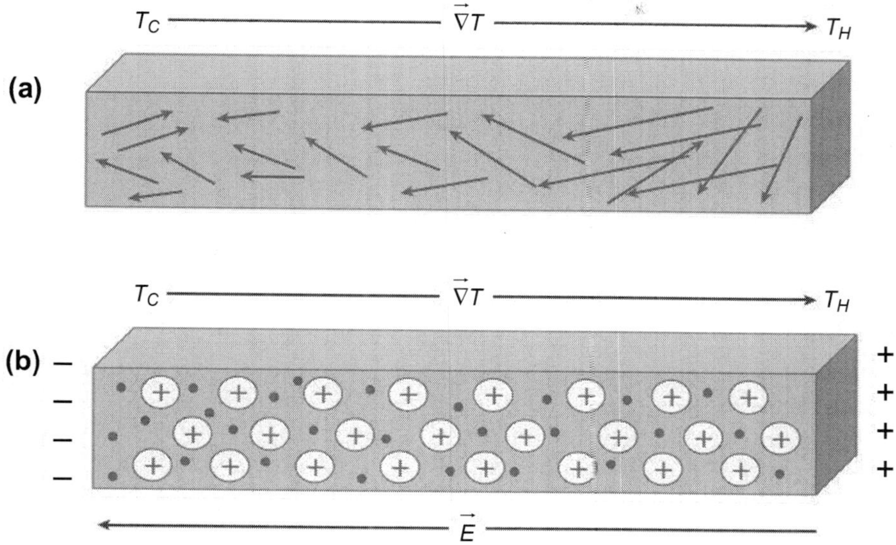

Fig. 2.1 Schematic illustration of the Seebeck effect: (a) Under temperature difference, charge carriers (electrons as an example) diffuse from the hot side to the cold side, as marked by arrows. (b) Mobile electrons (indicated by –) accumulate at the cold side to generate negative polar, while holes (indicated by +) are left behind at the hot side to form positive polar. Eventually, an electric field is generated between the two ends to balance the thermal gradient effects on electron movement.

2.1 Figure of Merit

The conversion efficiency of thermoelectric materials is defined the dimensionless figure of merit (ZT) and is given by (Hicks and Dresselhaus 1993).

$$ZT = \frac{S^2\sigma T}{\kappa} = \frac{S^2 T}{(\kappa_e + \kappa_l)\rho} \tag{1}$$

where σ is the electrical conductivity, ρ (= $1/\sigma$) is the electrical resistivity, T is the absolute temperature and κ is the thermal conductivity which is composed of the electronic contributions (κ_e) and lattice contributions (κ_l).

Categorized by their mobile charge carriers, the n-type thermoelectric materials are mainly governed by free electrons, while the p-type governed by holes. For real applications, both the n- and p-type thermoelectric components work together to form a thermoelectric couple with a combined ZT of (Lan et al. 2010).

$$ZT = \frac{(S_p - S_n)^2 T}{\left[\sqrt{k_p \rho_p} + \sqrt{k_n \rho_n}\right]^2} \tag{2}$$

In practice, a thermoelectric device contains many thermoelectric couples to meet the necessary output. For example, a thermoelectric power generator is a combination of numerous n- and p-type thermoelectric components that are connected electrically in series and thermally in parallel.

The efficiency (η_{TE}) of thermoelectric devices is a function of ZT and is presented by (Lan et al. 2010).

$$\eta_{TE} = \frac{W}{Q_n} = \eta_C \left(\frac{\sqrt{1+ZT} - 1}{\sqrt{1+ZT} + T_C/T_H} \right) \tag{3}$$

and the Carnot efficiency (η_C) is given by (Lan et al. 2010).

$$\eta_C = \frac{T_H - T_C}{T_H} \tag{4}$$

where Q_H is the net heat flow rate, W is the generated electric power, T_C is the temperature of the cold side and T_H is that of the hot side, respectively.

As can be seen from these equations, a significant difference in temperatures is needed to enhance η_C, so as to generate higher electrical energy. Currently, thermoelectric research has been focused on the optimization of a variety of conflicting parameters. To maximize ZT and in turn η_{TE}, of a material, high S and σ, as well as low κ, are required (Snyder and Toberer 2008). While there is no theoretical limit for maximizing ZT, the best bulk thermoelectric materials found so far, such as Bi_2Te_3, PbTe and $Si_{1-x}Ge_x$, have shown a peak ZT value of approximately 1 (Snyder and Toberer 2008), providing a moderated η_{TE} that is not sufficient for the large-scale applications of thermoelectric technology. A few remarkable progresses have been made to improve the thermoelectric properties, especially recent achievements to create nanostructured materials, such as superlattices, quantum dots, nanowires and nanocomposite, as illustrated in Fig. 2.2a, indicating the major milestones achieved for enhancing ZT over the past several decades.

Despite such high anticipation and achievements, the progress in developing thermoelectric materials has still been limited for practical applications due to lack of reliable, high performance, cost-effective thermoelectric materials (Tritt et al. 2008). Currently, the commercial thermoelectric generators on the market have a conversion efficiency of approximately 5% at the temperature of 100 K, as shown in Fig. 2.2b. As can be seen, there is a great potential to seek new materials with ZT values of 2-3 to provide desired conversion efficiencies to be competitive with traditional mechanical energy conversion systems. For a typical example, a thermoelectric power conversion device with $ZT = 3$ operating between 500°C and 30°C (room temperature) could provide η_C~30% (Tritt et al. 2008). Therefore, it is critical to develop thermoelectric materials with high ZT values in order to realize the practical applications of thermoelectric materials. The major activities in thermoelectric materials have been focused on the increase of the power factor ($S^2\sigma$) and the decrease of κ.

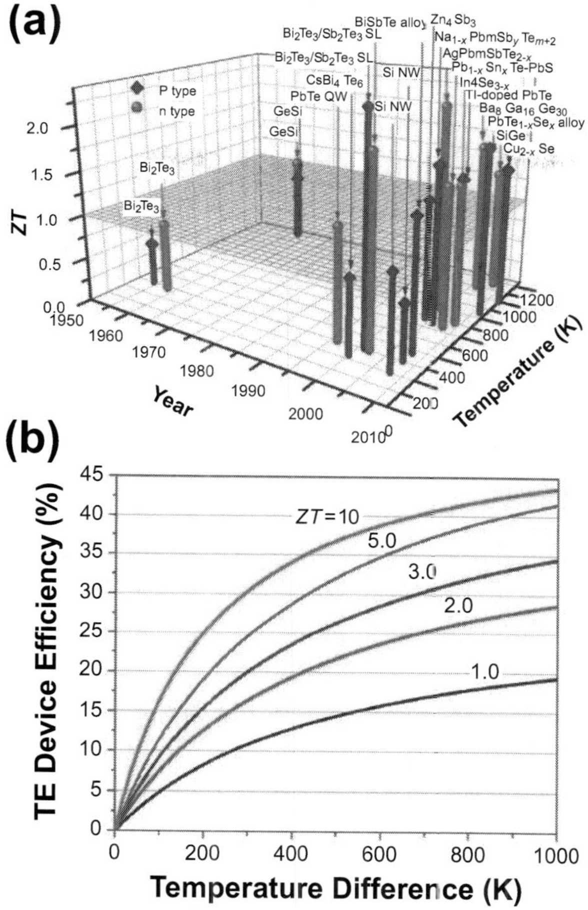

Fig. 2.2 (a) Peak ZT values as functions of temperature and year revealing the important development of thermoelectric materials. No material has yet achieved the target goal of $ZT \geq 3$ to date. (Bi_2Te_3 (Goldsmid and Douglas 1954), Bi_2Te_3 (Goldsmid et al. 1958), GeSi (Dismukes et al. 1964), GeSi (Dismukes et al. 1964), $Pb_{1-x}EuTe/PbTe$ MQWs (Harman et al. 1996), $CsBi_4Te_6$ (Chung et al. 2000), Bi_2Te_3/Sb_2Te_3 superlattices (Venkatasubramanian et al. 2001), PbSeTe-based QDSL (Harman et al. 2002), $AgPb_mSbTe_{2+m}$ (Hsu et al. 2004), $Na_{1-x}Pb_mSb_yTe_{m+2}$ (Poudeu et al. 2006), $Ba_8Ga_{16}Ge_{30}$ (Saramat et al. 2006), Si NWs (Boukai et al. 2008), Si NWs (Hochbaum et al. 2008), BiSbTe bulk alloy (Poudel et al. 2008), β-Zn_4Sb_3 (Snyder et al. 2004), $Pb_{1-x}Sn_xTe$-PbS (Androulakis et al. 2007), Tl-doped PbTe (Heremans et al. 2008), SiGe (Wang et al. 2008), $In_4Se_{3-\delta}$ (Rhyee et al. 2009b), Na-doped $PbTe_{1-x}Se_x$ alloy (Pei et al. 2011c) and nano β-$Cu_{2-x}Se$ (Yang et al. 2015). (b) Thermoelectric energy conversion as a function of ZT at the setting of T_C = 300 K.

2.2 Key Parameters for Controlling *ZT*

A material with high ZT needs to have a large S, often existing in low carrier concentration semiconductors or insulators and a high σ, often found in high carrier density (n) metals. Therefore, $S^2\sigma$ maximizes with n somewhere between semiconductors and metals, as revealed in Fig. 2.3 (Vaqueiro and Powell 2010). The transport properties for highly degenerate semiconductors is given by the Mott-Jones relations, expressed as (Snyder and Toberer 2008).

$$S = \frac{8\pi^2 k_B^2}{3eh^2} m^* T \left(\frac{\pi}{3n}\right)^{2/3} \tag{5}$$

$$\sigma = ne\mu \tag{6}$$

$$\kappa_e = L\sigma T \quad \text{and} \tag{7}$$

$$\kappa_e = 1/3\ (C_v v_s \lambda_{ph}) \tag{8}$$

where k_B is the Boltzmann constant, e is the carrier charge, h is the Plank's constant, m^* is the effective mass of the charge carrier, μ is the carrier mobility, L is Lorentz number, C_v is the constant-volume specific heat, v_s is the sound velocity and λ_{ph} is the phonon mean free path (Rowe 1995).

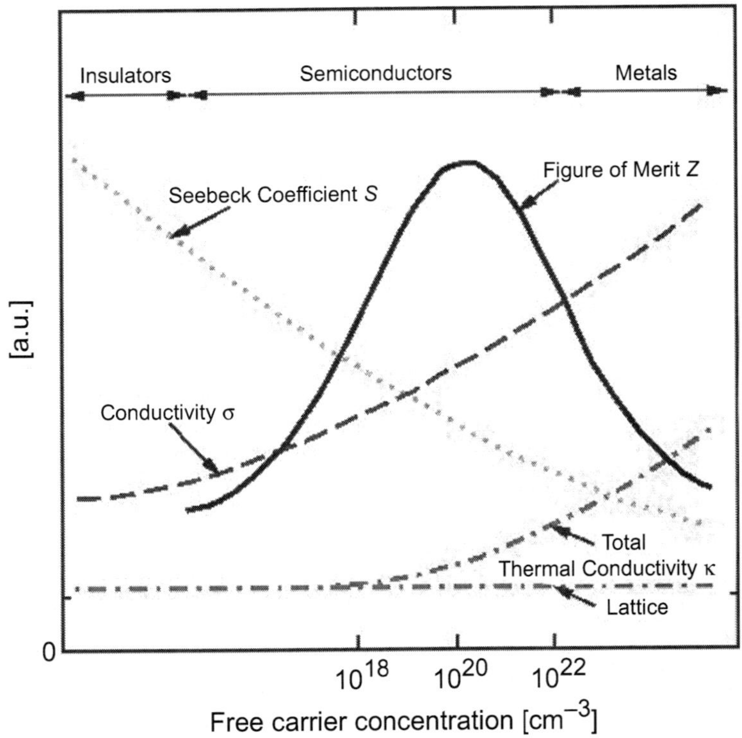

Fig. 2.3 Relations between S, σ, powder factor ($S^2\sigma$), as well as κ and n for a certain bulk thermoelectric material. Reproduced from ref. (Vaqueiro and Powell 2010).

Apparently, high S needs small n, which will in return decrease σ. Also, to obtain a high σ, electronic contribution to the thermal conductivity will increase as well. This dilemma among S, σ, $S^2\sigma$ and κ composed of κ_e and κ_l and n are presented in Fig. 2.3. It should be noted that, for the best thermoelectric materials, the maximized $S^2\sigma$ and minimized κ are desirable. Extensive studies have paid attention to explore new materials through increasing $S^2\sigma$ (by band convergence (Pei et al. 2011c, Wu et al. 2014, Wang et al. 2014), reversible phase transition (Zhao et al. 2014, Liu et al. 2012, Liu et al. 2013) and quantum confinement (Dresselhaus et al. 2007, Sun et al. 2012) and/or reducing κ by nanostructuring (Minnich et al. 2009, Lan et al. 2010, Kanatzidis 2010, Li et al. 2010), hierarchical architecturing (Chang et al. 2013, Han et al. 2014a, Zheng et al. 2015) and matrix with nano-precipitate (Poudel et al. 2008, Zhang et al. 2015a, Zhang et al. 2015b). However, the interdependence and coupling between these properties are still extremely challenging to enhance these parameters individually to realise peak *ZT*.

2.3 Material Requirements

Compared with metals and insulators, semiconductors with n-type or p-type are more eligible for thermoelectric applications. Despite their relatively low σ, their relatively high *S* and low κ make them become the most suitable thermoelectric systems. It is commonly believed that thermoelectric materials should be composed of heavy elements with large atomic number (Pichanusakorn and Bandaru 2010) with a suitable optimum band (Mahan 1989). Also, defects in alloys (Lan et al. 2010) and fabricating materials containing complex crystal structures (Snyder and Toberer 2008) can improve thermoelectric performance further.

Typically, good thermoelectric materials are heavily doped semiconductors with *n* in the range between 10^{19} and 10^{21} cm^{-3} (Vaqueiro and Powell 2010). To ensure a large *S*, a single type of carriers (n- or p-type) should be maintained, as mixed n-type/p-type charge carriers will lead to the opposite Seebeck effect and hence will lower the power factor (Heremans et al. 2012). In order to achieve single type of carriers, it is necessary to select materials with suitable energy bandgaps and appropriate doping, in which n- and p-type can be well separated. Therefore, effective thermoelectric materials are heavily doped semiconductors with energy bandgap within around between 6 k_BT and 10 k_BT (Sofo and Mahan 1994), in order to secure both a single carrier type and sufficiently high carrier mobility. Decoupling these thermal and electronic transport that determine *ZT*, including *S*, σ and κ, has been a key strategy to improve *ZT*. Heavy elements are always considered to use to reduce the phonon velocities to lower κ, while μ will not be affected significantly to reserve *S* and σ in thermoelectric system.

3. STRATEGIES TO ENHANCE THERMOELECTRIC PROPERTIES

For an ideal thermoelectric material, a high $S^2\sigma$ and a low κ are required to secure a high *ZT*. However, it is always a challenge to optimize the individual parameters of σ, *S* and κ for thermoelectric materials due to their interdependent and conflict (Rowe 2005). So far, besides using band engineering through tuning band convergence (Pei et al. 2011c), quantum confinement (Hicks et al. 1993, Molenkamp et al. 1993) and effective mass (Pei et al. 2012a) to maximizing $S^2\sigma$, most successful *ZT* enhancements have been achieved via nanostructural engineering (Snyder and Toberer 2008) to reduce κ.

3.1 Nanostructure Engineering

In conventional three-dimensional (3D) crystalline systems, quantities *S*, σ and κ are interrelated and very difficult to balance in a form that an increase in *S* usually results in a decrease in σ and a decrease in σ produces a decrease in κ_e. However, if the dimensionality of a material is decreased, new variables of length scale up to quantum effects become available for the control of material properties. With the size decrease, the density of electronic states can be changed dramatically, as illustrated in Fig. 2.4, which allows new opportunities to tune *S*, σ and κ. In addition, as the dimensionality is decreased from 3D crystalline solids to 2D (quantum wells) to 1D (quantum wires) and finally to 0D (quantum dots), new physical phenomena are also introduced and these phenomena may also create new opportunities to vary *S*, σ and κ independently (Dresselhaus et al. 2007, Bux et al. 2010, Szczech et al. 2011). Besides, the mean free path of electrons is much shorter than that of phonons in heavily doped semiconductors (Dresselhaus et al. 2007), nanostructuring serves to introduce a large density of interfaces in which phonons with a wide range of mean free path can be scattered more effectively and preferentially than electrons, as illustrated in Fig. 2.5. As a consequence, nanostructuring can reduce κ_l effectively while preserve μ and κ_e (Szczech et al. 2011). This can be satisfied by preparation of nanostructures with one or more dimensions smaller than the mean free path of phonons, while still larger than the mean free path of charge carriers. Therefore, the field of low-dimensional thermoelectricity should follow two strategies: the use of quantum confinement to enhance $S^2\sigma$ and to control *S* and σ somewhat independently and the use of numerous interfaces to effectively scatter phonons than electrons and to scatter preferentially those phonons that contribute most strongly to κ.

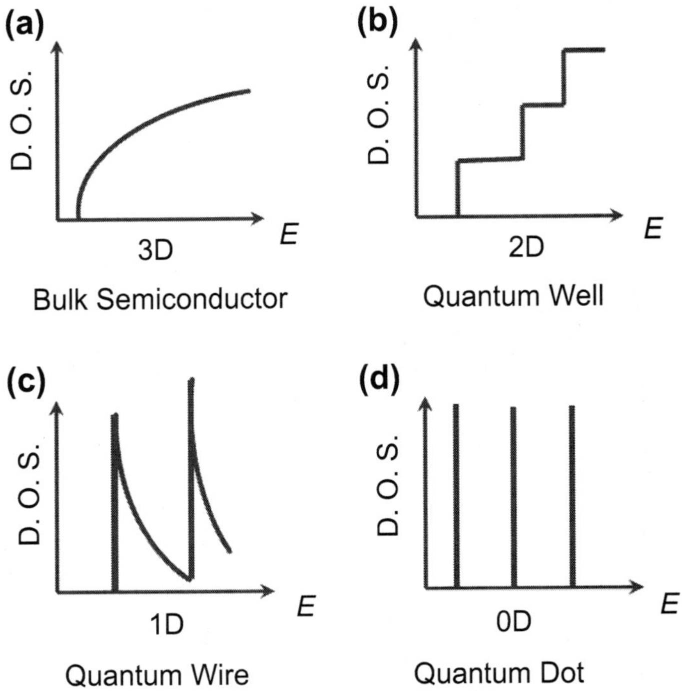

Fig. 2.4 Electronic density of states for (a) A bulk 3D crystalline semiconductor. (b) A 2D quantum well. (c) A 1D nanowire or nanotube. (d) A 0D quantum dot. Reproduced from ref. (Dresselhaus et al. 2007).

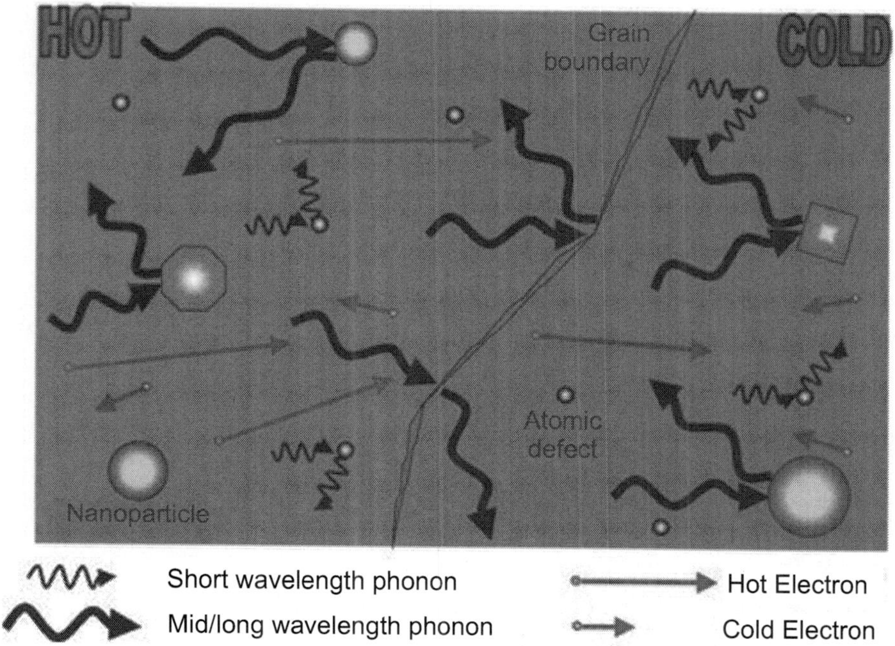

Fig. 2.5 Schematic diagram illustrating phonon scattering mechanisms and electronic transport of hot and cold electrons within a thermoelectric material. Reproduced from ref. (Vineis et al. 2010).

For most thermoelectrics, the bipolar κ_{bipolar} can be a significant component of the total κ at elevated high temperature and in turn S will be further considerably reduced. From the fundamentals, electrons in the valence band are excited to the conduction band, leaving behind an equal number of holes, which finally contribute according to the Eq. (7). The potential methodology to reduce the bipolar effect is to enhance the scattering of minor carriers by nanostructure engineering. A reduced bipolar effect was observed in p-type $Bi_xSb_{2-x}Te_3$ nanocomposites (Poudel et al. 2008). The electrons in p-type $Bi_xSb_{2-x}Te_3$ are more strongly scattered by the grain boundaries than the holes, resulting in a decreased electron contribution to the transport properties. Finally, a recorded high ZT of 1.4 is observed in such a $Bi_xSb_{2-x}Te_3$ system.

3.2 Band Engineering

Under favourable electron scattering by acoustic phonons for thermoelectrics, different band engineering strategies, such as resonate energy level tuning (Heremans et al. 2012), energy filtering (Zide et al. 2006, Vashaee and Shakouri 2004, Heremans et al. 2004, Paul et al. 2010, Prytz et al. 2009), increasing band divergence (Pei et al. 2011c) and lowering the effective mass (Pei et al. 2012b), have been used to increase in ZT, since $ZT \propto N_v/m_b^* \; \Xi^2$ (Pei et al. 2012a) and $S \propto g(E)$ (Heremans et al. 2008), where N_v is band degeneracy, m_b^* is band effective mass, Ξ is scattering strength and $g(E)$ is the electronic density of states. These strategies have led to *ZT* increases from 1.0 to 1.5 ~2.0 in several material systems (Pei et al. 2011a, b, Poudel et al. 2008, Liu et al. 2012, Snyder et al. 2004).

3.2.1 Resonant Energy Level Engineering

It has been found that the distortion of the density of states by a resonant impurity can be achieved by doping TI in PbTe, leading to a double *ZT* value than its parent semiconductor (Heremans et al. 2008), Fundamentally, according to the Mahan-Sofo theory (Mahan and Sofo 1996) increasing μ can magnify *S*, while the distortion of the density of states at the Fermi level can significantly boost *S*. Therefore, choosing a suitable dopant to manipulate energy levels near the Fermi level can lead to enhanced thermoelectric performance by increasing the density of electronic states near the Fermi level to enhance *S*. For example, Sn was used as a resonant impurity in Bi_2Te_3 to enlarge the overall *S* (Jaworski et al. 2009). Figure 2.6 shows the experimental result, in which an obvious enhanced *S* was observed in the Jaworski's study (Jaworski et al. 2009). This experimental result provided direct experimental indication for the presence of the resonant level in Sn doped Bi_2Te_3 system, which verified the resonant level can enlarge *S* (Kulbachinskii et al. 1988, Heremans et al. 2012).

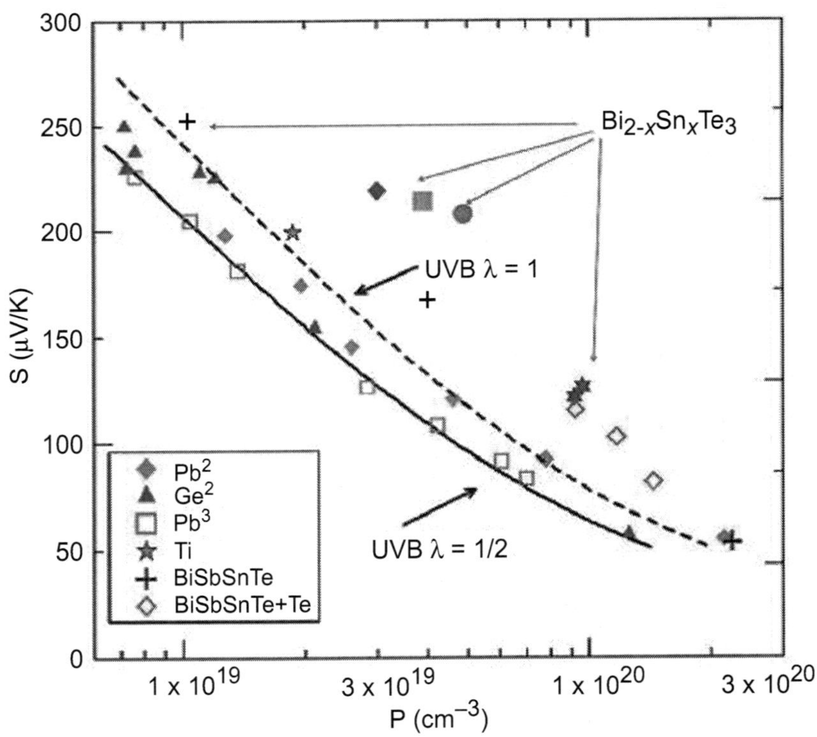

Fig. 2.6 Pisarenko relation (*S* versus carrier concentration) at 300 K for p-type Bi_2Te_3. The lines are calculated using a classical parabolic band with the effective mass for the UVB. The data points are obtained for the different acceptor impurities as marked. The points where Sn is used as a dopant differ significantly from the others and from the calculated lines. Reproduced from (Jaworski et al. 2009).

3.2.2 Carrier Energy Filter Engineering

The concept of carrier energy filtering is that, by introducing carrier barriers (~1-10 k_BT) in the conduction band (for n-type semiconductors) or valence band (for p-type semiconductors), only 'hot' carriers with high energy can be used for the materials conductivities. Such a carrier energy filtering can result in a substantially increased *S*, as *S* depends on the thermal energy transported by carriers (Martin et al. 2009). The corresponding enhanced *ZT* will be offset to some extent by decreased σ since fewer carriers participate in conduction and the extent of this reduction depends critically on whether energy or momentum filtering occurs. Thus, the overall impact on $S^2\sigma$ from this approach will be dependent on the specific material system but it is an exciting area worthy of further investigation, especially since it provides a possible avenue to enhanced $S^2\sigma$. For example, p-type Pt-doped Sb_2Te_3 nanocomposites showed an enlarged *S* by eliminating carriers with low energy via the energy filtering approach (Ko et al. 2011). Despite a little decrease in σ, the remarkable enhancement in *S* from 115.6 μV K^{-1} to 160 μV K^{-1}, resulting in a $S^2\sigma$ increase to 1.02 ± 0.36 μW cm^{-2}K^{-1} as a comparison with 0.96 ± 0.14 μW cm^{-2}K^{-1} measured by its Sb_2Te_3 counterpart.

Carrier trapping in the grain boundaries to form energy barriers is another approach and has been widely used in nanocomposites, especially in nanodot embedded nanocomposites (Zide et al. 2006, Vashaee and Shakouri 2004, Heremans et al. 2004, Paul et al. 2010, Prytz et al. 2009). These energy barriers can impede the conduction of carriers between grains and filter charge carriers with energy less than the barrier height to result in an enhanced *S*, which finally lead to an enhanced *ZT* value (Zide et al. 2006, Vashaee and Shakouri 2004, Heremans et al. 2004, Paul et al. 2010, Prytz et al. 2009).

3.2.3 Carrier-Pocket Engineering

Carrier-pocket engineering is used to design a superlattice structure, so that one type of carrier can be quantum confined in the quantum-well region and the opposite type of carrier can be quantum confined in the barrier region (Koga et al. 1998). This is accomplished by utilizing the differences in confinement energy experienced by different carrier pockets, to shift the carrier pocket energies to similar levels to enhance $S^2\sigma$.

The carrier-pocket engineering has been used in several different thermoelectric systems, such as cases of Bi_2Te_3-Sb_2Te_3 quantum well superlattices, Γ-point electrons for GaAs quantum wells and X-point electrons in the AlAs barriers in GaAs/AlAs quantum-well superlattices (Koga et al. 1998). Si/SiGe 2D superlattices (Koga et al. 1999). A significant enhancement of *ZT* has been observed.

4. PROGRESS IN LAYER-STRUCTURED THERMOELECTRIC MATERIALS

4.1 Layer-Structured Bi_2Te_3-Based Thermoelectric Materials

Layer-structured metal chalcogenide, mainly Bi_2Te_3,Bi_2Se_3 and Sb_2Te_3, the best thermoelectric materials around room temperature, have been extensively used for the first thermoelectric devices for commercial Peltier elements. Their typical crystal structure belongs to the rhombohedral system with a layered structure (Chen et al. 2012, Chen et al. 2014, Cheng et al. 2012, Cheng et al. 2013). As for a typical example, Bi_2Te_3 has a narrow bandgap (~0.15 eV) and its crystal structure is shown in Fig. 2.7. Under the hexagonal indexing, rhombohedral Bi_2Te_3 structure consists of stacked quintuple layers along its *c*-axis. These quintuple layers are bonded by van der Waals interactions between them. Within a quintuple layer, there are ionic and covalent bands exist between Bi and Te ions, which are much stronger than the van der Waals interaction. These structural characteristics result in very anisotropic properties and make Bi_2Te_3 easily cleave along their *c* planes. As for the band structure of Bi_2Te_3, there are six valleys for the valence band maximums, located in the mirror planes of the Brillouin Zone and the corresponding m^* is also highly anisotropic. Due to the narrow band gap, high valley degeneracy and the anisotropic effective mass, Bi_2Te_3 shows a relatively high σ and high S. Natively, stoichiometrically crystallized Bi_2Te_3 ingots have intrinsic p-type conductivity because their excessive Bi ions are easily able to form antisite defects. When an excess of Te was achieved in Bi_2Te_3 alloy, the conductivity type of Bi_2Te_3 can be tuned to an n-type due to the formation of Bi vacancies and antisite Te replaced Bi (Chen et al. 2014, Cheng et al. 2012, Cheng et al. 2013).

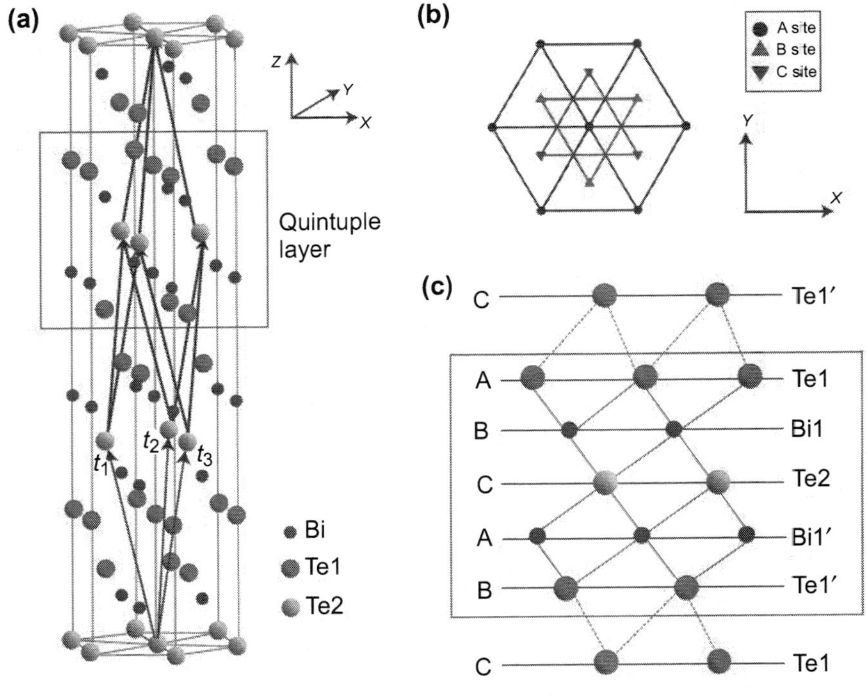

Fig. 2.7 Crystal structure: (a) Crystal structure of Bi_2Te_3 with three primitive lattice vectors denoted as t1; 2; 3. A quintuple layer with Te1–Bi1–Te2–Bi1′–Te1′ is indicated by the red square. (b) Top view along the *z*-direction. The triangle lattice in one quintuple layer has three different positions, denoted as A, B and C. (c) Side view of the quintuple layer structure. Along the *z*-direction, the stacking order of Te and Bi atomic layers is •••–C (Te1′)–A (Te1)–B (Bi1)–C (Te2)–A (Bi1′)–B (Te1′)–C (Te1)–•••. The Te1 (Bi1) layer can be related to the Te1′ (Bi1′) layer by an inversion operation in which the Te2 atoms have the role of inversion centres.

Table 2.1 State of the art Bi_2-Te_3-based nanocomposites.

Material systems	Carrier type	*ZT*	κ_l [$Wm^{-1}K^{-1}$]	*T*[K]	Synthetic method*	Ref.
BiSbTe	p	1.2	–	300	HEBM + HP	(Poudel et al. 2008)
BiSbTe	p	1.4	–	373	HEBM + HP	(Poudel et al. 2008)
BiSbTe	p	1.3	–	373	HEBM + HP	(Ma et al. 2008)
BiSbTe	p	1.4	–	373	HEBM + HP	(Lan et al. 2009)
$Bi_2Te_{2.7}Se_{0.3}$	n	1.04	–	400	HEBM + HP	(Yan et al. 2010)
$(Bi, Sb)_2Te_3$	p	1.5	–	390	MS + SPS	(Xie et al. 2010)
$(BiSb)_2Te_3$	p	1.47	–	440	HS + HP	(Cao et al. 2008)
$Bi_{0.52}Sb_{1.48}Te_3$	p	1.56	0.26	300	MS + SPS	(Xie et al. 2009b)
Bi_2Te_3	n	1	0.3	450	HS + HP	(Zhao et al. 2009)
$Bi_{0.4}Sb_{1.6}Te_3$	p	1.5	0.16	300	MS + HP	(Fan et al. 2010)
$Bi_{0.4}Sb_{1.6}Te_3$	p	1.8	–	315	MS + HP	(Fan et al. 2010)
$Bi_{0.5}Sb_{1.5}Te_3$	p	1.86		320	MS + HP	(Kim et al. 2015)

The abbreviations used in the column of the synthetic method represent the following meanings: HEBM = High Energy Ball Milling, HP = Hot Pressing, MS = Melting Solution, SPS = Spark Plasma Sintering.

With regard to the thermoelectric properties of Bi_2Te_3-based systems, there exists a wide variation in *ZT* values, ranging from ~0.4 to 1.85, which is summarized in Table 2.1. For example, the best *ZT* for single crystal Bi_2Te_3 is only 0.8, while polycrystalline p-type $Bi_{0.5}Sb_{1.5}Te_3$ bulk nanocomposites, fabricated by hot pressing of ball-milled nanopowders, exhibited *ZT* of 1.2 at room temperature and *ZT* of 1.4 at 373 K (Poudel et al. 2008). In this study, κ was significantly reduced to 1.0 $Wm^{-1}K^{-1}$ in the nanocomposite from 1.3 $Wm^{-1}K^{-1}$ in the bulk ingots with the same composition, although σ was slightly reduced in the nanocomposite; therefore, the maximum *ZT* of the nanocomposite was almost 30% higher than that of bulk ingots. Complex polycrystalline structures, with diameters ranging from a few micros down to a few nanometres were detected in the nanocomposite. In addition, Sb-rich nanodots ranging from 2 to 10 nm in diameter with diffuse boundaries and pure Te precipitates with diameter between 5 to 30 nm were also observed. These nanostructures can effectively scatter phonons with a broad range of wavelengths, which may account for the enhancement of thermoelectric properties in nanostructured Bi_2Te_3-based nanocomposites. In other follow-up researches, p-type $Bi_xSb_{2-x}Te_3$ nanocomposites produced by ball-milling showed a high *ZT*, about 1.3 between 75 and 100°C (Ma et al. 2008) and 1.4 at 100°C (Lan et al. 2009) and synthesized n-type Bi_2Te_3-based nanocomposite with a *ZT* of 1.04 at 125°C (Yan et al. 2010). A high *ZT* of about 1.5 at 390 K was achieved in $(Bi,Sb)_2Te_3$ nanocomposites by melt-spinning single elements of Bi, Sb and Te followed by s spark plasma sintering process. *ZT* = 1.47 (Cao et al.

2008) and $ZT = 1.56$ (Xie et al. 2009a, 2009b) were achieved in nanocomposites fabricated by hot pressing of Bi_2Te_3 and Sb_2Te_3 nanopowders and p-type $Bi_{0.52}Sb_{1.48}Te_3$ with 5 to 15 nm diameter precipitates produced by melt spinning and spark plasma sintering. Other Bi_2Te_3-based nanocomposites have revealed similar thermoelectric properties to bulk Bi_2Te_3 (Zhao et al. 2005, Fan et al. 2010), while others have poor performance in contrast (Fan et al. 2007). It can be deduced that the detailed structural and compositional characteristics of the nanostructures should play an important role in their represented thermoelectric performance.

Very recently, a significantly enhanced ZT in Te excess $Bi_{0.5}Sb_{1.5}Te_3$ nanocomposites has been realized by reducing κ_l through grain boundary and point-defect scattering, which target to scatter low- and high-frequency phonons (Kim et al. 2015). Dense dislocation arrays formed at low-energy grain boundaries by liquid-phase compaction in $Bi_{0.5}Sb_{1.5}Te_3$ effectively scatter mid-frequency phonons, leading to a substantially lower κ_l with a value of 0.33 $Wm^{-1}K^{-1}$ at 320 K, as shown in Fig. 2.8. Full-spectrum phonon scattering with minimal charge-carrier scattering dramatically improved its ZT up to 1.86 ± 0.15 at 320 K. A developed thermoelectric cooler using these Te excess $Bi_{0.5}Sb_{1.5}Te_3$ nanocomposites has a performance with a maximum temperature difference of 81 K, which is much higher than current commercial Peltier cooling devices (Kim et al. 2015).

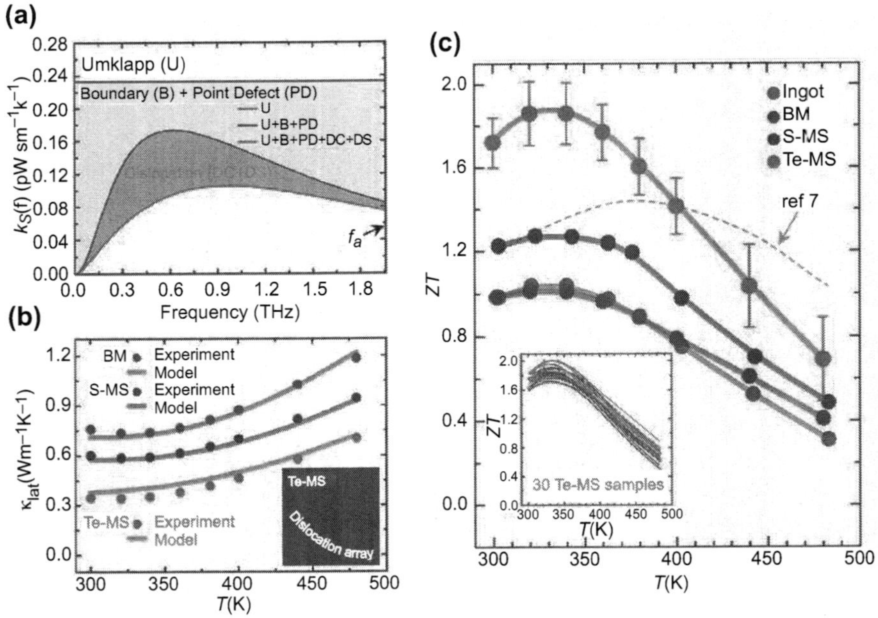

Fig. 2.8 Full-spectrum phonon scattering in high-performance bulk thermoelectrics: (a) The inclusion of dislocation scattering (DC + DS) is effective across the full frequency spectrum. Boundary (B) and point defect (PD) are effective only at low and high frequencies. The acoustic mode Debye frequency is fa. (b) k_L for $Bi_{0.5}Sb_{1.5}Te_3$ alloys produced by melt-solidification (ingot), solid-phase compaction (BM and S-MS) and liquid-phase compaction (Te-MS). The lowest k_L of Te-MS can be explained by the mid-frequency phonon scattering due to dislocation arrays embedded in grain boundaries. (c) The ZT as a function of temperature for $Bi_{0.5}Sb_{1.5}Te_3$ alloys. Reproduced from ref. (Kim et al. 2015).

4.2 Layer-Structured Metal Sulphide Thermoelectric Nanomaterials

Layer-structured metal sulphide nanomaterials, such as Bi_2S_3, Sb_2S_3 and TiS_2 (Wan et al. 2015), have recently attracted increasing interests due to their potential applications in the thermoelectrics, especially in the flexible thermoelectrics (Wan et al. 2015). Bi_2S_3 and Sb_2S_3 are semiconducting materials with a direct bandgap of 1.3 eV and 1.78 eV, respectively. They belong to the family of compounds M_2X_3 system (where M = Bi, Pb, Sb and X = S, Se, Te), which have the same crystal structure of the rhombohedral Bi_2Te_3 structure outlined above. The theoretical predictions (Wan et al. 2011) and subsequent experimental results (Wan et al. 2015) have proved the possibility of a significant enhancement in their *ZT* values. The reported $S^2\sigma$ values for these metal sulphide nanostructures is larger than that of their bulk counterparts (Lai et al. 2012). On these bases, the metal sulphide thermoelectric nanomaterials show great potential applications in solid-state power generation from heat sources (Lai et al. 2012).

It has been proposed (Wan et al. 2010, Wan et al. 2011) misfit $(MS)_{1+x}(TiS_2)_2$ (M = Pb, Bi, Sn) layer compounds as a new form of thermoelectric materials, due to the fact that misfit layer compounds compose of alternating rock-salt-type MS layers paired by the trigonal anti-prismatic TiS_2 layers with a van der Waals gap. Such naturally modulated structure shows low κ_l close to or even lower than the predicted minimum κ_l. The measured sound velocities shows that the ultra-low κ_l partially originates from the softening of the transverse modes of lattice wave due to weak interlayer bonding (Wan et al. 2010, Wan et al. 2011). Combined with a high $S^2\sigma$, the misfit layer compounds show a relatively high *ZT* value of 0.28~0.37 at 700 K. Table 2.2 summarizes the state of art metal sulphide thermoelectric materials.

Table 2.2 State of the art metal sulphide nanocomposites.

Material systems	Carrier type	*ZT*	κ_l [$Wm^{-1}K^{-1}$]	*T*[K]	Synthetic method*	Ref.
$(SnS)_{1.2}(TiS_2)_2$	n	0.37	0.3	700	SLV + SPS	(Wan et al. 2011)
$(BiS)_{1.2}(TiS_2)_2$	n	0.25	0.3	700	SLV + SPS	(Wan et al. 2010)
$(PbS)_{1.18}(TiS_2)_2$	n	0.30	1.1	775	SLV + SPS	(Wan et al. 2010)
TiS_2/[(hexylammonium) $x(H_2O)y(DMSO)z$]	n	0.28	0.12	373	CVD + solvent exchange	(Wan et al. 2015)

The abbreviations used in the column of the synthetic method represent the following meanings:
SLV: solid-liquid-vapour reaction.
SPS: Spark plasma sintering.
CVD: Chemical vapour deposition.

4.3 Layer-Structured Indium Selenide Thermoelectric Materials

Layer-structured In_4Se_3 shows relatively low bandgap of ~0.65 eV due to its In-In bonds exist in its crystal structure (Rhyee et al. 2009b), while In_3Se_4 is a self-doped n-type semiconductor with an optical gap of ~0.55 eV (Han et al. 2014b). In contrast, most of other indium selenides have bandgaps larger than 1 eV, such as 1.35 eV for α-In_2Se_3 (Julien and Balkanski 2003), 1.308 eV for β-In_2Se_3 (Julien et al. 1990), 1.812 eV for γ-In_2Se_3 (Julien et al. 1990), 1.25 eV for β-InSe (Erkoç et al. 1994) and 1.29 eV for γ-InSe (Julien and Balkanski 2003, Julien et al. 1985).

Charge Density Waves (CDWs) are periodic modulations of the electronic charge density – a standing wave in the electronic wave functions created by combining electron states moving in opposite directions (Rhyee et al. 2009b). CDWs and strong electron-phonon coupling and Peierls instabilities, have been used to lower κ of In_4Se_{3-x} crystal. In_4Se_{3-x} crystal shows very strong anisotropic thermoelectric properties. A very low κ of < 0.74 $Wm^{-1}K^{-1}$ at 705 k along the *b-c* plane is observed in the Se deficiency In_4Se_{3-x} crystal while the κ along *a-b* plane is higher than that along *b-c* plane, but still achieves a relatively low value of ~1.1 $Wm^{-1}K^{-1}$ at 705 K (Rhyee et al. 2009a). Through fundamental analysis, three aspects are responsible for the obtained low κ, including: (a) the in-plane Peierls lattice distortion and CDW instability result in the low κ along *b-c* plane in $In_4Se_{2.35}$, (b) the van der Waals interaction between In-Se layers along *a* axis leads to the low κ along *a-b* plane in $In_4Se_{2.35}$ and (c) the defect-induced phonon scattering at the Se-defect sites are further lowering κ with increasing the Se deficiency. Such a low κ finally results in a peak *ZT* of 1.48 at 705 K along the *b-c* plane, indicating In_4Se_{3-x} as a potential material for recycling waste heat in the mid-temperature range (500-900 K).

Doping with halogen substitutions, including F, Br, Cl and I, has also been employed in In_4Se_3 to maximize $S^2\sigma$, with single crystalline $In_4Se_{3-x}Cl_{0.03}$ and polycrystalline $In_4Pb_{0.01}Sn_ySe_3$ achieved maximum *ZT* of 1.53 (Rhyee et al. 2011). and 1.4 (Lin et al. 2013) at 698 K and 733 K, respectively, indicating their potentials in thermoelectric energy conversion.

Compared with the strong anisotropic thermoelectric properties of single crystal In_4Se_3, i.e. the *ZT* along the *b-c* plane (e.g., ~0.3 at 610 K for $In_4Se_{2.35}$) is only 1/5 of that along the *a-b* plane (e.g., 1.48 at 705 K for $In_4Se_{2.35}$), polycrystalline In_4Se_3 materials show much less anisotropic thermal and electrical transport properties (Rhyee et al. 2009a, Rhyee et al. 2010, Shi et al. 2010, Lim et al. 2010, Cho et al. 2011, Zhu et al. 2011). Polycrystalline Se vacancies and Cl doping were also introduced into the polycrystalline In_4Se_3, with $In_4Se_{2.7}Cl_{0.03}$ obtaining a peak *ZT* of 0.67 at 673 K.

As outlined above, nanostructuring is an effective approach to enhance *ZT* in polycrystalline In_4Se_3 materials as interfaces and surfaces introduced by nanostructuring can effectively enhance phonon scattering, leading to the decreased κ_l. With the embedded 10 nm inclusions between the grains of the high dense $In_4Se_{2.2}$ products, $In_4Se_{2.2}$ has a very low κ with only 0.41 $Wm^{-1}K^{-1}$ at 698 K, which is attributed to the enhanced phonon scattering by Se deficiency, defects and nanoscale inclusions (Zhu et al. 2011). Accordingly, a peak *ZT* of 0.97 at 425°C for $In_4Se_{2.2}$ are observed. It is believed that the thermoelectric property of nanostructured In_4Se_3 materials can be further enhanced by tuning the size and density control of their nanoscaled inclusions. Table 2.3 summarizes the thermoelectric properties obtained from In_4Se_3-based materials.

As suggested by theoretical modelling and sequent experimental measurements, In_3Se_4 is a self-doped n-type semiconductor (Han et al. 2014b). The calculated band structure and density of states of layer-structured rhombohedral In_3Se_4 structure indicates that it may be metallic or a heavily doped n-type semiconductor. Stoichiometric rhombohedral In_3Se_4 should have one extra electron per formula unit because of the octahedrally coordinated In^{+3} in the structure, which leads to a heavily doped n-type semiconductor. The n-type semiconducting nature was further confirmed by the optical absorption analysis and electrical property measurement of the synthesized In_3Se_4 nanostructures (Han et al. 2014b). The room temperature n of In_3Se_4 nanostructures was determined to be -1×10^{-17} cm^{-3} (Han et al. 2014b), suggesting that In-deficiency In_3Se_4 nanostructures is n-type semiconductors. The obtained n is comparable to that of In_4Se_3-based materials. With proper doping to enhance its n and σ, In_3Se_4 is anticipated to have reasonable thermoelectric properties.

Table 2.3 State of the art n-type indium selenides.

Material systems	Carrier type	ZT	κ_l [$Wm^{-1}K^{-1}$]	T(K)	Synthetic method*	Ref.
$In_4Se_{2.35}$ (b-c plane)	n	1.48	0.74	705	BM	(Rhyee et al. 2009b)
$In_4Se_{2.78}$ (b-c plane)	n	1.1	0.94	705	BM	(Rhyee et al. 2009b)
$In_4Se_{2.67}Cl_{0.03}$ (b-c plane)	n	1.53	~0.6	698	BM	(Rhyee et al. 2011)
$In_4Se_{2.67}Cl_{0.03}$ (b-c plane)	n	~1.05	~0.7	698	BM	(Rhyee et al. 2011)
$In_4Se_{2.35}$	n	~1.2	~0.7	660	BM	(Ahn et al. 2012)
$In_4Pb_{0.01}Sn_ySe_3$ (x = 0.03 or 0.04)	n	1.4	~0.56	733	SS	(Lin et al. 2013)

The abbreviations used in the column of the synthetic method represent the following meanings: BM = Bridgeman Method, SS = Solid State reaction + spark plasma sintering.

4.4 Two-Dimensional Thin Films and Superlattices

(Hicks and Dresselhaus (1993) pioneered improved ZT by two-dimensional Bi_2Te_3 based quantum well due to their enhanced $S^2\sigma$, as quantum confinement in the interlayer direction can increase the density of state near the Fermi level. The interfaces between quantum wells can effectively scatter phonons if the Bi_2Te_3 quantum well thickness is less than phonon mean free path, which can result in the decrease of κ_l (Hicks et al. 1993, Hicks and Dresselhaus 1993). As revealed in Fig. 2.9, the ZT values of Bi_2Te_3 quantum well structures are projected to be much higher than its bulk counterparts. Venkatasubramanian et al. 2001 observed the highest $ZT = 2.4$ using Bi_2Te_3-Sb_2Te_3 quantum well superlattices with a periodicity of 6 nm. Correspondingly, the highest ZT value for the bulk counterparts is only 1.1.

Inspired by the prediction that quantum confinement may lead to an increased S and therefore an enhanced ZT, Harman et al. (2002) developed quantum-dot superlattices in the PbTe-PbSeTe system, described as PbSe nanodots embedded in a PbTe matrix and showed $ZT = 1.6$, which is significantly higher than their bulk counterparts ($ZT = 0.34$). The enhanced thermoelectric properties have also been found in other two-dimensional thin films and quantum well systems, including Bi_2Te_3 superlattice-based thin film (Chowdhury et al. 2009), PbTe/Ag_2Te thin films (Urban et al. 2007), quantum well/barrier PbTe/$Pb_{1-x}Eu_xTe$ structures (Hicks et al. 1996) and n-PbTe/p-SnTe/n-PbTe quantum wells (Rogacheva et al. 2005), as summarized in Table 2.4. The early motivations for these investigations were based on the prediction that quantum confinement of in-plane carrier transport could substantially enhance $S^2\sigma$ over that of homogeneous materials, leading to ten-fold increases in *ZT* (Hicks and Dresselhaus 1993). Shakouri (2005) considered that such an enhancement can occur because sharp features in the electronic density of states of quantum-confined structures enable a doping-level-tuneable increase in the asymmetry between hot and cold electron transport, resulting in a large average transport energy and a large number of carriers moving in the material (i.e. a large *S* and a lager σ) (Vineis et al. 2010).

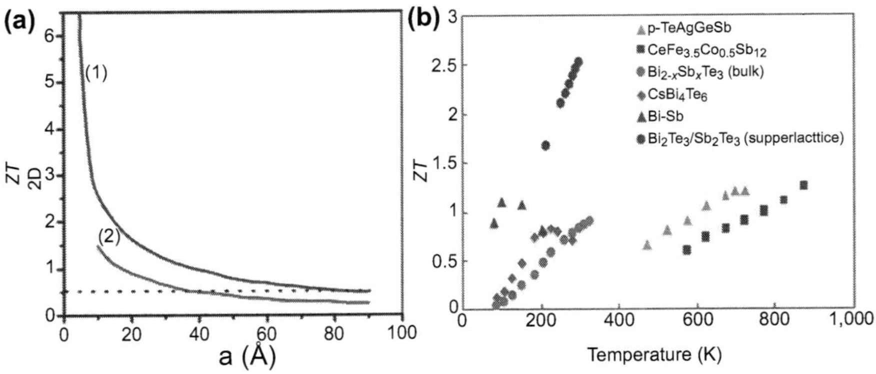

Fig. 2.9 (a) Calculated *ZT* as a function of layer thickness a in a quantum well structure for layers parallel to the *a-b* plane (1) and *b-c* plane (2) and the dashed line represents the optimized *ZT* for bulk Bi_2Te_3. Reproduced from ref. (Hicks and Dresselhaus 1993). (b) The improved *ZT* as a function of temperature for supperlattice Bi_2Te_3/Sb_2Te_3 compared with other TE materials (Venkatasubramanian et al. 2001).

Table 2.4 State of the art two-dimensional thin films and superlattices.

Material systems	Carrier type	*ZT*	κ_l [$Wm^{-1}K^{-1}$]	*T*(K)	Synthetic method*	Ref.
2D Materials: quantum well or superlattices						
PbTe/$Pb_{1-x}Eu_xTe$	–	2.0	–	300	MBE	(Hicks et al. 1996)
PbSeTe/PbTe	n	2.0	0.58	300	MBE	(Harman et al. 2002)
Bi_2Te_3/Sb_2Te_3	p	2.4	0.22	300	MBE	(Venkatasubramanian et al. 2001)
$Bi_2Te_3/Bi_2Te_{2.83}Se_{0.17}$	N	1.4	0.58	300	MBE	(Venkatasubramanian et al. 2001)

The abbreviations used in the column of the synthetic method represent the following meanings: MBE = molecular beam epitaxy.

Devices based on two-dimensional thermoelectric materials (such as thin film, quantum well and superlattices) can be used for small scale cooling, such as on-chip cooling (Chowdhury et al. 2009), however, their high costs and low scalabilities have limited their applications in large scale thermoelectric cooling and power generation devices. Especially, fully functional practical thermoelectric coolers have not been made from these nanomaterials due to the enormous difficulties in integrating nanoscale materials into microscale devices and packaged macroscale systems.

5. OUTLOOK AND CHALLENGE

Reducing the size and dimensionality of layer structured thermoelectric materials in nanoscale has been demonstrated as an effective approach to increase the thermoelectric performance, which confirmed the predicted *ZT* enhancement due to quantum confinement and nanostructure effects. These increased *ZT* values result primarily from lowered κ as interface density increases, as well as from possible quantum size effects including improved $S^2\sigma$ resulting from the increased electronic density of states at the Fermi level (Szczech et al. 2011). Obviously, a record highest *ZT* of 2.4 has been reported for Bi_2Te_3/Sb_2Te_3 superlattice system (Venkatasubramanian et al. 2001) and a remarkably low cross-plane value of 0.22 $Wm^{-1}K^{-1}$ was estimated in this Bi_2Te_3/Sb_2Te_3 system. Such super low cross-plane κ may be the reported lowest experimental and theoretical results in the present Bi_2Te_3/Sb_2Te_3 system (Venkatasubramanian et al. 2001).

It is of interest to note that the lowest κ is always observed in an amorphous material, since their average phonon mean free path is in the order of the lattice constant (~0.2-0.5 nm). The lowest κ can be estimated to be ~0.25-1 $Wm^{-1}K^{-1}$, which is confirmed by more sophisticated theories (Chiritescu et al. 2007, Zhao et al. 2014). The report in superlattices of WSe_2/W layers, however, is quite intriguing as they suggested cross-plane lattice κ values as low as 0.02 $Wm^{-1}K^{-1}$ (Chiritescu et al. 2007). Although the mechanism is not fully understood, it is likely that the layering creates large asymmetry in the directional phonon density of states and low coupling between phonons in different directions (Chiritescu et al. 2007, Zhao et al. 2014). Therefore, it is believed that new physical understanding should be developed to make further lowering κ becoming possible.

To reach *ZT* values of 3 or greater, seems to be feasible to date. It needs dramatic enhancements in $S^2\sigma$, which depend on significant reduction in κ and increasing σ. With regards to κ, any further reduction below the amorphous limit can only occur if one can actively change the phonon velocity or manipulate the wavelength number of phonon modes that propagate. This could result from coherent or correlated scattering effects, but so far this remains elusive for phonons (although widely known for electrons) and κ reduction through such mechanisms has never been conclusively demonstrated. Therefore, this may be an exciting scientific opportunity and creates a new open challenge to theorists and experimentalists alike to come up with new scattering mechanisms and concepts that will help achieve very large increases in $S^2\sigma$ and simultaneous decreases in κ.

High-performing thermoelectric properties appear to depend sensitively on both the nanostructure and band engineering with new synthesis approach. Various approaches will continue to study layer-structured thermoelectric nanomaterials with narrow bandgaps (less than 1 eV), heavy elements doping, defects loading and nanostructuring. Continued research to gain a more quantitative understanding is required to allow the rational design and preparation of layer-structured thermoelectric nanomaterials and accelerate the wide adoption of thermoelectric technologies in power generation and cooling applications.

6. ACKNOWLEDGEMENTS

This work was supported by the Australian Research Council. Dr Chen would like to thank QLD government for a state future fellowship and a UQ research foundation excellent award. The authors acknowledge the facilities and the scientific and technical assistance, of the Australian Microscopy & Microanalysis Research Facility (AMMRF) and the Centre for Microscopy and Microanalysis, The University of Queensland.

7. REFERENCES

Ahn, K., E. Cho, J. S. Rhyee, S. I. Kim, S. Hwang, H. S. Kim, et al. 2012. Improvement in the thermoelectric performance of the crystals of halogen-substituted $In_4Se_{3-x}H_{0.03}$ (H = F, Cl, Br, I): effect of halogen-substitution on the thermoelectric properties in In_4Se_{3-x}. J. Mater. Chem. 22: 5730-5736.

Androulakis, J., C.-H. Lin, H.-J. Kong, C. Uher, C.-I. Wu, T. Hogan, et al. 2007. Spinodal decomposition and nucleation and growth as a means to bulk nanostructured thermoelectrics: enhanced performance in $Pb_{1-x}Sn_xTe$-PbS. J. Am. Chem. Soc. 129: 9780-88.

Bell, L. E. 2008. Cooling, heating, generating power and recovering waste heat with thermoelectric systems. Science 321: 1457-1461.

Bottner, H., G. Chen and R. Venkatasubramanian. 2006. Aspects of thin-film superlattice thermoelectric materials, devices and applications. MRS. Bull. 31: 211-217.

Boukai, A. I., Y. Bunimovich, J. Tahir-Kheli, J.-K. Yu, W. A. Goddard, III and J. R. Heath. 2008. Silicon nanowires as efficient thermoelectric materials. Nature 451: 168-171.

Bux, S. K., J.-P. Fleurial and R. B. Kaner. 2010. Nanostructured materials for thermoelectric applications. Chem. Commun. 46: 8311-8324.

Cao, Y. Q., X. B. Zhao, T. J. Zhu, X. B. Zhang and J. P. Tu. 2008. Syntheses and thermoelectric properties of Bi_2Te_3/Sb_2Te_3 bulk nanocomposites with laminated nanostructure. Appl. Phys. Lett. 92: 143106.

Chang, H.-C., C.-H. Chen and Y.-K. Kuo. 2013. Great enhancements in the thermoelectric power factor of BiSbTe nanostructured films with well-ordered interfaces. Nanoscale 5: 7017-7025.

Chen, Z.-G., G. Han, L. Yang, L. Cheng and J. Zou. 2012. Nanostructured thermoelectric materials: current research and future challenge. Prog. Nat. Sci. 22: 535-549.

Chen, Z.-G., L. Yang, S. Ma, L. Cheng, G. Han, Z.-D. Zhang, et al. 2014. Paramagnetic Cu-doped Bi_2Te_3 nanoplates. Appl. Phys. Lett. 104: 053105.

Cheng, L., Z.-G. Chen, S. Ma, Z.-D. Zhang, Y. Wang, H.-Y. Xu, et al. 2012. High curie temperature Bi 1.85 Mn 0.15 Te_3 nanoplates. J. Am. Chem. Soc. 134: 18920-18923.

Cheng, L., Z.-G. Chen, L. Yang, G. Han, H.-Y. Xu, G. J. Snyder, et al. 2013. T-shaped Bi_2Te_3-Te heteronanojunctions: epitaxial growth, structural modeling and thermoelectric properties. J. Phys. Chem. C 117: 12458-12464.

Chiritescu, C., D. G. Cahill, N. Nguyen, D. Johnson, A. Bodapati, P. Keblinski, et al. 2007. Ultralow thermal conductivity in disordered, layered WSe_2 crystals. Science 315: 351-353.

Cho, J. Y., Y. S. Lim, S.-M. Choi, K. H. Kim, W.-S. Seo and H.-H. Park. 2011. Thermoelectric properties of spark plasma-sintered In_4Se_3-In_4Te_3. J. Elect. Mater. 40: 1024-1028.

Chowdhury, I., R. Prasher, K. Lofgreen, G. Chrysler, S. Narasimhan, R. Mahajan, et al. 2009. On-chip cooling by superlattice-based thin-film thermoelectrics. Nat. Nano. 4: 235-238.

Chung, D. Y., T. Hogan, P. Brazis, M. Rocci-Lane, C. Karnewurf, M. Bastea, et al. 2000. $CsBi_4Te_6$: a high-performance thermoelectric material for low-temperature applications. Science 287: 1024-1027.

Dismukes, J. P., E. Ekstrom, D. S. Beers, E. F. Steigmeier and I. Kudman. 1964. Thermal + electrical properties of heavily doped Ge-Si alloys up to 1300 degrees K. J. Appl. Phys. 35: 2899-2907.

Dresselhaus, M. S., G. Chen, M. Y. Tang, R. G. Yang, H. Lee, D. Z. Wang, et al. 2007. New directions for low-dimensional thermoelectric materials. Adv. Mater. 19: 1043-1053.

Erkoç, Ş., K. Allahverdi and Z. Ibrahim. 1994. Electronic states of InSe/GaSe superlattice. Solid State Comm. 90: 553-556.

Fan, S., J. Zhao, J. Guo, Q. Yan, J. Ma and H. H. Hng. 2010. P-type Bi(0.4)Sb(1.6)Te(3) nanocomposites with enhanced figure of merit. Appl. Phys. Lett. 96: 182104.

Fan, X. A., J. Y. Yang, Z. Xie, K. Li, W. Zhu, X. K. Duan, et al. 2007. Bi_2Te_3 hexagonal nanoplates and thermoelectric properties of n-type Bi_2Te_3 nanocomposites. J. Phys. D. Appl. Phys. 40: 5975-5979.

Goldsmid, H. J. and R. W. Douglas. 1954. The use of semiconductors in thermoelectric refrigeration. British J. Appl. Phys. 5: 386-390.

Goldsmid, H. J., A. R. Sheard and D. A. Wright. 1958. The use of semiconductors in thermoelectric refrigeration. British J. Appl. Phys. 9: 365-370.

Han, G., Z.-G. Chen, L. Yang, M. Hong, J. Drennan and J. Zou. 2014a. Rational design of Bi_2Te_3 polycrystalline whiskers for thermoelectric applications. ACS Appl. Mater. Interf. 7: 989-995.

Han, G., Z. G. Chen, C. Sun, L. Yang, L. Cheng, Z. Li, et al. 2014b. A new crystal: layer-structured rhombohedral In_3Se_4. CrystEngComm 16: 393-398.

Harman, T. C., D. L. Spears and M. J. Manfra. 1996. High thermoelectric Figures of merit in PbTe quantum wells. J. Electron. Mater. 25: 1121-1127.

Harman, T. C., P. J. Taylor, M. P. Walsh and B. E. LaForge. 2002. Quantum dot superlattice thermoelectric materials and devices. Science 297: 2229-2232.

Heremans, J. P., C. M. Thrush and D. T. Morelli. 2004. Thermopower enhancement in lead telluride nanostructures. Phys. Rev. B 70: 115334.

Heremans, J. P., V. Jovovic, E. S. Toberer, A. Saramat, K. Kurosaki, A. Charoenphakdee, et al. 2008. Enhancement of thermoelectric efficiency in PbTe by distortion of the electronic density of states. Science 321: 554-557.

Heremans, J. P., B. Wiendlocha and A. M. Chamoire. 2012. Resonant levels in bulk thermoelectric semiconductors. Energy Environ. Sci. 5: 5510-5530.

Hicks, L. D. and M. S. Dresselhaus. 1993. Effect of quantum-well structures on the thermoelectric figure of merit. Phys. Rev. B 47: 12727-12731.

Hicks, L. D., T. C. Harman and M. S. Dresselhaus. 1993. Use of quantum-well superlattices to obtain a high figure of merit from nonconventional thermoelectric materials. Appl. Phys. Lett. 63: 3230-3232.

Hicks, L. D., T. C. Harman, X. Sun and M. S. Dresselhaus. 1996. Experimental study of the effect of quantum-well structures on the thermoelectric figure of merit. Phys. Rev. B 53: R10493.

Hochbaum, A. I., R. Chen, R. D. Delgado, W. Liang, E. C. Garnett, M. Najarian, et al. 2008. Enhanced thermoelectric performance of rough silicon nanowires. Nature 451: 163-167.

Hsu, K. F., S. Loo, F. Guo, W. Chen, J. S. Dyck, C. Uher, et al. 2004. Cubic AgPbm $SbTe_{2+m}$: bulk thermoelectric materials with high figure of merit. Science 303: 818-821.

Jaworski, C. M., V. Kulbachinskii and J. P. Heremans. 2009. Resonant level formed by tin in Bi_2Te_3 and the enhancement of room-temperature thermoelectric power. Phys. Rev. B 80: 233201.

Julien, C., E. Hatzikraniotis, A. Chevy and K. Kambas. 1985. Electrical behavior of lithium intercalated layered In-Se compounds. Mater. Res. Bull. 20: 287-292.

Julien, C., A. Chevy and D. Siapkas. 1990. Optical properties of In_2Se_3 phases. Phys. Status Solidi A 118: 553-559.

Julien, C. M. and M. Balkanski. 2003. Lithium reactivity with III–VI layered compounds. Mater. Sci. Eng. B 100: 263-270.

Kanatzidis, M. G. 2010. Nanostructured thermoelectrics: the new paradigm? Chem. Mater. 22: 648-659.

Kim, S. I., K. H. Lee, H. A. Mun, H. S. Kim, S. W. Hwang, J. W. Roh, et al. 2015. Dense dislocation arrays embedded in grain boundaries for high-performance bulk thermoelectrics. Science 348: 109-114.

Ko, D. K., Y. J. Kang and C. B. Murray. 2011. Enhanced thermopower via carrier energy filtering in solution-processable Pt-Sb_2Te_3 nanocomposites. Nano Lett. 11: 2841-2844.

Koga, T., X. Sun, S. B. Cronin and M. S. Dresselhaus. 1998. Carrier pocket engineering to design superior thermoelectric materials using GaAs/AlAs superlattices. Appl. Phys. Lett. 73: 2950-2952.

Koga, T., X. Sun, S. B. Cronin and M. S. Dresselhaus. 1999. Carrier pocket engineering applied to "strained" Si/Ge superlattices to design useful thermoelectric materials. Appl. Phys. Lett. 75: 2438-2440.

Kulbachinskii, V. A., N. B. Brandt, P. A. Cheremnykh, S. A. Azou, J. Horak and P. Lostak. 1988. Magnetoresistance and hall-effect in Bi_2Te_3(Sn) in ultrahigh magnetic-fields and under pressure. Phys. Status Solidi B 150: 237-343.

Lai, C.-H., M.-Y. Lu and L.-J. Chen. 2012. Metal sulfide nanostructures: synthesis, properties and applications in energy conversion and storage. J. Mater. Chem. 22: 19-30.

Lan, Y., B. Poudel, Y. Ma, D. Wang, M. S. Dresselhaus, G. Chen, et al. 2009. Structure study of bulk nanograined thermoelectric bismuth antimony telluride. Nano Lett. 9: 1419-1422.

Lan, Y., A. J. Minnich, G. Chen and Z. Ren. 2010. Enhancement of thermoelectric figure of merit by a bulk nanostructuring approach. Adv. Funct. Mater. 20: 357-376.

Li, J. F., W. S. Liu, L. D. Zhao and M. Zhou. 2010. High-performance nanostructured thermoelectric materials. NPG Asia Mater. 2: 152-158.

Lim, Y. S., J. Y. Cho, J. K. Lee, S. M. Choi, K. H. Kim, W. S. Seo, et al. 2010. Microstructures and thermoelectric properties of spark plasma sintered In(4)Se(3). Elect. Mater. Lett. 6: 117-121.

Lin, Z.-S., L. Chen, L.-M. Wang, J.-T. Zhao and L.-M. Wu. 2013. A promising mid-temperature thermoelectric material candidate: Pb/Sn-Codoped $In_4Pb_xSn_ySe_3$. Adv. Mater. 25: 4800-4806.

Liu, H., X. Shi, F. Xu, L. Zhang, W. Zhang, L. Chen, et al. 2012. Copper ion liquid-like thermoelectrics. Nat. Mater. 11: 422-425.

Liu, H., X. Yuan, P. Lu, X. Shi, F. Xu, Y. He, et al. 2013. Ultrahigh thermoelectric performance by electron and phonon critical scattering in $Cu_2Se_{1-x}I_x$. Adv. Mater. 25: 6607-6612.

Ma, Y., Q. Hao, B. Poudel, Y. Lan, B. Yu, D. Wang, et al. 2008. Enhanced thermoelectric figure of merit in p-type nanostructured bismuth antimony tellurium alloys made from elemental chunks. Nano Lett. 8: 2580-2584.

Mahan, G. D. 1989. Figure of merit for thermoelectrics. J. Appl. Phys. 65: 1578-1583.

Mahan, G. D. and J. O. Sofo. 1996. The best thermoelectric. Proc. Natl. Acad. Sci. USA 93: 7436-7439.

Martin, J., L. Wang, L. Chen and G. Nolas. 2009. Enhanced seebeck coefficient through energy-barrier scattering in PbTe nanocomposites. Phys. Rev. B 79: 115311.

Minnich, A. J., M. S. Dresselhaus, Z. F. Ren and G. Chen. 2009. Bulk nanostructured thermoelectric materials: current research and future prospects. Energy Environ. Sci. 2: 466-479.

Molenkamp, L. W., H. Vanhouten, A. A. M. Staring and C. W. J. Beenakker. 1993. Quantum effects in thermal and thermoelectric transport in semiconductor nanostructures. Phys. Scripta. T49b: 441-445.

Paul, B., V. A. Kumar and P. Banerji. 2010. Embedded Ag-rich nanodots in PbTe: enhancement of thermoelectric properties through energy filtering of the carriers. J. Appl. Phys. 108: 064322.

Pei, Y., N. A. Heinz, A. LaLonde and G. J. Snyder. 2011a. Combination of large nanostructures and complex band structure for high performance thermoelectric lead telluride. Energy Environ. Sci. 4: 3640-3645.

Pei, Y., J. Lensch-Falk, E. S. Toberer, D. L. Medlin and G. J. Snyder. 2011b. High thermoelectric performance in PbTe due to large nanoscale Ag_2Te precipitates and La doping. Adv. Funct. Mater. 21: 241-249.

Pei, Y., X. Shi, A. LaLonde, H. Wang, L. Chen and G. J. Snyder. 2011c. Convergence of electronic bands for high performance bulk thermoelectrics. Nature 473: 66-69.

Pei, Y., H. Wang and G. J. Snyder. 2012a. Band engineering of thermoelectric materials. Adv. Mater. 24: 6125-6135.

Pei, Y. Z., A. D. LaLonde, H. Wang and G. J. Snyder. 2012b. Low effective mass leading to high thermoelectric performance. Energy Environ. Sci. 5: 7963-7969.

Pichanusakorn, P. and P. Bandaru. 2010. Nanostructured thermoelectrics. Mat. Sci. Eng. R. 67: 19-63.

Poudel, B., Q. Hao, Y. Ma, Y. Lan, A. Minnich, B. Yu, et al. 2008. High-thermoelectric performance of nanostructured bismuth antimony telluride bulk alloys. Science 320: 634-638.

Poudeu, P. F. R., J. D'Angelo, A. D. Downey, J. L. Short, T. P. Hogan and M. G. Kanatzidis. 2006. High thermoelectric figure of merit and nanostructuring in bulk p-type $Na_{1-x}Pb_mSbyTe_{m+2}$. Angew. Chem. Int. Edit. 45: 3835-3839.

Prytz, O., A. E. Gunnaes, O. B. Karlsen, T. H. Breivik, E. S. Toberer, G. J. Snyder, et al. 2009. Nanoscale inclusions in the phonon glass thermoelectric material Zn_4Sb_3. Phil. Mag. Lett. 89: 362-369.

Rhyee, J.-S., E. Cho, K. H. Lee, S. M. Lee, S. I. Kim, H.-S. Kim, et al. 2009a. Thermoelectric properties and anisotropic electronic band structure on the In [sub 4] Se [sub 3 – x] compounds. Appl. Phys. Lett. 95: 212106.

Rhyee, J.-S., K. H. Lee, S. M. Lee, E. Cho, S. I. Kim, E. Lee, et al. 2009b. Peierls distortion as a route to high thermoelectric performance in $In_4Se_{3-\delta}$ crystals. Nature 459: 965-968.

Rhyee, J.-S., E. Cho, K. Ahn, K. H. Lee and S. M. Lee. 2010. Thermoelectric properties of bipolar diffusion effect on In [sub 4] Se [sub 3 – x] Te [sub x] compounds. Appl. Phys. Lett. 97: 152104.

Rhyee, J.-S., K. Ahn, K. H. Lee, H. S. Ji and J.-H. Shim. 2011. Enhancement of the thermoelectric figure of merit in a wide temperature range in $In_4Se_{3-x}Cl_{0.03}$ bulk crystals. Adv. Mater. 23: 2191-2194.

Rogacheva, E. I., O. N. Nashchekina, A. V. Meriuts, S. G. Lyubchenko, M. S. Dresselhaus and G. Dresselhaus. 2005. Quantum size effects in n-PbTe/p-SnTe/n-PbTe heterostructures. Appl. Phys. Lett. 86: 063103.

Rowe, D. M. 1995. CRC Handbook of Thermoelectrics. CRC Press, New York.

Rowe, D. M. 2005. General principles and basic considerations. *In*: Thermoelectrics Handbook. CRC Press.

Sandoz-Rosado, E. J., S. J. Weinstein and R. J. Stevens. 2013. On the thomson effect in thermoelectric power devices. Int. J. Therm. Sci. 66: 1-7.

Saramat, A., G. Svensson, A. E. C. Palmqvist, C. Stiewe, E. Mueller, D. Platzek, et al. 2006. Large thermoelectric figure of merit at high temperature in czochralski-grown clathrate $Ba_8Ga_{16}Ge_{30}$. J. Appl. Phys. 99: 023708.

Shakouri, A. 2005. Thermoelectric, thermionic and thermophotovoltaic energy conversion. ICT 2005. 24th International Conference on Thermoelectrics 2005: 507-512.

Shi, X., J. Y. Cho, J. R. Salvador, J. Yang and H. Wang. 2010. Thermoelectric properties of polycrystalline In_4Se_3 and In_4Te_3. Appl. Phys. Lett. 96: 162108.

Snyder, G. J., M. Christensen, E. Nishibori, T. Caillat and B. B. Iversen. 2004. Disordered zinc in Zn_4Sb_3 with phonon-glass and electron-crystal thermoelectric properties. Nat. Mater. 3: 458-463.

Snyder, G. J. and E. S. Toberer. 2008. Complex thermoelectric materials. Nat. Mater. 7: 105-114.

Sofo, J. O. and G. D. Mahan. 1994. Optimum band gap of a thermoelectric material. Phys. Rev. B 49: 4565-4570.

Steele, W. H. 1893. A new thermoelectric phenomenon. Science (New York, N.Y.) 22: 256-256.

Sun, Y., H. Cheng, S. Gao, Q. Liu, Z. Sun, C. Xiao, et al. 2012. Atomically thick bismuth selenide freestanding single layers achieving enhanced thermoelectric energy harvesting. J. Am. Chem. Soc. 134: 20294-20297.

Szczech, J. R., J. M. Higgins and S. Jin. 2011. Enhancement of the thermoelectric properties in nanoscale and nanostructured materials. J. Mater. Chem. 21: 4037.

Tritt, T. M., H. Boettner and L. Chen. 2008. Thermoelectrics: direct solar thermal energy conversion. MRS. Bull. 33: 366-368.

Urban, J. J., D. V. Talapin, E. V. Shevchenko, C. R. Kagan and C. B. Murray. 2007. Synergism in binary nanocrystal superlattices leads to enhanced p-type conductivity in self-assembled PbTe/Ag_{-2}Te thin films. Nat. Mater. 6: 115-121.

Vaqueiro, P. and A. V. Powell. 2010. Recent developments in nanostructured materials for high-performance thermoelectrics. J. Mater. Chem. 20: 9577-9584.

Vashaee, D. and A. Shakouri. 2004. Improved thermoelectric power factor in metal-based superlattices. Phys. Rev. Lett. 92: 106103.

Venkatasubramanian, R., E. Siivola, T. Colpitts and B. O'Quinn. 2001. Thin-film thermoelectric devices with high room-temperature Figures of merit. Nature 413: 597-602.

Vineis, C. J., A. Shakouri, A. Majumdar and M. G. Kanatzidis. 2010. Nanostructured thermoelectrics: big efficiency gains from small features. Adv. Mater. 22: 3970-3980.

Wan, C., Y. Wang, N. Wang and K. Koumoto. 2010. Low-thermal-conductivity (MS)(1+x) (TiS2)(2) (M = Pb, Bi, Sn) misfit layer compounds for bulk thermoelectric materials. Materials 3: 2606-2617.

Wan, C., Y. Wang, N. Wang, W. Norimatsu, M. Kusunoki and K. Koumoto. 2011. Intercalation: building a natural superlattice for better thermoelectric performance in layered chalcogenides. J. Electron. Mater. 40: 1271-1280.

Wan, C., X. Gu, F. Dang, T. Itoh, Y. Wang, H. Sasaki, et al. 2015. Flexible n-type thermoelectric materials by organic intercalation of layered transition metal dichalcogenide TiS_2. Nat. Mater. 14: 622-627.

Wang, H., Z. M. Gibbs, Y. Takagiwa and G. J. Snyder. 2014. Tuning bands of PbSe for better thermoelectric efficiency. Energy Environ. Sci. 7: 804-811.

Wang, X. W., H. Lee, Y. C. Lan, G. H. Zhu, G. Joshi, D. Z. Wang, et al. 2008. Enhanced thermoelectric figure of merit in nanostructured n-type silicon germanium bulk alloy. Appl. Phys. Lett. 93: 193121.

Wu, D., L. Zhao, S. Hao, Q. Jiang, F. Zheng, J. W. Doak, et al. 2014. Origin of the high performance in GeTe-Based thermoelectric materials upon Bi_2Te_3 doping. J. Am. Chem. Soc. 136: 11412-11419.

Xie, W., X. Tang, Y. Yan, Q. Zhang and T. M. Tritt. 2009a. High thermoelectric performance BiSbTe alloy with unique low-dimensional structure. J Appl. Phys. 105: 113713.

Xie, W., X. Tang, Y. Yan, Q. Zhang and T. M. Tritt. 2009b. Unique nanostructures and enhanced thermoelectric performance of melt-spun BiSbTe alloys. Appl. Phys. Lett. 94: 102111.

Xie, W., J. He, H. J. Kang, X. Tang, S. Zhu, M. Laver, et al. 2010. Identifying the specific nanostructures responsible for the high thermoelectric performance of (Bi, Sb)2Te3 nanocomposites. Nano Lett. 10: 3283-3289.

Yamanashi, M. 1996. A new approach to optimum design in thermoelectric cooling systems. J. Appl. Phys. 80: 5494-5502.

Yan, X., B. Poudel, Y. Ma, W. S. Liu, G. Joshi, H. Wang, et al. 2010. Experimental studies on anisotropic thermoelectric properties and structures of n-type Bi(2)Te(2.7)Se(0.3). Nano Lett. 10: 3373-3378.

Yang, L., Z.-G. Chen, G. Han, M. Hong, Y. Zou and J. Zou. 2015. High-performance thermoelectric Cu_2Se nanoplates through nanostructure engineering. Nano Energy 16: 367-374.

Zhang, Q., X. Ai, L. Wang, Y. Chang, W. Luo, W. Jiang, et al. 2015a. Improved thermoelectric performance of silver nanoparticles-dispersed Bi_2Te_3 composites deriving from hierarchical two-phased heterostructure. Adv. Funct. Mater. 25: 966-976.

Zhang, Q., E. K. Chere, K. McEnaney, M. Yao, F. Cao, Y. Ni, et al. 2015b. Enhancement of thermoelectric performance of n-type PbSe by Cr doping with optimized carrier concentration. Adv. Energy Mater. 5: 1401977.

Zhao, L.-D., S.-H. Lo, Y. Zhang, H. Sun, G. Tan, C. Uher, et al. 2014. Ultralow thermal conductivity and high thermoelectric figure of merit in SnSe crystals. Nature 508: 373-377.

Zhao, X. B., X. H. Ji, Y. H. Zhang, T. J. Zhu, J. P. Tu and X. B. Zhang. 2005. Bismuth telluride nanotubes and the effects on the thermoelectric properties of nanotube-containing nanocomposites. Appl. Phys. Lett. 86: 062111.

Zhao, X. B., S. H. Yang, Y. Q. Cao, J. L. Mi, Q. Zhang and T. J. Zhu. 2009. Synthesis of nanocomposites with improved thermoelectric properties. J. Electr. Mater. 38: 1017-1024.

Zheng, Y., Q. Zhang, X. Su, H. Xie, S. Shu, T. Chen, et al. 2015. Mechanically robust BiSbTe alloys with superior thermoelectric performance: a case study of stable hierarchical nanostructured thermoelectric materials. Adv. Energy Mater. 5: 1401391.

Zhu, G. H., Y. C. Lan, H. Wang, G. Joshi, Q. Hao, G. Chen, et al. 2011. Effect of selenium deficiency on the thermoelectric properties of n-type In (4) Se (3 – x) compounds. Phys. Rev. B 83: 115201.

Zide, J. M. O., D. Vashaee, Z. X. Bian, G. Zeng, J. E. Bowers, A. Shakouri, et al. 2006. Demonstration of electron filtering to increase the seebeck coefficient in In (0.53) Ga (0.47) As/In (0.53) Ga (0.28) Al (0.19) As superlattices. Phys. Rev. B 74: 205335.

3

Thermoelectric Properties of Organic and Inorganic Materials and Cells

Khasan S. Karimov,[1,2,*] Muhammad Abid,[3,a]
Kuan Yew Cheong[4,b] and Muhammad Mehran Bashir[1,c]

ABSTRACT

In this chapter, design of materials and devices for low-power generation technologies based on thermoelectric effect i.e. photo-thermoelectric and thermo-chemical are introduced. Thermoelectric properties of quasi-one dimensional crystals of tetracyanoquinodimethane complexes such as tetracyanoquinodimethane-triethylammonium and tetracyanoquinodimethane-cesium, two-dimensional nanostructures of $SrTiO_3$, two-dimensional layered materials of graphene, hexagonal Boron nitride (hBN), phosphorene, transition metal dichalcogenides (e.g., MoS_2, WS_2 etc.), metal oxides (e.g., MoO_3) and composites of Carbon Nano-tubes (CNT) with silicone and polymer adhesives are discussed. Properties of photo-thermoelectric cells based on n-Si and p-Si single crystal strips, Cu/Orange dye aqueous solution consisting of flexible polymer tube casing and Cu electrodes as well as photo-thermoelectrochemical cells consisted of n-type InP/aqueous solution of orange dye/carbon cell and *n*-InAs orange dye/Zn are also being presented. It is concluded that these devices are environmental friendly, relatively easy to be recycled and can be used for low power applications.

Keywords: Thermoelectric, organic, inorganic, materials, cell, property, two-dimensional, three-dimensional, nanostructure, orange dye, polymer, power, complex.

[1] Ghulam Ishaq Khan Institute of Engineering Sciences and Technology, Topi, District Swabi, Pakistan.
[2] Center for Innovative Development of Science and New Technologies, Academy of Sciences, Rudaki St. 33, Dushanbe, 734025, Tajikistan.
[3] Interdisciplinary Research Center, COMSATS Institute of Information Technology, Wah Cantt, Pakistan.
[4] School of Materials & Mineral Resources Engineering, Engineering Campus, Universiti Sains Malaysia 14300 Nibong Tebal, Penang, Malaysia.
[a] E-mail: mabid19692000@gmail.com
[b] E-mail: srcheong@usm.my
[c] E-mail: mehran4gcu@gmail.com
* Corresponding author: khasansangink@gmail.com

1. INTRODUCTION

Energy is a key factor of life and energy consumption is the most significant feature influencing the life standards of human beings. With the increasing human population, the demand for energy is increasing continuously and the sustainable energy supply is arising as one of major problems of 21st century (Recalde et al. 2014). Cost effective and efficient conversion of energy is also the most important necessity for the development of any country (Markvart 2000). World Energy Council (WEC) estimated that, in 2020, world energy demand would be 50-80% more than the measured in 1990 (Omer 2008). Increasing demand of world's energy consumption has resulted in the reduction of oil reserves on the earth, while combustion of these fossil fuels has created environmental contamination and greenhouse effect (Narodoslawsky 2013).

Keeping in view these facts, it is realized that an efficient energy economy is based on a cost effective renewable and non-polluting source of energy (Deutsch et al. 2008). Therefore novel renewable technologies which are eco-friendly will become a long term resource of energy (Lamy et al. 2014). Because of its numerous advantages and inherent potential for power generation, solar energy is observed as the most efficient technology among all renewable energy forms such as tidal power, thermal, hydropower and biomass. Although the efficiency of photovoltaic (PV) system is low but such systems are capable to generate electrical power in a more economical way. Different constraints may apply that affect PV efficiency including climate condition, placement of modules i.e. angle of inclination, load matching, tracking system and reflector or concentrator (Karimov et al. 2002). Generally, a non-tracking system is preferable as it minimizes the maintenance cost and reduces the complexity of system. However, tracking modules are used where higher efficiency is preferable (Twidell and Weir 2005, Green 2004, TDFTE 1970).

At present, the scientific community is trying to explore other environmental friendly energy resources (Asif and Muneer 2007, Rodprasert et al. 2014). Currently the efficient production of clean and sustainable energy is the most provoking challenge of the coming few decades (Dresselhaus et al. 2007, Ondersma and Hamann 2013, Panda et al. 2012, Snyder and Toberer 2008).

To meet the future energy challenges, the thermoelectric phenomena can play an important role, which involve the conversion of heat to electricity and endow with the methods for materials heating and cooling (Dresselhaus et al. 2007). The sustainability of electricity can also be improved by scavenging of waste heat from industrial processes, factories, power plants, automotive exhaust computers, home heating and even from the human body by using thermoelectric generators (Choi et al. 2011, Snyder and Toberer 2008, Alam and Ramakrishna 2013, Elsheikh et al. 2014). Use of waste heat for the generation of electric power is of prime importance to meet the world's future energy requirements. The devices that are used for the conversion of heat energy into electricity are the semiconductor thermoelectric cells. These work on the base of Seebeck effect. Peltier effect is used for the cooling of thermoelectric refrigerators (Bell 2008).

Thermoelectricity has been known for a long time. The German-Estonian physicist Thomas Johann Seebeck discovered thermoelectricity in 1821; with the concept that a conductor can generate current if it is subjected to a thermal gradient. During the last years researchers have mostly concentrated on the conversion of solar energy into electric power using solar cells. Due to recent advancements in thermoelectricity, this technology has become complementary or even alternative to solar PV systems (Snyder 2008). Moreover thermoelectric cells are more universal compared to the solar cells as this technology allows conversion of solar and any thermal energy into electricity. Seebeck described the electromotive field (E_{emf}) using Eq. (1) as follows:

$$E_{emf} = -S\Delta T \tag{1}$$

S is Seebeck coefficient or thermo power and is usually in the range of –100 μV/K to + 1000 μV/K and ΔT is temperature gradient. The Seebeck coefficients depend on temperature and composition of material.

Thermoelectricity is produced by temperature on contact component, volume component and phonon component. Volume and phonon components of thermopower have the same sign whereas contact component as a rule has the opposite sign. Seebeck coefficient is observed low in metals or good conductive material, but is more than 1000 μV/K in semiconductors. The value of Seebeck coefficients depends on the impurities in the materials and is observed sensitive to mechanical and chemical properties. Due to any temperature gradient, average charge carrier energy varies than in the equilibrium stage and results in electron and phonons transport and is concluded as the main reason of Seebeck effect (Stilbans 1967). The electrons obey Bose-Einstein and Fermi-Dirac statistics. Electrons and phonons transport depends on the structure and state of the materials. Phonons participate in heat transport whereas electrons near of Fermi surface contribute to the transport. An increase in efficiency (Z) and figure of merit (ZT) of the thermoelectric materials and cells is concluded due to decrease in phonon conductivity.

Thermoelectric cells work on the principle of Seebeck effect and their efficiency (Z) is determined using Eq. (2) (LE 2008):

$$Z = \frac{\alpha^2 \sigma}{k_{\text{total}}} \tag{2}$$

where α and σ are the Seebeck coefficient and electrical conductivity respectively. k_{total} is the total thermal conductivity and is equal to the sum of thermal conductivities of electron (k_e) and phonon (k_{ph}). The increase in efficiency of thermoelectric generators depends, first of all on decrease in phonon thermal conductivity (k_{ph}). In this way, the layered chalcogenides with complex crystal structure are investigated intensively. During the last years, thermoelectric cells based on 11 mm thick layers of n-Si/SiGe-p-B4C/B9C deposited on the silicon substrate are fabricated, concluding a high efficiency of 15%. At the same time the thermoelectric effect is used not only for the conversion of energy, but also for the measurements of temperature gradient in instrumentation. It is also used for the measurement of concentration of gases (like CO, CH_4 and C_2H_5OH etc.) using thermoelectric cells on the base of oxides of tin and indium (Papadopoulos et al. 1996). In the reference (Walia et al. 2013a) the Bi_2Te_3–Sb_2Te_3 (p-type) and Bi_2Te_3–Bi_2Se_3 (n-type) based thermoelectric cells for the measurement of temperature gradient are presented. It is also reported that these cells have a high thermoelectric figure of merit (ZT) and can be used to determine the velocities of gas flow. In addition to chalcogenides, the transition metal oxides are also very attractive thermoelectric materials. These materials have excellent mechanical, chemical and electronic properties along with fascinating thermoelectric characteristics like tunable phonon and electronic transport properties, high electrical conductivity and Seebeck coefficient, high temperature stability and well known synthesis processes. Some representative thermoelectric metal oxides are MnO_2, TiO_2, ZnO and WO_3 (Walia et al. 2012, 2013a, b). Recently, Walia et al. (2011) implemented ZnO and MnO_2 for the fabrication of wave-based thermo-power energy generation devices. The concept of thermo-power waves demonstrates great potential for the miniaturization of power sources by maintaining their capabilities of energy generation. These devices have been fabricated by sequential deposition of thermoelectric material (ZnO or MnO_2) and solid fuel (nitrocellulose) on Al_2O_3 substrate. The thermo-power waves are generated by solid fuel's exothermic reaction and then propagated through thermoelectric material. This self-propagation of waves resulted in very high output voltage of 500 mV and 1.8 V in case of ZnO and MnO_2 based devices, respectively, while, their corresponding room temperature Seebeck coefficients are 2360 mV/K and 2460 mV/K (Walia et al. 2012, Sumino et al. 2011). Presently, not only inorganic but also organic materials based thermoelectric sensors and generators are investigated on the base of Seebeck effect. Sumino et al. investigated the properties of organic thin film thermoelectric cells based on semi conductive bilayer structures in which C60 and Cs_2CO_3 were used as n-type elements, while, pentacene and F4-TCNQ (tetracyanoquinodimethane) as a p-type elements. It is reported that the Seebeck coefficient for n-type and p-type elements was respectively measured as 0.19 and 0.39 mV/°C and it is also concluded that the bi-layer structures allow to increase conductivity and efficiency of the thermoelectric cells. Investigations on the thermoelectric properties of the nanomaterials show that their Seebeck coefficient and *ZT* as a rule are higher than that of traditional thermoelectric materials (Minnich et al. 2009). For example, maximum figure of merit of nano-BiSbTe and BiSbTe approximately were 1.5 and 1.0.

In this chapter thermoelectric properties of low-dimensional materials and nano-structures including quasi-one dimensional crystals of tetracyanoquinodimethane complexes, quasi-two dimensional crystals of organic complexes, thermoelectric cells based on CNT-silicone and polymer adhesive composites, silicon based thermoelectric cells with built-in solar cells, thermo-electrochemical cells based on organic semiconductor orange dye aqueous solution: n-InP/Orange dye aqueous solution/carbon cell and n-InAs/Orange dye aqueous solution/Zn cell are discussed.

2. LOW-DIMENSIONAL CRYSTALS

Low dimensional crystals are observed with high anisotropic electric parameters such as conductivity and thermoelectric power. Quasi-one dimensional crystals of tetracyanoquino-dimethane (TCNQ) are investigated and highest conductivity along the length of the crystals is concluded for elongated crystals. In addition, thermoelectric power is concluded higher in semiconducting crystals than the conductors or metals (Karimov 1993).

2.1 Thermoelectric Properties of Quasi-One Dimensional Crystals of Tetracyanoquino-Dimethanecomplexes

Quasi-one dimensional crystals of tetracyanoquinodimethane-triethylammonium (TEA(TCNQ)$_2$) and tetracyanoquinodimethane-cesium ($Cs_2(TCNQ)_3$) complexes show electric properties of conductors and semiconductors respectively. Figures 3.1 and 3.2 show molecular structures of (TEA) and (TCNQ) crystals. Temperature dependence of Seebeck coefficient of TEA (TCNQ)$_2$ crystals of size ($10 \times 3 \times 0.3$ mm^3) shows its behavior similar to the organic conductors and metals as shown in Fig. 3.3 (Karimov 1993). Seebeck coefficients of the TEA(TCNQ)$_2$ crystals measured at room temperature is (58-62) μV/K and is observed increasing slowly with temperature. Crystals of TEA(TCNQ)$_2$ show highly anisotropic conductivity of 5×10^2, 4 and 1×10^{-1} (Ω^{-1}cm^{-1}) along length ([001]), width ([010]) and thickness ([100]) of the crystal. Activation energy of conductivity-temperature relationship of the TEA (TCNQ)$_2$ crystals measured at room temperature is 0.12 eV.

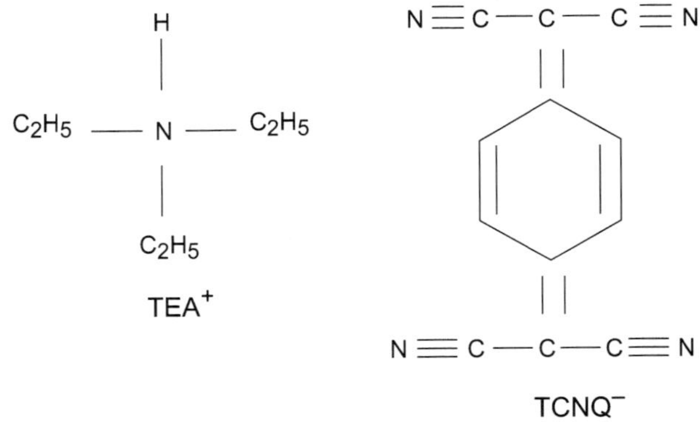

Fig. 3.1 Molecular structures of (TEA)$^+$ and (TCNQ)$^-$.

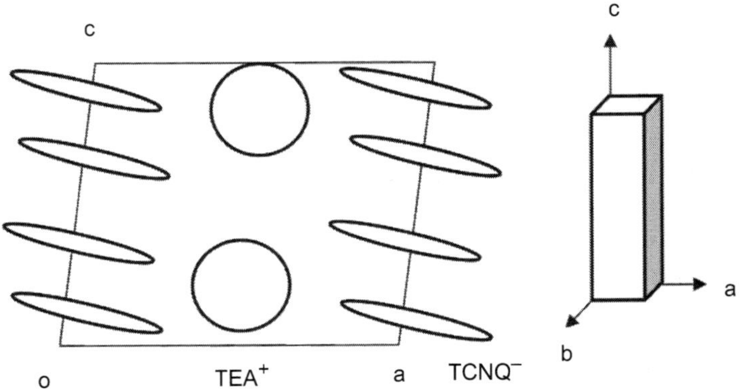

Fig. 3.2 Orientation of crystal and Projection of TEA (TCNQ)$_2$ structure on [010] plane.

Crystal structure of $Cs_2(TCNQ)_3$ is observed approximately similar to the TEA(TCNQ)$_2$ structure as shown in Fig. 3.2. Unlike TEA (TCNQ)$_2$, the $Cs_2(TCNQ)_3$ crystals temperature dependence of the Seebeck coefficient (Fig. 3.4) shows semiconductive behavior. In addition based on the measured thermo-power along the length ([010]) of the crystals of sizes 6 × 1.2 × 0.8 mm³ its behavior is observed similar to the organic conductors and metals (Karimov 1993). Seebeck coefficients of the $Cs_2(TCNQ)_3$ crystals measured at room temperature is (760-850) μV/K. Crystals of $Cs_2(TCNQ)_3$ show highly anisotropic conductivity of 1×10^{-1}, 4×10^{-3} and 4×10^{-3} ($\Omega^{-1}cm^{-1}$) along length ([010]), width ([100]) and thickness ([001]) of the crystal (Karimov 1993). Activation energy of conductivity-temperature relationship of $Cs_2(TCNQ)_3$ crystals measured at room temperature 0.30 eV.

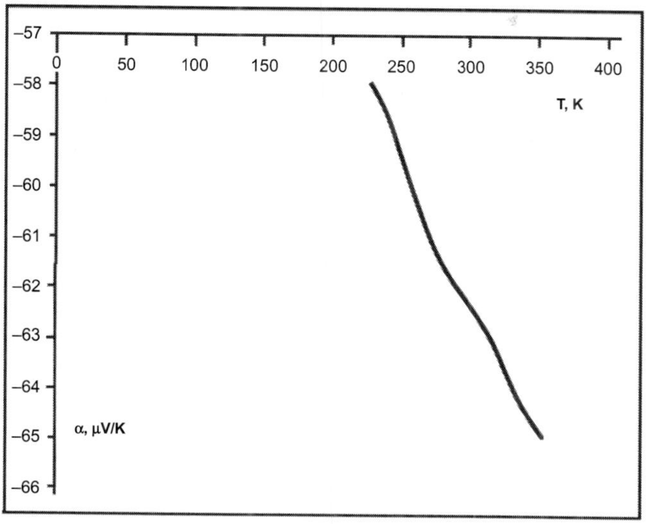

Fig. 3.3 Dependence of Seebeck coefficient of $TEA(TCNQ)_2$ on temperature along length ([001]) of the crystal (Karimov 1993).

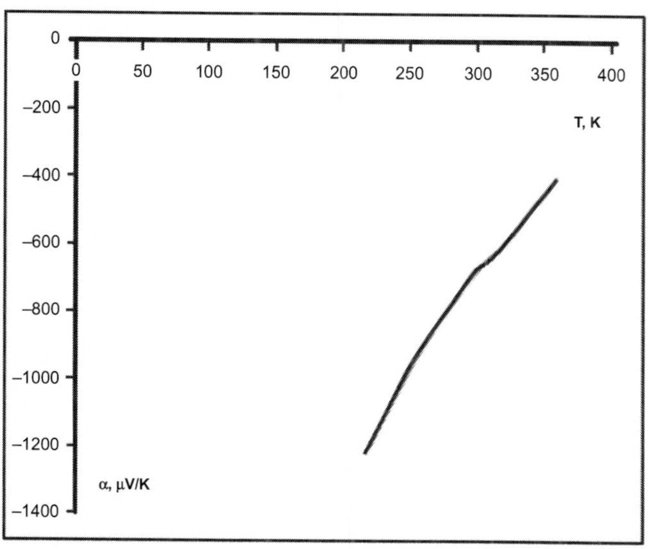

Fig. 3.4 Dependence of Seebeck coefficient of $Cs_2(TCNQ)_3$ on temperature along length ([010]) of the crystal (Karimov 1993).

3. THERMOELECTRIC PROPERTIES OF TWO-DIMENSIONAL NANOSTRUCTURES

Low-dimensional materials, in particular, zero, one and two-dimensional nanostructures schematically can be represented as shown in Fig. 3.5. Sizes of the particles less than 100 μm are considered as critical and concerned to nanoscale. These materials potentially can be used in light-emitting diodes, solar cells, photoelectrochemistry and fuel cells (Reed et al. 1998, Chattopadhyay et al. 2011). Nanostructures show very promising results for the fabrication of thermoelectric cells, based on Seebeck effect and are used for energy conversion technologies.

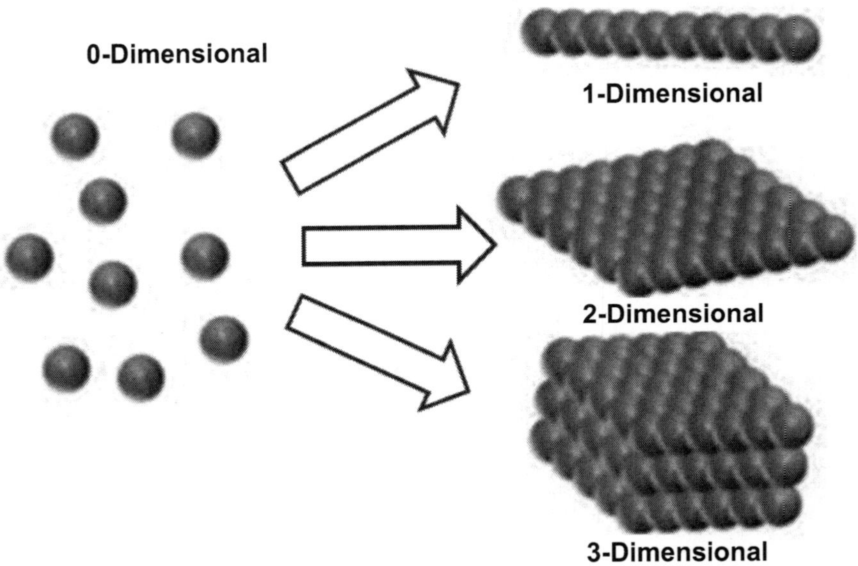

Fig. 3.5 Schematic representation of the zero (0D), one (1D) and two (2D) dimensional nanostructures where 0D, 1D and 2D represent concerned sizes that should be less than 100 nm.

Large thermoelectric Seebeck coefficient of two-dimensional electron gas in $SrTiO_3$ is observed by (Wang et al. 2009). Figure of merit (*ZT*) of more than 2 observed was approximately five times more than the bulky $SrTiO_3$ materials and thermoelectric power calculated was 850 μV/K.

Metal sulfide nanomaterials such as bismuth sulfide (Bi_2S_3), antimony sulfide (Sb_2S_3) and lead sulfide (PbS) nanostructures potentially can be used for thermoelectric applications (Lai et al. 2012). Bi_2S_3 and Sb_2S_3 are semiconducting materials with a direct bandgap of 1.3 eV and 1.78 eV respectively. In these materials a significant enhancement in the thermoelectric figure of merit (*ZT*) is expected due to the presence of superlattices based on lead chalcogenide compounds.

Thermoelectric properties of two-dimensional (2D) materials, such as graphene, hexagonal boron nitride (hBN), phosphorene, transition metal dichalcogenides (e.g., MoS_2, WS_2, etc.), metal oxides (e.g., MoO_3) were investigated (Zhang and Zhang

2015). It was found that figure of merit (*ZT*) depends on strain and can be increased at strain condition.

It was found that a newly synthesized family of two-dimensional transition metals such as carbides and nitrides, so-called MXenes, can be used for thermoelectric energy conversion applications at high temperatures (Khazaei et al. 2014). Thermoelectric properties of monolayer and multilayer M_2C (M = Sc, Ti, V, Zr, Nb, Mo, Hf and Ta) and M_2N (M = Ti, Zr and Hf) MXenes functionalized with F, OH and O groups were calculated and it was observed that Mo_2C may have a large Seebeck coefficient and simultaneously a good electrical conductivity at low carrier concentrations.

A theoretical study on the thermoelectric properties of two-dimensional topological insulators (2DTIs) based on CdTe/HgTe quantum wells showed that by increasing the doping concentration of nonmagnetic impurity, the edge states dominate the thermoelectric transport and the bulk-state conduction is largely suppressed (Li and Xu 2014). It was found that the thermoelectric figure of merit (*ZT*) can be much larger than 1 and is much higher compared to the conventional thermoelectric materials. It was also observed that by decreasing the 2DTI ribbon width or the Hall-bar width, *ZT* can be considerably improved. The CdTe/HgTe 2DTIs doped with nonmagnetic impurities can be potentially used as thermoelectric materials for thermoelectric devices.

Using density-functional theory and Boltmann transport theory, electronic band structures and electronic transport coefficients for AMN_2 (A = Sr, Ba; M = Ti, Zr, Hf) with a $KCoO_2$ crystal structure was calculated (Ohkubo and Mori 2014). Electronic band structures and electronic transport properties were observed highly anisotropic. It is considered that the two-dimensional layered materials are promising candidates for high performance thermoelectric materials which are due to the two-dimensional electronic structures in AMN_2 layered complex metal nitrides.

4. THERMOELECTRIC CELLS BASED ON CNT COMPOSITES

4.1 Thermoelectric Cells Based on CNT-Silicone Adhesive Composite

Theoretical investigations show that the *ZT* of CNT based thermoelectric cells can be larger than 2 (Sun 2008, Zhao et al. 2012). However, experimental results show that *ZT* is in the range of 1.023 to 1.202, Seebeck coefficient is around of 40 μV/°C (at room temperature) (Zhao et al. 2012). The value of *ZT* can be increased up to 0.4 by plasma treatment of CNT in argon atmosphere, which results in increase in Seebeck coefficient and decrease in thermal conductivity of the material.

Nevertheless the figure of merit is low for utilization of the CNT for conversion of the heat energy into electricity (Zhao et al. 2012). The thermoelectric properties of ultra-small single-wall carbon nanotubes showed that *ZT* can be increased by surface design, formation of bundles, increasing the tube length and so on, which significantly reduce the phonon and electron-derived thermal conductance (Tan et al. 2011).

Investigation of thermoelectric properties of single-wall carbon nanotube/ceramic nanocomposites (3Y-TZP/Al_2O_3) produced by spark-plasma-sintering showed that the thermoelectric power and *ZT* of the composites heighten with increase in temperature. The thermoelectric power changed from 28.5 μV/K to 50.4 μV/K on change in temperature from 345 K to 644 K, while the value of *ZT* is 0.02 at 850 K, which is double than that of SWCNT (in pure form). These factors indicate the potential of CNTS for their use as a thermoelectric material (Zhan and Mukherjee 2005). Seebeck coefficient increased linearly with increase in temperature. The conductivity of the composite decreased with temperature showing the metallic behavior. The effect of single (SWCNTs), few (FWCNTs) and multi-walled CNTs (MWCNTs) on the thermoelectric performance of CNT/polymer (Nafion) nanocomposites was studied by Choi et al. (2011). It was found that the electrical properties of the CNT/Nafion nanocomposites were primarily affected by the CNTs since the Nafion acts as an electrically non-conducting matrix. The thermal conductivity of the nanocomposites was dominated by the Nafion mainly due to weak van der Waals interaction. The electrical conductivity and Seebeck coefficient increased as the concentration of CNTs was increased. It was found that for thermoelectric applications FWCNTs and MWCNTs are preferred over SWCNTs in CNT/Nafion nanocomposites. In polyaniline/carbon nanotube (PANI/CNT) composites in which PANI coats CNT networks, the enhanced Seebeck coefficients and figure of merits were obtained (Meng et al. 2010). It was found that the thermoelectric parameters are several times larger than those of either of the individual components. It is also considered that this new approach has potential for synthesizing high-performance thermoelectric materials. Therefore, it would be reasonable to investigate the possibilities to use CNTs in the thermoelectric cells because of their relatively low cost and commercial availability. Here the results of the investigation of thermoelectric cells fabricated on the base of CNT-silicone adhesive composites are presented.

Commercially available (Sun Nanotech Co Ltd., China) multi-walled carbon nanotubes (MWNTs) powder and liquid silicon adhesives (Hero Gum) were used for the fabrication of composites. The diameter of MWNTs varied between 10-30 nm. For the fabrication of thermoelectric cells, the composite was prepared by mixing the multi-walled carbon nanotubes powder with silicone adhesive. The ratio of components was 1:1 by weight. The medical glass slides were used as substrate. Before deposition of composite layer on glass substrates, the substrates were cleaned by methanol and dried. The layers of composite were deposited by sequential use of drop-casting and doctor blade technologies. The length, width and total thickness of the composites layers were equal to 45 mm, 10 mm and 100 mm respectively. The thickness of the CNT composite layers was controlled by screen and measured by optical microscope. After deposition, the samples were dried for one day in room temperature conditions and then for 2 hours at 90°C. Figure 3.6 shows SEM image of the CNT-silicone adhesive composite layer at various magnification. The composite layer consisted of grains, which are in the range of 1 to 4 micron. For the measurement of temperature, thermocouples were used, which also played the role of electrodes as well for the measurement of voltage. The thermocouples were fixed at the cell's surface by silver paste. Figure 3.7 shows schematic diagram of the thermoelectric cell. The temperature, voltage and current were measured by using FLUKE 87 multimeter, while, the Seebeck coefficients were obtained as a ratio of the voltage developed between hot and cold thermocouples and temperature gradient.

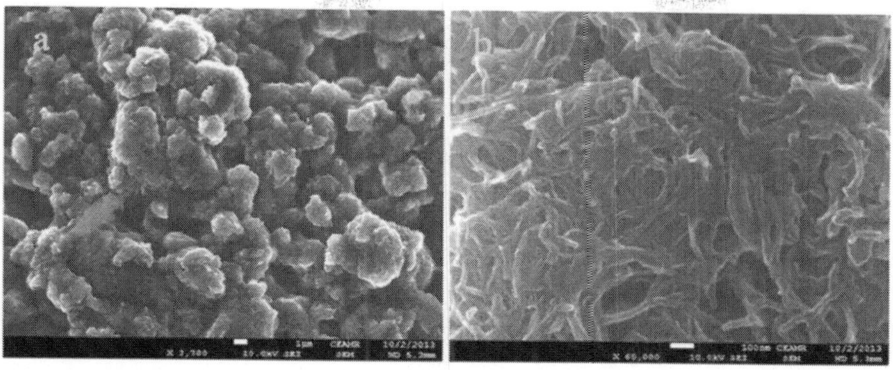

Fig. 3.6 SEM image of CNT-silicone adhesive composite layer at lower (a) and higher (b) magnification (Chani et al. 2014).

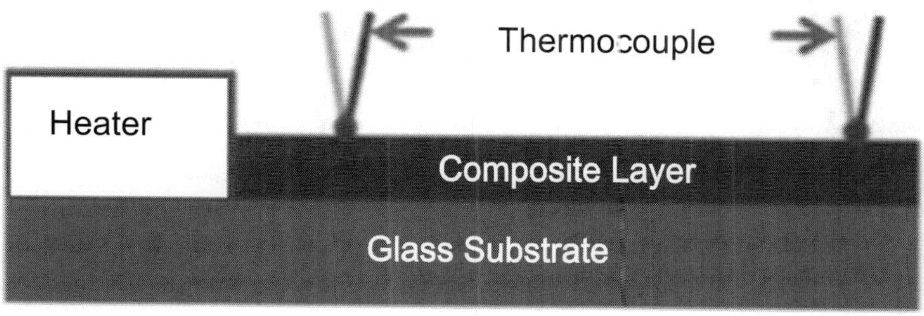

Fig. 3.7 Schematic diagram of the thermoelectric cell based on CNT-silicone adhesive composite (Chani et al. 2014).

Figure 3.8 shows current-voltage characteristics of the thermoelectric cell at different values of temperature gradient (ΔT) and at different values of load resistance of the cell. It can be seen that as ΔT increases the open circuit voltages and short circuit currents of the cell also increase. Figure 3.9 shows open-circuit voltage and short circuit current-temperature gradient relationships. These relationships are quasi-linear, that make application of the cells more suitable.

As the temperature gradient increases approximately seven times, the open circuit voltage and short circuit current increase up to 40 and five times respectively. The Seebeck coefficient (a)-temperature relationship is shown in Fig. 3.10. It can be seen that Seebeck coefficient initially increases with temperature and shows saturation behavior. Usually, the CNT composite samples are the blend of semiconductor and metallic phases. The temperature dependence of the Seebeck coefficient of the CNTs can be described by Eq. (3) (Bandaru 2007):

$$\alpha = \frac{\pi^2 K_B^2}{3e} \frac{Td\left[DOS(E^F)\right]}{dE^F} \tag{3}$$

where T is temperature, DOS (E^F) is density of states on the Fermi Energy level (E^F). In the case of metallic and semi conductive nanotubes, the derivative d (DOS (E^F))/dE^F is equal to zero or nonequal, respectively (Bandaru 2007). In the case of the metallic or degenerate semiconductor behavior of the CNTs, the following Eq. (4) may be more relevant (Snyder and Toberer 2008, Walia et al. 2013b, Cutler and Leavy 1964):

$$\alpha = \frac{8m^* \pi^2 K_B^2}{3eh^2} T \left(\frac{\pi}{3n}\right)^{2/3} \tag{4}$$

where n and m^* are the carrier concentration and the effective mass of the carriers, respectively. It means that metallic behavior of the investigated CNT composite observed in Seebeck coefficient α-temperature relationship shown in Fig. 3.10 shows that metallic phase is dominating over of semiconducting phase in this case. Under the temperature gradient, the measurements of the polarity of open-circuit voltage, in particular, positive potential of the cold side of the sample with respect to the hot side revealed that the CNT-silicone adhesive composite is a p-type material. Previously, it was stated that limitation in the sizes of particles led to the increase in efficiency of the thermoelectric cells. The cells fabricated on the base of nanomaterials were more efficient as compared to the ordinary materials of the same composition. However, it was observed that in the case of intrinsic undoped nano- materials, the thermoelectric effect was relatively low due to symmetrical contribution of the electrons and holes (Minnich et al. 2009). Therefore, due to the increase in thermoelectric parameters like figure of merit and efficiency, the practical application of the nanomaterials may be realized by doping them with n-type and p-type impurities.

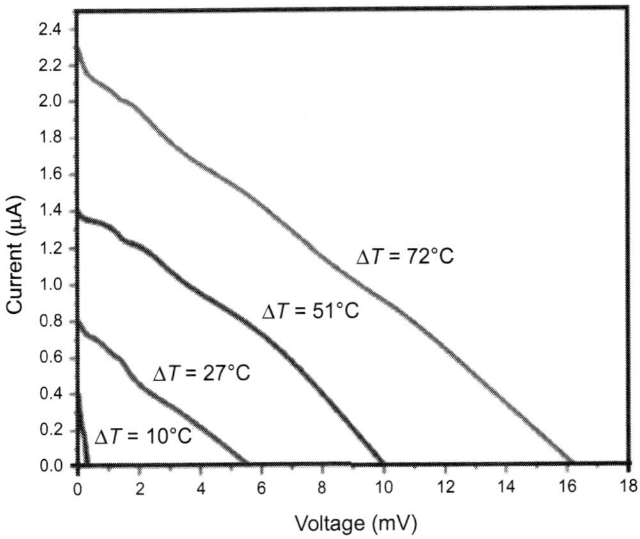

Fig. 3.8 Current-voltage characteristics of the thermoelectric cell at different values of temperature gradient (ΔT) (Chani et al. 2014).

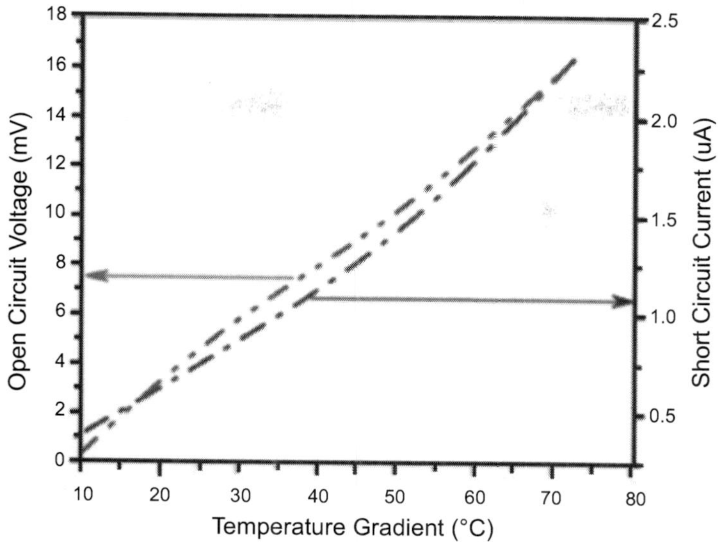

Fig. 3.9 Open-circuit voltage and short circuit-current temperature gradient relationships for the thermoelectric cell based on CNT-silicone adhesive composite (Chani et al. 2014).

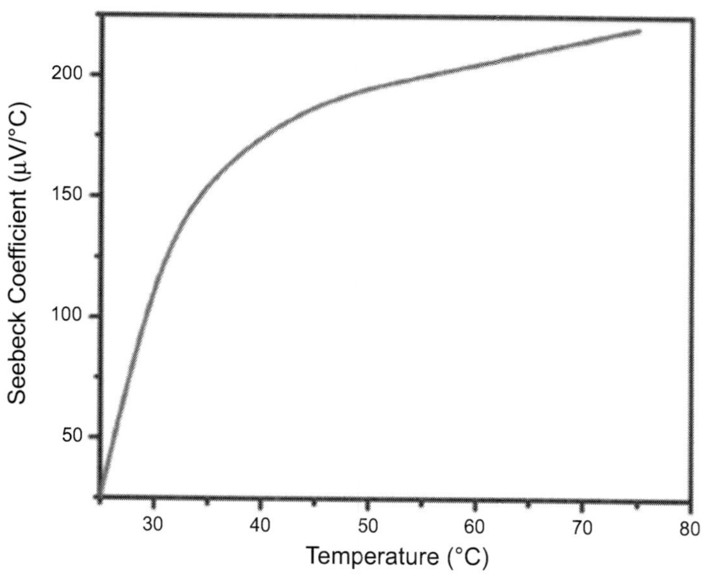

Fig. 3.10 Seebeck coefficient (α)-temperature relationship of the cell (Chani et al. 2014).

As far as the electronic properties are concerned, it is well known that CNTs show metallic or semi-conductive (small bandgap) behavior depending on the orientation of the graphene lattice with respect of CNT axis (Grow et al. 2005). This idea is also supported by the resistance temperature relationships investigated and it is found that with increase in temperature from 29°C to 72°C, the resistance of the cells decreases from 662 V to 586 V, i.e. temperature coefficient of the resistance is 20.27%/°C. The conductivity of charges in the composite can be attributed to the percolation theory (Böttger and Bryksin 1985, Dyakonov and Sariciftci 2003). According to this theory, the average conductivity can be calculated using Eq. (5):

$$\sigma = \frac{1}{LZ} \tag{5}$$

where L is the characteristic length between sites and Z is the average resistance of the connecting path between sites. Here total conduction can be considered as a conduction of both layers. Probably, the contribution of the CNT layer is considerably larger, as the change in the thickness of the CNT layer and concentration of CNT influences much to the total conduction of the composite. The adhesive in the composite plays a vital role in making the CNT layers firm, so that these layers could provide stable properties. This effect was practically observed during the current study.

Another expected role of adhesive is that it may affect the thermoelectric properties of the composite. But, the comparison of the Seebeck coefficients of the investigated composite with different CNTs and CNT-composites (Choi et al. 2011, Zhao et al. 2012, Tan et al. 2011, Zhan and Mukherjee 2005, Meng et al. 2010) shows that the effect of adhesive to the thermoelectric properties was really negligible as the value of the main parameter, Seebeck coefficient, measured by us was in the similar range as presented in the relevant references. Unlike thermoelectric generators, where the efficiency or figure of merit are most important parameters, the most important parameters for the temperature gradient sensors are the Seebeck coefficient, linearity of the voltage-temperature gradient and the range of the temperature gradient. In principle, both organic and inorganic semiconductors can be used for the temperature gradient sensors. For example, as organic semiconductors the quasi-one dimensional crystals of tetracyanoquino-dimethane complexes could be used, where the Seebeck coefficient is around of 1000 mV/°C. At the same time, for practical applications, the growth of sufficiently large sized crystals is difficult. Therefore, utilization of the thin film thermoelectric cells on the base of the CNT composites seems reasonable. During last few years, the properties of thin films based cells have been improved (Minnich et al. 2009). The simulation of current voltage behavior is carried out by using mathematical relationship in Eq. (6) (Croft and Davison 1992):

$$f(x) = ax + b \tag{6}$$

The modified form of above relationship for the current-voltage is the following Eq. (7):

$$I = kV + C \tag{7}$$

where I is the short circuit current, V is open circuit voltage and k is current-voltage factor. The value of k for the data presented in Fig. 3.8 is calculated as 21.46 A/V. The comparison of experimental and simulated results is given in Fig. 3.11 and it is evident from the graphs that the experimental and simulated results are in good agreement. For the simulation of open-circuit voltage and short circuit current-temperature gradient relationships the following exponential function in Eq. (8) is used (Croft and Davison 1992):

$$f(x) = e^x \tag{8}$$

Eq. (8) has been modified for open circuit voltage-temperature gradient relationship given in Eq. (9) as follows:

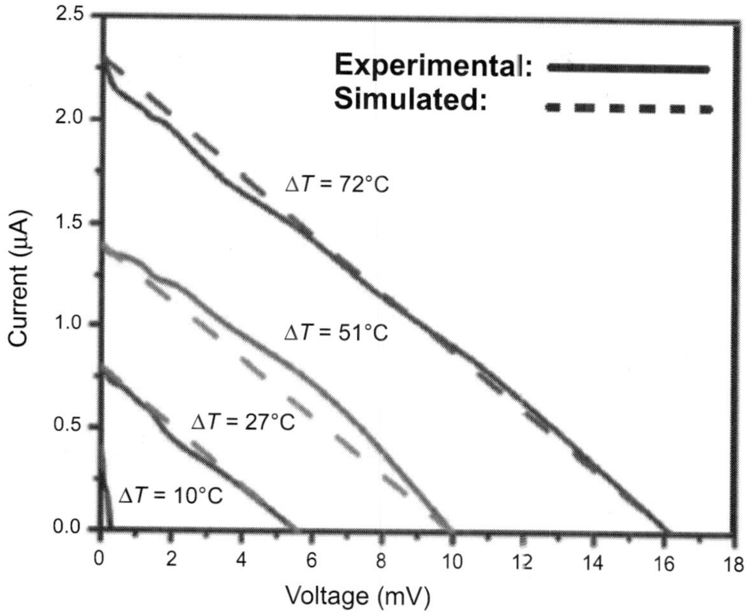

Fig. 3.11 Comparison of experimental and simulated results of the current-voltage behavior of the cell (Chani et al. 2014).

$$\frac{V_{oc}}{(V_{oc})_o} = e^{k_1\Delta(\Delta T)(4\Delta Tm + \Delta T)/(5\Delta T} \tag{9}$$

where $(V_{oc})_o$ is initial open circuit voltage at minimal temperature gradient (ΔT = 10°C), V_{oc} is open circuit voltage at instantaneous temperature gradients (ΔT) and k_1 is voltage-temperature gradient factor.

The value of voltage-temperature gradient factor is 5.46/°C. For the simulation of short circuit current temperature gradient relationship the Eq. (8) is modified as:

$$\frac{I_{sc}}{(I_{sc})_o} = e^{k_2\Delta(\Delta T)(\Delta Tm + 2\Delta T)/(3\Delta T)} \tag{10}$$

where $(I_{sc})_o$ and I_{sc} are initial short circuit current at minimal temperature gradient (ΔT = 10°C) and instantaneous values of short circuit current, respectively, while ΔT and ΔT_m are temperature gradient and maximum temperature gradient accordingly. The k_2 is current-temperature gradient factor and its value is calculated as 2.86/°C. The comparison of experimental and simulated results of voltage-temperature gradient and current-temperature gradient relationship is given in Fig. 3.12(a) and (b) respectively. It is evident from the graphs that the simulated results are in agreement with experimental results. Further improvement in the simulation is possible, which may be carried out by our group as a future work.

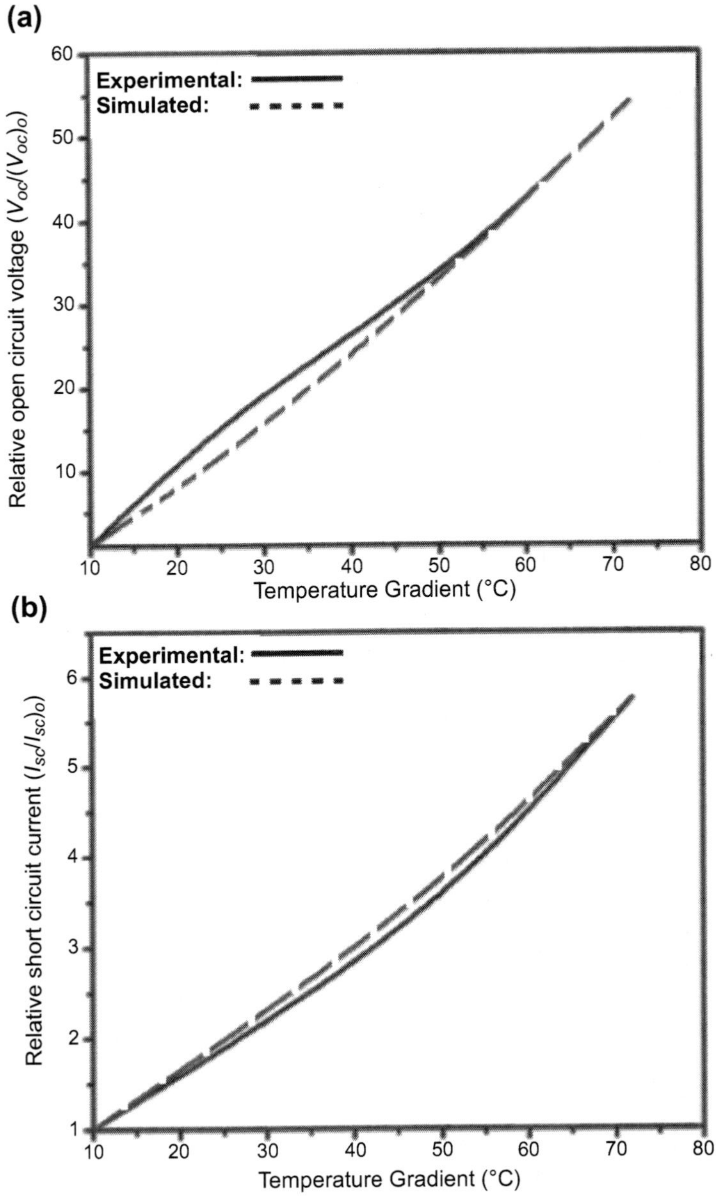

Fig. 3.12 Experimental and simulated results of voltage-temperature gradient (a) and current -temperature gradient (b) behavior of the cells (Chani et al. 2014).

The investigation of the thermoelectric cells based on composites of carbon nano-tubes (CNT) and silicone adhesive showed that the cells have good stability of performance at large temperature gradient as well. It may allow the fabrication of less energy consuming (from the heater) temperature gradient sensors that can be used in instrumentation for the measurement of temperature gradient or for the generation of electric energy for low power applications. Moreover the efficiency of the cells can be increased by the modification of composites. The simulated results are in good agreement with the experimental results.

4.2 Thermoelectric Cells Based on CNT-Polymer Adhesive Composite

Here the results of investigation of temperature gradient cells based on thermoelectric cells, fabricated on the base of CNT composites with polymer adhesive are presented. Commercially produced (Sun Nanotech Co Ltd., China) multi-walled carbon nanotubes powder was used for fabrication of the composite. Fabrication of the thermoelectric cells was similar to the cells described earlier. The diameter of MWCNTs in the powder particles varied between 10-30 nm. Thin layer of the composite of CNT (50 wt.%) and polymer (GMSA) adhesive were deposited on the glass substrate. The liquid adhesives are commercially available. The length, width and thickness of the composite layer were equal to 45 mm, 10 mm and 100 μm, respectively. Thermocouples, which played the role of electrodes for the measurement of the voltage and current, were used for the measurement of temperature. The thermocouples were fixed at the cell's surface by the silver paste as shown in Fig. 3.7. Figure 3.13 shows open-circuit voltage and short-circuits current-temperature gradient relationships for the CNT-adhesive composite. Seebeck coefficient-temperature gradient relationships for the CNT-adhesive composite are shown in Fig. 3.14.

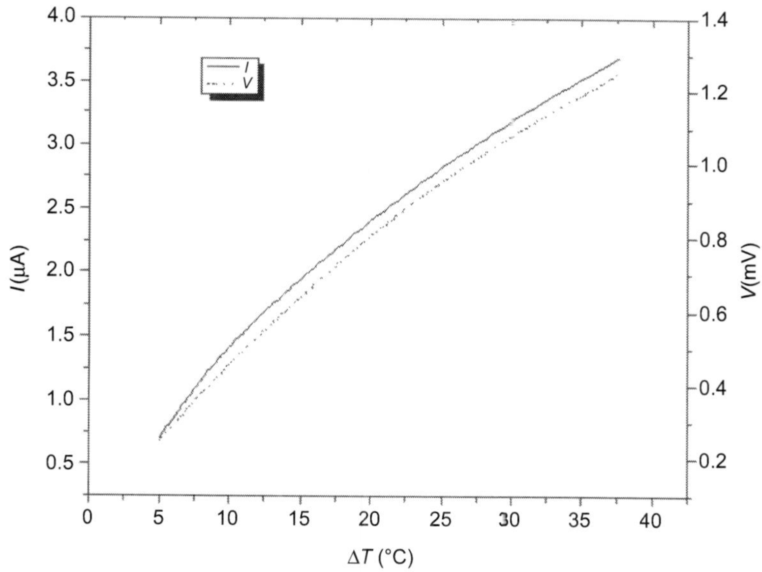

Fig. 3.13 Open-circuit voltage (V) and short-circuit current (I)-temperature gradient (ΔT) relationships for the CNT-adhesive composite (Chani et al. 2015a).

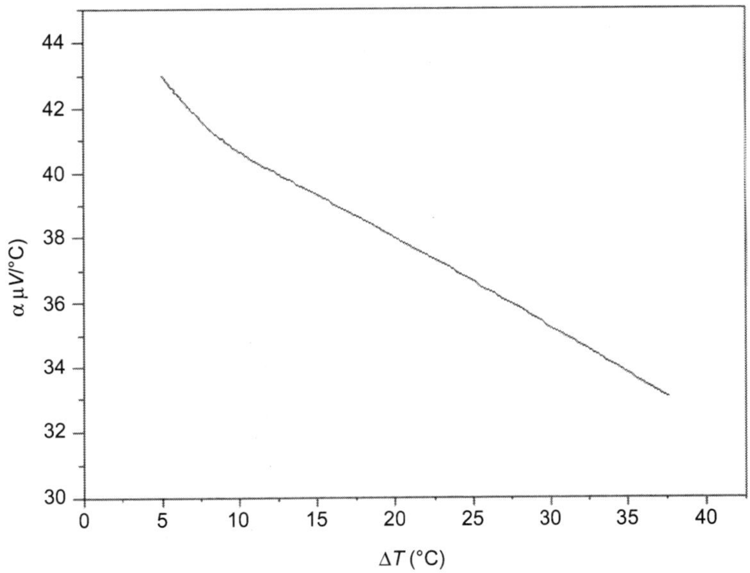

Fig. 3.14 Seebeck coefficient (α)-temperature gradient (Δ*T*) relationship for the CNT-adhesive composite (Chani et al. 2015a).

It is seen that open-circuit voltage and short-circuit current increased by 5 and 5.3 times, respectively and Seebeck coefficient decreased by 1.3 times as temperature gradient increased by 7.5 times. Measurements of the open-circuit voltage and of Seebeck coefficient showed that the composite is p-type material. Figures 3.15 and 3.16 exhibit resistance-temperature and Seebeck coefficient-temperature relationships for the CNT-adhesive composite. It is observed that resistance and Seebeck coefficient decrease with the increase of temperature by 1.13 and 1.43 times, respectively. At room temperature conditions ($T = 25°C$), Seebeck coefficient of the different samples were in the range from 40 $\mu V°C^{-1}$ to 45 $\mu V°C^{-1}$.

As it was shown in (Minnich et al. 2009) and earlier, the limitation on the size of the particles, as it was observed in the case of nano-particles, have led to the increase of efficiency of the thermoelectric cells fabricated on the base of these materials in comparison with the ordinary materials of the same composition. In the case of intrinsic and undoped nano-materials, the thermoelectric effect is relatively low due to symmetrical contribution of the electron and holes (Minnich et al. 2009). For the increase of such thermoelectric parameters (Figure of merit and efficiency), practical application of the nano-materials may be realized by doping of the materials by n-type and p-type impurities.

Electrical conductivities of the CNTs showed that they had properties of metals or semiconductors with a small gap, depending on the orientation of the graphene lattice with respect to the CNT axis (Grow et al. 2005). Probably, the investigated CNT composite samples were a blend of semiconductor and metallic phases. The temperature dependence of the Seebeck coefficient of the CNTs can be described by Eq. (3) (Bandaru 2007). The conductivity charges in the CNT-adhesive composite can be attributed to Percolation theory as it was described earlier.

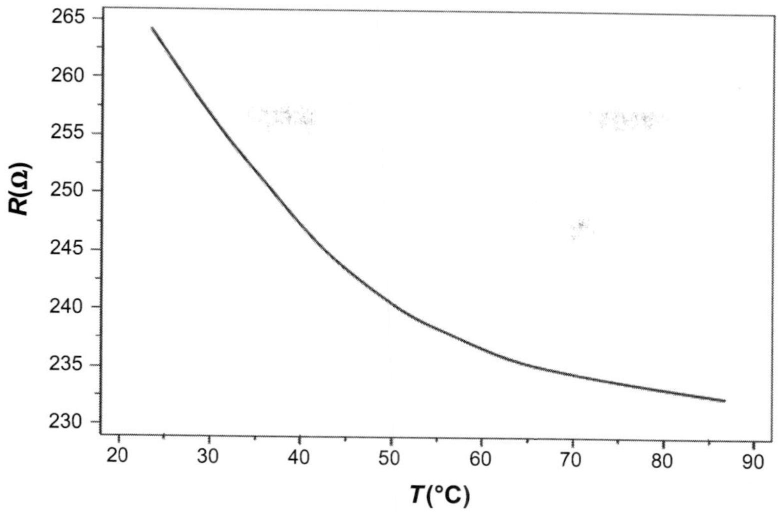

Fig. 3.15 Resistance (R)-temperature (T) relationship for the CNT-adhesive composite (Chani et al. 2015a).

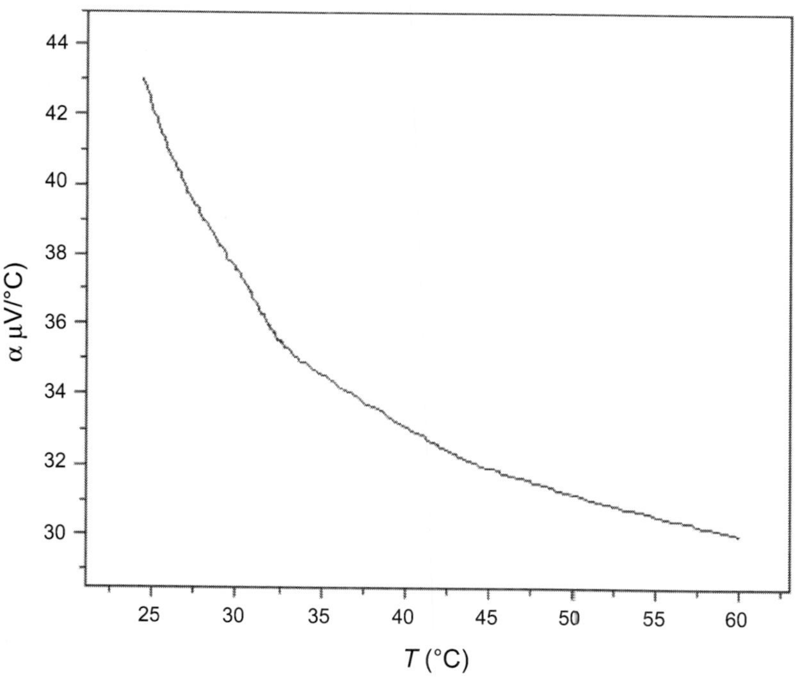

Fig. 3.16 Seebeck coefficient (α)-temperature (T) relationship for the CNT-adhesive composite (Chani et al. 2015a).

Selection of adhesive is very important. The primary role of the adhesive in the composite was in making the CNTs layer firm that could provide the stable properties which were observed practically. Secondly, the adhesive could partly influence the thermoelectric properties of the composite. The comparison of the Seebeck coefficients of the investigated CNT-adhesive composite with different CNTs and CNT-composites (Zhao et al. 2012, Tan et al. 2011, Zhan and Mukherjee 2005, Choi et al. 2011, Meng et al. 2010) shows that really the effect of the adhesive to the thermoelectric properties was negligible as the value of the main parameter, Seebeck coefficient, measured by us was in the range of the values presented in the relevant references.

Unlike the thermoelectric generators, where the efficiency or figure of merit are the most important parameters, for the temperature gradient sensors, the value of Seebeck coefficient, linearity of the voltage-temperature gradient and the range of the temperature gradient are the most important parameters. By using the approach described in (Gorodetsky and Kravchenko 1986), the open-circuit voltage-temperature gradient relationships (Fig. 3.2) can be represented by the following Eq. (11):

$$V = c(\Delta T)^B \qquad (11)$$

where c is proportional factor and B is nonlinearity coefficient that may be determined from the Eq. (12):

$$B = \frac{\text{In} V_2 - \text{In} V_1}{\text{In}\Delta T_2 - \text{In}\Delta T_1} \qquad (12)$$

where V_1 and V_2 are measured voltages at ΔT_2 and ΔT_1 temperature gradients in the sensor.

It was found that nonlinearity coefficient is relatively low, it is equal to 0.74. By means of modern electronics, non-linearity of the characteristics can be linearized by use of special non-linear op-amp circuits (Irvine 1994). For the temperature gradient sensors, inorganic semiconductors or organic semiconductors can be used. For example, as organic semiconductor temperature gradient sensor as described earlier, the quasi-one dimensional crystals of tetracyanoquinodimethane complexes could be used, where Seebeck coefficient is around of 1000 $\mu V°C^{-1}$ (Karimov 1993). At the same time, it is difficult to grow the crystals of sufficiently large sizes for practical applications. Therefore utilization of the thin film thermoelectric cells on the basis of the CNTs composites seems reasonable as their properties have improved in past few years (Minnich et al. 2009).

5. SILICON BASED PHOTO-THERMOELECTRIC CELLS WITH BUILT-IN SOLAR CELL

Photothermoelectric effect was observed at graphene interface field-effect transistors (Xu et al. 2009). Potentially it can allow designing the graphene-based optoelectronics, such as photothermocouples and photovoltaics. In (Baglio et al. 2003, Xu et al. 2009) actually light energy was converted into heat energy that in turn created a difference in temperature between illuminated and non-illuminated parts of the sample and that was the reason of generation of Seebeck voltage. In (Basko 2011) grapheme based samples having the structure of the field effect transistor with source, drain and gate were fabricated. Under effect of the laser light photothermoelectric effect was observed that was the sum of thermoelectric or

Seebeck effect and photovoltaic effect due to charges separation by the electric field of the applied to gate voltage. It may allow developing graphene-based optoelectronic devices (Basko 2011).

The photo-thermoelectric generator integrating Dye-Sensitized Solar Cells (DSSC) with thermoelectric modules was fabricated and investigated. Actually the sunlight illuminating the solar cell produces electric power and at the same time heats the cell that reduces the photoelectric conversion efficiency. This heat energy can be converted into electric power by thermoelectric cells. For fabrication of the DSSC nano-TiO_2 powder was used. Commercial nano-Cu powder was used as the medium with high thermal conductivity that can effectively transfer heat produced by sunlight in DSSC to the thermoelectric generator (TEG). Intensity of light illumination was 100 mW/cm^2. It was observed that photoelectric conversion efficiency of DSSC was 4.83% and power output was equal to 4.83 mW/cm^2. The thermoelectric conversion efficiency of the thermoelectric modules was 1.53%. The total output power was 4.97 mW/cm^2, i.e. it increased on 2.87% output with respect of the use of DSSCs alone (Chang et al. 2010). The results of investigations of the experimental model of the photo-thermoelectric cells integrating solar cells with built-in thermoelectric cells based on n-Si and p-Si were discussed.

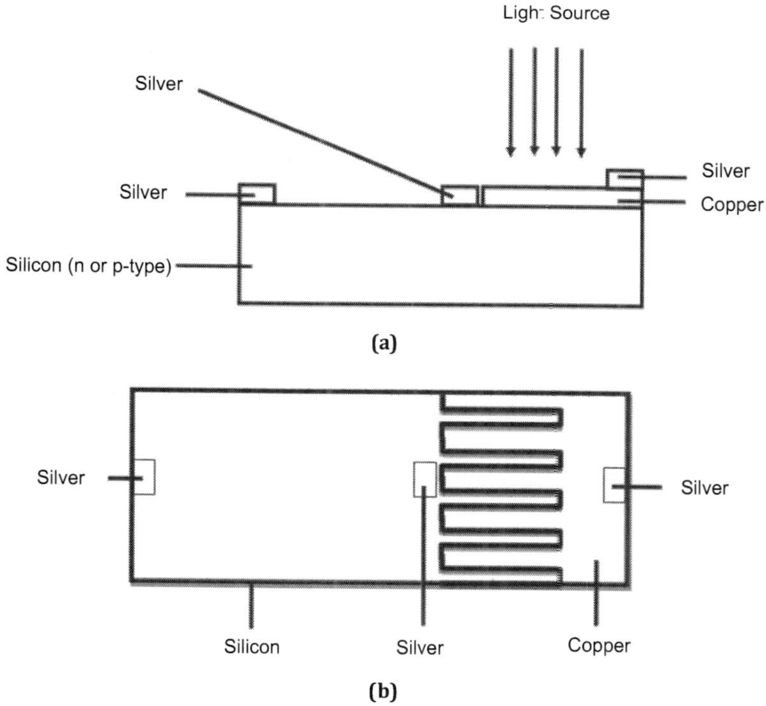

Fig. 3.17 Side view (a) and top view (b) of single photo-thermoelectric cell (Karimov et al. 2014).

Single crystal strips of planes of (100) of p-type and n-type samples with dopants concentrations of 10^{18} cm^{-3} and 10^{17} cm^{-3} respectively were tested. The sizes of strips were 3.5 cm and 2 cm in length and width. On the strip's area of 2 cm^2 Schottky junction solar cells (photo-voltaic cell) by deposition of the thin films 100 nm of Cu (Fig. 3.17a) was fabricated. Copper film comprised of the fingers of 1 mm in width, the distances between fingers were 0.5 mm (Fig. 3.17b). The contacts to the strip shaped photo-thermoelectric (PTE) cells were made by silver paste.

Photo-thermoelectric module contained thermoelectric cells of n-Si and p-Si connected in series (Fig. 3.18). Non-tracking Combined Parabolic Concentrator (CPC) with concentration ratio of 3.8, designed in (Mahammed et al. 2012) was used to increase intensity of illumination of light source (Fig. 3.19). The CPC played a double role, first of all for concentration of light to the solar cells, secondly, for shading of thermoelectric cells from the light source. As a light source the filament bulb of power of 100 W was used.

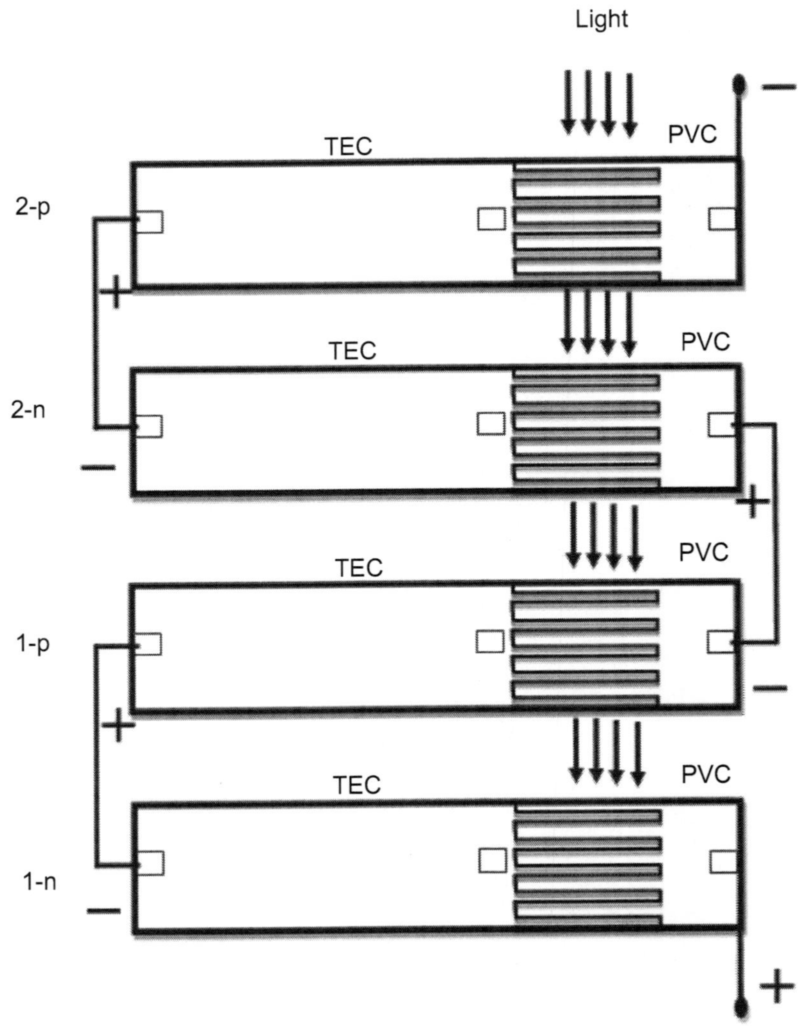

Fig. 3.18 Photo-thermoelectric module containing photo-thermoelectric cells of n-Si and p-Si connected in series (Karimov et al. 2014).

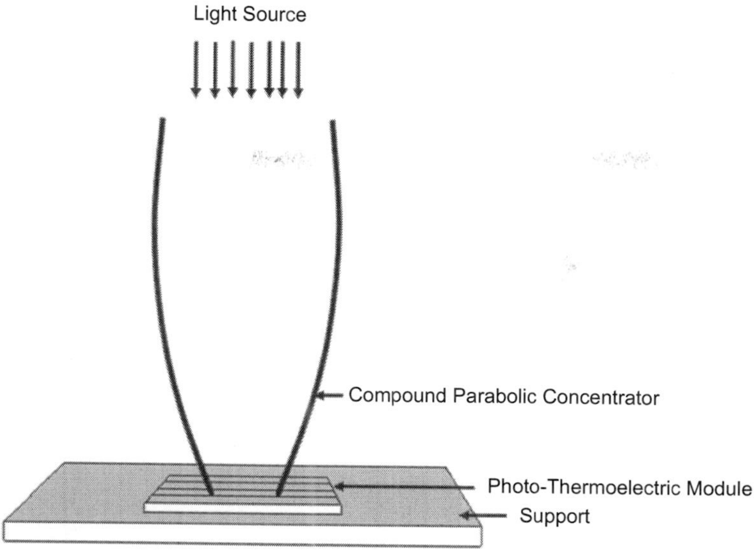

Fig. 3.19 Non-tracking combined parabolic concentrator (CPC) with concentration ratio of 3.8 integrated with photo-thermoelectric module (Karimov et al. 2014).

Table 3.1 shows the results obtained for the investigation of the properties of the photo-thermoelectric cells on the base of n-Si and p-Si strips without using combined parabolic concentrator. An estimate showed that in the case of using the CPC in average the short circuit currents (I_{sc}) and voltage of the thermoelectric cells (V_{tec}) increases to three times and voltage of the solar cells (V_{sc}) increased to 1.3 times. Total open circuit voltage and short circuit current of the four photo-electric cells in the module with CPC calculated was 3.7 mV and 0.6 μA respectively. As it is known thermal conductivity of the bulk Si is high (150 $Wm^{-1}K^{-1}$ at room temperature) and gives figure of merit (ZT) < 0.01 at 300 K (Reboun et al. 2010). Due to low ZT, actually efficiency of the thermoelectric cells fabricated on the base of bulky Si is much lower than 1%. In (Joseph et al. 2008) the nanowire with an average diameter of approximately 100 nm was investigated. It was found that by using roughened Si nanowires, the thermal conductivity related to the phonons can be reduced to 1.6 W $m^{-1}K^{-1}$, due to decrease of the mean free path lengths of phonons (Joseph et al. 2008). It can allow increasing ZT up to 1 if diameter of nanowires will be decreased. This will make Si nanowires attractive for applications in the thermoelectric generators.

Table 3.1 The properties of the photo-thermoelectric cells on the base of n-Si and p-Si (Karimov et al. 2014).

Cell material	Intensity mW/cm^2	V_{tec} (mV)	V_{sc} (mV)	V_{tot} (mV)	I_{sc} (μA)	ΔT (°C)	Seebeck coefficient (μV/°C)
n-Si	12	−0.2	−0.1	−0.3	0.2	1.3	−152
p-Si	12	0.2	0.4	0.6	0.2	1.2	167

6. PHOTO-THERMOELECTRIC CELLS BASED ON ORGANIC SEMICONDUCTOR ORANGE DYE AQUEOUS SOLUTION

6.1 Photo-Thermoelectric Cell Based on Cu/Orange Dye Aqueous Solution/Cu

For the conversion of solar and heat energies mostly made from solid semiconductors solar cells and the thermoelectric cells based on Seebeck effect are used in practice (Twidell and Weir 2015, LE 2008, Cutler and Mallon 1965). Along with solid state semiconductors a lot of research work is devoted to the investigation of liquid cells that can be potentially used for the solar and heat energy conversion into electric power. Cutler and Mallon (1965) investigated the Seebeck coefficient of thallium and tellurium liquid compounds of n-type and p-type in the temperature range of 200°C to 800°C. It was found that the figure of merit (*ZT*) was in the range of 0.1 to 0.85, which is a considerably high range. Like buoyancy and thermocapillary the thermoelectric mechanism related to convective motion in liquid semiconductors (semimetals) was analyzed by Eidel (1993). Thermopower in the Ni-Te alloys liquid semiconductors was investigated by Newport et al. (1982) and it was found that depending on concentration of Ni and Te in the alloys the Seebeck coefficient was in the range of –38 to +22.7 μV/°C. A thermoelectric cell consisted of cylindrical casing with hollow central annulus member was fabricated by (Hed 1993). In this cell the heated fluid is pumped in the central tube from the outer periphery. The gradient of temperature between the inner member and the peripheral surface of the cell results in the generation of the electric power. Male investigated glasses and liquids of chalcogenide semiconductors and observed an unpredictable phenomenon in the temperature range of 20 to 500°C e.g., in amorphous As-Se-Te and As-Se-Te-Tl systems (Male 1967). The signs of the thermoelectric coefficient and Hall Effect voltage were the opposite: the Hall coefficients were negative while the thermoelectric coefficients were positive. Usually in semiconductive materials the Hall coefficient and thermoelectric coefficient have the same sign because the sign shows type of dominating charge carriers i.e. electrons or holes.

Recently, a liquid selenium based Schottky diode has been fabricated for thermal energy harvesting. This diode shows high open circuit voltage (2.1 V). The Seebeck coefficients of non-aqueous electrolytes like tetrabutylammonium nitrate, tetraoctylphosphonium bromide and tetradodecylammonium nitrate in 1-octanol, 1-dodecanol and ethylene-glycol were investigated in the temperature range of 30 to 45°C. It was found that Seebeck coefficient was 7 mV/K at 0.1 M concentration for tetrabutylammonium nitrate in 1-dodecanol (Bonetti et al. 2011). During the last few years, a number of papers have been published on electric properties of Orange Dye (OD) and the electrochemical elements based on it. The electrical parameters of orange dye semiconductor diode were estimated by the investigation of I-V characteristics (Moiz et al. 2005). Electrochemical properties of the cell based on orange dye aqueous solution (Zn/orange dye aqueous solution/carbon) are discussed in ref. (Karimov et al. 2006). The photo-electrical behavior of electrochemical sensors based on n-Si/orange dye, vinylethynyl-trimethyl-piperidole/conductive glass was studied in ref. (Elahi et al. 2007). The investigation of electrical properties of orange dye aqueous solution was also presented in ref. (Karimov et al. 2008), where it was concluded that the conductivity of aqueous solution of OD depends upon its concentration in the water. In continuation of the efforts for the fabrication and investigation of organic semiconductor power devices (Chani et al. 2014, 2015a and 2015b), fabrication and investigation of orange dye based flexible photo-thermo electrochemical cells is discussed.

The photo thermoelectric cells were fabricated using commercially available organic semiconductor material orange dye $C_{17}H_{17}N_5O_2$) with molecular weight of 323.35 g/mol and density 0.9 g/cm^3. The IUPAC name of OD is 3-[N-Ethyl-4-(4-nitrophenylazo) phenylamino]propionitrile. The type of conduction of OD used for experiments was confirmed as p-type by the hot-probe method. The molecular structure of OD is shown in Fig. 3.20.

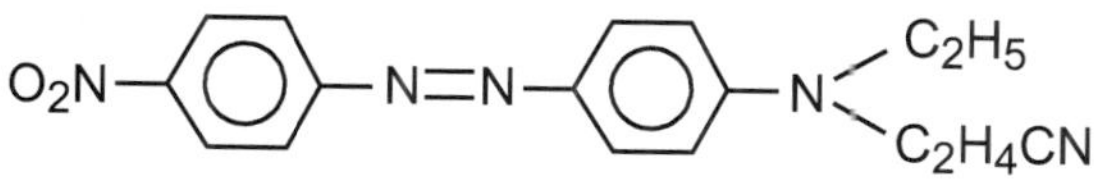

Fig. 3.20 Molecular Structure of orange dye ($C_{17}H_{17}N_5O_2$).

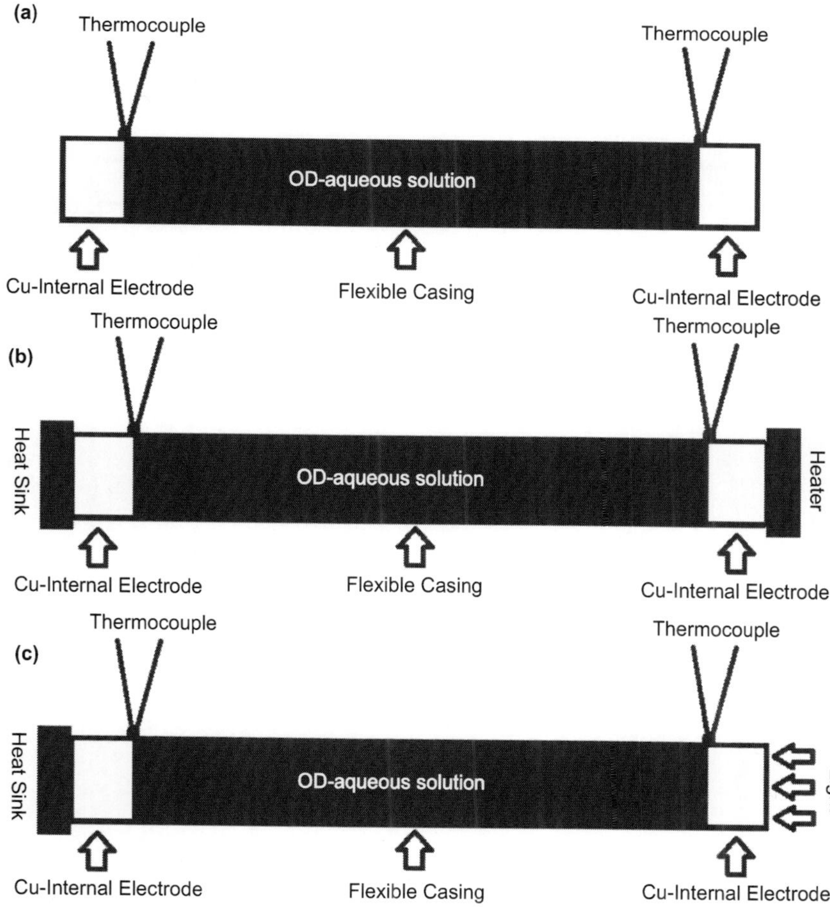

Fig. 3.21 Flexible Cu/orange dye aqueous solution/Cu photo-thermo electrochemical cell (a) the cell at experiments with low power electric heater (b) and the cell at experiments with heating by light source. (c) (Chani et al. 2015b).

The cells were fabricated by using flexible polymer tubes of internal diameter 4 mm and length 4-5 cm. The 5 wt.% aqueous solution of the OD was used as an electrolyte. The copper plates having 4 mm diameter and 2 mm thickness were used as internal electrodes. Schematic diagrams of the fabricated flexible Cu/orange dye aqueous solution/Cu photo and thermo electrochemical cells are shown in Fig. 3.21.

The voltage and currents were measured by HIOKI 3256 Digital HiTESTER. Gradient of temperature was created by low power electric heater or light and measured by FLUKE 87 by using thermocouples. The filament lamp was used as a light source, while the intensity of light was measured by KYOCERA JIM-100.

It can be seen that V_{oc} and I_{sc} increase with increase in ΔT up to 10°C approximately quasi-linearly (Fig. 3.22). Figure 3.23 shows relationships of open-circuit voltage (V_{oc}), short-circuit current (I_{sc}), versus intensity of light (G) for the Cu/orange dye aqueous solution/Cu cell. Figure 3.23 also reveals the dependence of ΔT on G. It can be seen that relationships presented in Figs. 3.22 and 3.23 have many similarities, except the values of largest temperature gradient that was achieved at particular experimental conditions. The temperature gradient created by small power electric heater and light was 10°C and 7°C, respectively. The comparison of Seebeck coefficient-temperature gradient and Seebeck coefficient-light intensity relationships of the cell are given in Fig. 3.24.

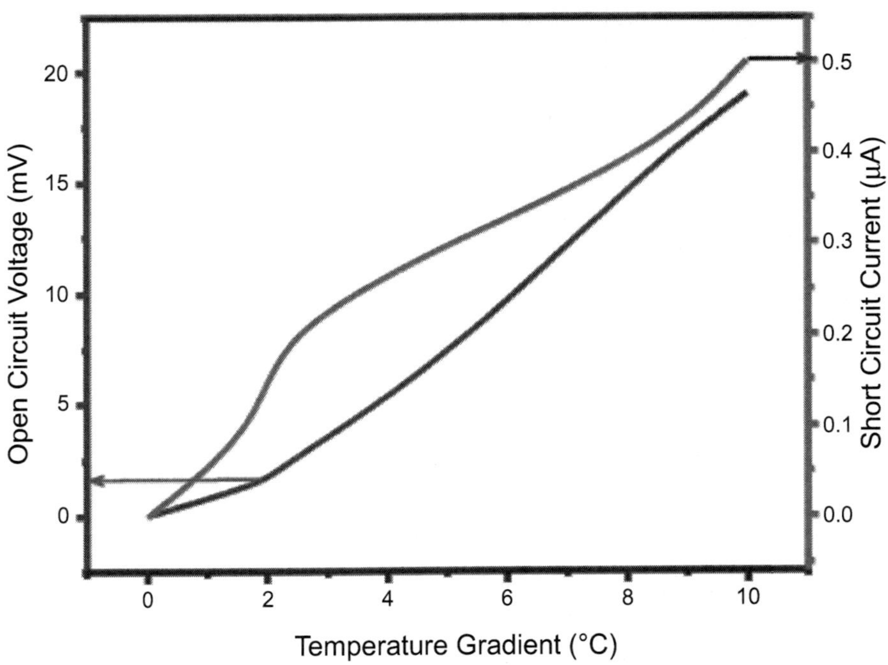

Fig. 3.22 Open-circuit voltage (V_{oc}) and short-circuit current (I_{sc}) versus temperature gradient (ΔT) relationships of the Cu/orange dye aqueous solution/Cu cell (Chani et al. 2015b).

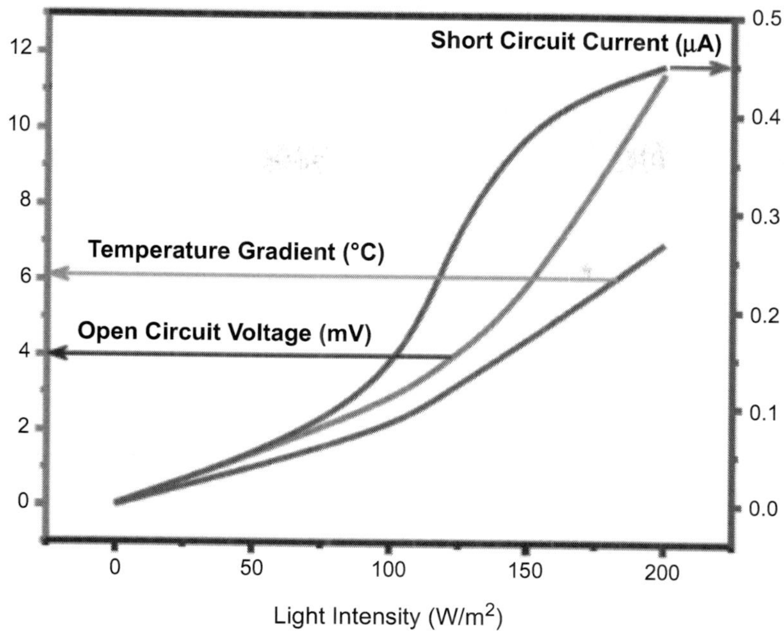

Fig. 3.23 Open-circuit voltage (V_{oc}), short-circuit current (I_{sc}) and temperature gradient (ΔT) versus light intensity relationships of the Cu/orange dye aqueous solution/Cu cell (Chani et al. 2015b).

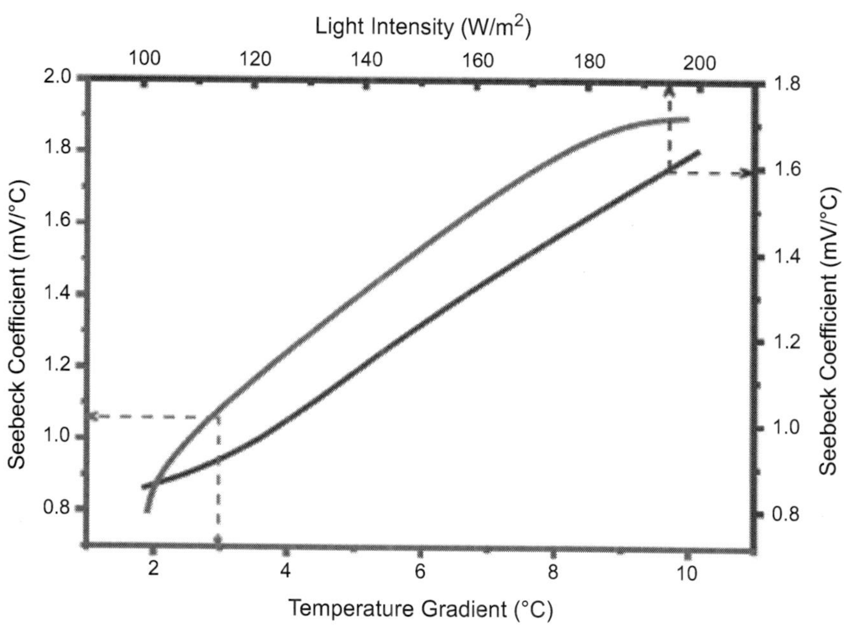

Fig. 3.24 Thermoelectric voltage-temperature gradient and thermoelectric voltage-light intensity relationships of the Cu/orange dye aqueous solution/Cu cell (Chani et al. 2015b).

The experimentally obtained maximum values of Seebeck coefficient were sufficiently large (1.9-2.2 mV/°C) as compared to the Seebeck coefficient (α) of traditional and even new thermoelectric materials; for example, for the C60 thin film n-type and p-type thermoelectric cells, plasma treated carbon nanotubes (CNTs) and Few Layers Graphene (FLG) the α is equal to 190-390 µV/°C, 350 µV/°C at 670 K and 700 µV/°C, respectively (Sumino et al. 2011, Xiao et al. 2011, Zhao et al. 2012).

The last achievements were obtained in the new materials mostly due to enhancement of phonon scattering and bandgap opening. The thermoelectric effects that were observed in a number of liquid (melted) semiconductors, for example presented in the introduction, are actually Seebeck effect because all these inorganic semiconductors are the electronic semiconductors where electrons and holes are charge carriers. Unlike these materials in electrolytes ions play the role of charges usually at electrode-electrolyte interface chemical reactions take place (Hibbert 1993). These reactions result in generation of electrode potentials that are temperature dependent as well (Hibbert 1993). Therefore we can guess that measured value of Seebeck coefficient is also affected by the temperature dependency of electrode potentials and the values of electrode potentials as well. In the case of identical electrodes probably the contribution of electrochemical processes is minimized i.e. the values of electrode potential 'theoretically' should be equal to zero, 'practically' can be low and/or constant.

The electrochemical reactions related to OD aqueous solution were presented in refs. (Karimov et al. 2006, Karimov et al. 2013). The reactions concerned to Cu electrode in electrochemical cell (Daniel cell) are described in ref. (Hibbert 1993). Electrode potential and the temperature coefficient at 298 K in aqueous solution for Cu electrode are equal to +0.337 V and +0.01 mV/°C (Hibbert 1993). Therefore it can be considered that the Seebeck coefficient of the cell will depend on the nature of metallic electrode as well. The influence of the electrodes to the value of Seebeck coefficient can be increased if two properly selected different kinds of metallic electrodes will be used in the electrochemical cell. The investigated cells potentially can be used for low power application in medicine and in instrumentation for the measurement of gradient of temperature as well. The simulation of the experimental results shown in Fig. 3.24 is carried out by using the following exponential function (Irwin and Nelms. 2007):

$$f(x) = e^x \tag{13}$$

The modified form of the above function for the simulation of Seebeck coefficient-temperature gradient relationships of the cell is the following:

$$\frac{S}{S_o} = e^{10k_1\Delta(\Delta T)/(\Delta T)} \tag{14}$$

While for the Seebeck coefficient-light intensity relationship the modified form of the function is the following:

$$\frac{S}{S_o} = e^{k_2\Delta G} \tag{15}$$

where S_o and S are the initial and instantaneous values of Seebeck coefficients, ΔT is the temperature gradient, $\Delta(\Delta T)$ is the change in temperature gradient ($\Delta T = \Delta T - \Delta T_o$) and ΔG is the change in light intensity. The k_1 is the Seebeck coefficient-temperature gradient factor and k_2 is the thermoelectric voltage-light intensity factor. The values of k_1 and k_2 are 1.08×10^{-1} and 6.6×10^{-3}, respectively. The normalized experimental and simulated results of the cells are shown in Fig. 3.25. It is apparent from Fig. 3.25 that the simulated results are in good comparison with the experimental results.

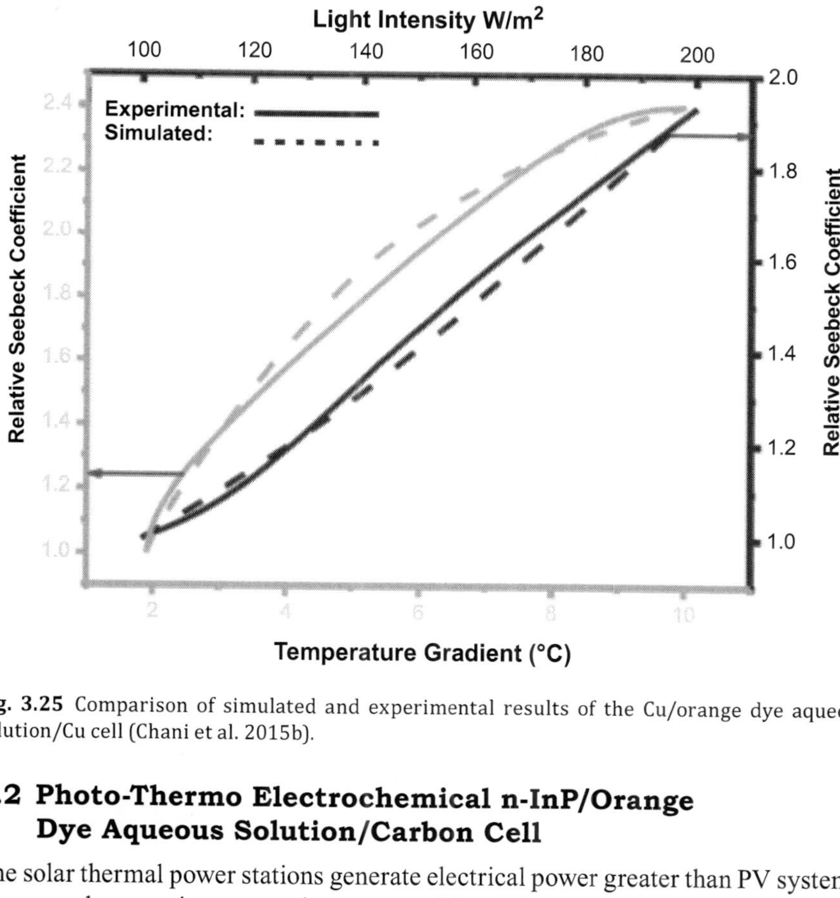

Fig. 3.25 Comparison of simulated and experimental results of the Cu/orange dye aqueous solution/Cu cell (Chani et al. 2015b).

6.2 Photo-Thermo Electrochemical n-InP/Orange Dye Aqueous Solution/Carbon Cell

The solar thermal power stations generate electrical power greater than PV systems; however, they require non-stationary, movable tracking systems to achieve a higher level of efficiency. In contrast to these power stations, investigations are made on such systems that do not require movable parts. For this, the photo electrochemical cell has prime importance in which electrolysis take places in presence of light.

Faith of almost all types of energies is heat and a larger amount of heat is wasted globally. Various thermo devices have been studied and proposed. Among them the thermo electrochemical cell is also capable of utilizing this waste heat into electrical power (Lamy 2014, Loutfy et al. 1983, Ludwig and Rowlette 1966, Shizawa et al. 1994, Stachurski 1967, Kim et al. 2014). The main principle of thermo electrochemical cell is the conversion of heat into electrical energy using chemical reaction and is a reliable technique to take advantage of thermo electrochemical cell by involving the solar light which is the source of heat (Shizawa et al. 1994).

The utilization of solar energy in organic semiconductor devices gives a cost effective, reliable and non-polluting energy device. Organic semiconductor based electronic devices have been of a greater interest and have been used for the last several decades (Elkington et al. 2014). A number of organic semiconductor based electronics and optoelectronic devices have been studied including sensors (Fukuda et al. 2014) flexible electronic devices (Karzazi 2014), OLEDs (Moiz et al. 2005) and solar cells. (Fang and Haataja 2014, Maennig et al. 2004). The advantage of organic based semiconductor devices over an inorganic semiconductor is low cost materials and fabrication technology, easy to prepare, low processing temperature and flexible devices (Chuan et al. 2011). However, the efficiency of organic semiconductor based solar cells is much lesser than inorganic solar cells (Karimov et al. 2006).

In last few years, organic semiconductor based electrochemical cell have been investigated (Saleem et al. 2007, 2009a, Elahi et al. 2007, Karimov et al. 2008). Among them, orange dye based electrochemical cell were studied (Saleem et al. 2007, 2009a, Elahi et al. 2007, Karimov et al. 2008). Orange dye ($C_{17}H_{17}N_5O_2$) is p-type organic semiconductor and shows good electrical conductivity in distilled water. The electrical conductivity versus concentration of aqueous solution of orange dye (OD) was investigated and showed maximum conductivity at 5 wt.% of OD. OD is stable at room temperature and has excellent solubility in H_2O. It also shows good light absorption and is harmless. Due to this, it is being used in many electrochemical cells and sensors (Saleem et al. 2009b, Elahi et al. 2007). In 2006, Karimov et al. studied the electrochemical properties of Zn/OD/Carbon based electrochemical cell and observed the charge and discharge of the cell was found rechargeable (Saleem et al. 2007). Here an investigation is made to explore the effect of heat due to light on n-InP/aqueous solution of orange dye/carbon cell.

The n-type InP/aqueous solution of orange dye/carbon cell is fabricated using the organic semiconductor material orange dye, OD, ($C_{17}H_{17}N_5O_2$) which is commercially available. The molecular structure of OD is shown in Fig. 3.20. The nature of OD used in the experiment was confirmed as p-type using the 'hot-probe' method. The organic orange dye with molecular weight 323 g and density 0.9 g/cm^3 is studied as photo-thermo-electrochemical cell. The cell is fabricated using an organic glass box having a width, height and depth as 30, 35 and 14 mm, respectively. Carbon and n-type indium phosphide plates with dimensions $26 \times 34 \times 3$ mm^3 and $22 \times 12 \times 0.5$ mm^3, respectively were used to serve as electrodes. Tin is used as dopant material at [100] plane of n-type indium phosphide. The doping concentration is 10^{18} cm^3 Te. The carbon electrode serves as a cathode whereas n-type indium phosphide is anode. The electrode separation is 11 mm and the complete assembly is properly sealed to avoid contamination from the environment and to minimize the evaporation of electrolyte. The schematic and 3D view of cell is shown in Fig. 3.26.

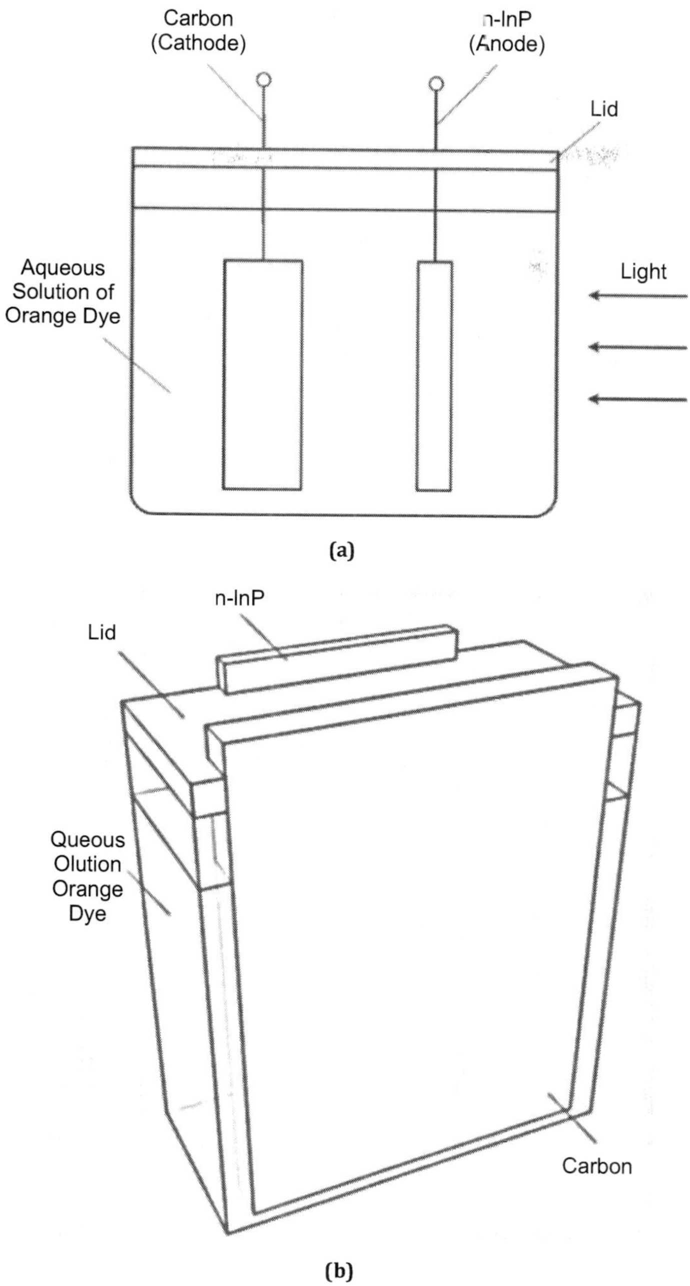

Fig. 3.26 (a) The schematic. (b) 3D view of n-InP/aqueous solut on of orange dye/carbon cell (Ali et al. 2015).

The photo-thermo-electrochemical is studied at three different levels of concentrations of electrolyte OD i.e. 1, 3 and 5 wt.%. The cell was observed under the presence of heat and light of bulb. The cell was illuminated at side of n-InP.

The open circuit voltage (V_{oc}) and short circuit current (I_{sc}) of the cell is observed by increasing the light intensity and hence the temperature of the cell. The short circuit current (I_{sc}), open circuit voltage (V_{oc}) and temperature measurements were taken using the Fluke-87 digital meter whereas the light intensity is precisely measured using the lux meter Amprove LM-80 and intensity meter Kyocera JIM-100. The light is concentrated at the cell using a non-tracking Compound Parabolic Concentrator (CPC) (Mahammed et al. 2012). Figure 3.27 shows the schematic diagram of the experimental setup of n-InP/aqueous solution of orange dye/carbon cell using compound parabolic concentrator.

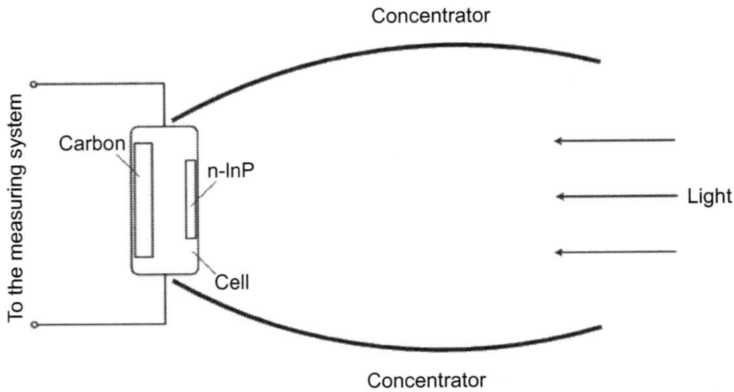

Fig. 3.27 The top view of schematic diagram of experimental setup (Ali et al. 2015).

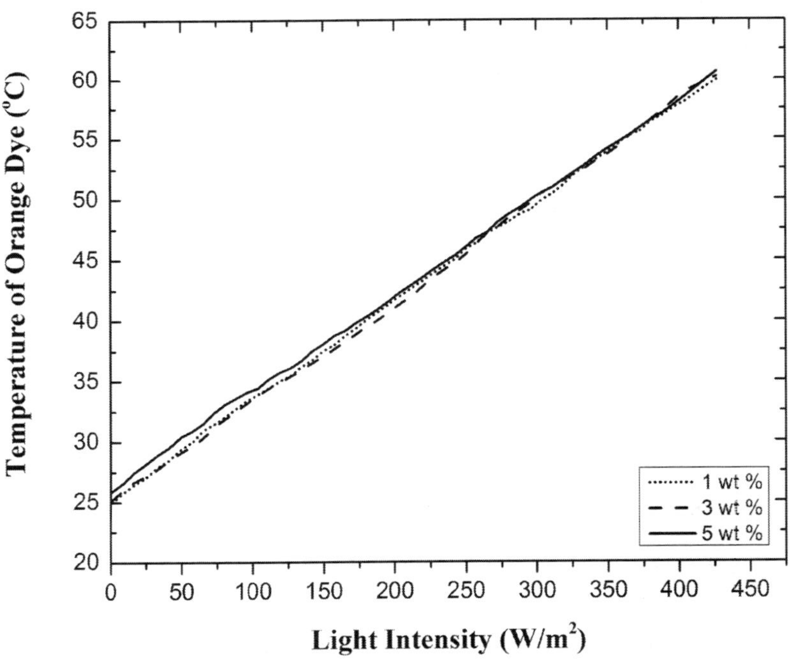

Fig. 3.28 Light intensity versus temperature of orange dye in n-InP/orange dye aqueous solution/carbon cell at 1, 3 and 5 wt.% concentration (Ali et al. 2015).

The photo-thermo-electrochemical effect of the cell is studied for n-type InP/aqueous solution orange dye/carbon cell. The effect of light on temperature is observed and depicted as in Fig. 3.28 for 1, 3 and 5 wt.% of aqueous solution of OD. It can be seen that the relationship between light and temperature is quasi linear.

The electrical properties of the cell are observed by varying light and temperature. The effect of temperature at open circuit voltage (V_{oc}) of the cell at 1, 3 and 5 wt.% aqueous solution of OD is recorded in Fig. 3.29. It can be seen that the open circuit voltage (V_{oc}) increased approximately linearly with temperature for all concentrations of OD.

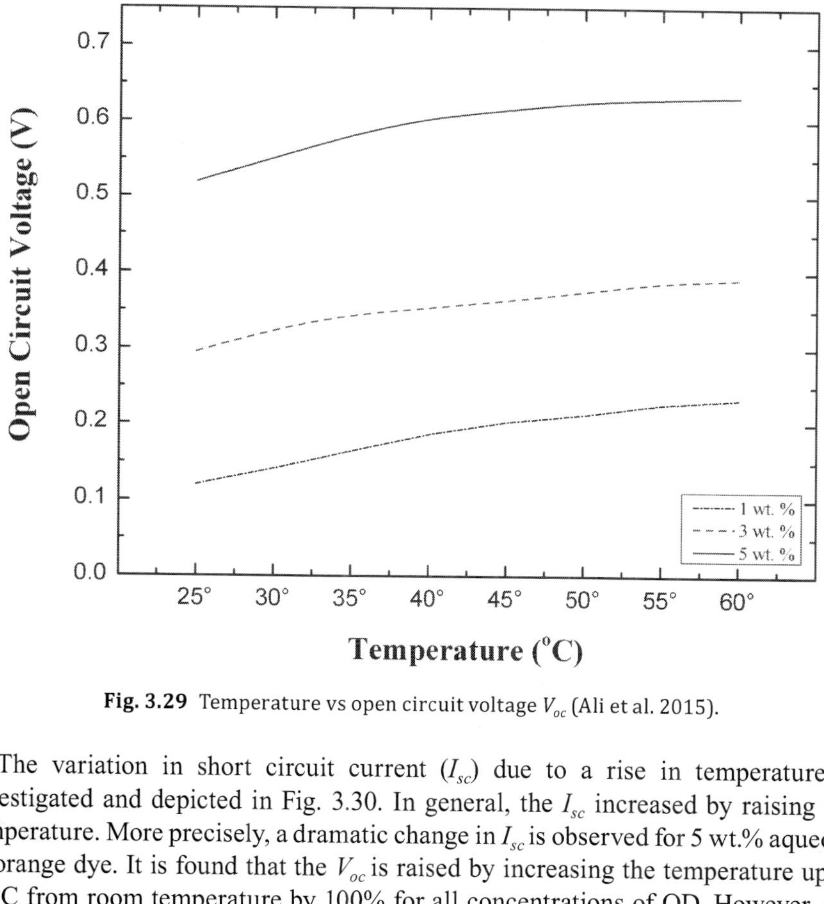

Fig. 3.29 Temperature vs open circuit voltage V_{oc} (Ali et al. 2015).

The variation in short circuit current (I_{sc}) due to a rise in temperature is investigated and depicted in Fig. 3.30. In general, the I_{sc} increased by raising the temperature. More precisely, a dramatic change in I_{sc} is observed for 5 wt.% aqueous of orange dye. It is found that the V_{oc} is raised by increasing the temperature up to 60°C from room temperature by 100% for all concentrations of OD. However, the % change in I_{sc} is 300% for 5 wt.%.

The effect of 'light' on n-InP/aqueous solution of orange dye/carbon electrochemical cell at 1, 3 and 5% is also investigated. The V_{oc} and I_{sc} due the variation of light are observed. The effect of light on V_{oc} and I_{sc} is shown in Figs. 3.31 and 3.32, respectively.

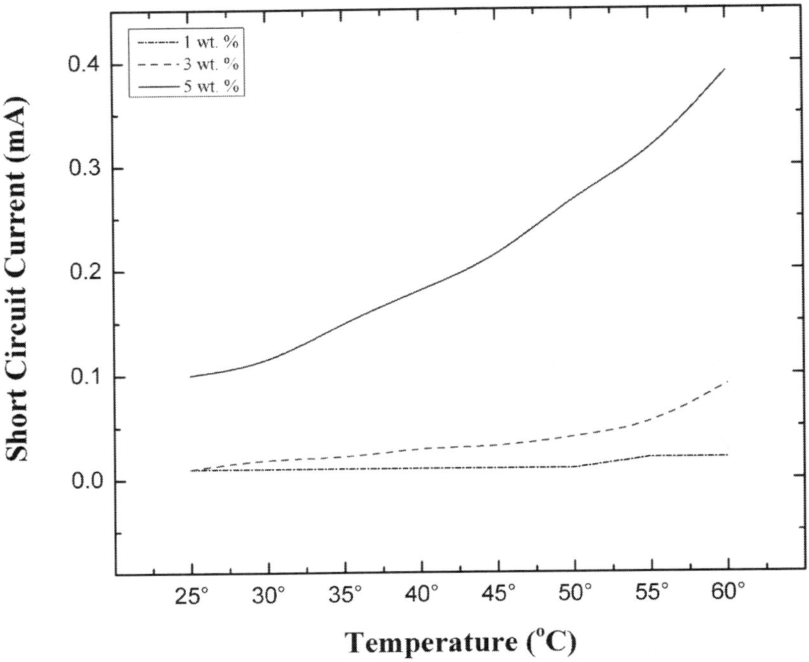

Fig. 3.30 Temperature vs short circuit current I_{sc} (Ali et al. 2015).

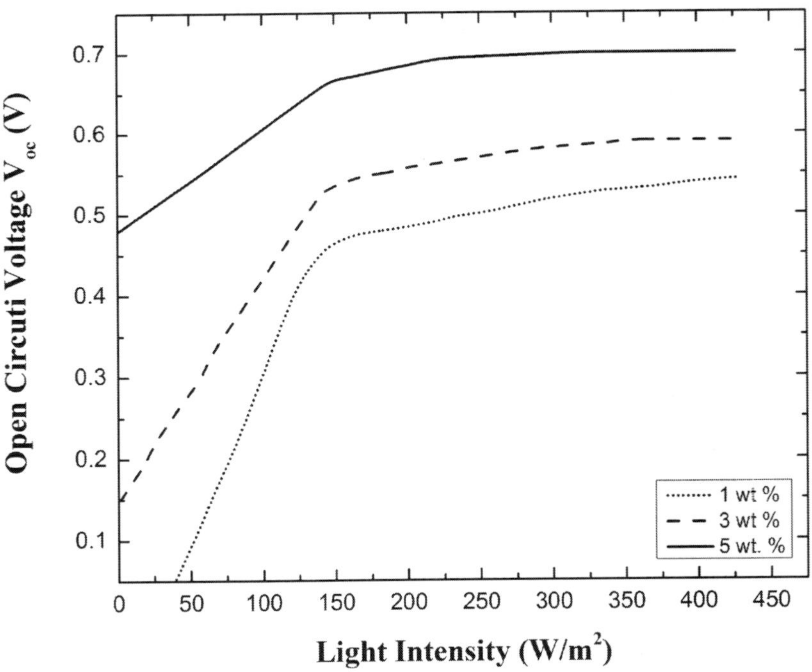

Fig. 3.31 Light intensity vs open circuit voltage of n-InP/orange dye aqueous solution/carbon cell at 1, 3 and 5 wt.% concentration (Ali et al. 2015).

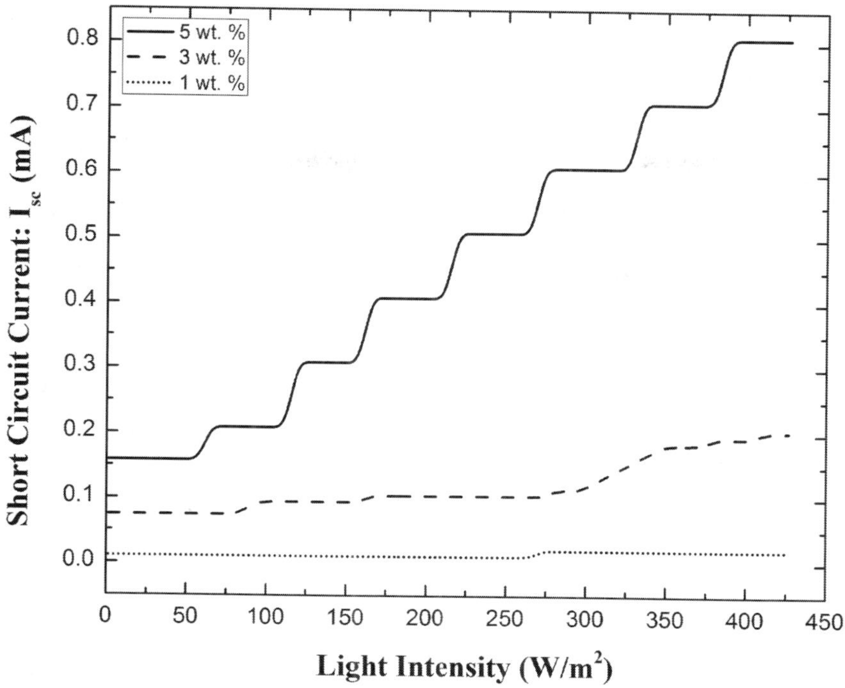

Fig. 3.32 Light intensity with short circuit current of n-InP/orange dye aqueous solution/ carbon cell at 1, 3 and 5 wt.% concentration (Ali et al. 2015).

The maximum value of V_{oc} and I_{sc} under temperature only, however, is less than the maximum value of V_{oc} and I_{sc} when observed under light. This can be used to justify that the light and temperature are both key responsible phenomena for the generation of power.

In this experiment the light intensity is increased up to 425 W/m² from dark conditions which results the temperature of the cell increased from 30°C to 60°C. It is observed that the open circuit voltages of the cell increases sharply for variation in lower values of light intensity however, it is approximately constant for increasing the higher values of light intensity. The maximum value of V_{oc} is observed 0.7 V for 5 wt.% concentration of OD. The short-circuit currents of the cell are also raised by increasing the light intensity and the maximum increment is observed for 5 wt.% concentration of OD. The change in I_{sc} by rising the light intensity and temperature at higher concentration of OD can be explained by the conductivity of ions. The ion concentration is associated with the concentration of solute i.e. orange dye molecules and their dissociation in solution. The electrolyte conductivity is given by (Mehmood et al. 2015):

$$\sigma = 10^{-3} c\alpha F(v_1 + v_2) \tag{16}$$

where c is concentration of solute in solution (mol dm³), α is dissociation constant for the molecules of OD, F is constant v_1 and v_2 are the velocities of cations and anions, respectively. However, the conductivity and temperature relationship is given as:

$$\sigma_2 = \sigma_1[1 + A(T_2 - T_1)] \quad (17)$$

where is the temperature conductance coefficient, σ_1 and σ_2 are the conductivities at temperature T_1 and T_2, respectively.

The light intensity is increased from a dark condition to 425 W/m^2 as a result of the open circuit voltage and short circuit currents are also increased. A dominate increase in short circuit current (I_{sc}) is observed from 0.15 mA to 0.8 mA for 5 wt.% of OD. Therefore the output power (ΔP_o) generated by the cell also increased and can be determined as:

$$\Delta P_o = \Delta V_{oc} \times \Delta I_{sc} \times FF \quad (18)$$

where ΔI_{sc} and ΔV_{oc} is change in short circuit current and open circuit voltage as light intensity is increased and FF called the fill factor which is 0.5 calculated form the I-V curve presented in (Mehmood et al. 2014). In Table 3.2 output power is presented at different solutions of OD.

Table 3.2 Output power (P_{out}) of 1, 3 and 5 wt.% solution of OD.

($\times 10^{-6}$ W)	1 wt.%	3 wt.%	5 wt.%
P_{out}	2.265	28.61	71.33

As the area of light exposed electrode is 264 mm^2 therefore power received (P_i) by the cell is 40.8 × 10^{-3} W at light intensity of 425 W/m^2. The power efficiency (η) is calculated as:

$$\eta = \frac{\Delta P_o}{P_i} \times 100\% \quad (19)$$

The three different samples of cell using 1, 3 and 5 wt.% of OD showed different efficiencies. It is calculated that the efficiency of the conversion of light energy into electric power by the electro chemical cell is equal to 0.55 × 10^{-2}%, 7.009 × 10^{-2}% and 14.48 × 10^{-2}% of 1, 3 and 5 wt.%. The results are depicted in Fig. 3.33.

The rise in V_{oc} and I_{sc} due to increase in light and hence temperature can be explained on the basis of the fact that as increasing temperature decreases the resistance of n-InP and OD by this current increases. The open circuit voltage increases due to dependence of electrochemical potential from temperature. Therefore, light affects the resistance of n-InP and decreases it and eventually increases the output power of electrochemical cell. Note that in both cases current increases more than voltage.

The efficiency of the proposed electrochemical cell can be further increased by using the optimal concentration of OD and choosing the electrode that are more sensitive towards light absorption. If the efficiency of the cell is increased furthermore then the cell can be practically realized as light energy convertor to electrical power generator (Ali et al. 2015).

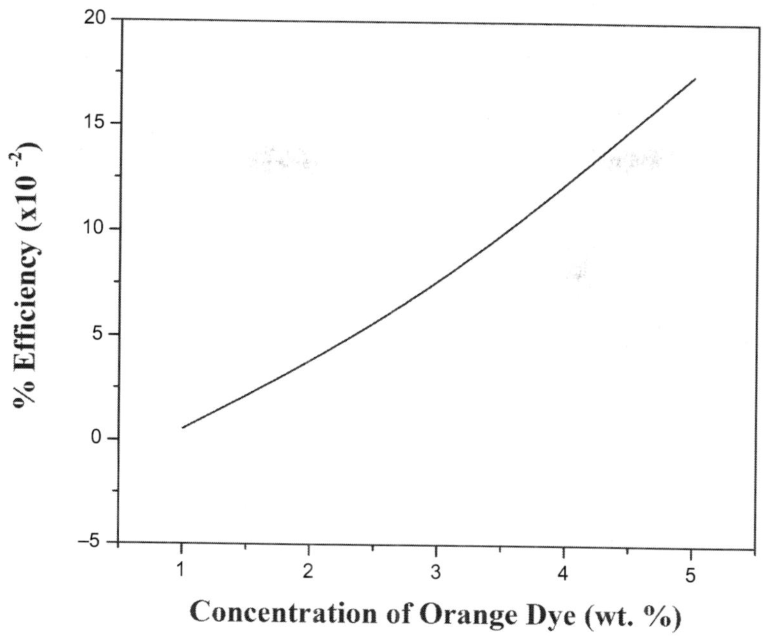

Fig. 3.33 Concentration of orange dye and % efficiency of Carbon/Orange dye aqueous solution/n-InP cell (Ali et al. 2015).

6.3 Photo-Thermo Electrochemical n-InAs/Orange Dye Aqueous Solution/Zn cell

The first practical solar cell was fabricated in 1954 using diffused silicon p-n junction with an efficiency of 6%. Although efficiency has reached 15 to 20% for silicon-based solar cells, silicon based solar cells have been restricted from local and industrial application because of their high cost, complex manufacturing techniques and rigid structures. These limitations encouraged researchers toward cheap and eco-friendly solar devices. In last 10 years, organic solar cells have been designed and investigated for the purpose of energy harvesting. Organic solar cells become the most promising alternative to silicon-based solar cells because they are cost-effective, lightweight and require easy fabrication as well as low processing temperatures (Ludin et al. 2014, Lizin et al. 2012, Shelke et al. 2014, Panozzo et al. 2002). Research is being carried on the fabrication of organic solar cells because of the previously mentioned advantages over silicon based solar cells. Nevertheless, organic solar cells are emerging in terms of efficiency compared to inorganic semiconductor solar cells.

Research has proved that fabrication and investigation of electrochemical cells based on organic electrolytes is a promising technology on the basis of cost and efficiency. Panozzo et al. (2002) reported high efficiency for light-emitting electrochemical cell based on organic salt. Bartic et al. (2003) developed organic based transistors for charge detection. Organic semiconductor material is OD that is highly soluble in water and good light absorption in a visible spectrum. Based on these advantages, OD has been used in many electrochemical cells and sensors. Previously, Karimov et al. (2006) studied electrochemical properties of Zn/orange dye/C based electrochemical cell and it was observed that the cell was rechargeable and charging/discharging current efficiency of cell was 67%. Saleem et al. (2009a) studied electrochemical properties of Zn/orange dye/C cell with varying concentration of OD and reported an increase in V_{oc} and I_{sc} at higher concentration of OD. Recently, thermo-photo electrochemical effect in n-InP/orange dye/C cell was studied with different wt.% concentration of OD (Ali et al. 2015). Maximum power efficiency of n-InP/orange dye/C cell was obtained at high concentrations of OD. These results showed that use of OD in electrochemical cell for conversion of light into electricity was useful. In the current study, an investigation of photo-electrochemical properties and power generation efficiency for n-InAs/orange dye/Zn cell has been carried out.

At the outset of the investigational procedure of fabricated Zn/orange dye/n-InAs cell, the aqueous solution of dye as electrolyte was prepared in distilled water at 1, 3 and 5 wt.% concentrations of organic semiconductor OD ($C_{17}H_{17}N_5O_2$) having molecular weight of 323 g and density of 0.9 g cm^3, which was obtained from Sigma-Aldrich, USA. The Hot-probe technique was used to confirm OD as p-type semiconductor. The molecular structure of OD is shown in Fig. 3.20.

The cell was assembled within an organic glass box having dimensions of 30, 40 and 14 mm. Figure 3.34 shows a schematic view of cell assembly. In the proposed cell design, Zn was employed as counter electrode whereas n-InAs was used as photo-electrode because of its electrochemical and semiconducting properties given in Table 3.3.

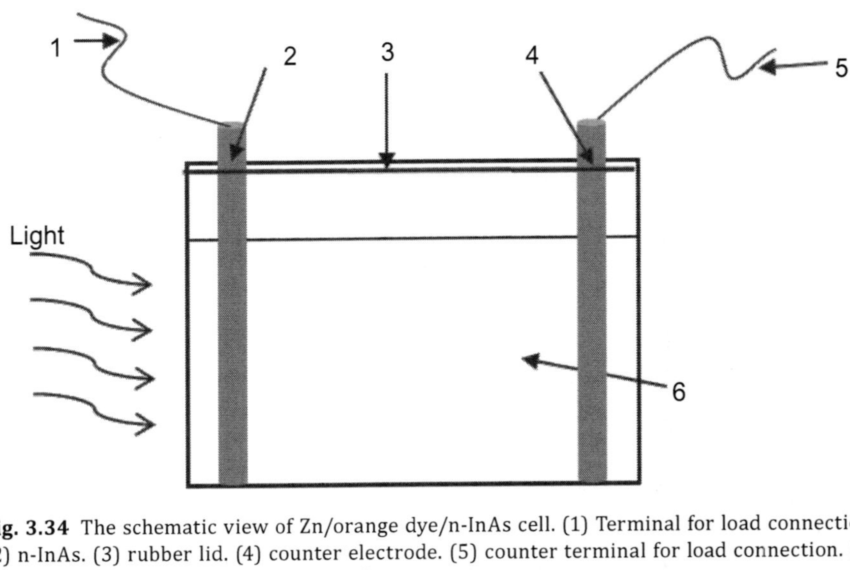

Fig. 3.34 The schematic view of Zn/orange dye/n-InAs cell. (1) Terminal for load connection. (2) n-InAs. (3) rubber lid. (4) counter electrode. (5) counter terminal for load connection. (6) aqueous solution of OD (Ajmal et al. 2015).

Table 3.3 Semiconducting properties of n-InAs, n-InP and GaAs.

Property	n-InAs	n-InP	GaAs
Electron mobility ($cm^2v^{-1}s^{-1}$)	4×10^4	5400	8500
Hole mobility ($cm^2v^{-1}s^{-1}$)	5×10^2	200	400
Electron diffusion coefficient (cm^2s^{-1})	10^3	130	200
Hole diffusion coefficient (cm^2s^{-1})	13	5	10^3
Electron thermal velocity (ms^{-1})	7.7×10^5	3.9×10^5	4.4×10^5
Hole thermal velocity (ms^{-1})	2×10^5	1.7×10	1.8×10^5

The fabricated cell is principally photo electro chemical in which thermo-galvanic effect was also observed. Owing to direct bandgap and high electron mobility, metals and semiconductor of group III-V are considered as potential candidates for power generation from renewable energy aspect (Bak et al. 2002, Kim et al. 2011). n-InAs has high electron mobility as compared to n-InP as well as conventional semiconductors i.e. Si and GaAs (Bratsch 1989). The Czochralski method was employed for synthesis and crystal growth of n-InAs photo electrode in Academy of Sciences of the Tajik SSR.

Tellurium (Te) was used as dopant material at (111) plane in n-InAs. The doping concentration was 1.5×10^{18} cm^{-3}. In the designed cell, n-InAs served as photo cathode and Zn as anode. Microscopic analysis of n-InAs photo-electrode and Zn counter electrode was carried out by utilizing PHILIPS XL 30 scanning electron microscope, while Energy Dispersive X-ray Spectroscopy (EDS) was performed for compositional analysis of both n-InAs photo-electrode and Zn counter electrode using Oxford detector. Figure 3.35 shows the SEM images of Zn and photo electrode n-InAs, Fig. 3.36 shows EDS results of n-InAs and Zn.

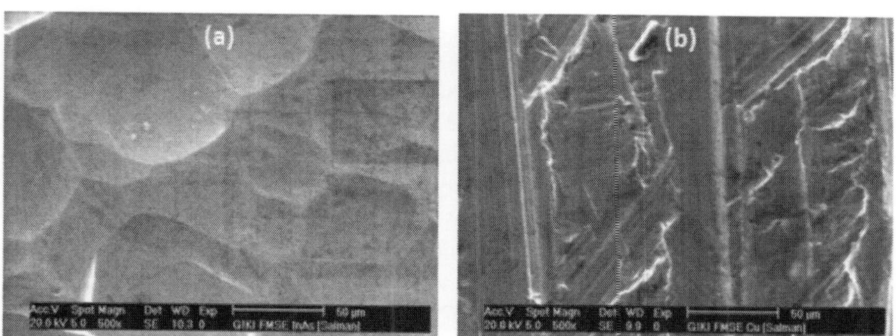

Fig. 3.35 SEM images of (a) n-InAs and (b) Zn (Ajmal et al. 2015).

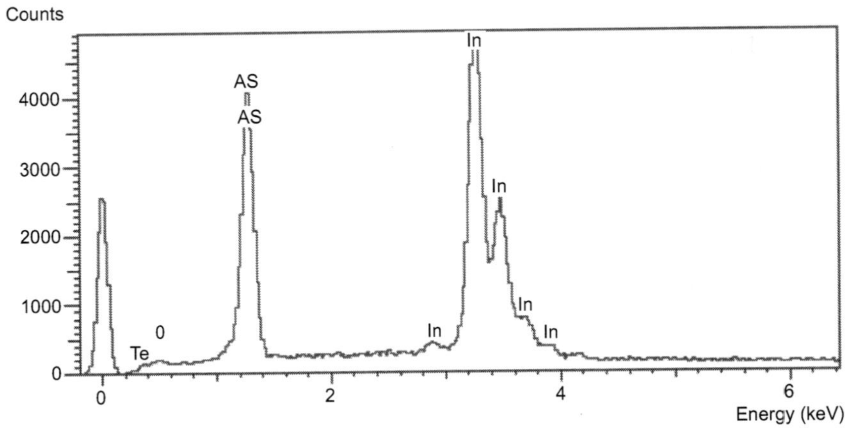

(a)

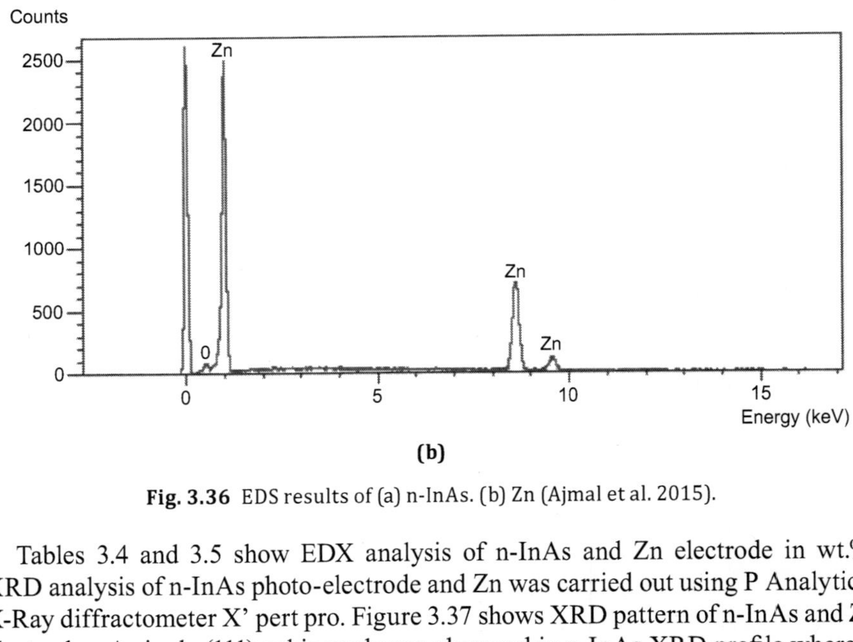

(b)

Fig. 3.36 EDS results of (a) n-InAs. (b) Zn (Ajmal et al. 2015).

Tables 3.4 and 3.5 show EDX analysis of n-InAs and Zn electrode in wt.%. XRD analysis of n-InAs photo-electrode and Zn was carried out using P Analytical X-Ray diffractometer X' pert pro. Figure 3.37 shows XRD pattern of n-InAs and Zn electrodes. A single (111) cubic peak was observed in n-InAs XRD profile whereas an orthorhombic crystal structure was identified in Zn Xrd profile using JCPDS standard database.

To avoid the externalities from surroundings and to reduce electrolyte evaporation during the experiment, the cell assembly was appropriately sealed. The cell was illuminated from n-InAs side firstly. From the side of counter electrode Zn, the effect of light was also examined. By increasing light intensity and the temperature of the cell, V_{oc} and I_{sc} of the cell were observed. The I_{sc}, V_{oc} readings and temperature (T) measurements were taken by means of the Fluke-87 digital meter. However light intensity was measured by using the intensity meter Kyocera JIM-100 and lux meter Amprove LM-80.

Table 3.4 Chemical composition of n-InAs photo electrode.

Element	In	As	Te
wt.%	60.19	39.33	0.48

Table 3.5 Chemical composition of Zn counter electrode.

Element	Zn	O
wt.%	95.19	4.14

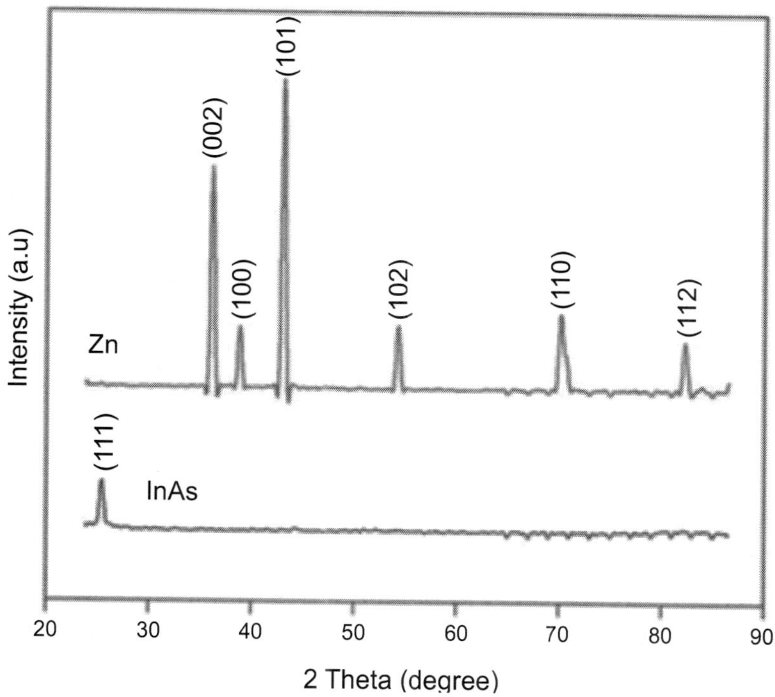

Fig. 3.37 XRD patterns of n-InAs and Zn electrodes (Ajmal et al. 2015).

The core drive is to study the photo thermo electrochemical effect at three (1, 3 and 5 wt.%) levels of concentration of OD for n-InAs/orange dye/Zn cell. The electrical properties of photo-thermo electrochemical cell were observed by varying light intensity and average temperature of the cell. It can be observed that V_{oc} and I_{sc} rised correspondingly with increasing light intensity and average temperature for all concentrations of OD. The dependency of V_{oc} and I_{sc} on light and average temperature is shown in Fig. 3.38.

The average temperature of the cells increased from 22°C to 46°C and gradient of temperature between electrodes increased from 0 to 4°C as intensity of solar radiation increased from 0 to 884 W/m^2 as shown in Fig. 3.39.

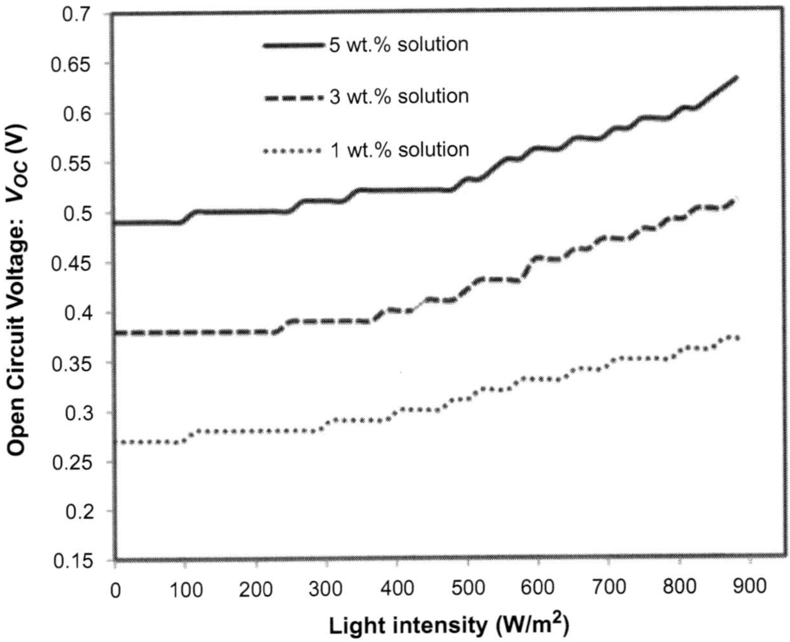

(a)

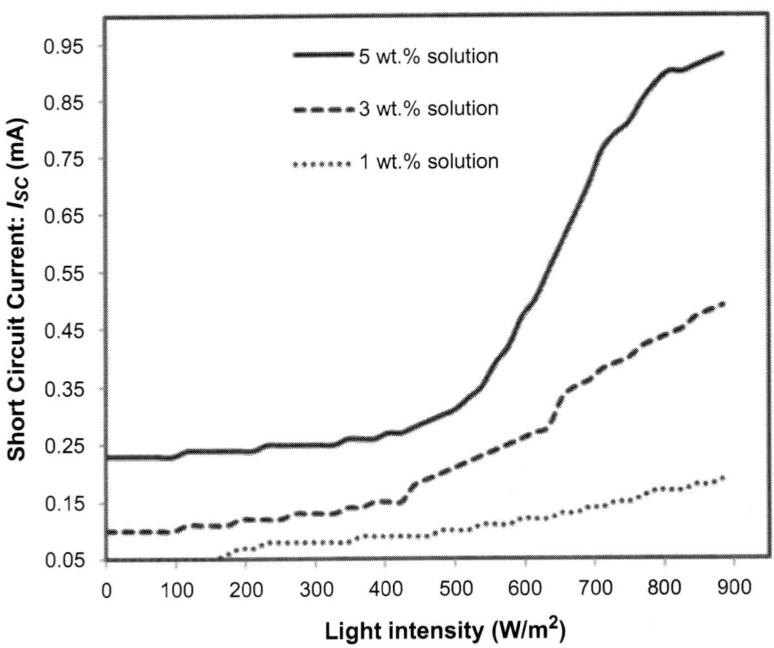

(b)

Fig. 3.38 (a) V_{oc} vs light intensity. (b) I_{sc} vs light intensity at three different concentrations of OD by illuminating the n-InA (Ajmal et al. 2015).

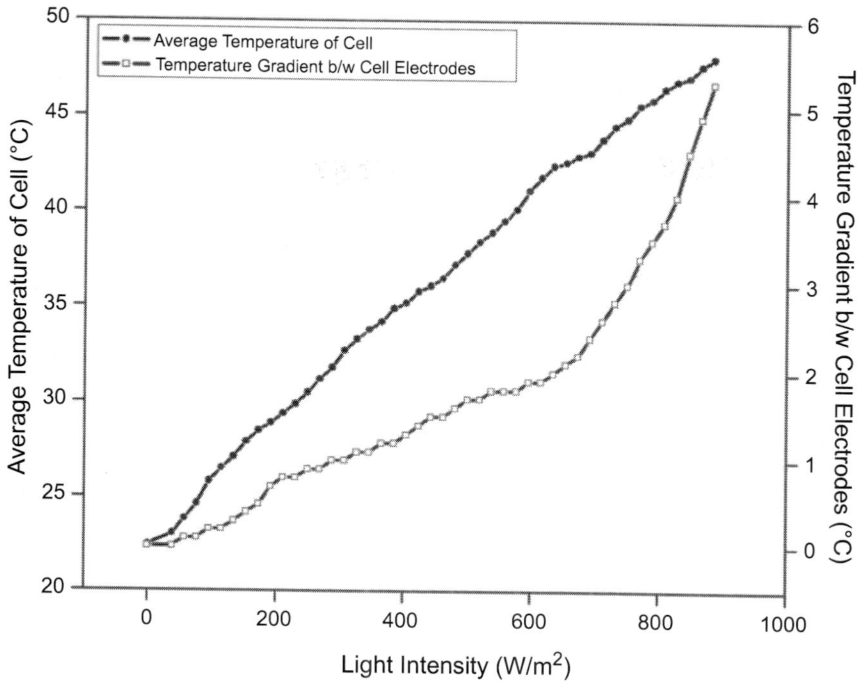

Fig. 3.39 Average temperature and gradient of temperature of the cell vs light intensity (Ajmal et al. 2015).

The maximum value of I_{sc} of single cell was observed 0.93 mA for 5 wt.% concentration of OD. Similarly, the maximal change in the value of V_{oc} of the fabricated cell was also observed for 5 wt.% concentration of OD under the illumination. The maximal change in V_{oc} and I_{sc} for 5 wt.% concentration of electrolyte was credibly higher due to the higher ions density as compared to the other wt.% concentrations. At higher concentration of OD, the variation in current and voltage by increasing light and temperature is explained by the fact that the conductivity of charged particles in the electrolyte solution increases. The charge carrier concentration is directly related with the concentration of solute. Eventually, the I_{sc} increased more than 300 times. The electrolyte conductivity related to concentration and temperature and is given by Eq. (16) and Eq. (17).

Photo electric effect certainly occurs at n-InAs electrode because of its photosensitive nature as a semiconductor and thermo-galvanic effect that is the sum of Seebeck effect in the electrolyte as well as the voltage generated due to the dependence of electrode potentials on temperature. Owing to presence of incident light, the decline in the internal resistance of n-InAs occurs. Principally, the reduction in overall resistance takes place due to high conductivity (σ) of charge carriers in the 5 wt.% concentration of electrolyte at high temperature and photosensitivity of semiconductors when it is exposed to light. Thus the output power increases. If the light falls on the counter electrode (Zn) side instead of n-InAs side, the effect observed is several times lesser as shown in the Fig. 3.40.

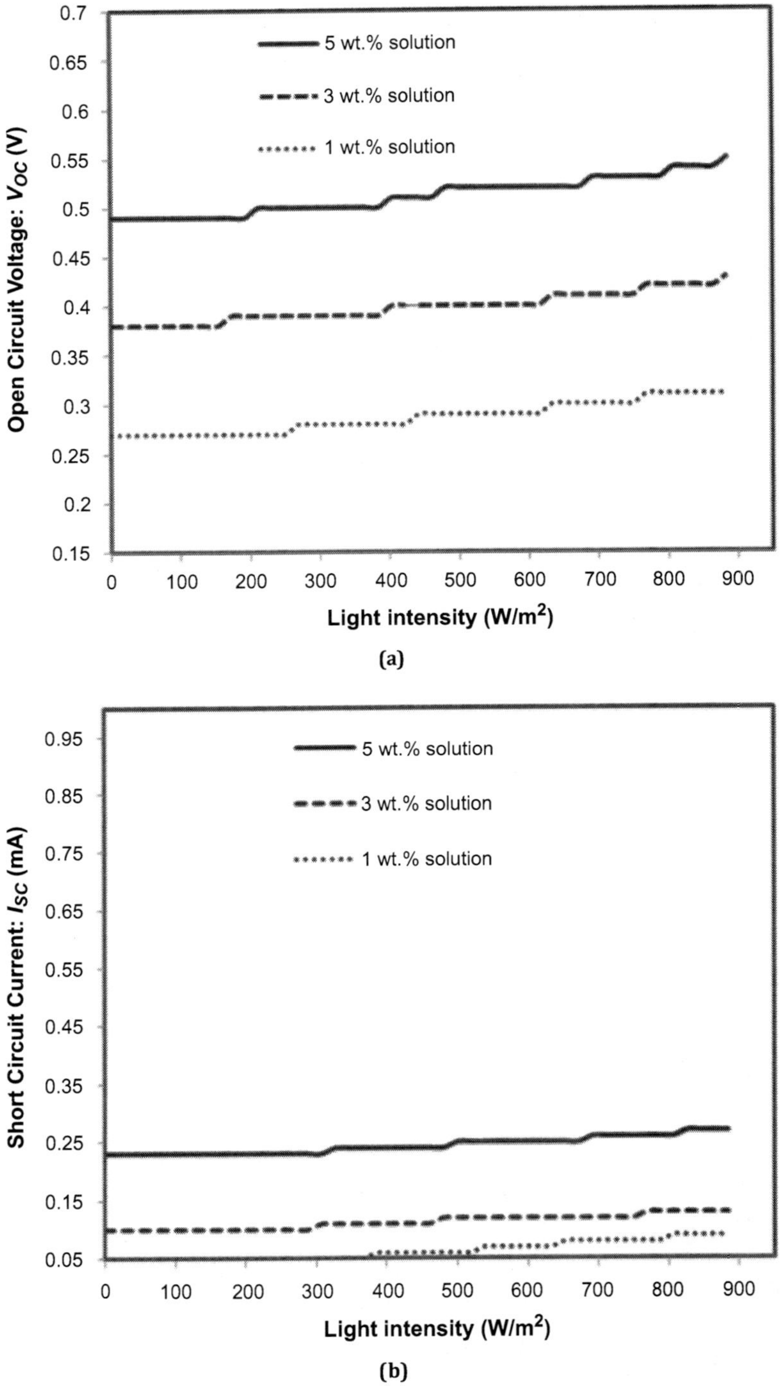

Fig. 3.40 (a) V_{oc} vs light intensity. (b) I_{sc} vs light intensity at three different concentrations of OD by illuminating Zn side (Ajmal et al. 2015).

In dark conditions, the output power was measured because of only electrochemical process that occurs in the cell, while in the presence of light and gradient of temperature the increase in the output power was due to the effects mentioned earlier. Output power can be estimated by use of Eq. (18). The value of $FF = 0.5$ was calculated from the I-V curve.

The area of photo electrode under illumination was 15 mm^2. Therefore, the input power received (P_i) by the cell was 0.13 W at maximum light intensity of 884 W/m^2. The power efficiency was calculated by use of Eq. (19). It is estimated and found that the cell having 5 wt.% electrolytic solution has the highest efficiency, as shown in Fig. 3.41, for different concentration level of OD when n-InAs was illuminated.

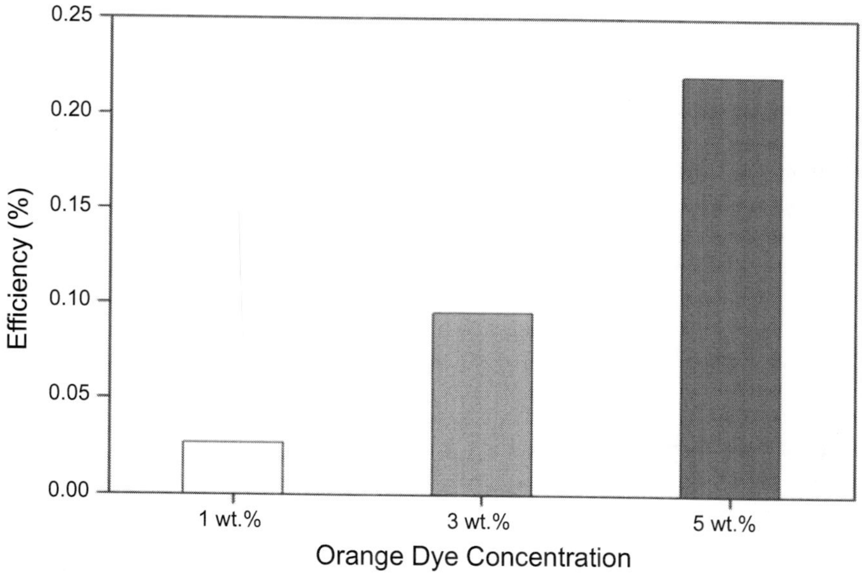

Fig. 3.41 Efficiencies of 1, 3 and 5 wt.% solution of OD under illumination of n-InAs as photo electrode (Ajmal et al. 2015).

The conversion efficiencies of light energy into electric power of the reported electrochemical cells are 0.027, 0.095 and 0.22% for 1, 3 and 5 wt.% concentrations of OD respectively. It is noticed that the efficiency can be improved further by using the OD solution of optimal concentration and selecting a suitable combination of light sensitive photo electrode with the counter electrode.

7. CONCLUSIONS

Based on the detailed study and analysis of discussed materials and devices for power generation, it is concluded that these can be potentially used for low power applications. As the main parameter for practical application of these devices is the power conversion efficiency; therefore, it is concluded that necessary efforts are required to increase Seebeck coefficient and electrical conductivity and decrease thermal conductivity of materials by selecting nanomaterials as basic ingredients or as additions. In order to increase figure of merit of thermoelectric materials at micro level, requirement of more scattering of phonons and less scattering of electrons is observed (Bulusu and Walker 2008). This is achieved by the investigation of thermoelectric effect in low dimensional materials (Zheng 2008) by the fabrication of a new carbon allotrope graphdiyne material having high electrical and low thermal conductivity and large Seebeck coefficient (Sun 2015). The room temperature *ZT* value of 3.0-4.8 makes this material one of the new candidates for thermoelectric applications. In addition, in this chapter, properties of thermoelectric cells based on solid and liquid materials (George and Kaye, 1962, George and Psarouthakis, 1964), devices such as thermo electron engines based on combination of vacuum-tube with electron gas and thermoelectric cell that can be used for the conversion of thermal energy into electric energy were also discussed. It is also concluded that a further development of materials sciences and engineering can achieve practically sufficiently high energy conversion efficiency of thermoelectron engines.

8. ACKNOWLEDGEMENTS

The authors are grateful to Ghulam Ishaq Khan Institute of Engineering Sciences and Technology of Pakistan, Center for Innovative Development of Science and New Technologies of Academy of Sciences of Tajikistan, Interdisciplinary Research Center, COMSATS Institute of Information Technology, Wah Cantt. Pakistan, School of Materials & Mineral Resources Engineering, Engineering Campus, Universiti Sains, Penang, Malaysia for support of this work.

9. REFERENCES

Ajmal, M. S., K. S. Karimov and M. Mehran Bashir. 2015. Zn/OD/n-InAs photo-electrochemical cell (unpublished).

Alam, H. and S. Ramakrishna. 2013. A review on the enhancement of figure of merit from bulk to nano-thermoelectric materials. Nano Energy 2(2): 190-212.

Ali, T., K. S. Karimov, K. M. Akhmedov, K. Kabutov and A. Farooq. 2015. Thermo photo-electrochemical effect in n-InP/aqueous solution of orange dye/C cell. Electron. Mater. Lett. 11(2): 259-265.

Al-Khalfioui M., A. Michez, A. Giani, A. Boyer and A. Foucaran. 2003. Anemometer based on Seebeck effect. Sens. Actuators, A 107(1): 36-41.

Asif, M. and T. Muneer. 2007. Energy supply, its demand and security issues for developed and emerging economies. Renewable Sustainable Energy Rev. 11(7): 1388-1413.

Baglio, S., S. Castorina, L. Fortuna and N. Savalli. 2003. Photo-thermoelectric power generation for autonomous microsystems. *In*: Circuits and Systems, 2003. ISCAS'03. Proceedings of the 2003 International Symposium on (Vol. 4: pp. IV-760). IEEE.

Bak, T., J. Nowotny, M. Rekas and C. C. Sorrell. 2002. Photo- electrochemical hydrogen generation from water using solar energy. Materials-related aspects. Int. J. Hydrogen Energy 27(10): 991-1022.

Bandaru, P. R. 2007. Electrical properties and applications of carbon nanotube structures. J. Nanosci. Nanotechnol. 7(4-5): 1239-1267.

Bartic, C., A. Campitelli and S. Borghs. 2003. Field-effect detection of chemical species with hybrid organic/inorganic transistors. Appl. Phys. Lett. 82(3): 475-477.

Basko, D. 2011. A photothermoelectric effect in graphene. Science 334: 610-616.

Bell, L. E. 2008. Cooling, heating, generating power and recovering waste heat with thermoelectric systems. Science 321(5895): 1457-1461.

Bonaccorso, F., Z. Sun, T. Hasan and A. C. Ferrari. 2010. Graphene photonics and optoelectronics. Nat. Photonics 4(9): 611-622.

Bonetti, M., S. Nakamae, M. Roger and P. Guenoun. 2011. Huge Seebeck coefficients in nonaqueous electrolytes. J. Chem. Phys. 134(11): 114513.

Böttger, H. and V. V. Bryksin. 1985. Hopping Conduction in Solids. VCH, Deerfield Beach, FL.

Brabec, C. J., V. Dyakonov, J. Parisi and N. S. Sariciftci. 2003. Organic Photovoltaics: Concepts and Realization. Springer-Verlag Berlin Heidelberg.

Bratsch, S. G. 1989. Standard electrode potentials and temperature coefficients in water at 298.15 K. J. Phys. Chem. Ref. Data 18(1): 1-21.

Bryce-Smith, D. 1982. Photochemistry. Royal Society of Chemistry, London.

Bulusu, A. and D. G. Walker. 2008. Review of electronic transport models for thermoelectric materials. Superlattices Microstruct. 44(1): 1-36.

Chang, H., M. J. Kao, K. D. Huang, S. L. Chen and Z. R. Yu. 2010. A novel photo-thermoelectric generator integrating dye-sensitized solar cells with thermoelectric modules. Jpn. J. Appl. Phys. 49(6S): 06GG08.

Chani, M. T. S., K. S. Karimov, A. M. Asiri, N. Ahmad, M. M. Bashir, S. B. Khan, et al. 2014. Temperature gradient measurements by using thermoelectric effect in CNTs-Silicone adhesive composite. PLoS One 9(4): e95287.

Chani, M. T. S., K. S. Karimov, S. B. Khan and A. M. Asiri. 2015a. Fabrication and investigation of flexible photo-thermo electrochemical cells based on Cu/orange dye aqueous solution/Cu. Int. J. Electrochem. Sci. 10: 5694-6701.

Chani, M. T. S., S. B. Khan, A. M. Asiri, K. S. Karimov and M. A. Rub. 2015b. Photo-thermoelectric cells based on pristine α-Al_2O_3 co-doped CdO, CNTs and their single and bi-layer composites with silicone adhesive. Journal of the Taiwan Institute of Chemical Engineers (in press).

Chattopadhyay, S., L. C. Chen and K. H. Chen. 2011. Energy production and conversion applications of one-dimensional semiconductor nanostructures. NPG Asia Mater. 3(8): 74-81.

Cheng, M. and Y. Zhu. 2014. The state of the art of wind energy conversion systems and technologies: a review. Energy Convers. Manage. 88: 332-347.

Choi, Y., Y. Kim, S. G. Park, Y. G. Kim, B. J. Sung, S. Y. Jang, et al. 2011. Effect of the carbon nanotube type on the thermoelectric properties of CNT/Nafion nanocomposites. Org. Electron. 12(12): 2120-2125.

Chuan, J., L. Tianze, H. Luan and Z. Xia. 2011. Research on the characteristics of organic solar cells. J. Phys. Conf. Ser. 276(1): 012169). IOP Publishing.

Chung, D.-Ym, T. P. Hogan, M. Rocci-Lane, P. Brazis, J. R. Ireland, C. R. Kannefurf, et al. 2013. A new thermoelectric material: $CsBi_4Te_6$. JACS (in press).

Croft, A. and R. Davison. 1992. Engineering Mathematics: A Modern Foundation for Electrical Electronic and Control Engineering. Addison-Wasley-Publishing Company.

Cutler, M. and J. F. Leavy. 1964. Electronic transport in high-resistivity cerium sulfide. Phys. Rev. 133(4A): A1153-A1163.

Cutler, M. and C. E. Mallon. 1965. Thermoelectric properties of liquid semiconductor solutions of thallium and tellurium. J. Appl. Phys. 36(1): 201-205.

Deutsch, T. G., J. L. Head and J. A. Turner. 2008. Photoelectrochemical characterization and durability analysis of GaInPN epilayers. J. Electrochem. Soc. 155(9): B903-B907.

Dresselhaus, M. S., G. Chen, M. Y. Tang, R. G. Yang, H. Lee, D. Z. Wang, et al. 2007. New directions for low-dimensional thermoelectric materials. Adv. Mater. 19(8): 1043-1053.

Dyakonov, V. and N. S. Sariciftci. 2003. Organic Photovoltaics: Concepts and Realization. Heidelberg, Germany.

Eidel, M. 1993. Thermoelectric convection in liquid semiconductors. J. Exp. Theor. Phys. 76(5): 802-807.

Elahi, A., M. N. Sayyad, K. S. Karimov, K. Zakaullah and M. Saleem. 2007. The photoelectrical behavior of n-Si/orange dye, vinyl-ethynyl-trimethyl-piperidole/conductive glass electrochemical sensor. Optoelectron. Adv. Mater. Rapid Commun. 1(7): 333-338.

Elkington, D., N. Cooling, W. Belcher, P. C. Dastoor and X. Zhou. 2014. Organic thin-film transistor (OTFT)-based sensors. Electronics 3(2): 234-254.

Elsheikh, M. H., D. A. Shnawah, M. F. M. Sabri, S. B. M. Said, M. H. Hassan, M. B. A. Bashir, et al. 2014. A review on thermoelectric renewable energy: principle parameters that affect their performance. Renewable Sustainable Energy Rev. 30: 337-355.

Fang, A. and M. Haataja. 2014. Crystallization in organic semiconductor thin films: a diffuse-interface approach. Phys. Rev. E 89(2): 022407.

Fukuda, K., Y. Takeda, M. Mizukami, D. Kumaki and S. Tokito. 2014. Fully solution-processed flexible organic thin film transistor arrays with high mobility and exceptional uniformity. Scientific Reports, 4.

George, H. and J. Kaye. 1962. Process and apparatus for converting thermal energy into electrical energy, US Patent 3054914 A.

George, H. and J. Psarouthakis. 1964. Process and apparatus for converting thermal energy into electrical energy, US Patent 3129345 A.

Godfrey, B. 1996. Renewable Energy: Power for a Sustainable Future. Oxford University Press, USA.

Gorodetsky, A. F. and A. F. Kravchenko. 1986. Semiconductor Devices. Visshaya Shkola, Moscow.

Green, M. A. 2004. Recent developments in photovoltaics. Sol. Energy 76(1): 3-8.

Grow, R. J., Q. Wang, J. Cao, D. Wang and H. Dai. 2005. Piezoresistance of carbon nanotubes on deformable thin-film membranes. Appl. Phys. Lett. 86(9): 093104.

Hed, A. Z. 1993. U.S. Patent No. 5,228,923. Washington, DC: U.S. Patent and Trademark Office.

Hibbert, D. B. 1993. Introduction to Electrochemistry. Macmillan, UK.

Huang, D., B. Yang and T. Tran. 2014. A nanocomposite approach: lowering thermal conductivity of Si-Ge thermoelectric materials for power generation applications. J. Appl. Mech. Eng. 3(1): 1-6.

Irvine, R. G. 1994. Operational Amplifier Characteristics and Applications. Prentice Hall, USA.

Irwin, J. D. and R. M. Nelms. 2007. Basic Engineering Circuit Analysis. John Wiley, New York, USA.

Joseph, J., V. Mathew and K. E. Abraham. 2008. Structural, photoconductivity and electroluminescence studies of Mn doped Barium Tartrate crystals. Optoelectron. Adv. Mater. Rapid Commun. 2(11): 707-713.

Karimov, K. S. 1993. Electrophysical Properties of Low-dimensional Organic Materials at Deformation. D.Sc. Thesis Academy of Sciences, Tashkent, Uzbekistan.

Karimov, K. S., J. A. Chattha and M. M. Ahmed. 2002. Investigations of parameters and characteristics of solar modules at different conditions. J. Acad. Sci. TJ. 9: 75-83.

Karimov, K. S., M. H. Sayyad, M. Ali, M. N. Khan, S. A. Moiz, K. B. Khan, et al. 2006. Electrochemical properties of Zn/orange dye aqueous solution/carbon cell. J. Power Sources 155(2): 475-477.

Karimov, K. S., I. Qazi, Z. M. Karieva, T. A. Khan and I. Murtaza. 2008. Electrical properties of orange dye aqueous solution. Kuwait J. Sci. Eng. 35: 27-35.

Karimov, K. S., M. Abid and Z. M. Karieva. 2013. Application of Nano and Organic Composites Based Devices. LAMBERT Academic Publishing, Germany.

Karimov, K. S., M. Abid, C. K. Yew, N. Ahmed, M. M. Bashir and Z. Abbas. 2014. Silicon based thermoelectric cells with built in solar cells and CNT composite based thermoelectric cells for measurement of gradient of temperature. Adv. Mater. Res. 1024: 385-392.

Karzazi, Y. 2014. Organic light emitting diodes: devices and applications. J. Mater. Environ. Sci. 5(1): 1-12.

Khazaei, M., M. Arai, T. Sasaki, M. Estili and Y. Sakka. 2014. Two-dimensional molybdenum carbides: potential thermoelectric materials of the MXene family. Phys. Chem. Chem. Phys. 16(17): 7841-7849.

Kim, S. Y., J. D. Song and T. W. Kim. 2011. Effect of InAs thickness on the structural and the electrical properties of InAs layers grown on GaAs substrates with an AlAs 0.32 Sb 0.68 buffer layer.

Kim, T. S., S. H. Kim, M. Jang, H. Yang and T. W. Lee. 2014. Charge transport and morphology of pentacene films confined in nano-patterned region. NPG Asia Mater. 6(3): e91.

Krasnov, K. S. 1982. Physical Chemistry. Published by Visshaya Shkola, Moscow, Russia.

Lai, C. H., M. Y. Lu and L. J. Chen. 2012. Metal sulfide nanostructures: synthesis, properties and applications in energy conversion and storage. J. Mater. Chem. 22(1): 19-30.

Lamy, J., I. L. Azevedo and P. Jaramillo. 2014. The role of energy storage in accessing remote wind resources in the Midwest. Energy Policy 68: 123-131.

LE, S. 2008. Anisotropic thermoelectric materials for thermoelectric generators based on layered chalcogenides. Perspect. Mater. 2: 28-38.

Li, L. and W. Xu. 2014. Thermoelectric properties of two-dimensional topological insulators doped with nonmagnetic impurities. J. Appl. Phys. 116(1): 013706.

Lizin, S., S. Van Passel, E. De Schepper and L. Vranken. 2012. The future of organic photovoltaic solar cells as a direct power source for consumer electronics. Sol. Energy Mater. Sol. Cells 103: 1-10.

Loutfy, R. O., A. P. Brown and N. P. Yao. 1983. U.S. Patent No. 4,410,606. Washington, DC: U.S. Patent and Trademark Office.

Ludin, N. A., A. A. A. Mahmoud, A. B. Mohamad, A. A. H. Kadhum, K. Sopian and N. S. A. Karim. 2014. Review on the development of natural dye photosensitizer for dye-sensitized solar cells. Renewable Sustainable Energy Rev. 31: 386-396.

Ludwig, F. A. and J. J. Rowlette. 1966. U.S. Patent No. 3,231,426. Washington, DC: U.S. Patent and Trademark Office.

Maennig, B., J. Drechsel, D. Gebeyehu, P. Simon, F. Kozlowski, A. Werner, et al. 2004. Organic pin solar cells. Appl. Phys. A 79(1): 1-14.

Mahammed, S. S., H. J. Khalaf and T. A. Yassen. 2012. Theoretical study of the compound parabolic trough solar collector. Tikrit J. Eng. Sci. (TJES) 19(2): 1-9.

Male, J. C. 1967. Hall effect measurement in semiconducting chalcogenide glasses and liquids. Br. J. Appl. Phys. 18(11): 1543.

Markvart, T. 2000. Solar Electricity, Vol. 6. John Wiley and Sons, New York City, USA.
Mehmood, U., S. U. Rahman, K. Harrabi, I. A Hussein and B. V. S. Reddy. 2014. Recent advances in dye sensitized solar cells. Adv. Mater. Sci. Eng. 2014: 1-12.
Mehmood, U., I. A. Hussein, K. Harrabi, M. B. Mekki, S. Ahmed and N. Tabet. 2015. Hybrid TiO_2–multiwall carbon nanotube (MWCNTs) photoanodes for efficient dye sensitized solar cells (DSSCs). Sol. Energy Mater. Sol. Cells 140: 174-179.
Meng, C., C. Liu and S. Fan. 2010. A promising approach to enhanced thermoelectric properties using carbon nanotube networks. Adv. Mater. 22(4): 535-539.
Minnich, A., M. S. Dresselhaus, Z. F. Ren and G. Chen. 2009. Bulk nanostructured thermoelectric materials: current research and future prospects. Energy Env. Sci. 2(5): 466-479.
Moiz, S. A., M. M. Ahmed and K. S. Karimov. 2005. Estimation of electrical parameters of OD organic semiconductor diode from measured IV characteristics. ETRI J. 27(3): 319-325.
Narodoslawsky, M. 2013. Chemical engineering in a sustainable economy. Chem. Eng. Res. Des. 91(10): 2021-2028.
Newport, R. J., R. A. Howe and J. E. Enderby. 1982. The structure and electrical properties of liquid semiconductors. II. Electron transport in liquid Ni-Te alloys. J. Phys. C: Solid State Phys. 15(22): 4635.
Ohkubo, I. and T. Mori. 2014. Two-dimensional layered complex nitrides as a new class of thermoelectric materials. Chem. Mater. 26(8): 2532-2536.
Omer, A. M. 2008. Energy, environment and sustainable development. Renewable Sustainable Energy Rev. 12(9): 2265-2300.
Ondersma, J. W. and T. W. Hamann. 2013. Recombination and redox couples in dye-sensitized solar cells. Coord. Chem. Rev. 257(9): 1533-1543.
Ostroverkhova, O. (Ed.). 2013. Handbook of Organic Materials for Optical and (Opto) Electronic Devices: Properties and Applications. Elsevier, Amsterdam, Netherlands.
Paish, O. 2002. Small hydro power: technology and current status. Renewable Sustainable Energy Rev. 6(6): 537-556.
Panda, M. K., K. Ladomenou and A. G. Coutsolelos. 2012. Porphyrins in bio-inspired transformations: light-harvesting to solar cell. Coord. Chem. Rev. 256(21): 2601-2627.
Panozzo, S., M. Armand and O. Stephan. 2002. Light-emitting electrochemical cells using a molten delocalized salt. Appl. Phys. Lett. 80(4): 679-681.
Papadopoulos, C. A., D. S. Vlachos and J. N. Avaritsiotis. 1996. A new planar device based on seebeck effect for gas sensing applications. Sens. Actuators, B 34(1): 524-527.
Reboun, J., A. Hamáček, T. Džugan and M. Kroupa. 2010. Stabilization of organic materials for sensors. In Electronics Technology (ISSE), 2010 33rd International Spring Seminar, IEEE: 40-44.
Recalde, M. Y., C. Guzowski and M. I. Zilio. 2014. Are modern economies following a sustainable energy consumption path. Energy Sus. Dev. 19: 151-161.
Reed, M. A., J. N. Randall, R. J. Aggarwal, R. J. Matyi, T. M. Moore and A. E. Wetsel. 1988. Observation of discrete electronic states in a zero-dimensional semiconductor nanostructure. Phys. Rev. Lett. 60(6): 535-542.
Rodprasert, R., T. Chandarasupsang, N. Chakpitak and P. P. Yupapin. 2014. Green energy community with smart society for sustainable living. Energy Procedia 56: 678-689.
Saleem, M., M. H. Sayyad, Z. Ahmad and K. S. Karimov. 2007. Fabrication and investigation of the charge/discharge characteristics of zinc/PVA-KOH-H_2O-Iodine/carbon cell. Optoelectron. Adv. Mater. Rapid Commun. 1(9): 477-479.
Saleem, M., M. H. Sayyad, Z. Ahmad, K. S. Karimov and Z. M. Karieva. 2009a. Effects of orange dye concentration on electrochemical properties of zinc/orange dye aqueous solution/carbon cells. Kuwait J. Sci. Eng. 36(1A): 77-86.

Saleem, M., M. H. Sayyad, K. S. Karimov and Z. Ahmad. 2009b. Fabrication and investigation of the charge/discharge characteristics of zinc/PVA-KOH/carbon cell. Acta Phys. Pol. A 116(6): 1021-1024.

Sharma, A. K. and P. J. Reddy. 1984. Electrical conductivity of InAs films prepared at different temperatures. Zech. J. Phys. B 34(7): 705-711.

Shelke, R. S., S. B. Thombre and S. R. Patrikar. 2014. Cobal sulphate as an alternative counter electrode material in dye sensitized solar cells. J. Sol. Energy Eng. 136(4): 41006.

Shizawa, M., T. Ogata, K. Shindo, S. Kitada and M. Koyama. 1994. U.S. Patent No. 5,310,608. Washington, DC: U.S. Patent and Trademark Office.

Snyder, G. J. and E. S. Toberer. 2008. Complex thermoelectric materials. Nat. Mater. 7(2): 105-114.

Snyder, J. 2008. Small thermoelectric generators. J. Electrochem. Soc. 17(3): 54-56.

Stachurski, Z. O. 1967. U.S. Patent No. 3,357,860. Washington, DC: U.S. Patent and Trademark Office.

Stilbans, L.S. 1967. Physics of Semiconductors, Moscow.

Sumino, M., K. Harada, M. Ikeda, S. Tanaka, K. Miyazaki and C. Adachi. 2011. Thermoelectric properties of n-type C60 thin films and their application in organic thermovoltaic devices. Appl. Phys. Lett. 99(9): 093308.

Sun, K. 2008. Charge transport in organic/semiconductor quantum dot ensembles and phonon scattering in carbon nanotubes. ProQuest, USA.

Sun, L., P. H. Jiang, H. J. Liu, D. D. Fan, J. H. Liang, J. Wei, et al. 2015. Graphdiyne: a two-dimensional thermoelectric material with high figure of merit. Carbon 90: 255-259.

Tan, X. J., H. J. Liu, Y. W. Wen, H. Y. Lv, L. Pan, J. Shi, et al. 2011. Thermoelectric properties of ultrasmall single-wall carbon nanotubes. J. Phys. Chem. C 115(44): 21996-22001.

Thermoelectric device with fluid thermoelectric element. U.S. Patent 3,508,974, issued April 28, 1970.

Twidell, J. and T. Weir. 2005. Renewable Energy Resources. Taylor & Francis, Second Edition, Spon.

Walia, S., R. Weber, S. Sriram, M. Bhaskaran, K. Latham, S. Zhuiykov, et al. 2011. Sb_2Te_3 and Bi_2Te_3 based thermopower wave sources. Energy Env. Sci. 4(9): 3558-3564.

Walia, S., R. Weber, S. Balendhran, D. Yao, J. T. Abrahamson, S. Zhuiykov, et al. 2012. ZnO based thermopower wave sources. Chem. Commun. 48(60): 7462-7464.

Walia, S., S. Balendhran, H. Nili, S. Zhuiykov, G. Rosengarten, Q. H. Wang, et al. 2013a. Transition metal oxides–thermoelectric properties. Prog. Mater. Sci. 58(8): 1443-1489.

Walia, S., S. Balendhran, P. Yi, D. Yao, S. Zhuiykov, M. Pannirselvam, et al. 2013b. MnO_2-based thermopower wave sources with exceptionally large output voltages. J. Phys. Chem. C 117(18): 9137-9142.

Wang, Y., Y. Sui, H. Fan, X. Wang, Y. Su, W. Su, et al. 2009. High temperature thermoelectric response of electron-doped $CaMnO_3$. Chem. Mater. 21(19): 4653-4660.

Wei, C. 2006. Adhesion and reinforcement in carbon nanotube polymer composite. Appl. Phys. Lett. 88(9): 093108.

Wrixon, G. T., A. M. E. Rooney and W. Palz. 1993. Conclusions: Prospects for the Development of Renewable Energy in Europe to the year 2000. In Renewable Energy-2000. Springer Berlin Heidelberg: 93-111.

Xiao, N., X. Dong, L. Song, D. Liu, Y. Tay, S. Wu, et al. 2011. Enhanced thermopower of graphene films with oxygen plasma treatment. ACS Nano 5(4): 2749-2755.

Xu, X., N. M. Gabor, J. S. Alden, A. M. van der Zande and P. L. McEuen. 2009. Photo-thermoelectric effect at a graphene interface junction. Nano Lett. 10(2): 562-566.

Zalba, B., J. M. Marin, L. F. Cabeza and H. Mehling. 2003. Review on thermal energy storage with phase change: materials, heat transfer analysis and applications. Appl. Therm. Eng. 23: 251-283.

Zhan, G. D. and A. Mukherjee. 2005. Processing and characterization of nanoceramic composites with interesting structural and functional properties. Rev. Adv. Mater. Sci. 10(3): 185-196.

Zhang, G. and Y. W. Zhang. 2015. Strain effects on thermoelectric properties of two-dimensional materials. Mech. Mater. 91: 382-398.

Zhao, W., S. Fan, N. Xiao, D. Liu, Y. Y. Tay, C. Yu, et al. 2012. Flexible carbon nanotube papers with improved thermoelectric properties. Energy Env. Sci. 5(1): 5364-5369.

Zheng, J.-C. 2008. Recent advances on thermoelectric materials. Front. Phys. China 3(3): 269-279.

4

Graphene Applications in Optoelectronic Devices

Lung-Chien Chen* and Jia-Ren Wu[a]

ABSTRACT

This chapter summarizes recent researches on graphene-based optoelectronic devices. Graphene is the simplest and strongest material and it is one of the most promising candidates for the fabrication of high mobility electrodes and transparent semiconductors for the future. This chapter reviews the effects of graphene uses for different applications and devices such as electrodes for light-emitting diodes, electron transport layers for dye-sensitized solar cells and field emitters for field emission devices. Finally, recent and substantial developments of graphene for the use of optoelectronic devices are summarized and the advantages and disadvantages of the different nanostructures are analyzed.

Keywords: Graphene, light-emitting diodes, solar cells, field emission devices.

1. INTRODUCTION

Graphene is a potential material with a zero-bandgap. Therefore, its behavior is somewhat like a metal and somewhat like a semiconductor. It is a two-dimensional material, consisting of a single layer of carbon atoms forming a hexagonal lattice. The unique advantages of graphene include its high thermal conductivity (~5,000 $Wm^{-1}K^{-1}$), high carrier mobility (200,000 $cm^2V^{-1}s^{-1}$), chemical and mechanical stability, excellent mechanical properties (~1,100 GPa) and optical transparency (Lee et al. 2008, Geim 2009, Falkovsky and Pershoguba 2007).

Graphene can be considered as the base of other graphitic forms with different structures, including bilayer graphene, Multi-Layer Graphene (MLG), 3D graphite, 2D carbon nanotubes and 0D fullerenes. The methods of growing graphene layers include micro-mechanical cleavage, reduction of exfoliated graphene and Chemical

Department of Electro-optical Engineering, National Taipei University of Technology, 1, 3 Sec., Chung-Hsiao E. Rd., Taipei 106, Taiwan.

[a] E-mail: t9659012@ntut.edu.tw

* Corresponding author: ocean@ntut.edu.tw

Vapor Deposition (CVD) (Inagaki et al. 2011). Graphene is a material with potential applications ranging from electronics, optics, plasmonics, spintronics, biomedical uses and sensors.

2. OPTOELECTRONIC PROPERTIES OF GRAPHENE

2.1 Electronic Properties

Single layer graphene has an ideal two-dimensional hexagonal lattice structure assembled of sp^2 orbitals. No hybrid orbitals of $2p_z$ neighboring atoms form the $2p_z$ orbital localized π bond. Since these π electrons are bound by weak nuclei, after thermal excitation they can be moved along the planar direction and therefore the graphene plane can be conductive. Near the first Brillouin zone, the relationship of kinetic energy and momentum appears linear. This spot is called the Dirac point and the formula is:

$$E^{\pm}(k) = \pm h v_F |k| \tag{1}$$

where $k = (k_x, k_y)$ is the wave number and $v_F \approx 10^6$ m/s is the Fermi velocity which is 300 times smaller than the velocity of light c (Bonaccorso et al. 2010). This is why a lot of uncommon properties can be observed in graphene.

At this point, the linear dispersion in which the Conduction Band (CB) and the Valence Band (VB) intersect is shown in Fig. 4.1. The bandgap at the Dirac point is zero. The structure is symmetric at this c point. The electron mobility of graphene is higher than 15,000 $cm^2/V{\cdot}s$, the resistivity is 10^{-6} $\Omega{\cdot}cm$ and even lower than Ag, indicating that the graphene is an outstanding conductor (Geim 2009).

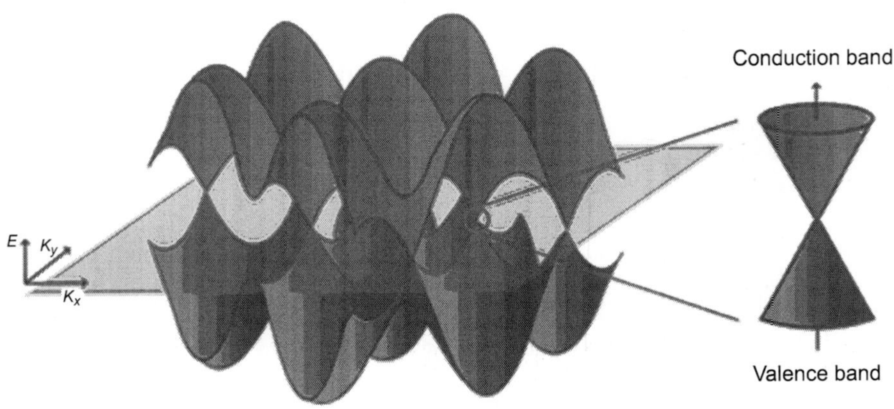

Fig. 4.1 The structure of graphene with the conducting band and the valence band approaching six points at the Brillouin zone (Geim 2009).

Another feature of the Dirac point is insensitivity of the external electrostatic potentials. According to the Klein paradox, the Dirac fermions can be transmitted through a forbidden region with a probability of 1 (Bonaccorso et al. 2010, Castro Neto et al. 2009, Avouris 2010, Lu et al. 2006).

2.2 Optical Properties

The transmittance (T) of a single graphene layer can be derived by applying the Fresnel equations in the thin film limit for a material with a fixed universal optical conductance $G_0 = e^{24}/\sim 6.08 \times 10^{-5}\,\Omega^{-1}$, to give:

$$T = (1 + \pi\alpha/2)^{-2} \approx 1 - \pi\alpha \approx 97.7\%, \tag{2}$$

where $\alpha = e^2/4\pi c = G_0/\pi c \approx 1/137$ is the fine structure constant (Lee et al. 2008) and it is independent of wavelength. Graphene only reflects < 0.1% of the incident light in the visible region, absorbing A = 1-T = $\pi\alpha$ = 2.3% over the visible spectrum. A single atomic layer of material should not have such a high opacity. It is the unique electronic properties of graphene that cause its extremely high opacity. Because graphene has an unusually low energy electronic structure in the Dirac point, electrons and holes will meet in a conical band, resulting in the high opacity. The high conductivity and excellent light transmission of graphene make it an excellent conductive electrode for applications in solar cells, flat panel displays, touch screens and Organic Light-Emitting Diodes (OLEDs).

2.3 Saturated Absorption

Interband excitation of the graphene by ultrafast optical pulse has been studied. The relaxation duration of the electrons is shorter than the pulse duration. This behavior is known as the saturated nonlinear optical absorption threshold called saturation fluency. When there is stronger visible or near-infrared excitation due to the overall absorption of light and zero graphene bandgap properties, graphene very easily becomes saturated. The limiting velocity is acquired by carriers when the kinetic energy is enough to excite the phonons of the material. The coupling of surface phonons can affect the value of the saturation velocity (Bonaccorso et al. 2010, Zhang et al. 2008, Lu et al. 2006).

3. TRANSPARENT CONDUCTIVE LAYER IN GaN-BASED LIGHT-EMITTING DIODES

The Transparent Conductive Layer (TCL) applied in the GaN-based LEDs has several characteristics: (1) transparent for emitted light, (2) ohmic contact to the p-GaN layer and (3) conductivity matched to that of the p-GaN layer. Graphene can potentially be used in GaN-based light-emitting diodes to replace the conventional indium tin oxide (ITO) TCL or Ni/Au TCL due to its superior optical and electrical properties.

The GaN-based UV light-emitting diode (LED) was used to study the graphene TCL. The LED consists of a 2 μm GaN:Si, a AlGaN/GaN/AlGaN Single Quantum Well (SQW) with thickness of 8 nm/5 nm/8 nm and a 200 nm GaN:Mg contact layer. The graphene layer is manufactured using a simple mechanical exfoliation of highly ordered pyrolytic graphite (HOPG) (ZYA grade). The graphene layer is deposited on a GaN-based UV LED. The desired thickness is obtained by the scotch-tape method and measured by micro-Raman spectroscopy using a 514 nm Ar-ion laser with a power at the sample of 0.5 mW for 60 seconds. Measurements are obtained using a back scattering geometry. Finally, Ti/Au (40/100 nm) bi-layer is used as the back contact on the surface of the SiC substrate by evaporator. Then, rapid thermal annealing (RTA) is performed on the sample at 450°C for 40 seconds in an N_2 atmosphere (Kim et al. 2010).

UV light is emitted from the graphene layer with light input power of 30 mW. The graphene layer with various thicknesses produced various EL intensities under UV irradiation. Comparing the brightness of the HOPG, it can be seen that the greatest brightness is in the region consisting of four layers of graphene. The electronic properties of the graphene-based electrodes were measured by applied voltage. The EL spectrum of the graphene layer exhibits the strongest peak located at 368 nm. The weak EL spectrum is due to the small emission area. The I-V characteristics shows typical diode-like curve. The forward voltage at the injection current of 1 mA is around 26.5 V.

Specifically, graphene-based transparent conductive electrodes have two advantages for the optimization of UV LEDs, one is their excellent thermal conductivity and the other is better UV transmission if compared with other transparent conductive electrodes such as ITO or the standard Ni/Au metallized electrodes. We compare the graphene layer and ITO in terms of thickness and thermal conductivity. The 1-2 μm-thick graphene layer has very good thermal conductivity, about 5000 W/m·K. In contrast, the thermal conductivity of the 100-200 nm-thick ITO is only about 11-12 W/m·K.

Comparison shows that Ni/Au metallized layer improves current spreading in p-GaN but it also absorbs a lot of light which interacts with surface of the LED. A graphene-based contact not only reduces this photon blocking effect but also increases long-term stability, due to the improved heat distribution through the graphene contact. The typical forward bias at an injection current of 20 mA is around 3.5 V for conventional GaN-based LEDs with Ni/Au TCL. In the case of graphene TCL, the forward bias is over 5V. Several studies have reported on the use of doped graphene TCL for Ohmic contacts in GaN-based LEDs. (Kim et al. 2013, Youn et al. 2013).

Youn et al. (2013) reported an improvement in optical power and thermal stability of GaN LEDs fabricated using a chemically doped graphene TCL. A patterned graphene TCL with Cr/Au electrodes is suggested as being able to obtain a low contact resistance between the TCL and p-GaN layer. A bi-layer patterning method for a graphene TCL was utilized to prevent the graphene from peeling-off from the p-GaN surface. The preparation of the graphene was conducted using CVD and the graphene was doped onto Au/Cr by a chemical treatment using a HNO_3 solution (Youn et al. 2013).

The GaN-based LEDs with single graphene layers and the Au/Cr/graphene contacts yielded forward biases of 5 and 4.2 V, respectively, at an injection current of 20 mA. The forward bias of the GaN-based LEDs with the Au/Cr/graphene TCL is close to that of the GaN-based LEDs with Cr/Au TCL (3.9 V) and that of the GaN-based LEDs with Ni/Au TCL (3.5 V). The light output power of the GaN-based LEDs with the single graphene layers and the Au/Cr/graphene contacts is 5 and 7.5

mW, respectively. The improvement in the light output power is due to the current spreading effect on the surface of the p-GaN layer (Youn et al. 2013).

4. REFLECTIVE CONDUCTIVITY LAYER IN GaN-BASED LIGHT EMITTING DIODES

To prepare a reflective conductivity layer in GaN LEDs, a 2 um-thick p-GaN epitaxial layer with a GaN buffer layer was grown on a sapphire substrate by metal-organic chemical vapor deposition (MOCVD) method.

The resultant GaN LED layer was measured by Hall measurement. The hole concentration and mobility of the p-GaN were approximately 8-10 $cm^2/V{\cdot}s$. Next, radio-frequency magnetron sputtering was used to deposit the reflective layer (Ag/Ni/MLG tri-layer) on the p-GaN epitaxial layer. The typical deposition conditions of various RF-sputtered films are shown in Table 4.1 (Chen and Chiang 2014a). The samples were measured by Circular Transmission Line Method (CTLM) patterns to obtain the specific contact resistance ρc (King et al. 1997, Marlow and Das 1982). Next, the samples were annealed by thermal treatment in the range 500-700°C for 10 minutes in ambient nitrogen and measured the current-voltage (I–V) characteristics by CTLM used a spacing of 5 μm.

Table 4.1 Typical deposition conditions for various films (Chen and Chiang 2014a).

Target	Silver	Nickel	Graphite
Target diameter (cm)	5	5	5
Substrate-to-target distance (cm)	4	4	4
RF power (W)	30	70	90
Sputtering ambient	Ar	Ar	Ar
Gas flow rate (sccm)	40	40	90
Residual pressure × 10^{-6} torr micro/μ torr	<5	<5	<5
Deposition time (min)	4	60	40

Next, a thermal treatment at 650°C for 10 minutes in nitrogen ambient was used to activate the p-type dopant. Then etching was performed on the surface of the p-type GaN layer until the n-type GaN layer was exposed. The process was followed to grow a Ag/Ni/MLG (160/90/200 nm) reflective conductivity layer on the p-type GaN LED. Next, a Ti/Al/Ti/Au (15/50/100/1000 nm) electrode was placed on the reflective conductivity surface. Finally, Raman spectroscopy was used to obtain the Raman scattering spectra under the following conditions: 5145 Å-line argon-ion laser with a power of 80 mW and a spot diameter of ~100 μm. Figure 4.2a shows an SEM cross-sectional image of the Ag/Ni/MLG on p-GaN and Fig. 4.2b shows the structure of Ag/Ni/multilayer graphene reflective ohmic contacts with p-type GaN.

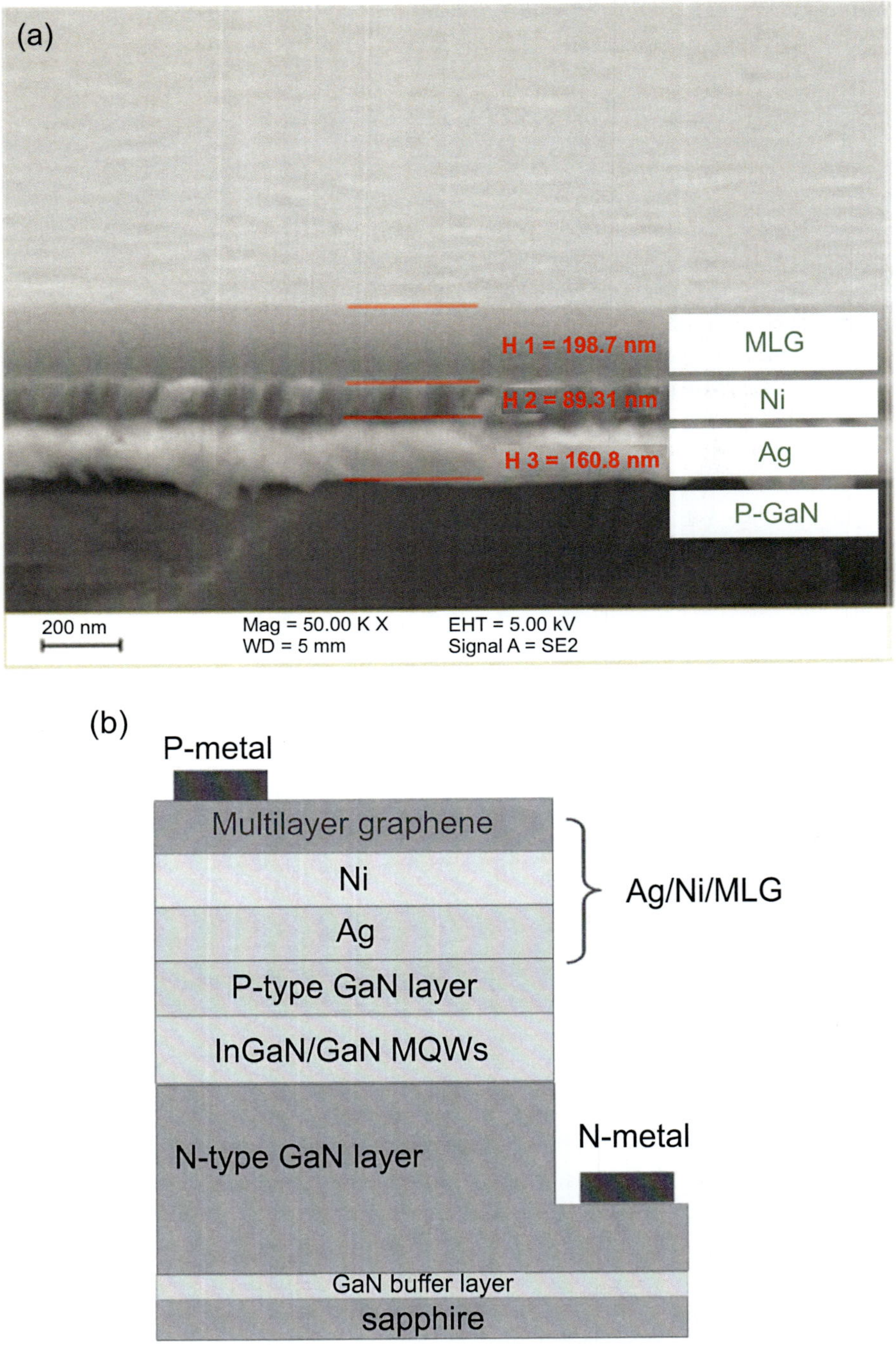

Fig. 4.2 (a) SEM cross sectional image of the Ag/Ni/MLG on *p*-GaN. (b) the structure on AgNi multilayer graphene reflective ohmic contacts with p-type GaN (Chen and Chiang 2014a).

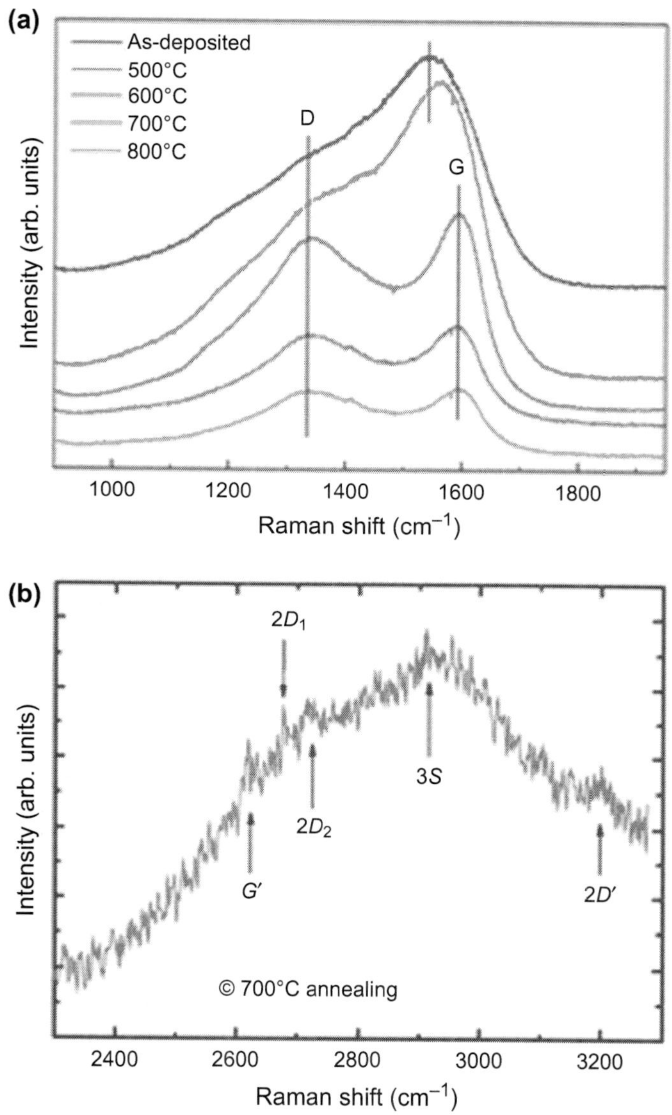

Fig. 4.3 Typical Raman spectra of MLG films at different annealing temperatures. (a) MLG films have first and second Raman-active modes in the *G* and *D* band. (b) double-resonance Raman process in 2D band graphene (Chen and Chiang 2014a).

Figure 4.3 shows typical Raman spectra of MLG films at different annealing temperatures. It is easy to analyze and read by using the *D* bands and *G* bands which are the main features in the Raman spectra of graphitic carbons. In Fig. 4.3a, the obtained MLG films have the first (*G* band at 1563 cm^{-1}) and the second (*D* bands at 1350 cm^{-1}) Raman active modes at different annealing temperatures. The Raman spectra reveal first order scattering where the graphite-like zone is called the *G* band at the Brillouin zone center. The figure reveals the degree of graphitization.

It represents the stretching of sp^2-bonds between both rings and chains for all pairs of atoms. The *D* band (second order) is very weak which represents the disordering of hybridized carbon atoms (Marlow and Das 1982, Castiglioni et al. 2001, Ferrari and Robertson 2000). Many defects in the materials can be overcome by annealing. In the Raman spectra, it can be seen that the *G* bands of multi-layer graphene films are shifted to 1592 cm^{-1} and the intensity decreases with increasing annealing temperatures. The charge impurities in the graphene unintentionally doped onto the substrate could cause the blue-shift at the *G* peak. Another possibility may be the self-doping effects. There are some defects in the graphene that cause the intensity and *D* bands at 1350 cm^{-1} to become obvious. The results indicate the creation of sp^3 domains because of extensive oxidation. Figure 4.3b shows the observed double-resonance Raman which causes the *G′* band (2618 cm^{-1}) in defect-free graphene (Sidorov et al. 2008). The second-order modes of the *D* band are at the $2D_1$ (2674 cm^{-1}) and $2D_2$ (2710 cm^{-1}) modes. The 2*D* band represents the multilayer graphene spilt into multiple components (Das et al. 2008). Then, the 3*S* bands (~2910 cm^{-1}) from Fig. 4.3b are formed by the second-order that combines the *D* and *G* peaks and the 2*D′* band (~3200 cm^{-1}) is the overtone of the 2*D* peak.

Figure 4.4 shows the current-voltage (*I–V*) characteristics of a Ag/Ni/MLG contact that is deposited on p-GaN layers before and after annealing at temperatures of 500-700°C. The graph reveals that samples annealed at 700°C have the best I-V characteristics while at 800°C the sample breaks. The sample annealed at 700°C had the lowest resistance of Ag/Ni/MLG on p-GaN which was 1.02×10^{-3} Ω·cm². From this we reach the conclusion that the formation of the gallium oxide by annealing improved the properties of the contact, especially at 700°C (Sohn et al. 2011).

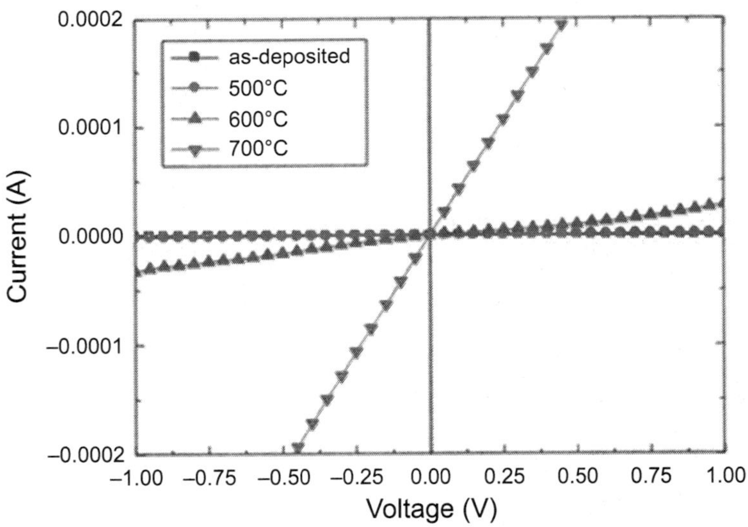

Fig. 4.4 The *I–V* characteristics of Ag/Ni/MLG deposited on *p*-GaN layers after annealing at a temperature range from 500-700°C for 10 min (Chen and Chiang 2014a).

Figure 4.5 shows typical *I-V* curves at a forward bias that reveal the characteristics of GaN-based LEDs. When Ag/Ni/MLG contacts are injected with a current of 20 mA a forward bias of 3.51 V was produced after annealing at 700°C. The traditional GaN-based LED with ITO transparent contact layer shows a forward bias of 3.35 V. However, the forward voltages of Ag/Ni/MLG and ITO were quite close. The illustration shows the Ag/Ni/MLG structure of the GaN LED.

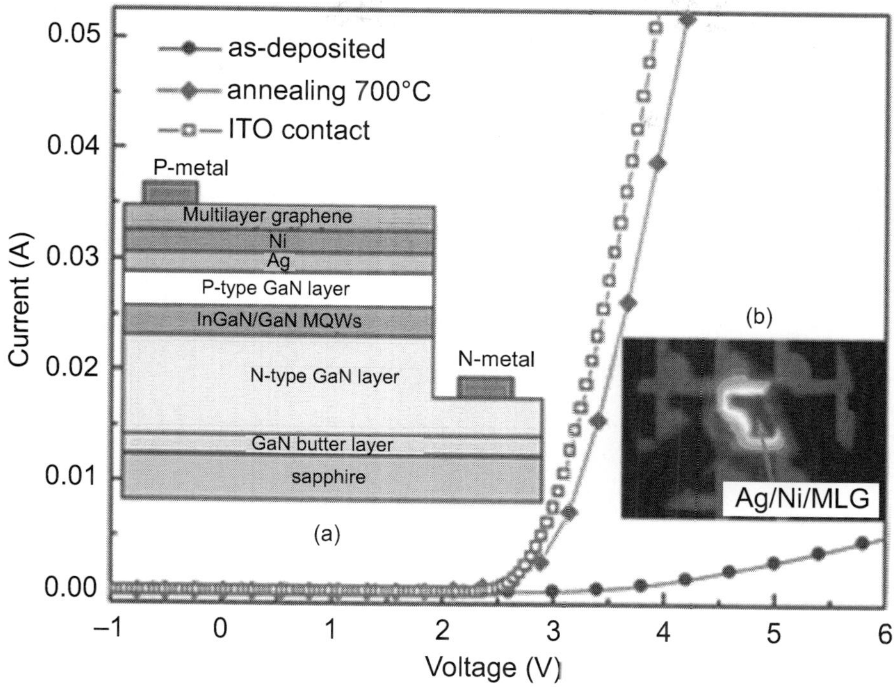

Fig. 4.5 (a) Typical *I–V* characteristics of the GaN-based LEDs with Ag/Ni/MLG reflective contacts. (b) The illustration shows the operating image of an LED at 20 mA (Chen and Chiang 2014a).

Figure 4.6 shows the curves indicating the external quantum efficiency and light output between a GaN LED with a Ag/Ni/MLG reflective layer and the traditional GaN LED with the ITO contact as the control sample with a size of 10 mil × 12 mil operating under a forward bias at a forward current of 20 mA. The light output and EQE were estimated to be 6.25 mW and 12.4%, respectively, whereas in the traditional GaN LED the light output and EQE were estimated to be only 5.2 mW and 9.8%, respectively. Therefore, the light output and the external quantum efficiency were increased by using the Ag/Ni/MLG reflective layer in the GaN LED. The rate of increment of the light output power was almost 25%. However, when the injection current was 100 mA, the light output power of the Ag/Ni/MLG reflective layer on GaN LED and the traditional GaN LED with ITO was increased up to 16 mW and 11.8 mW, respectively and the quantum efficiency decreases to 5.1 and 5.9%, respectively. The GaN LED with the Ag/Ni/MLG reflective layer shows a lower light output power and higher external quantum efficiency when it operated at a high injection current. The GaN LED with the Ag/Ni/MLG reflective layer demonstrated the best optical properties owing to improvement of the thermal characteristics.

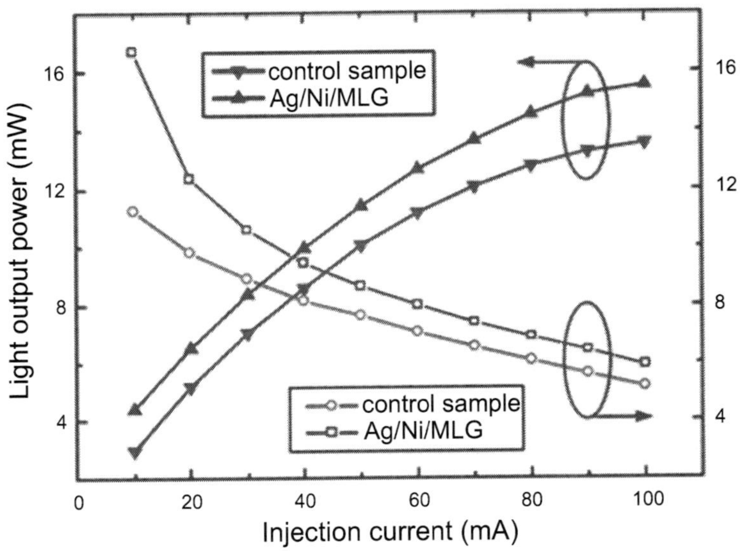

Fig. 4.6 Light output power and external quantum efficiency of the GaN-based LED with the Ag/Ni/MLG reflective contact and the control sample against the forward current (Chen and Chiang 2014a).

5. HOLE TRANSPORT LAYER IN ORGANIC SOLAR CELLS

The most commonly used Hole Transport Layer (HTL) in organic solar cell structures is poly(3, 4-ethylenedioxythiophene) (PEDOT): poly(4-styrenesulfonate)(PSS) although graphene has the potential to replace this material because of its excellent reproducibility and much better stability. Several articles reported the results of using graphene HTLs in organic solar cells (Zheng et al. 2013, Li et al. 2013, Park et al. 2012a, b, Ho et al. 2014).

For example, Zheng et al. (2013) prepared Graphene Oxide (GO) from a GO-Mo solution and then applied it to form the structures of poly(3-hexylthiophene) (P3HT)/phenyl-C61-butyric acid methyl ester (PCBM) Organic Solar Cells (OSC) for use as an HTL. In this study, different weights of ammonium heptamolybdate ($(NH_4)_6Mo_7O_{24}$-$4H_2O$) were added to the aqueous solution to evaluate the performance of the P3HT/PCBM OSCs. The light transmittance of the film was prepared by spin coating. The films show good transmission of light which was in the visible wavelength range. The transmittance of all GO-Mo films exceeded 90% in the range from 400 to 800 nm. Therefore, they are suitable for the window layers of solar cells.

Table 4.2 lists the average short-circuit current density (J_{sc}), open-circuit voltage (V_{oc}), Fill Factor (*FF*) and the maximum and average values of the power conversion efficiency (PCE or η) for the sets of devices. The measurement results are listed for the conventional PEDOT: PSS and evaporated MoO_3 as HTLs. The best performance is obtained for the device with the HTL, which is prepared from 0.10 g Mo-precursor in a GO solution. The values of J_{sc}, V_{oc} and *FF* are 9.02 mA-cm^{-2}, 0.59 V and 49%, respectively. The maximum value of the PCE is 2.61%. The results suggest that the performance of the device with GO-Mo as the HTL demonstrates significantly more improvement than the cells with GO as the HTL. When the Mo-precursor content exceeds 0.10 g in the GO sample, the PCE will decrease. We can see a significant change of J_{sc} and *FF* in the different sets of cells. With the increase in the Mo-precursor content in the GO, J_{sc} will initially increase and then decrease. The R_s can be derived from the slope of the I-V curve of a solar cell at $V = V_{oc}$ (at $I = 0$) (Wang et al. 2011).

Table 4.2 Detailed parameters of device performance obtained using GO and GO-Mo from solutions with different Mo-precursor weights (Zheng et al. 2013).

Mo-precursor weight (g)	J_{sc} (mA cm^{-2})	V_{oc}(V)	*FF* (%)	PCE (%) max/average	Rs (Ω cm^2)
0.00	2.70	0.24	29	0.19/0.17	48
0.10	9.02	0.59	49	2.61/2.59	19
0.20	7.60	0.56	39	1.67/1.62	37
PEDOT:PSS	8.51	0.57	54	2.62	16
Thermally evaporated MoO_3	8.6	0.52	53	1.94	41

The performance of OSCs with graphene HTLs can be enhanced by employing Graphene Quantum Dots (GQDs) to replace graphene (GP), GO or PEDOT:PSS in the OSC structure (Sidorov et al. 2008). The GQDs were synthesized form carbon fibers by a facile one-step acid treatment (Peng et al. 2012, Luo et al. 2010). The current density-voltage of P3HT/PC_{61} BM OSCs with the different thicknesses of GQD HTL. The optimal measured parameters of the sample with GQD HTLs were the V_{oc}, J_{sc}, *FF* and η, which had values of 0.92 V, 11.36 mA/cm^2, 0.652 and 6.82%, respectively. The results matched those of OSC with PEDOT:PSS HTL. This means the GQDs have great potential to replace the conventional PEDOT:PSS HTL (Li et al. 2013).

6. ELECTRON TRANSPORT LAYER IN DYE-SENSITIZED SOLAR CELLS

Graphene has the potential to be used as an electron transport layer in dye-sensitized solar cells (Ho et al. 2014). In this study, two structures were demonstrated. First titanium dioxide (TiO_2) with added graphene was prepared, a graphene layer 60 nm was sputtered onto an ITO conductive glass substrate by radio-frequency magnetron sputtering using a graphite target as an electron transport layer to improve the electron transfer in the Dye-Sensitized Solar Cell (DSSC) structure. Next 1 g of TiO_2 nanocrystalline powder with a diameter ~25 nm, 1 mL Triton X-100, acetic acid and deionized water were mixed together to make the TiO_2 solution. The solution was then stirred for 24 hours before being coated on the ITO glass substrate. Then the film was annealed at 450°C for 10 minutes. The photoelectrode with the graphene layer was immersed in an (($Bu_4N)_2$-[Ru(dcbpyH)$_2$(NCS)$_2$] complex) (N719) dye solution to absorb the dye. Next an electrolyte composed of 0.05 M iodide, 0.5 M lithium iodide (LiI) and 0.5 M 4-tert-butylpyridine (TBP) propylene carbonate was prepared and platinum was sputtered onto the ITO substrate for the counter electrode. Finally, cells were fabricated by sealing (SX1170-60, SOLARONIX) the two electrodes, leaving two via-holes to inject the electrolyte. Figure 4.7 shows the completed graphene/TiO_2 structure. The other structure is fabricated like the first one, with the difference being the addition of a TiO_2 layer between the graphene and the ITO substrate (Hsu et al. 2014). Figure 4.8 shows the TiO_2/graphene/TiO_2 sandwich structure. The thickness of the first TiO_2 layer was 4 μm and the thickness of the second layer was 6.5 μm (Chen et al. 2014).

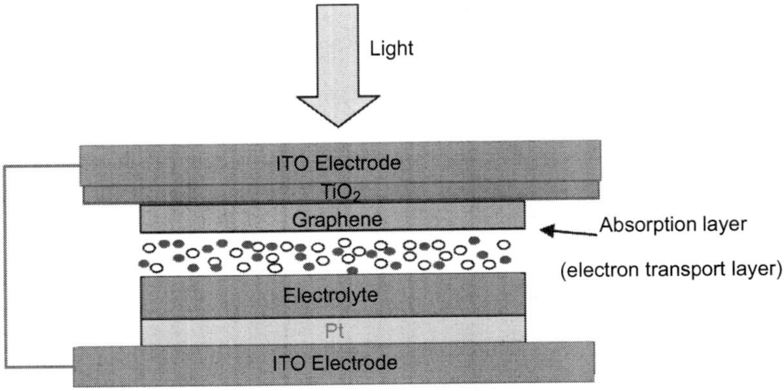

Fig. 4.7 Completed graphene/TiO_2 structure (Hsu et al. 2014).

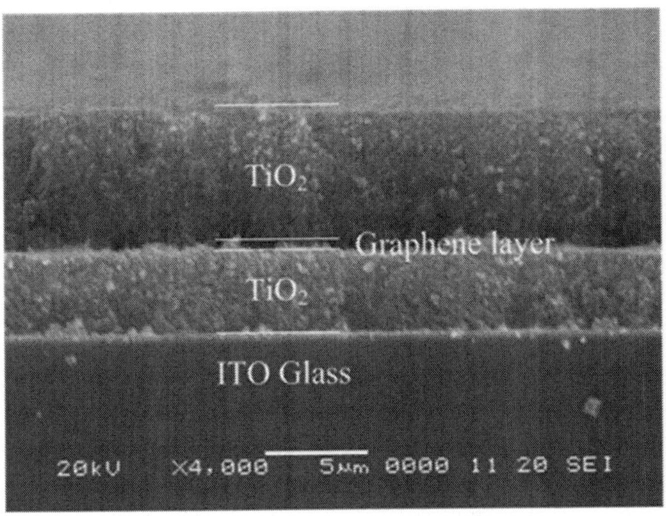

Fig. 4.8 SEM image of TiO_2/graphene/TiO_2 sandwich structure (Chen et al. 2014).

Figure 4.9 shows the absorbance spectra for the DSSCs with and without the graphene electron transfer layer. Absorption of the graphene with a DSSC of 60 nm is higher than for 100 nm of graphene and without a graphene electron transfer layer, in the range of 310-400 nm. The mean graphene electron transfer layer can increase absorption coefficient and also improves the absorption of the DSSC. Figure 4.10 shows the *J-V* curves of DSSCs with and without the graphene electron transfer layer. The cell performance was measured under AM 1.5 illumination, with a solar intensity of 1000 W xenon lamp at room temperature. From the spectra we can obtain the V_{oc}, J_{sc}, FF and η. It can be seen in Fig. 4.10 that the DSSC with graphene shows increased electron transport reach and has a higher short circuit current.

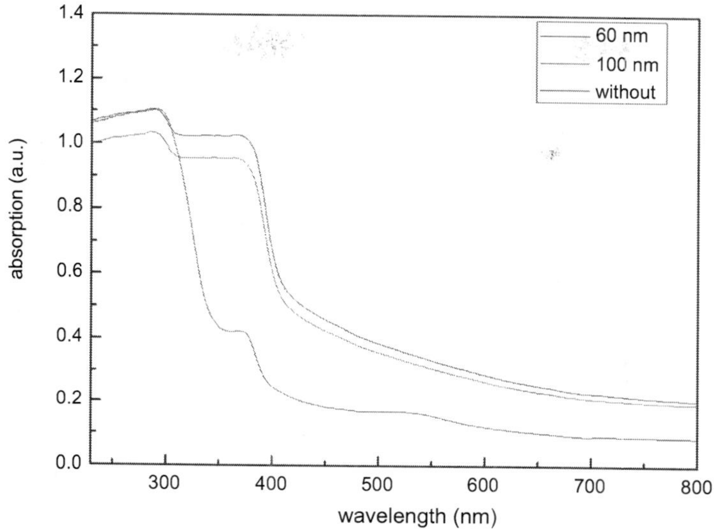

Fig. 4.9 Absorbance spectra with and without the graphene electron transfer layer (Hsu et al. 2014).

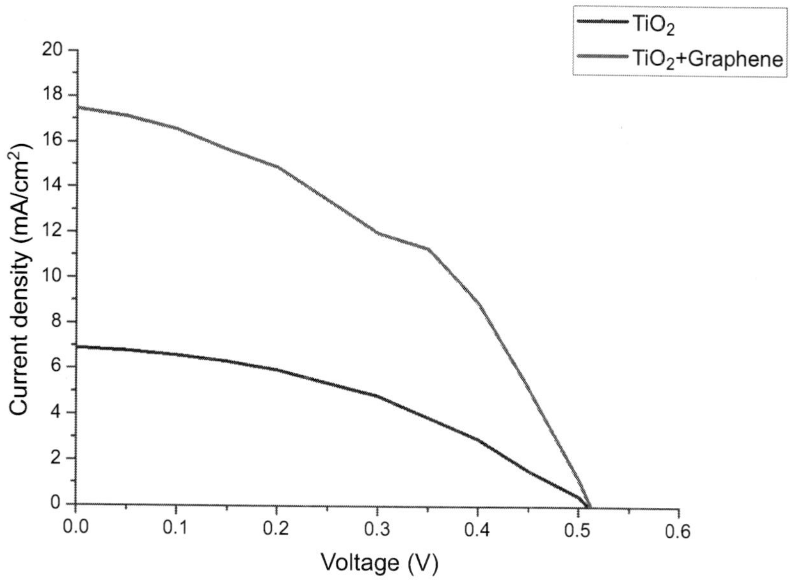

Fig. 4.10 *J–V* curves of DSSCs with different transfer layers under illumination (Hsu et al. 2014).

Figure 4.11 shows a plot of the UV-vis spectra of photoelectrodes with a TiO_2/graphene/TiO_2 sandwich structure on ITO glass and a traditional structure with a single TiO_2 using different dye coating speeds. The TiO_2/graphene/TiO_2 sandwich structure exhibited higher absorption between the TiO_2 and graphene, mainly in the upper photoelectrode, with more N719 dye being absorbed by the porous structure. Figure 4.12 shows the photovoltaic performance of the three structures. The TiO_2/graphene/TiO_2 sandwich structure has much better photovoltaic performance than traditional structures. From Figs. 4.11 and 4.13 it can be seen that the DSSCs with the TiO_2/graphene/TiO_2 sandwich structure, mainly through the transport of photoexcited dye to the conduction band because this structure, exhibited more via holes and higher absorption of the N719 dye, therefore increasing the absorption intensity.

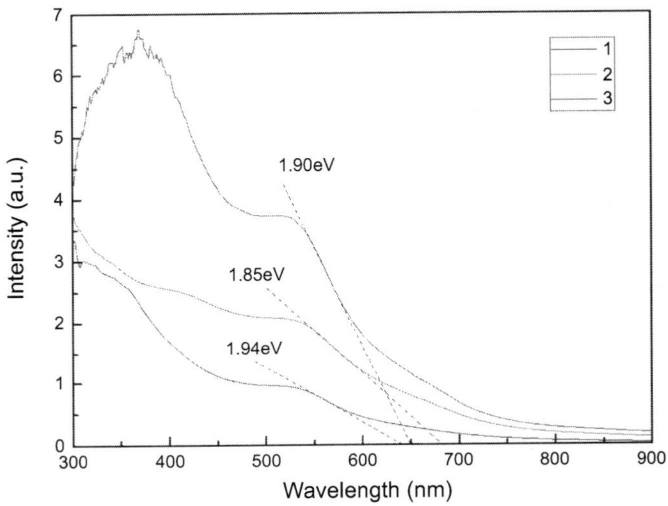

Fig. 4.11 UV-vis absorption spectra of DSSCs in dye with the: (1) traditional structure with a single TiO_2 4000 rpm; (2) traditional structure with a single TiO_2 2000 rpm; (3) TiO_2/graphene/TiO_2 sandwich structure with 4000 rpm (Chen et al. 2014).

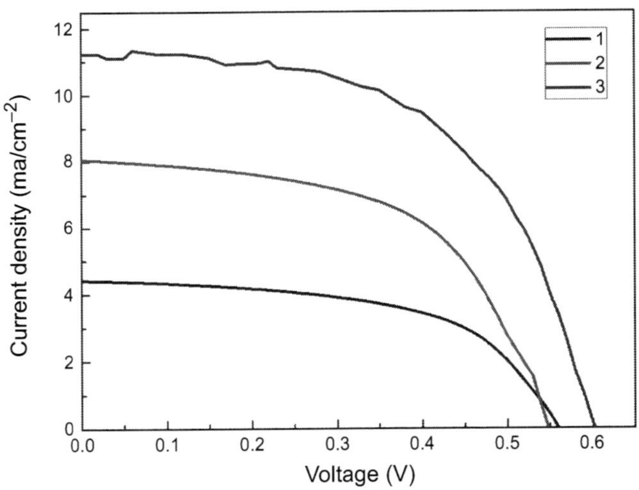

Fig. 4.12 Photovoltaic performance of DSSCs manufactured with different structures: (1) traditional structure with a single TiO_2 4000 rpm; (2) traditional structure with a single TiO_2 2000 rpm; (3) TiO_2/graphene/TiO_2 sandwich structure with 4000 rpm (Chen et al. 2014).

Figure 4.13 shows the energy level diagrams for graphene/TiO_2 and TiO_2/graphene/TiO_2 sandwich structures, where CB and VB indicate the conduction band and the valence band, respectively. Electrons in the photoexcited dye are transported from the Highest Occupied Molecular Orbital (HOMO) to the Lowest Unoccupied Molecular Orbital (LUMO) because of the graphene's excellent electrical conduction. The graphene allowed photoexcited electrons to be rapidly transported from the CB of TiO_2 to the graphene electron transfer layer.

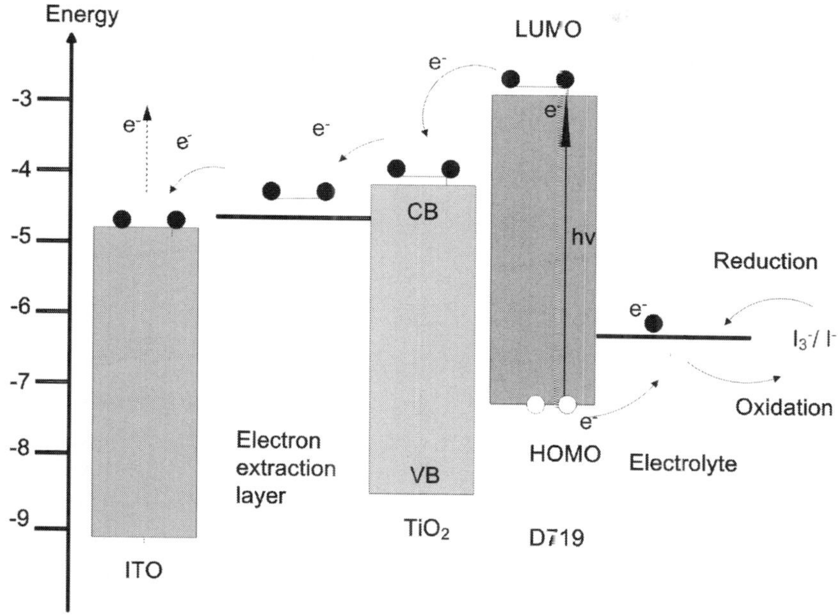

Fig. 4.13 Energy level diagram and mechanism of photocurrent in DSSCs with a graphene/TiO_2 structured graphene electron transfer layer (Chen et al. 2014).

Finally, cell parameters with two different structures were measured. The V_{oc}, J_{sc}, *FF* and η are summarized in Table 4.3. Through the addition of the graphene electron transfer layer there is obvious improvement in the J_{sc}, *FF* and η. The TiO_2/grapheme/TiO_2 sandwich structure is just higher than the graphene/TiO_2 in terms of V_{oc} and *FF*, however the energy conversion efficiency is close to the graphene/TiO_2 conversion efficiency. The TiO_2/grapheme/TiO_2 sandwich structure can still be improved in the future.

Table 4.3 The parameters for DSSCs with different structures (Hsu et al. 2014, Chen et al. 2014).

	TiO_2	Graphene/TiO_2	TiO_2/grapheme/TiO_2
J_{sc} (mA/cm^2)	6.9	17.5	11.22
V_{oc} (V)	0.5	0.5	0.6
FF	0.419	0.456	0.58
η (%)	1.45	3.98	3.93

7. EMITTERS IN FIELD-EMISSION DEVICES

A Field Emission Device (FED) is a device that functions based on electron extraction from the material surface by quantum mechanical tunneling (Fowler and Nordheim 1928). This simple principle has been widely used in field emission displays and electron guns (Brodie and Schwoebel 1994, De Jonge and Bonard 2004). Figure 4.14 shows a basic diagram of an FED with a graphene emitter.

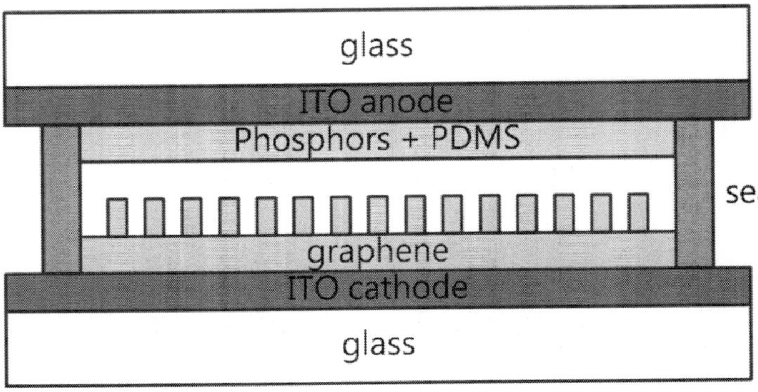

Fig. 4.14 Cross-section of the completed structure of the FED with the graphene emitter.

The process for fabrication of the FED with graphene emitters is as follows: (1) mixing of polydimethylsiloxane (PDMS) with a hardening agent, (2) calculation of the proportion of Yttrium Aluminium Garnet (YAG) fluorescent powder to be doped into the PDMS solution, (3) annealing of the mixed solution on ITO and solidification of the PDMS with RTA, (4) deposition of graphene on ITO glass by sputtering and formation of conical structures by dipping into the etchant, (5) assembly of ITO anode glass, insulating sealer and ITO cathode glass, as shown in Fig. 4.15, after pumping to 10^{-3} Torr in a vacuum apparatus and (6) application of the electrical field and observation of the lamination, as shown in Fig. 4.16.

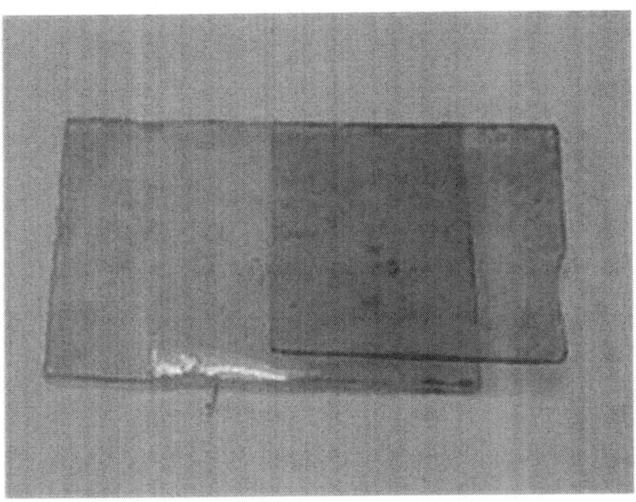

Fig. 4.15 Assembly of the ITO anode glass, insulating sealer, and ITO cathode glass.

Fig. 4.16 Picture of the luminance of the FED with the graphene emitter.

The electron field emission characteristics of a graphene (GP)/carbon nanotube (CNT) hybrid field emitter are also studied (Fowler and Nordheim 1928, Brodie and Schwoebel 1994). The multi-walled carbon nanotubes (MCNTs) and GP are manufacture by the CVD method. Some of the MCNT and GP solution was diluted in 200 ml of isopropyl alcohol. The mixture was placed in a sonic bath for 5 hours for electrophoresis suspension. Then the Si wafer acting as a cathode was immersed into the MCNT/GP electrophoresis suspension. Afterwards the MCNT/GP film covered the Si wafer. Phosphor-coated ITO glass was used as the anode (Chen et al. 2014). Xu et al. (2013) prepared GP by chemical reduction from Graphene Oxide (GO). The as-prepared GP and MCNT powder was mixed with different weight ratios of GP and Sodium Dodecyl Sulfate (SDS) dissolved in Deionized water (DI water). The graphene-MCNT hybrid films which on the membrane can be transferred to substrates (Xu et al. 2013).

The MCNTs are interlinked with the GP film. The MCNTs and the edges of the GP are stability field emitters. It can be seen that most of the vacant space of the MCNTs is covered with GP film. In the hybrid structure, the MCNTs and the GP are entangled with each other. This structure has low resistance and low contact resistance. The current densities and electric field curves (J-E) of the hybrid, GP and MCNTs. The turn-on electric fields for the hybrid, GP and MCNTs at 10 $\mu A/cm^2$ are 0.93, 2.18 and 2.62 V/μm, respectively (Chen et al. 2014).

8. CONCLUSIONS

Based on the above discussion, the following conclusions concerning graphene in optoelectronic devices can be drawn:

(1) Current spreading over the entire p-GaN layer can be significantly improved by using a doped graphene TCL to replace the single graphene TCL. The Au/Cr/graphene electrode structure leads to low contact resistance with p-GaN owing to the penetration of the Cr into the graphene

and improvement of the thermal stability in high-power GaN LEDs can be achieved. The GaN-based LEDs with a single graphene layer and the Au/Cr/graphene contacts yielded forward biases of 5 and 4.2 V at an injection current of 20 mA, respectively. The Au/Cr/graphene contacts to the p-GaN layer demonstrated good potential electrical, optical and thermo-stability performance.

(2) Ag/Ni/graphene was used as a p-type ohmic contact with p-GaN that was formed following heat-treatment. Optimal conditions adopted for the formation of Ag/Ni/ graphene contacts to minimize the specific contact resistance to around $1 \times 10^{-3}\ \Omega\,cm^2$ were obtained after heat treatment at alloying temperatures of 700°C for 10 minutes in nitrogen. The Ag/Ni/graphene contacts had a reflectance of approximately 60% at a wavelength of 470 nm. GaN-based LEDs with Ag/Ni/multilayer graphene contacts produced forward biases of 3.51 V at an injection current of 20 mA. These parameters are similar to the traditional Ni/Ag/Au reflective electrodes. Therefore, based on its reflectivity and specific contact resistance, the Ag/Ni/MLG contact is suitable for use in a reflective electrode in a GaN-based LED.

(3) The optimal measured parameters of samples with GQD HTL were the V_{oc}, the J_{sc}, the *FF* and η, which had values of 0.92 V, 11.36 mA/cm^2, 0.652 and 6.82%, respectively. The results matched those obtained for the OSC with PEDOT:PSS HTL. This means the GQDs have good potential to replace the conventional PEDOT:PSS HTL in P3HT/PCBM-based OSCs.

(4) The DSSCs with the graphene electron transport layer had a higher photocurrent than that of traditional DSSCs without a graphene electron transport layer. This improvement in performance arose from an increase in the photocurrent owing to the inserted graphene layer which collected electrons, acting as a transporter in the effective separation of charge and rapid transport of the photogenerated electrons. This occurs because graphene has a work function of approximately 4.5 eV, which is between that of the ITO and TiO_2 photoelectrodes.

(5) Field emission devices with graphene-MCNT hybrid field emitters were fabricated by vacuum filtration and screen printing processes. The hybrids displayed mechanical and electrical properties that were better than those of either nanotube-only or graphene-only field emitters. These properties suggest that such hybrid field emitters have the potential to replace conventional CNT field emitters. The addition of graphene improves the physical properties of such materials. The use of graphene-MCNT hybrid field emitters would significantly reduce the cost.

9. ACKNOWLEDGMENTS

Financial support for this research was provided by the Ministry of Science and Technology of the Republic of China under Contract No. NSC103-2221-E-027-029-MY2.

10. REFERENCES

Avouris, P. 2010. Graphene: electronic and photonic properties and devices. Nano Lett. 10: 4285-4294.

Bonaccorso, F., Z. Sun, T. Hasan and A. C. Ferrari. 2010. Graphene photonics and optoelectronics. Nat. Photonics 4: 611-622.

Brodie I. and P. R. Schwoebel. 1994. Vacuum microelectronic devices. Proc. IEEE 82: 1006-1034.

Castiglioni, C., F. Negri, M. Rigolio and G. Zerbi. 2001. Raman activation in disordered graphites of the A′1 symmetry forbidden $k \neq 0$ phonon: the origin of the D line. J. Chem. Phys. 115: 3769-3778.

Castro Neto, A. H., F. Guinea, N. M. R. Peres, K. S. Novoselov and A. K. Geim. 2009. The electronic properties of graphene. Rev. Mod. Phys. 81: 109-162.

Chen, L., H. He, H. Yu, Y. Cao, D. Lei, Q. Menggen, et al. 2014. Electron field emission characteristics of graphene/carbon nanotubes hybrid field emitter. J. Alloy. Compd. 610: 659-664.

Chen, L. C. and M. H. Chiang. 2014a. Ag/Ni/Multilayer graphene reflective ohmic contacts with p-type GaN. Sci. Adv. Mater. 6: 159-163.

Das, A., B. Chakraborty and A. K. Sood. 2008. Raman spectroscopy of graphene on different substrates and influence of defects. Bull. Mat. Sci. 31: 579-584.

De Jonge, N. and J. M. Bonard. 2004. Carbon nanotube electron sources and applications. Philos. Trans. R. Soc. A-Math. Phys. Eng. Sci. 362: 2239-2266.

Falkovsky, L. A. and S. S. Pershoguba. 2007. Optical far-infrared properties of a graphene monolayer and multilayer. Phys. Rev. B 76: 153410.

Ferrari, A. C. and J. Robertson. 2000. Interpretation of Raman spectra of disordered and amorphous carbon. Phys. Rev. B 61: 14095-14107.

Fowler, R. H. and L. Nordheim. 1928. Electron emission in intense electric fields. Proc. R. Soc. A-Math. Phys. Eng. Sci. 119: 173-181.

Geim, A. K. 2009. Graphene: status and prospects. Science 324: 1530-1534.

Ho, N. T., V. Senthilkumar, H. S. Cho, S. H. Nho, S. Cho, M.C. Jung, et al. 2014. Reliability improvement of bulk-heterojunction organic solar cell by using reduced graphene oxide as hole-transport layer. Phys. Status Solidi A-Appl. Mat. 211: 1873-1876.

Hsu, C. H., J. R. Wu, L. C. Chen, P. S. Chan and C. C. Chen. 2014. Enhanced performance of dye-sensitized solar cells with nanostructure graphene electron transfer layer. Adv. Mater. Sci. Eng. 2014: 107352.

Inagaki, M., Y. A. Kim and M. Endo. 2011. Graphene: preparation and structural perfection. J. Mater. Chem. 21: 3280-3294.

Kim, B. J., M.A. Mastro, J. Hite, C. R. Eddy and J. Kim. 2010. Transparent conductive graphene electrode in GaN-based ultra-violet light emitting diodes. Opt. Express 18: 23030-23034.

Kim, B. J., G. Yang, H. Y. Kim, K. H. Baik, M. A. Mastro, J. K. Hite, et al. 2013. GaN-based ultraviolet light-emitting diodes with $AuCl_3$-doped graphene electrodes. Opt. Express 21: 29025-29030.

King, D. J., L. Zhang, J. C. Ramer, S. D. Hersee and L. F. Lester. 1997. Temperature behavior of Pt/Au ohmic contacts to p-GaN. Mat. Res. Soc. Symp. Proc. 468: 421-426.

Lee, C., X. Wei, J. W. Kysar and J. Hone. 2008. Measurement of the elastic properties and intrinsic strength of monolayer graphene. Science 321: 385-388.

Li, M., W. Ni, B. Kan, X. Wan, L. Zhang, Q. Zhang, et al. 2013. Graphene quantum dots as the hole transport layer material for high-performance organic solar cells. Phys. Chem. Chem. Phys. 15: 18793-18798.

Lu, C. L., C. P. Chang, Y. C. Huang, R. B. Chen and M. L. Lin. 2006. Influence of an electric field on the optical properties of few-layer graphene with AB stacking. Phys. Rev. B 73: 144427.

Luo, J., L. J. Cote, V. C. Tung, A. T. L. Tan, P. E. Goins, J. Wu, et al. 2010. Graphene oxide nanocolloids. J. Am. Chem. Soc. 132: 17667-17669.

Marlow, G. S. and M. B. Das. 1982. The effects of contact size and non-zero metal resistance on the determination of specific contact resistance. Solid-State Electron. 25: 91-94.

Park, H., P. R. Brown, V. Bulović and J. Kong. 2012a. Graphene as transparent conducting electrodes in organic photovoltaics: studies in graphene morphology, hole transporting layers and counter electrodes. Nano Lett. 12: 133-140.

Park, H., R. M. Howden, M. C. Barr, V. Bulović, K. Gleason and J. Kong. 2012b. Organic solar cells with graphene electrodes and vapor printed poly (3, 4-ethylenedioxythiophene) as the hole transporting layers. ACS Nano 6: 6370-6377.

Peng, J., W. Gao, B. K. Gupta, Z. Liu, R. Romero-Aburto, L. Ge, et al. 2012. Graphene quantum dots derived from carbon fibers. Nano Lett. 12: 844-849.

Sidorov, A. N., S. Pabba, K. P. Hewaparakrama, R. W. Cohn and G. U. Sumanasekera. 2008. Side-by-side comparison of Raman spectra of anchored and suspended carbon nanomaterials. Nanotechnology 19: 195708.

Sohn, Y., S. K. Kim and B. K. Min. 2011. Continuous reduced graphene oxide film prepared by stitching of nanosheets at the interface of two immiscible solutions. Bull. Korean Chem. Soc. 32: 713-715.

Wang, M., Y. Li, H. Huang, E. D. Peterson, W. Nie, W. Zhou, et al. 2011. Thickness dependence of the MoO_3 blocking layers on ZnO nanorod-inverted organic photovoltaic devices. Appl. Phys. Lett. 98: 103305.

Xu, J., T. Feng, Y. Chen and Z. Sun. 2013. Field emission properties of the graphene double-walled carbon nanotube hybrid films prepared by vacuum filtration and screen printing. J. Nanomater. 2013: 536302.

Youn, D. H., Y. J. Yu, H. Choi, S. H. Kim, S. Y. Choi and C. G. Choi. 2013. Graphene transparent electrode for enhanced optical power and thermal stability in GaN light-emitting diodes. Nanotechnol. 24: 075202.

Zhang, Z. Z., K. Chang and F. M. Peeters. 2008. Tuning of energy levels and optical properties of graphene quantum dots. Phys. Rev. B 77: 235411.

Zheng, Q., G. Fang, F. Cheng, H. Lei, P. Qin and C. Zhan. 2013. Low-temperature solution-processed graphene oxide derivative hole transport layer for organic solar cells. J. Phys. D: Appl. Phys. 46: 135101.

5

Epitaxial Growth of III-Nitrides on Si Substrates for Highly-Efficient LED Application

Guoqiang Li,[1,2,*] Wenliang Wang[1,a] and Yunhao Lin[1,b]

ABSTRACT

GaN and its related III-nitride with outstanding optoelectronic properties are suitable for the fabrication of Light-Emitting Diodes (LEDs), laser diodes, etc. So far, GaN-based LEDs prepared on sapphire substrates have been commercialized. However, due to the lack of large size and low thermal conductivity of sapphire, it hampers the further development of high-power low-cost GaN-based LEDs. In this regard, Si substrate which has superior properties, such as high thermal conductivity, available of large size, etc. is a promising candidate for the fabrication of high-power lowcost GaN-based LED devices. This chapter focuses on the recent progress of the fabrication of GaN-based LEDs on Si substrates. The perspectives for the fabrication of highly-efficient LEDs on Si substrates are also discussed.

Keywords: III-nitride, LEDs, Si, epitaxial growth, perspectives.

1. INTRODUCTION

1.1 Advantages of Si Substrates

Nowadays, GaN and its relevant III-nitrides have attracted considerable attention in the fabrication of GaN-based devices, for example, light-emitting diodes (LEDs), due to their superior properties (Tsao et al. 2010, Shur and Zukauskas 2005, Laubsch et al. 2010). After a lot of hard work, significant progress in the area of GaN-based LEDs on sapphire substrates has been achieved and GaN-based LEDs prepared

[1] State Key Laboratory of Luminescent Materials and Devices, South China University of Technology, Guangzhou 510640, China.

[2] Engineering Research Center on Solid-State Lighting and its Informationisation of Guangdong Province, South China University of Technology, Guangzhou 510640, China.

[a] E-mail: 784812649@qq.com

[b] E-mail: 297501154@qq.com

* Corresponding author: msgli@scut.edu.cn

on sapphires have been commercialized (Laubsch et al. 2010, Shur and Zukauskas 2005, Tsao et al. 2010). Even then, conventional sapphire substrates still present several problems in the preparation of highpower, low-cost LEDs. On the one hand, the large-size low-cost sapphire substrates are still unavailable (Liu and Edgar 2002). On the other hand, the low thermal conductivity of sapphire makes it hard to fabricate high-power LEDs (Liu and Edgar 2002, Wang et al. 2014c).

Si substrates, on the contrary, can overcome these disadvantages. As we know, Si is one of the most abundant elements on the earth. Due to its excellent semiconducting properties and good thermal conductivity, Si is widely used in the electronic industry (Ando et al. 1982, Dadgar et al. 2003, Liu and Edgar 2002, Swanepoel 1983, Zhu et al. 2013). After years of development, the preparation of Si single crystal has become a mature industry. In comparison to sapphire substrates, Si substrates are available in large sizes at lowcost with high quality, which would greatly reduce the cost of LED fabrication (Guha and Bojarczuk 1998, Liu and Edgar 2002, Wang et al. 2015, Zhu et al. 2013). Furthermore, Si substrates have much higher electronic and thermal conductivity than sapphire, which is superior for fabricating highpower LEDs (Dadgar et al. 2003, Kukushkin et al. 2008). All these advantages of Si make it a very promising substrate to replace sapphire for fabrication of lowcost and high efficiency LEDs.

Si has diamond cubic crystal structure with lattice parameter a = 0.5431 nm and the space group Fd3m (Liu and Edgar 2002). Because the Si (111) plane has similar hexagonal arrangement with GaN (0002) plane, it is deemed as the most suitable plane for fabrication of GaN-based LEDs (Lei and Moustakas 1992, Kung et al. 1995).

1.2 Issues Involved in The Growth of GaN-Based LED Structures on Si Substrates

As mentioned above, the advantages for the growth of GaN-based LEDs on Si substrates are obvious. However, there are still some challenges for growth of GaN-based LED structures on Si substrates.

First, there exists a so-called melt-back etching phenomenon between Si substrates and Ga atoms during the direct growth of GaN on Si due to the serious reaction between Si and Ga, which leads to the formation of Si-Ga alloys at elevated temperatures (Ishikawa et al. 1998, Krost and Dadgar 2002). The melt-back etching is a serious and fast etching reaction taking place in the interface of GaN and Si, resulting in the destruction of the quality of GaN films. Figure 5.1 shows the plan-view and cross-sectional images of Ga atoms etching into Si substrates if no proper actions are taken during the growth of GaN on Si (Krost and Dadgar 2002). Therefore, it is urgent to avoid the direct contact between Ga atoms and Si substrates during the growth.

Second, during the cooling process, cracks may be caused due to the large tensile stress in the GaN layers grown on Si substrate. This tensile stress originates from the mismatch in Coefficient of Thermal Expansion (CTE) (–115%) between Si (2.6×10^{-6} K^{-1}) and GaN (5.59×10 K^{-1}) (Fu et al. 2000). It is reported that, for cracks on GaN grown on Si substrates, the primary cracks direction is along <$11\bar{2}0$>, which normally results in {1$\bar{1}$00} cleavage planes (Ulrici 2004). These cracks seriously impede the fabrication of LED devices. As an example, Fig. 5.2a shows an optical microscopy of a surface of GaN film grown on Si with a high density of cracks. The performance of the devices with different crack densities is shown in Figs. 5.2b-d, from which, we can find that the higher the crack density, the more non-uniform illumination. Evidently, cracks would lead to a series of problems, such as non-uniform illumination, current leakage, short lifetime, etc. for LEDs (Zhu et al. 2013).

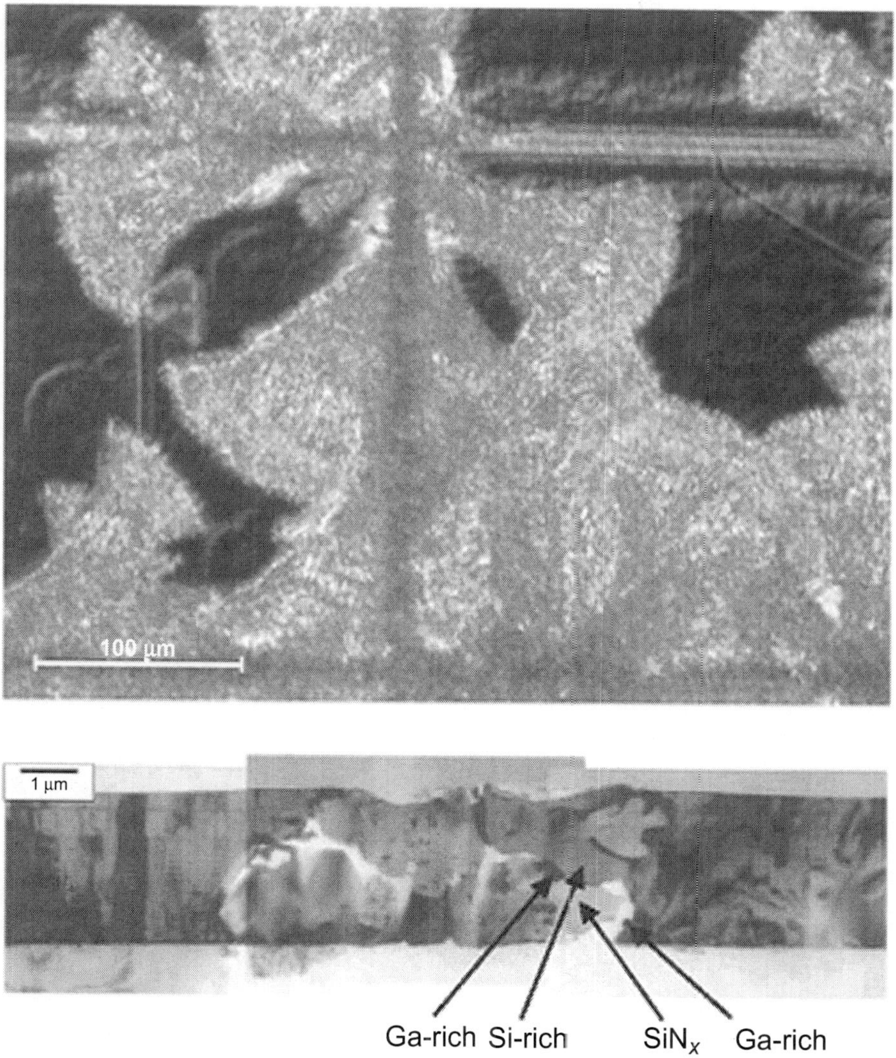

Fig. 5.1 Optical microscopy and cross-sectional TEM (bottom) images of melt-back etching (Krost and Dadgar 2002).

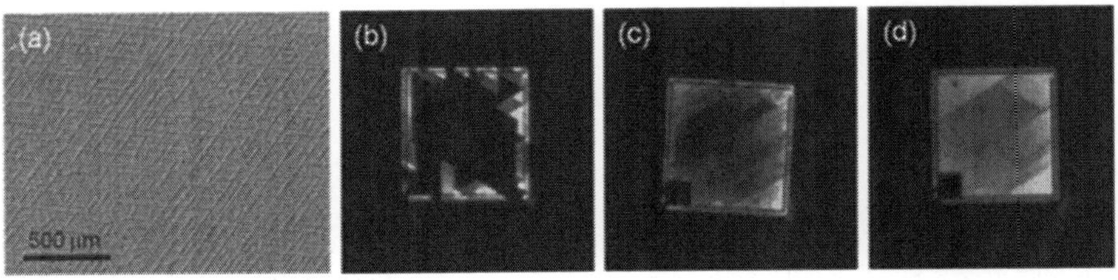

Fig. 5.2 (a) Optical microscopy images of a heavily cracked GaN surface on Si. (b)-(d) EL test for different magnitudes of cracks of GaN based LED on Si (Zhu et al. 2013).

Third, the lattice mismatch between GaN (0002) and Si (111) is as high as 17%. This massive lattice mismatch would lead to a high density of Threading Dislocations (TDs) in as-grown GaN films. As we know, thanks to this large lattice mismatch, a typical dislocation density of 5×10^9 cm^{-2} will extend into InGaN/GaN multiple quantum wells if no methods of dislocation reduction are employed (Dadgar et al. 2003), which would eventually deteriorate the performance of LEDs.

Fourth, the energy gap of Si is 1.12 eV, which is smaller than the wavelength of visible light and results in the fact that more than half of the visible light generated from the active region is absorbed by Si substrates, which leads to a serious reduction of the EQE of LEDs on Si (Butte et al. 2005, Smeeton et al. 2003, Zhang et al. 2005a, Zhou et al. 2007).

These four issues involved in the growth of GaN-based LED structures on Si substrates make it tough to fabricate high-efficiency GaN-based LEDs on Si substrates.

2. EPITAXIAL GROWTH OF GaN-BASED EPILAYERS ON Si SUBSTRATES

2.1 Buffer Layer Technology

Because of the melt-back etching between Si and Ga, as mentioned above, GaN films are very hard to be grown on Si substrates directly. Therefore, a buffer layer is deemed to be required on this occasion. Besides, the buffer layer is vital in the defect control and the stress control during the growth process of GaN-based LEDs on Si substrates. Several research groups have deployed various buffer layer technologies to achieve high-quality GaN films or GaN-based LEDs on Si substrates.

Nishimura et al. (2004, 2002a, b) used Boron Phosphide (BP) as buffer layer to grow GaN film on a 2-inch Si (100) substrate by MOCVD, due to the very small lattice mismatch of 0.6% between BP and cubic GaN. The triethy boron (TEB) and t-butyl phosphine (TBP) are used as precursors for boron and phosphorus. The 30-50 nm-thick low temperature BP layers are deposited on Si substrates, followed by a 100 nm-thick BP. The cubic GaN films are thus successfully grown. There are no voids or large density of defects existing at the interface of BP and Si, as observed from the SEM image in Fig. 5.3.

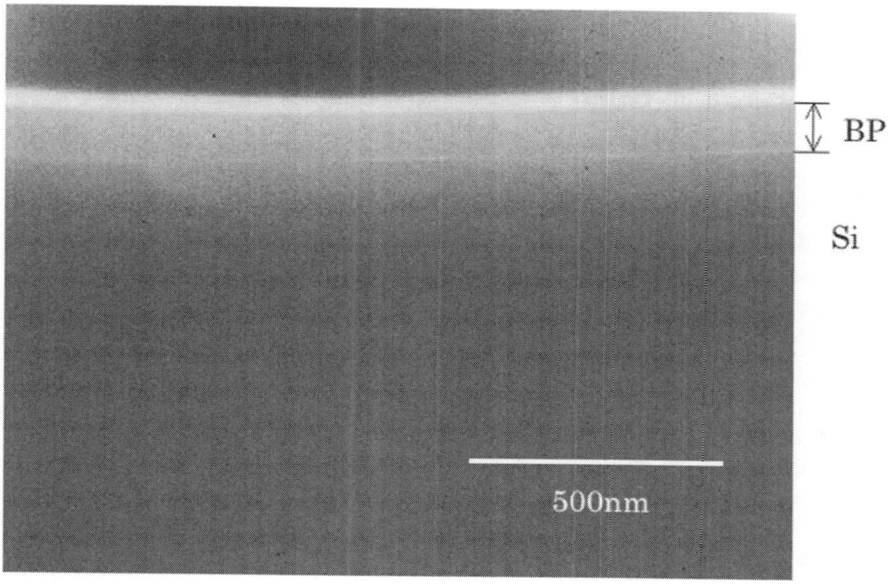

Fig. 5.3 SEM image of abrupt interface of BP and Si (Nishimura et al. 2004).

Zirconium diboride (ZrB_2) (Blake et al. 2012, Fleurence et al. 2013, Roucka et al. 2008), Scandium Nitride (ScN) (Moram et al. 2006, Norenberg et al. 2006), Al_2O_3 (Fenwick et al. 2009a, b), SiC (Abe et al. 2012, Fang et al. 2014, Komiyama et al. 2009) and SiN (Huang et al. 2002) are also applied as buffer layers. Nevertheless, all of these materials cannot be grown by *in situ* process, which adds the additional deposition process and increases the cost of GaN-based LEDs on Si substrates (Zhu et al. 2013). However AlN and step graded $Al_xGa_{1-x}N$ buffer layer, etc. can be grown by *in situ* process and have been adopted in the epitaxial growth of GaN-based LED epitaxial materials on Si substrates (Chen et al. 2001, Feng et al. 2014, Kim et al. 2001, Cheng et al. 2006, Lahreche et al. 2000, Lu et al. 2004, Marchand et al. 2001, Ng et al. 2015, Pan et al. 2011, Xiang et al. 2011, Zhu et al. 2010) and significant progress has been achieved.

2.1.1 AlN Buffer Layer Technology

AlN is regarded as the one of the most suitable buffer layer materials in consideration of the cost and the quality for growth of GaN-based LEDs (Chen et al. 2001, Feng et al. 2014, Lahreche et al. 2000, Lu et al. 2004, Marchand et al. 2001, Ng et al. 2015, Pan et al. 2011). On the one hand, AlN buffer layer can be grown within the same MOCVD system, being naturally incorporated into the entire growth process. On the other hand, the in-plane lattice of AlN is smaller than that of GaN and this lattice mismatch between AlN and GaN can induce the compressive stress in GaN layer, which can compensate the tensile stress originated from mismatch of CTE. Therefore, AlN buffer layer has been widely used in the growth of GaN films on Si substrates.

Raghavan and Redwing (2004) studied the stress during the MOCVD growth of AlN buffer layer and GaN layer on Si substrate by *in situ* stress measurements. They reported that the stress of AlN layer on Si substrate is under tensile stress and there is a sharp increment in stress level from 0.5 to >1 GPa between 800 and 900°C. This increase is ascribed to the transition in crystal structure from polycrystalline to epitaxially oriented as growth temperature increases from 800 to 900°C. Meanwhile, the compressive stress of 0.5 GPa exists in 400 nm-thick GaN layer on 200 nm-thick AlN buffer layer grown at 1100°C during growth process. In this case, the FWHM for X-ray Rocking Curves (XRCs) of GaN (0002) is 0.4°. To understand the effect of AlN layer, Luo et al. (2008) reported on the effect of thickness of the High-Temperature (HT) AlN buffer layer on the properties of GaN grown on Si (111). Their studies demonstrate that suitable thickness of HT-AlN buffer layer is beneficial in achieving high crystalline quality of GaN films with a reduction of cracks.

It is known that the structural and chemical properties for interfaces between AlN buffer layer and Si substrate affect the properties of GaN and its relevant devices. Therefore, significant attention has been paid to the study of the interface between AlN buffer layer and Si substrate. Liu et al. (2003) reported the atomic arrangement at low-temperature AlN nucleation layer on Si (111) substrate with crystallographically abrupt interface, which reveals that the AlN/Si interface is thermodynamically stable and is good for the growth of high crystalline quality AlN nucleation layer. The large lattice mismatch between Si (111) and AlN (0001) is released by the formation of a regular network of misfit dislocations with average separation of about 5.3 $\{1\bar{1}00\}$ AlN planes, which is equal to 4.3 {111} Si planes, as shown in Fig. 5.4. Although the abrupt interface is revealed, the bonding configurations cannot be distinguished (Liu et al. 2003).

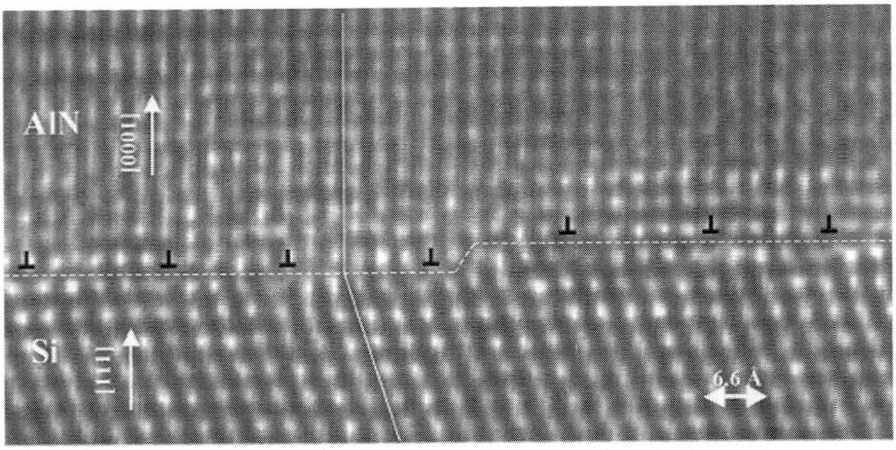

Fig. 5.4 High-resolution electron microscopy images of the abrupt cubic/hexagonal lattice interface and regular misfit dislocations are indicated by "⊥" (Liu et al. 2003).

Radtke et al. (2010, 2012) studied the structure and chemistry of the Si (111)/AlN interface. It is found that there is an intermixing region existing between Si and AlN, where the Si, N and Al exist in and extent to each other and N atoms go further toward Si substrate than Al atoms in comparable width, as shown in Fig. 5.5. It demonstrates that a SiN_x interlayer exists between AlN film and Si substrate and Si-N bonds instead of Al-Si bonds form in the interface region. In addition, they also obtained a much thinner SiN_x interfacial layer at lower growth temperature, which can be attributed to the weak kinetics of inter-diffusion for atoms at lower growth temperature.

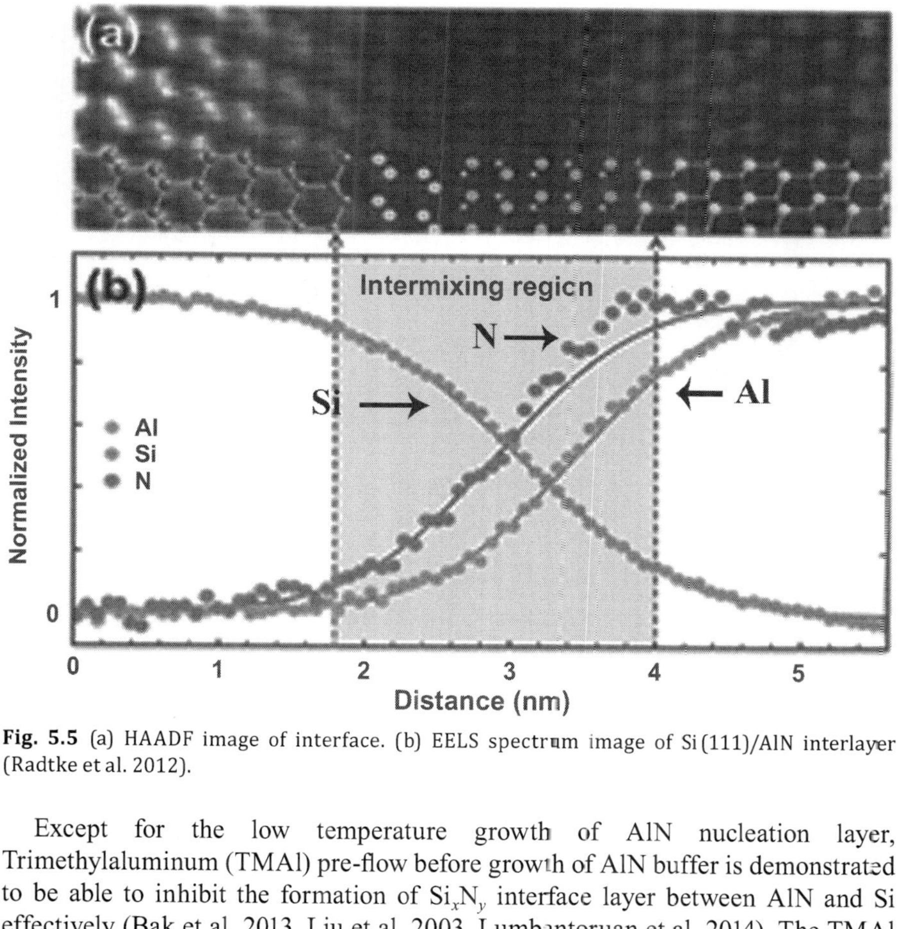

Fig. 5.5 (a) HAADF image of interface. (b) EELS spectrum image of Si(111)/AlN interlayer (Radtke et al. 2012).

Except for the low temperature growth of AlN nucleation layer, Trimethylaluminum (TMAl) pre-flow before growth of AlN buffer is demonstrated to be able to inhibit the formation of Si_xN_y interface layer between AlN and Si effectively (Bak et al. 2013, Liu et al. 2003, Lumbantoruan et al. 2014). The TMAl pre-flow is achieved by exposing the substrate on a single TMAl flow at the absence of NH_3 for a certain duration. It could deposit monolayer Al atoms in advance to prevent the reaction between Si and NH_3 which would produce the Si_xN_y layer (Liu et al. 2003). Lumbantoruan et al. (2014) studied the interface layer between AlN and Si with and without the TMAl pre-flow by high-resolution TEM, as shown in Fig. 5.6, which confirms a thinner Si_xN_y interface layer with the TMAl pre-flow than that without TMAl pre-flow. Besides, it is found that the TMAl pre-flow could significantly improve the quality of AlN and GaN on Si substrates (Bak et al. 2013, Clames et al. 2006, Kim 2007, Ni et al. 2005). Cao et al. (2010) studied the influence of duration for TMAl pre-flow ranging from 0 to 56 seconds on the properties of AlN buffer layer and GaN films on Si substrates. They demonstrated a moderate duration of TMAl pre-flow, which is 36 seconds in their work, could improve the crystalline quality and surface morphology of AlN buffer layer. In this regard, the GaN grown on the top of improved AlN buffer layer also reveals better crystalline quality and flatter surface morphology. In addition, the moderate duration of TMAl pre-flow can inhibit the formation of cracks.

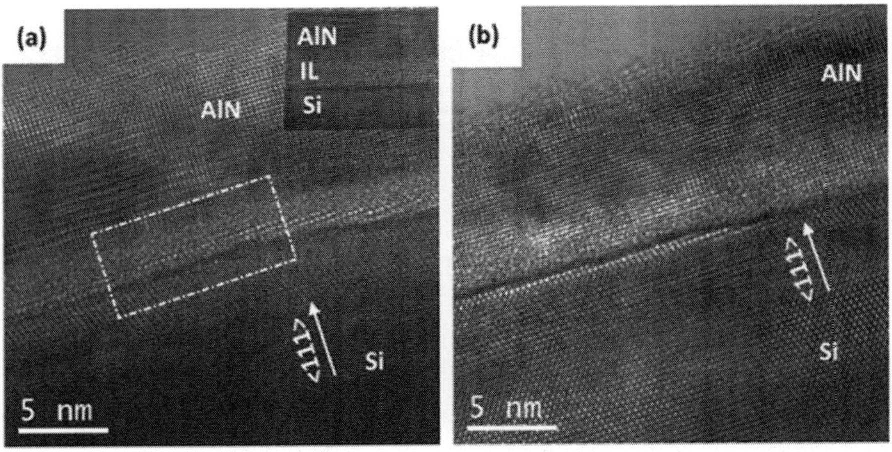

Fig. 5.6 High-resolution cross-sectional TEM images of AlN on Si (a) with and (b) without TMAl pre-flow (Lumbantoruan et al. 2014).

2.1.2 Step Graded Al_xGa_{1-x} N Buffer Layer Technology

Although various kinds of materials are used as the buffer layer, the crack and defect problems still could not be overcome effectively. Therefore, other structures for epitaxial growth of GaN on Si with multilayer buffer layers are proposed (Xiang et al. 2011, Kim et al. 2001, Cheng et al. 2006, Zhu et al. 2010).

Kim et al. (2001) showed that step-graded $Al_xGa_{1-x}N$ buffer in Ultrahigh Vacuum Chemical Vapor Deposition (UVCVD) growth of GaN layers on Si (111) substrates is a very effective method to suppress the crack formation. 2 μm-thick GaN layers are grown on Si substrates with and without the graded $Al_xGa_{1-x}N$ buffer layer, respectively. As compared with the GaN layer without the graded $Al_xGa_{1-x}N$ buffer layer, GaN grown on the graded $Al_xGa_{1-x}N$ buffer layer shows higher quality with less cracks on the surface, smaller FWHMs of XRCs for GaN (0002) and GaN ($10\bar{1}2$), smaller RMS surface roughness and narrower FWHMs for PL peaks. These results demonstrate that the graded $Al_xGa_{1-x}N$ buffer layer can reduce the residual tensile stress and eventually improve the properties of as-grown GaN films. Nevertheless, the stress relaxation mechanisms have not been systematically studied in their work. Cheng et al. (2006) discussed the condition of stress for GaN grown on $Al_xGa_{1-x}N$/AlN buffer layer. It is deduced that initial compressive stress of −11.3 GPa at room temperature originates from the lattice mismatch between GaN and AlN due to the fact that the in-plane lattice constant of GaN (0.3189 nm) is larger than that of

AlN (0.3112 nm) buffer layer. However, this compressive stress is released by the severe stress relaxation mechanism. And this stress relaxation mechanism results from that the epitaxial GaN layer suffers tensile stress from coalescence and misorientation of GaN islands in the 3D growth process of GaN. Therefore, the initial compressive stress of −11.3 GPa is weakened and the tensile is enhanced. For growth of GaN on $Al_xGa_{1-x}N$/AlN buffer layers on Si substrates, the step-graded $Al_xGa_{1-x}N$ layers can impose the compressive stress. As a consequence, the compressive stress can balance the tensile stress from the CTE mismatch and eventually suppress the formation of cracks.

Zhu et al. (2010) studied the propagation of dislocations through the $Al_xGa_{1-x}N$/AlN buffer layers on Si substrates. The screw, edge and mixed type dislocations are investigated by weak-beam dark-field TEM, as shown in Fig. 5.7. It is found that the step-graded AlGaN layers can lead to significant bending of edge type dislocations and the screw type dislocations can be annihilated by the formation of loop. The results confirm the effect of $Al_xGa_{1-x}N$ buffer layer on crystalline quality of GaN layer. Meanwhile, the dislocation density is further reduced by inserting a SiN_x interlayer. Afterwards, through optimizing the process, a highquality 1.8 μm-thick crack-free GaN film with threading dislocation density of 6.2×10^8 cm^{-2} is achieved (Zhu et al. 2013).

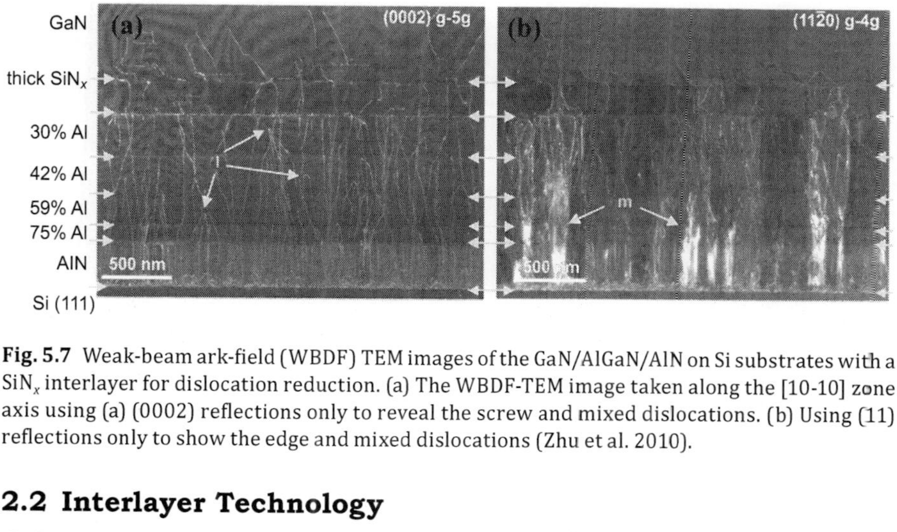

Fig. 5.7 Weak-beam ark-field (WBDF) TEM images of the GaN/AlGaN/AlN on Si substrates with a SiN_x interlayer for dislocation reduction. (a) The WBDF-TEM image taken along the [10-10] zone axis using (a) (0002) reflections only to reveal the screw and mixed dislocations. (b) Using (11) reflections only to show the edge and mixed dislocations (Zhu et al. 2010).

2.2 Interlayer Technology

2.2.1 Low Temperature AlN Interlayer Technology

Another classical structure of AlN interlayers for the growth of GaN layer on Si substrate is also proposed. Dadgar et al. (2006) successfully grew the 5.4 μm-thick crack-free GaN-based LED structures on 150 mm Si (111) with the Low Temperature (LT) AlN interlayers. Afterwards, the crystalline quality of GaN film with AlN interlayers is further improved, the FWHM for XRCs of GaN (0002) and (10$\bar{1}$0) are reduced to 380 and 372 arcsec, respectively. Furthermore, a 14.6 μm-thick GaN-on-Si layer is achieved by LT-AlN interlayers and the FWHM for GaN (0002) XRC is as low as 252 arcsec (Dadgar et al. 2011a). The influence of the thickness and the layer number of the LT-AlN interlayers on the stress and the crystalline quality of GaN film on Si substrates is discussed by Cong et al. (2005). It is found that the increase of AlN thickness and the layer number leads to the decrease of tensile stress in the GaN epilayer at room temperature. Besides, the GaN layer with 16 nm-thick AlN interlayer reveals the best crystalline quality of GaN layer. As for the layer numbers of AlN interlayers, dislocations are reduced at the increase of the interlayer numbers. From the multi-beam dark-field cross-sectional TEM, it is clearly identified that both screw and edge dislocations can be reduced by AlN interlayers (Dadgar et al. 2002, Drechsel and Riechert 2011), as shown in Fig. 5.8.

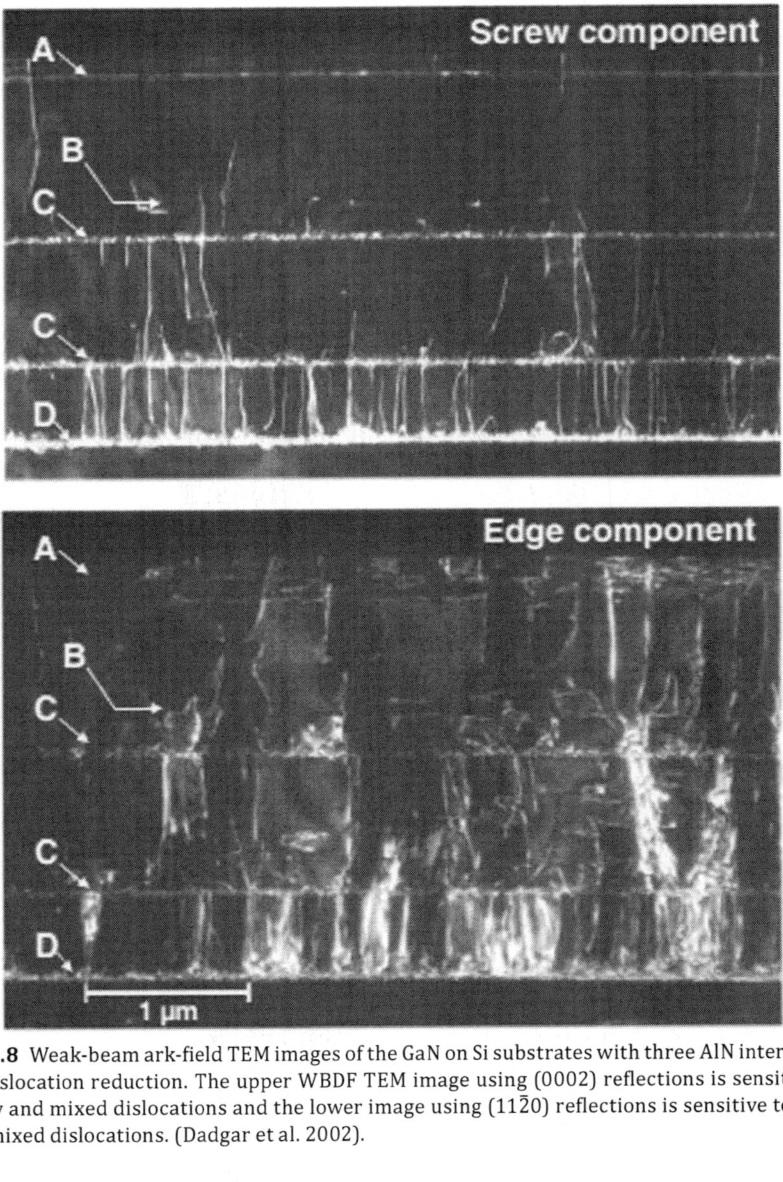

Fig. 5.8 Weak-beam ark-field TEM images of the GaN on Si substrates with three AlN interlayers for dislocation reduction. The upper WBDF TEM image using (0002) reflections is sensitive to screw and mixed dislocations and the lower image using $(11\bar{2}0)$ reflections is sensitive to edge and mixed dislocations. (Dadgar et al. 2002).

As for the stress in GaN layers grown on Si substrates, on the one hand, LT-AlN interlayer can destroy the coherence between subsequent layers and the underlying buffer layer which can weaken the tensile stress caused by the lattice mismatch between the Si and AlN/AlGaN/GaN. On the other hand, these interlayers can induce the strong pre-stress during the growth process of GaN layer and this pre-stress can compensate the tensile stress during the cooling process (Krost et al. 2003). Therefore, much thicker crack-free GaN layers on Si substrates can be obtained by AlN interlayers in comparison with the AlGaN buffer.

Liu et al. (2013) utilized the lattice mismatch between AlN interlayer and GaN layer to induce 3D growth mode for GaN nucleation layer. They inserted an AlN interlayer between step graded $Al_xGa_{1-x}N$ buffer layer and GaN layer and investigated the influence of the thickness of the AlN interlayer ranging from 0 to 30 nm on the growth of GaN. According to the SEM images of the GaN nucleation layer on the AlN interlayer with various interlayer thicknesses, Fig. 5.9, obvious increase of void density and decrease in GaN seed sizes can be observed. It indicates enhanced 3D growth mode for GaN nucleation layer, with increasing AlN interlayer thickness. They attributed this phenomenon to the recovery of lattice content for AlN interlayer on the $Al_{0.3}Ga_{0.7}N$ buffer, which restores the lattice mismatch between AlN interlayer and GaN layer, as the AlN interlayer thickness increases. While compared to conventional AlN interlayer, a larger improvement in terms of crystalline quality of GaN could be obtained. Li's group investigated the influence of growth reactor pressure on the GaN nucleation layer which is grown on AlN interlayer (Lin et al. 2015). They demonstrated that, not only the thickness of AlN interlayer, but also the reactor pressure for GaN nucleation layer plays a significant role on the formation of 3D growth mode for GaN nucleation layer on AlN interlayer. In comparison to the low reactor pressure, the high reactor pressure is advantageous in forming 3D growth mode for GaN. In addition, they deployed the TEM to confirm that the 3D growth mode of GaN on AlN interlayer could accelerate the bending and annihilation of dislocations effectively shown in Fig. 5.10, which improves the crystalline quality of GaN drastically. They also found that the 3D growth mode for GaN nucleation layer would raise the residual tensile stress and because of that the compensated compressive stress, which originates from the mismatch between AlN and GaN is weakened by the coalescence of GaN islands.

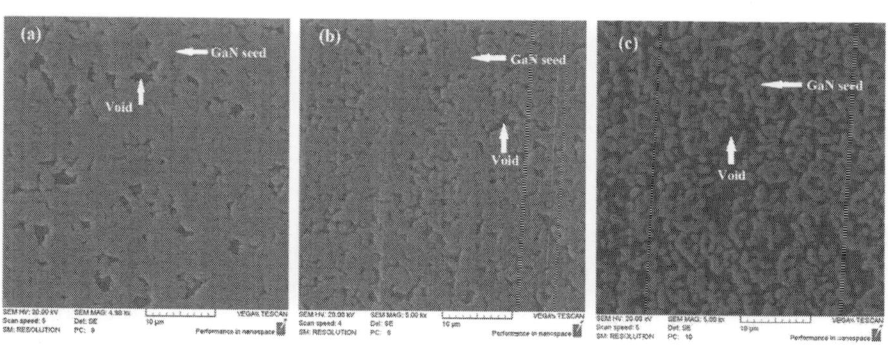

Fig. 5.9 SEM images of GaN nucleation layer on the AlN thicknesses with a thickness of (a) 0. (b) 10. (c) 30 nm (Liu et al. 2013).

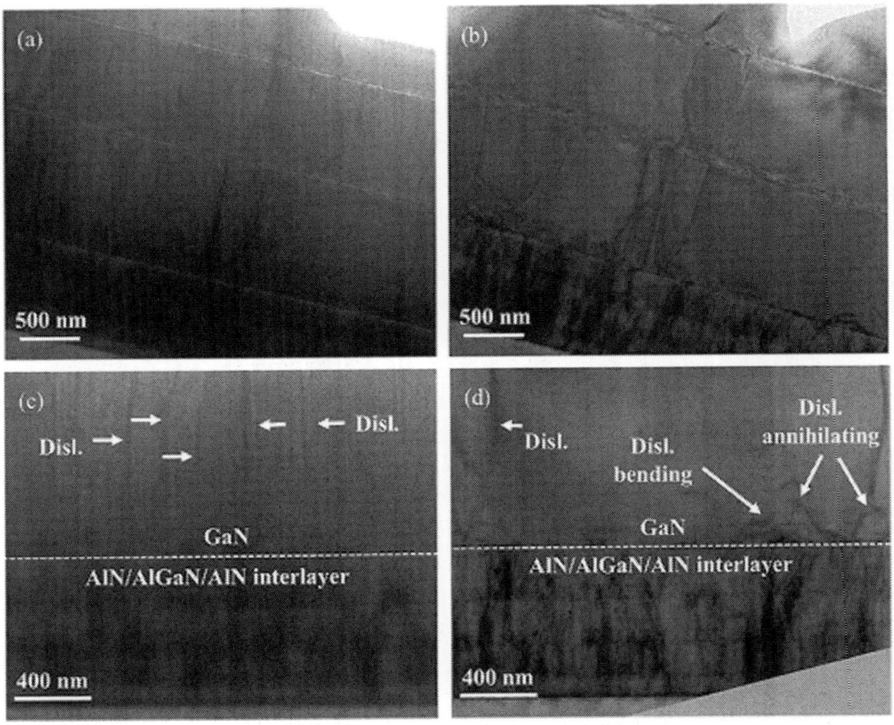

Fig. 5.10 Cross-sectional TEM images of (a) Sample A with the GaN nucleation layer grown at 200 Torr. (b) Sample B with the GaN nucleation layer grown at 500 Torr and the detailed cross-sectional BF-TEM images of the GaN layer on the AlN interlayer for (c) Sample A. (d) Sample B (Disl. = dislocation) (Lin et al. 2015).

2.2.2 In Situ S_xN_y Interlayer Technology

A greater reduction in the Threading Dislocations (TDs) density can be achieved by *in situ* Si_xN_y interlayer. The technology of Si_xN_y interlayer has been widely employed and the mechanism for the suppression of the TDs has also been studied by many groups (Arslan et al. 2009, Contreras et al. 2002, Wang 2014a, 2014b). The *in situ* Si_xN_y interlayer is usually deposited by the reaction between NH_3 and SiH_4 for a short period with an interruption of trimethylgallium (TMGa) in MOCVD system. There are two kinds of suppression for TDs and corresponding mechanisms induced by Si_xN_y interlayer, which have been reported. The first kind of suppression for TDs is that the TDs are terminated or annihilated on the interface of Si_xN_y (Contreras et al. 2002, Wang et al. 2014b). In this regard, the form of dislocations can be classified into three types, as shown in Fig. 5.11 (Wang et al. 2014b). Type I shows that Si_xN_y acts as a surface dislocation to induce a line bending of the dislocation to the basal plane. For type II, after being bent by the Si_xN_y, the dislocation bends back into the direction of growth and leaves behind a kink. Besides, two dislocations belonging to type II with the opposite Burgers vectors encounter each other to form an annihilation loop which is classified into the type III. The dislocations in type I and type III could be suppressed successfully. The corresponding mechanism can be explained to rely on the effect of the mask that Si atoms prefer to binding at GaN dislocation cores to accelerate the annihilation of dislocations (Riemann et al. 2006, Tanaka et al. 2000, Wang et al. 2014b, Zang et al. 2007). This tropism of Si atoms is attributed to the presence of N-dangling bonds at dislocation pits can trap the Si atoms more easily.

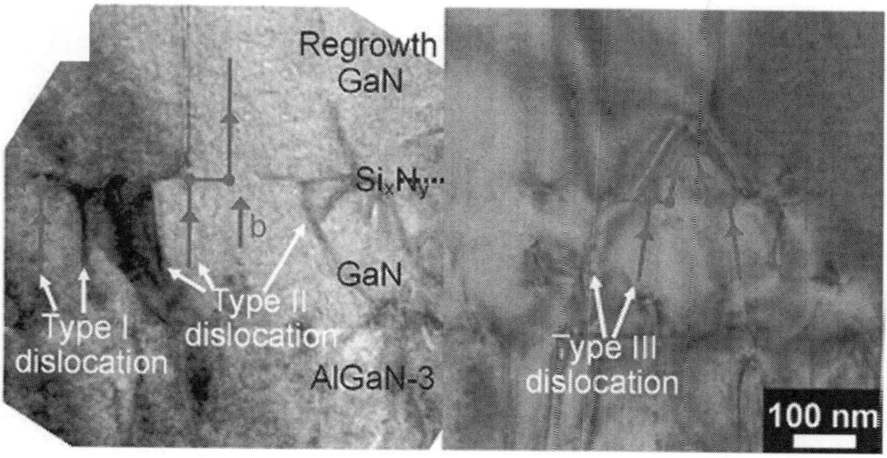

Fig. 5.11 Cross-section TEM image of three types of dislocation induce by *in situ* Si_xN_y interlayer in GaN epilayer (Wang et al. 2014b).

Another suppression of TDs is to utilize the nanoporous Si_xN_y network to induce a nano-epiaxial lateral overgrow (nano-ELOG) (Datta et al. 2004, Haeberlen et al. 2010, Kappers et al. 2007). It is interpreted that the Si_xN_y interlayer deposited on the GaN forms $x^{a\ no}$ non porous SN_y network. Due to the fact that the anti-surfactant effect of the Si-N mask weakens the GaN growth kinetics, the GaN just grows through the nanopores. At this stage, the growth of GaN transfers into 3D mode. Then, the transition from 3D to 2D of GaN is intentionally delayed by adjusting the growth process such as increasing the reactor pressure and using lower V/III ratio. As a result, the dislocations can be promoted to bend during the lateral overgrowth, eventually, annihilating the dislocations by bending dislocations in the GaN layer above the Si_xN_y interlayer, as shown in Fig. 5.12 (Kappers et al. 2007).

Markurt et al. (2013) studied and explained the mechanism of anti-surfactant effect of Si_xN_y interlayer in the growth of GaN. Firstly, they demonstrated that the Si_xN_y interlayer deposited on the GaN layer is a crystalline $SiGaN_3$ monolayer consisting of an ordered Ga vacancy (V_{Ga}), Si and Ga atoms with a form of a $\sqrt{3} \times \sqrt{3}R30°$ structure, as shown in Fig. 5.13. The Ga vacancies would induce a strong competition between chemistry and charge compensation. The growth of GaN on top of $SiGaN_3$ layer will result in the change transferring from the surface cation dangling bonds to the Ga vacancy. This process leads to the formation of an electrical dipole between the $SiGaN_3$ layer and the surface. The more the layers of GaN that are deposited, the more dipole moment would be. It would result in an energetically highly unfavorable surface which prevents the further growth of

GaN. Therefore, the GaN is exclusively grown at the areal without covering by the $SiGaN_3$ layer.

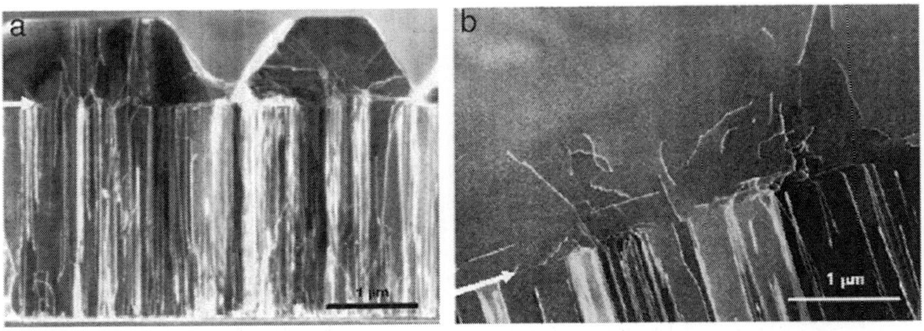

Fig. 5.12 Cross-section weak-beam dark-field TEM images of (a) GaN islands grown on the Si_xN_y which is covered on under GaN surface. The TDs penetrate the Si_xN_y interlayer and then bend over at different angles in the GaN islands. (b) A coalesced GaN layer with a Si_xN_y interlayer. The TDs penetrating the Si_xN_y interlayer are bent again. (In this image, only the edge and mixed TDs are visible) (Kappers et al. 2007).

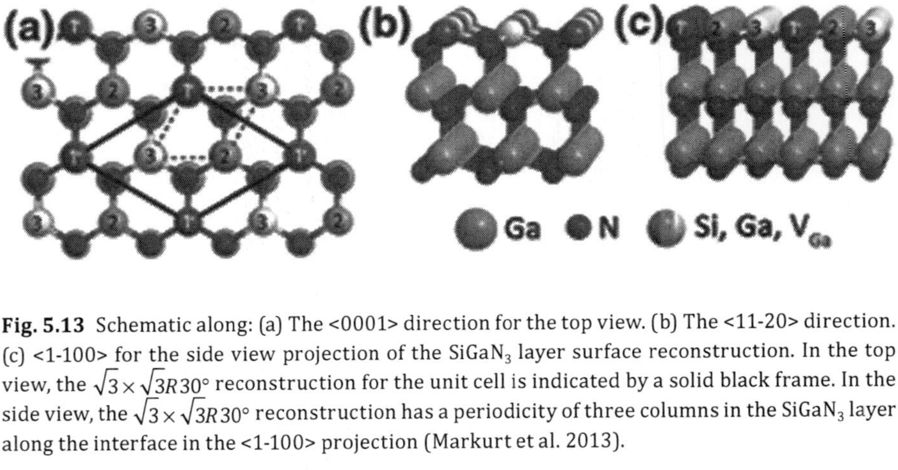

Fig. 5.13 Schematic along: (a) The <0001> direction for the top view. (b) The <11-20> direction. (c) <1-100> for the side view projection of the $SiGaN_3$ layer surface reconstruction. In the top view, the $\sqrt{3}\times\sqrt{3}R30°$ reconstruction for the unit cell is indicated by a solid black frame. In the side view, the $\sqrt{3}\times\sqrt{3}R30°$ reconstruction has a periodicity of three columns in the $SiGaN_3$ layer along the interface in the <1-100> projection (Markurt et al. 2013).

However, this method of *in situ* Si_xN_y interlayer still has a shortcoming. The GaN inserted by *in situ* Si_xN_y interlayer requires a long coalesced thickness, while the thickness of GaN grown on Si is limited due to the mismatch in CTE (Cheng et al. 2008, Zhu et al. 2013). Zhu et al. (2013) investigated the influence of deposited time of Si_xN_y on the crystalline quality and surface morphology of GaN films on Si substrates. It is found that the additional time for Si_xN_y deposition increases the thickness of the GaN layer required to achieve a fully coalesced surface and the long deposited time might induce cracks on the surface of GaN films on Si substrates. Therefore, a moderate time for Si_xN_y is crucial on the growth of GaN on Si. Cheng et al. (2008) reduced the coalesced thickness of GaN by increasing the growth temperature of GaN layer after inserting Si_xN_y interlayer. At high temperature, an accelerated coalescence is obtained and the crystalline quality is also improved.

In general, the *in situ* Si_xN_y interlayer plays an important role on the growth of GaN on Si substrate owing to the effect in reducing the TDs in the GaN layer. More advanced technology such as double Si_xN_y interlayer (Hikosaka et al. 2014), multilayers Si_xN_y interlayer (Zang et al. 2007), etc. have been reported. Actually, the great reduction in TDs by the Si_xN_y interlayer enables the development of GaN-based LEDs on Si substrates.

2.3 Patterned Si Substrate Technology

Patterning Si substrate is another effective method to overcome the problem of cracks (Honda et al. 2002, Lee et al. 2008, Liu et al. 2013, Zamir et al. 2002). Dadgar et al. (2001) reported the growth of crack-free GaN-based LEDs on Si (111) patterned substrates with sputtered SiN_x masks. Due to the amorphous SiN_x, the crystalline seed layer cannot be grown on it, resulting in GaN grown on restricted region and separated by SiN_x mask, as shown in Fig. 5.14a and b. When compared with the unpatterned samples, cracks within the GaN films grown on patterned Si substrates disappear, which confirms that patterning the Si substrate can effectively reduce the overall stress of GaN layer.

Fig. 5.14 SEM images of a GaN based LED on Si (111) patterned with SiN_x mask at (a) low- and (b) high-magnification (Dadgar et al. 2001).

Lau et al. (2007) also achieved 2 μm-thick crack-free GaN-based blue LEDs on patterned Si substrates. The difference is that the Si substrates are patterned with square islands separated by 3 μm deep and 20 μm wide trenches and the trenches are etched along the crystalline orientations of <$1\bar{1}0$> and <$11\bar{2}$>, as shown in Fig. 5.15. Although the GaN layer is grown on trenches, the deepness of trenches can guarantee the separation of the GaN layer on platforms and GaN layer on trenches. Meanwhile, the SEM measurement reveals that cracks only exist in the trench mask.

It is known that the impurity concentration, especially Si-doping concentration, may cause the tensile stress in GaN layer grown on Si substrates (Dadgar et al. 2003). For GaN with a doping concentration of 1×10^{18} cm^{-3}, about 0.1 GPa/μm of additional tensile stress is introduced. Up to about 0.5 GPa/μm of tensile stress is induced when the Si-doping concentration in GaN reaches 5×10^{18} cm^{-3}. Hence, the method to reduce the tensile stress originated from Si-doping needs to be carefully investigated.

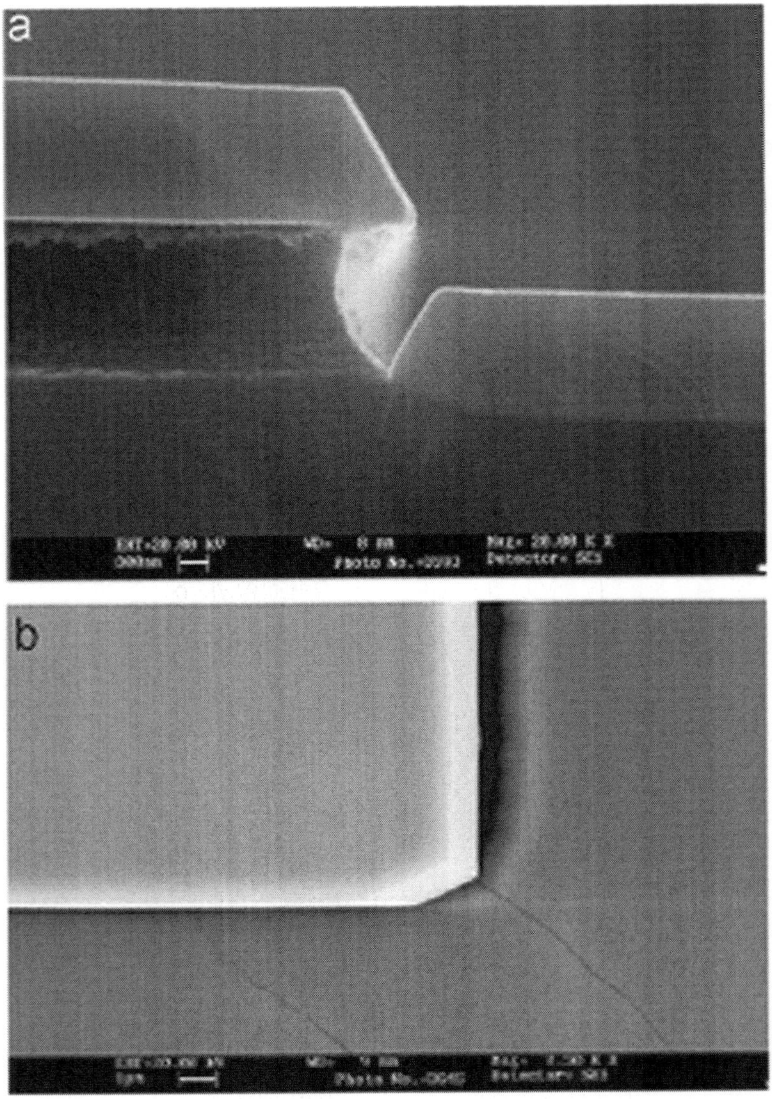

Fig. 5.15 (a) Cross-sectional SEM images. (b) Planform SEM images of GaN on patterned Si substrates (Lau et al. 2007).

2.4 Ge Doping Technology

Si is widely adopted as the n-doping for GaN and it is reported that Si-doping would induce edge-type dislocation climbs which lead to a decrease in compressive stress and finally an increase in tensile stress for tensely stressed layers (Krost et al. 2003,

Dadgar et al. 2011b). Therefore, it is necessary to search for a new doping source to reduce the tensile stress on GaN epitaxil films.

Dadgar et al. (2011b) and Fritze et al. (2012) studied the effect of Ge doping instead of Si doping as the n-GaN doping. The wafer curvature caused by the stress is shown in Fig. 5.16. The larger curvature means the greater tensile stress above the 0 m^{-1}. From Fig. 5.17, the GaN with Ge-doping has a smaller curvature than that with Si-doping, which means the tensile stress induced by Si-doping can be alleviated by Ge-doping. This is ascribed to the lower impact for Ge-doping on dislocation climbing in comparison with that of Si-doping. This dislocation climbing is originated from the enhancing vacancy formation by combining SiN at the dislocation coherence when the concentrations of Si-doping is high. In contrast to the formation of SiN, the GeN (Ge_3N_4) is unstable at the elevated temperature and is decomposed around 900°C, which is lower than the growth temperature of GaN in MOCVD. Meanwhile, Ge can be doped up to 2.9×10^{20} cm^{-3}, which is much higher than the concentration of Si-doping, For GaN with high Si-doping levels, the incorporation is in homogeneous for GaN with high Si-doping levels and it would cause 3D growth mode in GaN layer with Si-doping at concentrations around 1.9×10^{19} cm^{-3}, while it is 2D growth mode in GaN layer with Ge doping at concentrations around 6×10^{19} cm^{-3} (Fritze et al. 2012). Therefore, Ge-doping is a promising technology for the growth of high-quality GaN-based LEDs on Si.

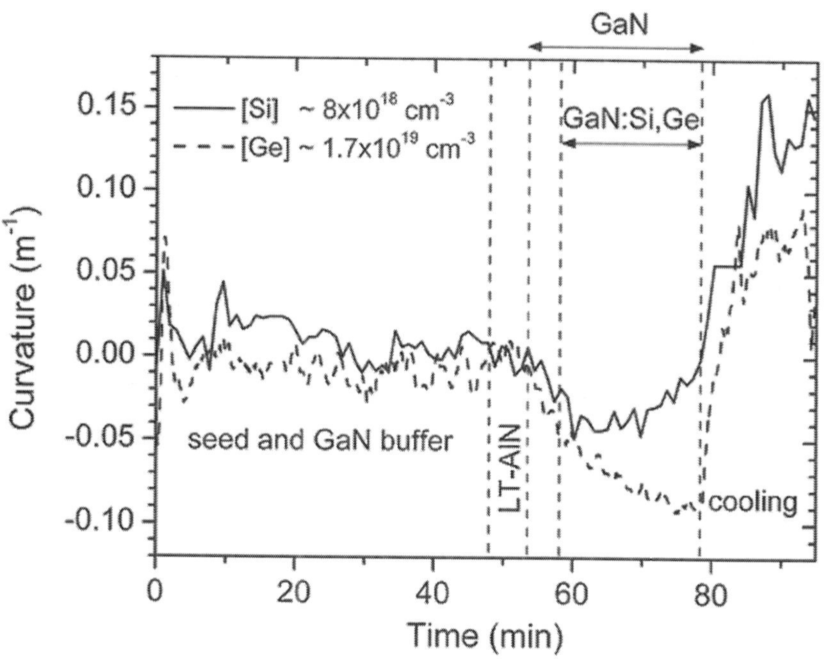

Fig. 5.16 *In situ* curvature measurement of GaN layer with LT-AlN interlayers, the GaN layers are doped with Si and Ge. The compressive stress induced by LT-AlN interlayers is significantly weakened by Si-doping and transform to tensile stress (solid line). The GaN doped by Ge stayed compressive stress all the time during the growth process of GaN layer. (Dadgar et al. 2011b).

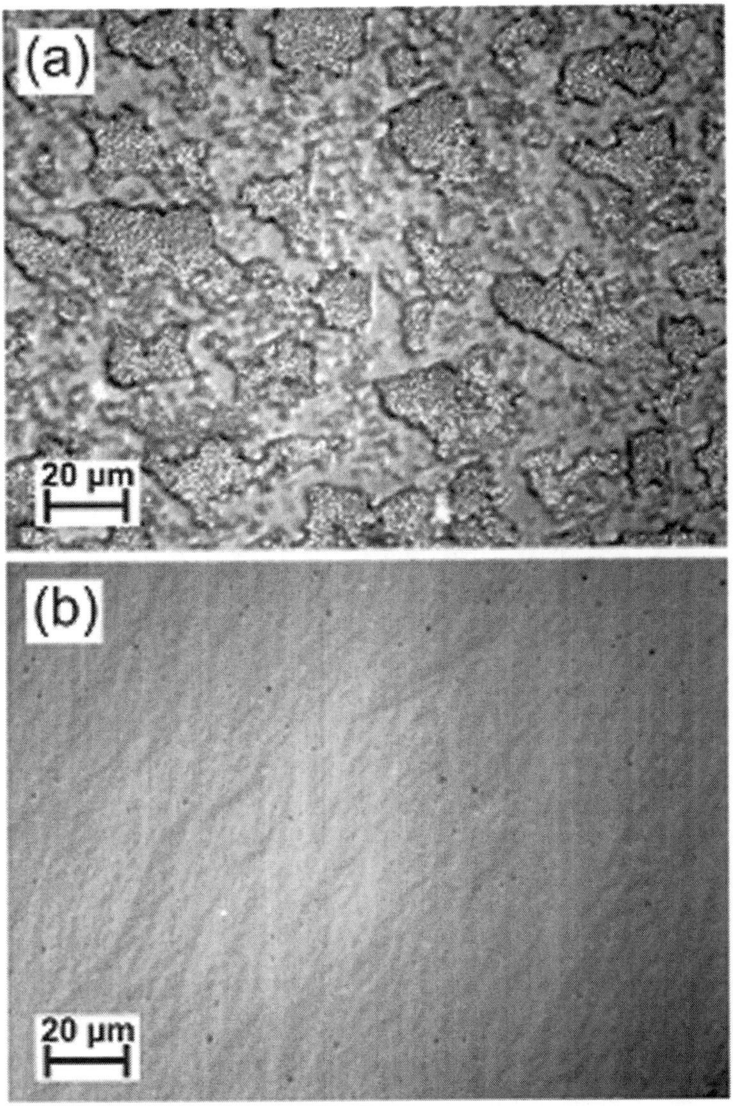

Fig. 5.17 Nomarski microscope images of (a) GaN surface with Si-doping at concentrations around 1.9×10^{19} cm^{-3}. (b) GaN surface with a Ge-doping at concentrations of 6×10^{19} cm^{-3}. (Fritze et al. 2012).

3. FABRICATION AND PROPERTIES OF LED DEVICES ON Si SUBSTRATES

After the stress and the crack in the as-grown GaN films on Si substrates are well controlled, naturally, whole LED epitaxial structures including the InGaN/GaN MQWs and Mg-doped GaN layer could be grown subsequently. Furthermore, the distinct properties of Si from sapphire and the sophisticated automated processing for Si processing lines offer more possibilities for the more advanced device structures. Researchers have tried to fabricate LED devices with various structures.

3.1 Fabrication of Lateral LED Devices on Original Si Substrates and The Corresponding Properties

Zhu et al. (2010) successfully grew crack-free high-quality LED structures on Si substrates and fabricated the lateral-structure GaN-based LED chips on Si substrates, as shown in Fig. 5.18a and b. This GaN-based LED structure includes AlN nucleation layer, AlGaN buffer layers, SiN_x interlayers, Si-doped GaN films, InGaN/GaN MQWs and Mg-doped GaN layer. While for GaN-based LEDs with the size of 500 × 500 μm² mesa, Ti/Al/Pt/Au alloys are deposited on the exposed n-GaN layer as the n-type contact and semi-transparent annealed Ni/Au is used as the p-type contact.

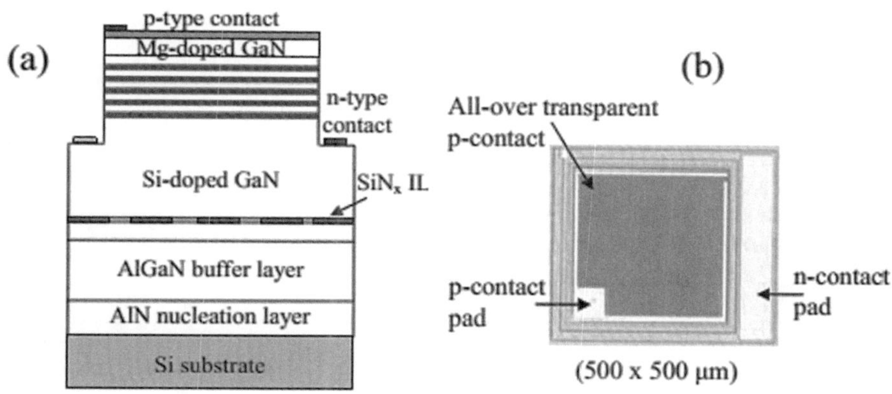

Fig. 5.18 (a) Schemiatic diagram of fabricated lateral LEDs on Si. (b) platform of device (Zhu et al. 2010).

After preparation of lateral LEDs on Si substrates, the relationship between Internal Quantum Efficiency (IQE) and crystalline quality of GaN has been discussed by Zhu and Humphreys (2013), as shown in Fig. 5.19a. It reveals that higher IQE is obtained in the sample with a narrower GaN $(10\bar{1}1)$ omega peak by low-temperature photoluminescence (LT-PL). Lower IQE is also observed for the sample with a longer emission wavelength. Furthermore, the EL-output is also improved as the FWHM of XRC for GaN $(10\bar{1}1)$ is decreased, Fig. 5.19b. These trends are mainly relevant to the reduction in threading dislocations that act as the non-radiation centers and are detrimental to the IQE (Cherns et al. 2001, Zhang et al. 2005b, Monemar and Sernelius 2007). Therefore, the reduction of threading dislocations is significant for the enhancement in light output power of LEDs on Si substrates.

Lin's group also confirmed the effect of crystalline quality of as-grown GaN films on the LEDs performance (Lin et al. 2015). The crystalline quality of GaN is controlled by the reactor pressure of GaN nucleation layers. The LEDs with high crystalline quality and low threading dislocation density are grown with GaN nucleation layer at high reactor pressure and show stronger light output power

(LED C), Fig. 5.20a. The LED chips of 500 × 500 μm² mesa with the best crystalline quality show the light output power of 4.22 mW and the forward voltage of 3.47 V under 20 mA, Fig. 5.20b.

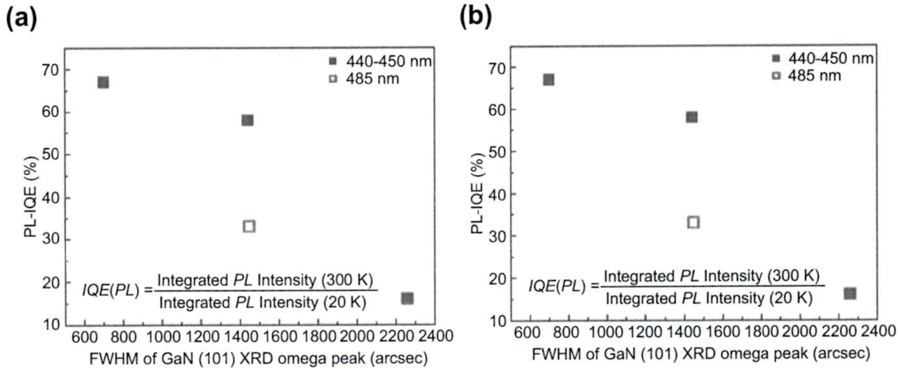

Fig. 5.19 The dependence of (a) PL-IQE. (b) EL-output (20 mA) on FWHM of GaN (10$\bar{1}$1) omega peak, respectively, for LED devices grown on Si substrate (Zhu and Humphreys 2013).

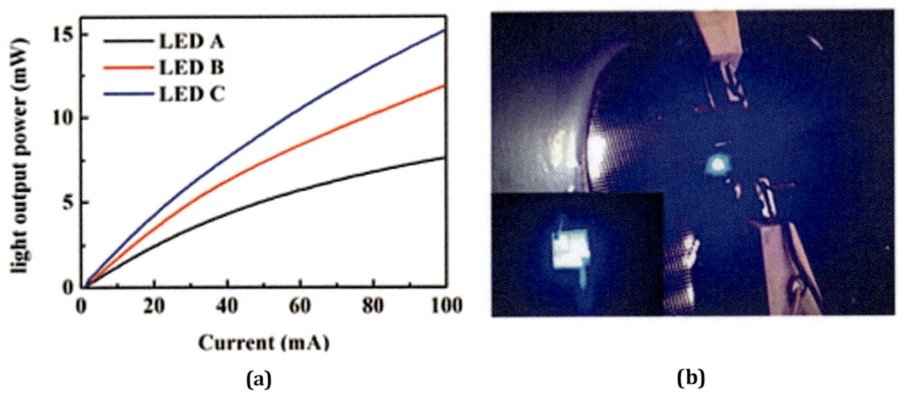

Fig. 5.20 (a) Light output power vs current. (b) The light emission image of LED chips from LED C (Lin et al. 2015).

In order to understand the concreted effect of the Si substrate on the performance of LEDs, Zhu et al. (2011) calculated the theoretical Light Extraction Efficiency (LEE) for LEDs on both Si and sapphire, which is 3.74 and 5.4%, respectively, Fig. 5.21 (Zhu et al. 2011). Taking the transmission of the p-contact metal (75%) into consideration, the final LEE for LEDs on Si and sapphire are 2.8 and 4.05%, respectively. This result confirms that the decline in LEE results from the absorption of visible light by Si substrates.

In order to minimize the absorption of light by Si substrate, a reflector can be inserted, which can reflect the light downward from the active region. The reflected light can escape from the top surface of LEDs. Bragg reflector (DBR) is a popular interlayer as the reflector (Butte et al. 2005, Arkun et al. 2012, Ishikawa et al. 2008). However, the DBR interlayers may induce more tensile stress or defects.

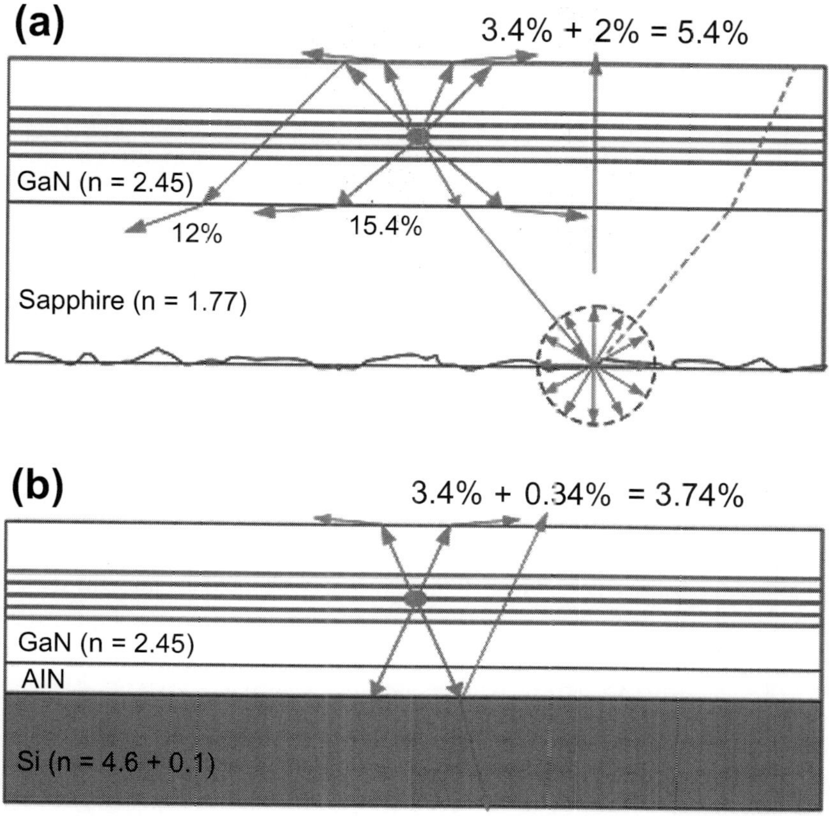

Fig. 5.21 Diagram of the calculation of extraction efficiency of LED devices on (a) Sapphire. (b) Si (Zhu et al. 2011).

Apart from the adoption of the reflector, removing the Si substrates during the chips fabrication process can also improve the LEE for LEDs on Si substrates and fortunately, Si can be easily dislodged by chemical etching (Jiang et al. 2007, Lau et al. 2011, Zhang et al. 2005a, Shi et al. 2014).

3.2 Fabrication of Lateral LED Devices on Transferred Substrates and The Corresponding Properties

Lau et al. (2011) improved the performance of LEDs by transferring the Si substrate to copper, the preparation process is described as follows. First, standard procedure is used to prepare the lateral LED device on patterned Si substrates. Second, the wafer is temporarily bonded to a sapphire substrate. Before the bonding process, a polyimide layer SDR-PI-5 was spin-coated to protect the LEDs during Si wet etching. Third, the Si substrate is removed by chemical etching. Fourth, Ti (5 nm)/Al (150 nm)/Ti (10 nm)/Au (100 nm) metal layers are deposited on the exposed n-GaN surface and then the 100 μm-thick Cu is electroplated to replace the Si as the new substrate. Finally, temporary substrate sapphire and protective layer (polyimide layer SDR-PI-5) are removed by trichloroethylene (TCE) and organic resist stripper, respectively. The final structure of LEDs on Cu with 160 μm in radius circular mesas is shown in Fig. 5.22 (Lau et al. 2011).

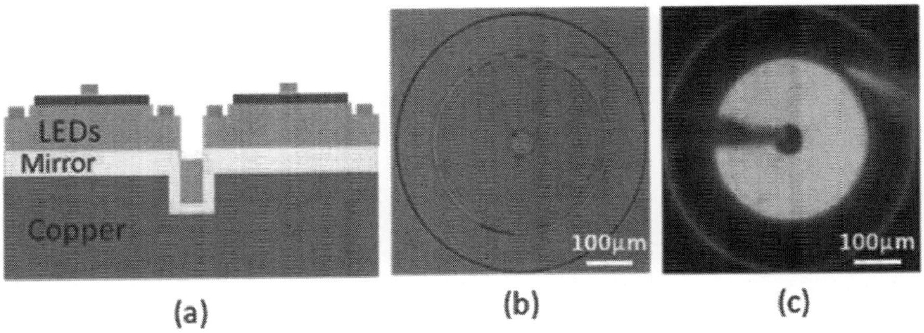

Fig. 5.22 (a) Cross-section diagram of LEDs on Cu with a mirror layer. (b) Microscope images of the circular LED mesa on Cu. (c) Emission image of the LED at 5 mA (Lau et al. 2011).

As a comparison, the forward voltages of LEDs on Si and Cu substrates are 3.7 and 4.0 V, respectively, Fig. 5.23 (Lau et al. 2011). The higher forward voltage for LEDs on Cu is ascribed to the deterioration of the metal contacts during the polyimide solidification process at 210°C. Nevertheless, the reverse leakage current of LEDs increases. The similar phenomenon is also observed in LED chips transferred from sapphire to Cu. According to Sun et al. (2008, 2010), this increase in the reverse leakage current may be related to the defects which are induced by the fabrication process of chips. The light output power of LEDs on Si and Cu is 1.4 and 2.5 mW, respectively, as shown in Fig. 5.24 (Lau et al. 2011). This increase in light output power is attributed to the higher reflectivity of Cu metal layer than that of Si. Meanwhile, the higher thermal conductivity of Cu (395 W/(m·K)), in comparison with that of Si (148 W/(m·K)), can more effectively alleviate the heat accumulation problem in LEDs. Therefore, better performance at high injection currents and longer life-time can be expected for the LEDs on Cu substrates.

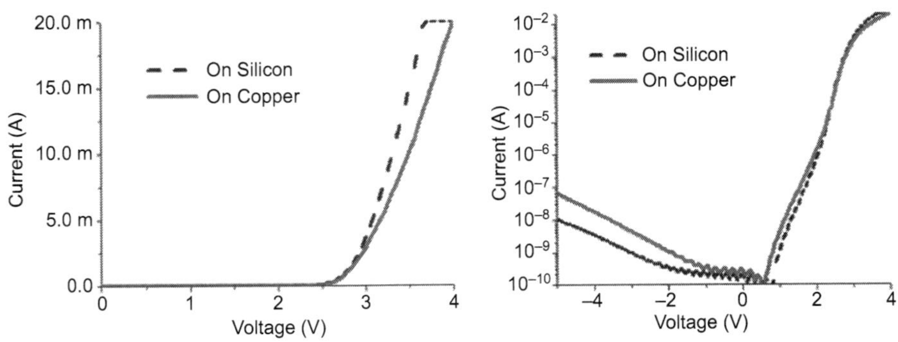

Fig. 5.23 The dependence of current on voltage, respectively, for LEDs on Si and on Cu (Lau et al. 2011).

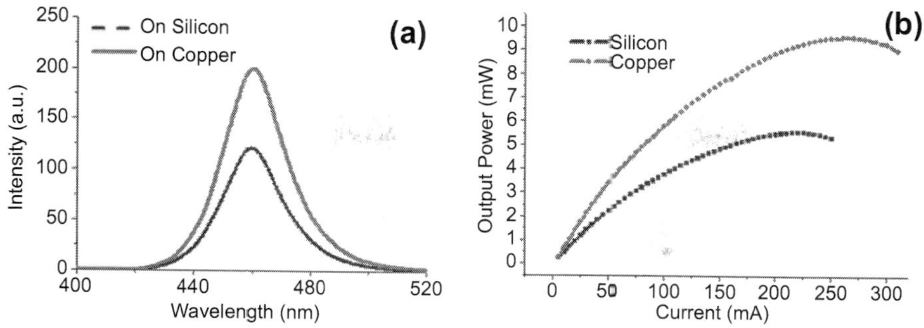

Fig. 5.24 (a) EL spectra of LEDs on Si and Cu substrates at 20 mA. (b) Light output power as a function of current (L-I) of LEDs on Si and Cu substrates (Lau et al. 2011).

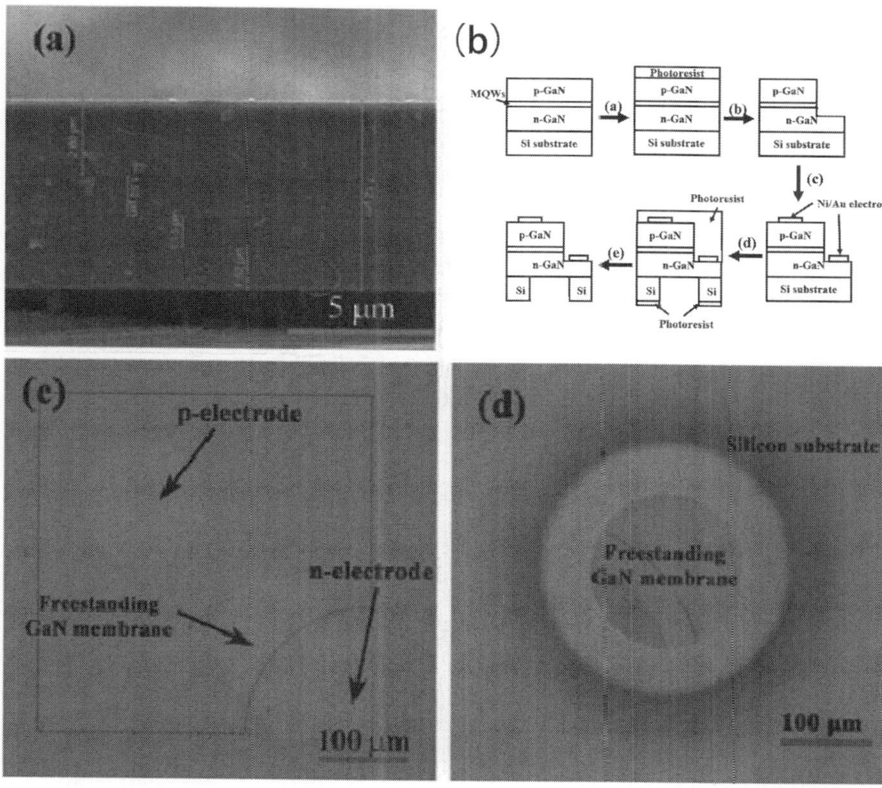

Fig. 5.25 (a) Cross-sectional SEM image of LED structure on Si substrate. (b) Processing steps for fabrication of freestanding membrane LED chips. (c) Optical Microscope image of a freestanding membrane LED chip. (d) Optical Microscope image of the backside of freestanding membrane LED chip (Shi et al. 2014).

Shi et al. (2014) fabricated a kind of lateral LED chips with a free standing membrane structure by removing a segment of Si substrates. The epitaxial structure of LEDs on Si substrates is shown in Fig. 5.25a. The thicknesses of the buffer layers, n-GaN layer, InGaN/GaN MQWs and p-GaN are 4.29 μm, 2.8 μm, 125 nm and 120 nm, respectively. Fig. 5.25b depicts the fabrication process of the freestanding membrane LEDs. Step a-c: An standard procedure for lateral LED chips is performed on the LED wafers on Si substrates. Step d: an AZ5214 photoresist is first pin-coated on the surface to protect the LEDs. Afterwards, the Si substrate is patterned by backside alignment photolithography and then is partly removed by deep reactive ion etching which is conducted by alternating steps of SF_6 etching and C_3F_8/O_2 passivation. Herein, the buffer layer could stop the etching to protect the structure of LEDs. Step e: Finally, the photoresist is removed. In this case, the size of the LED is designed to be 400×400 μm^2 and the layout of p-electrode and n-electrode are circular with radiuses of 100 and 150 μm, respectively. The etching areal is also designed to be circular and the diameter of the etching circular hole for the freestanding membrane is 200 μm. Figs. 5.25c and d reveal the microscope images of the LEDs from the top and backside view, respectively.

In comparison to the LED chips fabricated on Si substrates, approximately six times higher light output power can be obtained for this freestanding membrane LED by removing Si substrate. However, due to the standing membrane segment is suspended, the size of LED chips, especially the membrane size, is difficult to be enlarged as the freestanding membrane may not be sufficiently strong to sustain.

3.3 Fabrication of Vertical LED Devices and The Corresponding Properties

Great improvement can be achieved in lateral structure LED chips on Si substrates by removing the Si substrates. The inherent imperfections of lateral structure LED chips such as the weak current expanding and the reduction in light-emitting area inhibit the further improvement in performance of LEDs on Si substrates (Horng et al. 2014).

It is well known that the vertical-structure LED chips have incomparable advantages compared to the lateral structure LED chips. Fortunately, the easy removal of Si substrate enables the fabrication of vertical structure LED chips.

Jiang et al. (2007) reported the fabrication of vertical-structure GaN-based LED chips by wafer bonding and chemical etching to transfer the GaN-based LEDs from Si substrate to a new Si substrate. Firstly, the p-type contact is fabricated and then a thick Au layer was deposited as the bonding metal layer. Subsequently, a Au-coated new Si is bonded with the wafer. Afterwards, the Si substrate is removed by the chemical etching process. Finally, the buffer layer was removed by Inductively Coupled Plasma (ICP) etching. In addition, a lateral LEDs are fabricated with the same structure when compared with the vertical-structure LEDs.

The measured light output powers at 20 mA are 2.8 and 0.76 mW, respectively, Fig. 5.26. The improvement of the light output power can be attributed to the reflection of light downward from the active region. The forward voltages are 3.2 and 4.0 V for the vertical- and lateral-structure LEDs at 20 mA, respectively, Fig. 5.37b. Meanwhile, the I–V characteristics reveal the series resistance of the vertical-structure LEDs is lower than that of the lateral-structure LEDs. This lower series resistance in the vertical-structure LEDs can be attributed to its lower perpendicular resistance compared to that of the lateral-structure LEDs. Besides, the tensile stress in the vertical-structure LEDs is smaller than that in lateral-structure LEDs. The smaller tensile stress leads to larger energy gap and smaller piezoelectricity-induced quantum-confined Stark effect for vertical-structure LEDs in comparison with those of the lateral LEDs, which is also a significant reason for the performance improvement of the vertical-structure LEDs on Si.

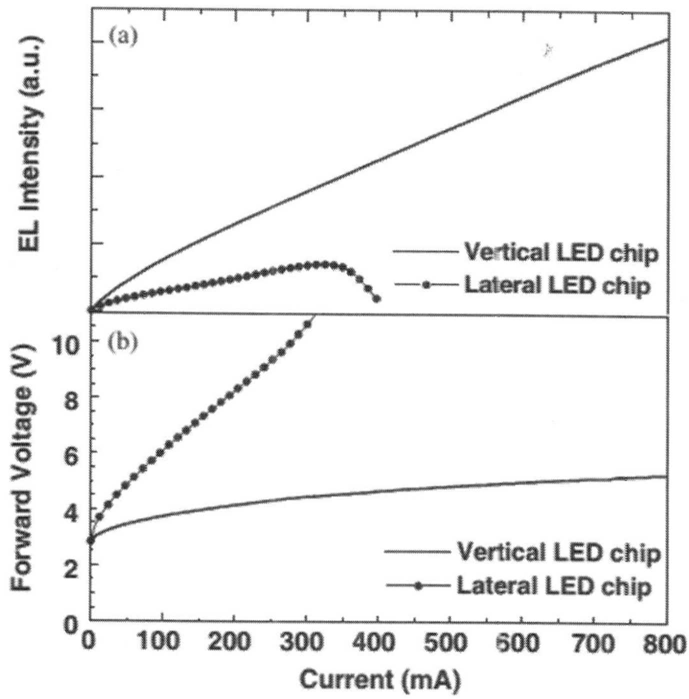

Fig. 5.26 The dependence of (a) EL intensity. (b) Forward voltage on injection current, respectively, for vertical- and lateral-structure LED chips (Jiang et al. 2007).

Except for the conventional vertical structure LED devices mentioned before, Zhang et al. (2013) reported another more advantageous LED device structure of thin-film flip chip fabricated from the LED wafer on Si. Compared to the conventional vertical structure LED chips, the thin-film flip chip could offer higher light extraction efficiency by reducing the areal of n-electrode pattern. The fabrication process is depicted in Fig. 5.27 (Zhu et al. 2013). (1) Growth of LED epitaxial films on Si substrates. (2) The wafer is etched to form a hole array, which is nonstop to n-GaN layer, by photoetching and ICP etching, then a layer of Ag alloy is deposited on the surface of p-GaN as the p-type ohmic contact by electron beam evaporation. On the top of p-type contact, a passivation layer is deposited by the PECVD, then the part of passivation grown on the n-GaN layer is removed by photo-etching and ICP etching. Afterwards, a layer of n-GaN Ohmic contact metal for the current dispersion and bonding layer is deposited in the hole array and on the surface of passivation layer by electron beam evaporation. (3) The wafer is bonded to another Si with a back metal which acts as the n-contract. (4) The original Si substrate is removed by wet etching. (5) The p-GaN is exposed by photo-etching and ICP etching, then the p-type contract is deposited by electron beam evaporation. Afterwards, the surface of n-GaN is roughened by submerging the sample in the potassium hydroxide solution. The cross-sectional schematic diagram and plane view microscope image of thin-film flip chip are shown in Figs. 5.28a and b (Zhang et al. 2013), respectively.

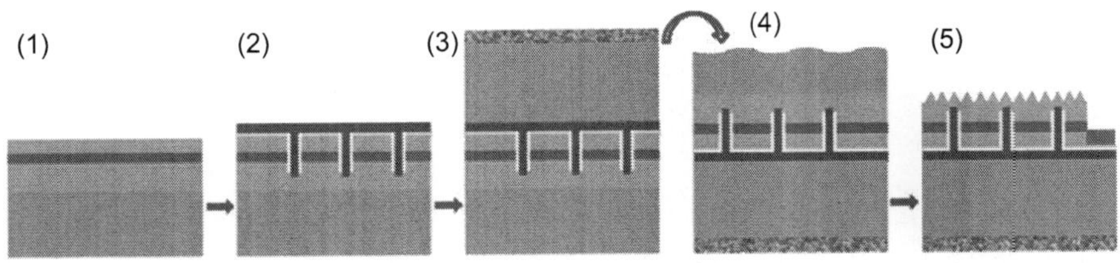

Fig. 5.27 Process steps for fabrication of thin-film flip LED chips from a GaN based LEDs on Si substrates: (1) GaN-based LEDs wafer on Si substrates. (2) Deposition of highly reflective mirror layer and the fabrication of the current spreading pillar for n-contact. (3) Wafer bonding by another Si substrates with a back metal for n-contract. (4) Removal of Si substrates by wet etching. (5) ICP etching and deposition of p-contact metal (P-pad) and surface roughening (Zhu et al. 2013).

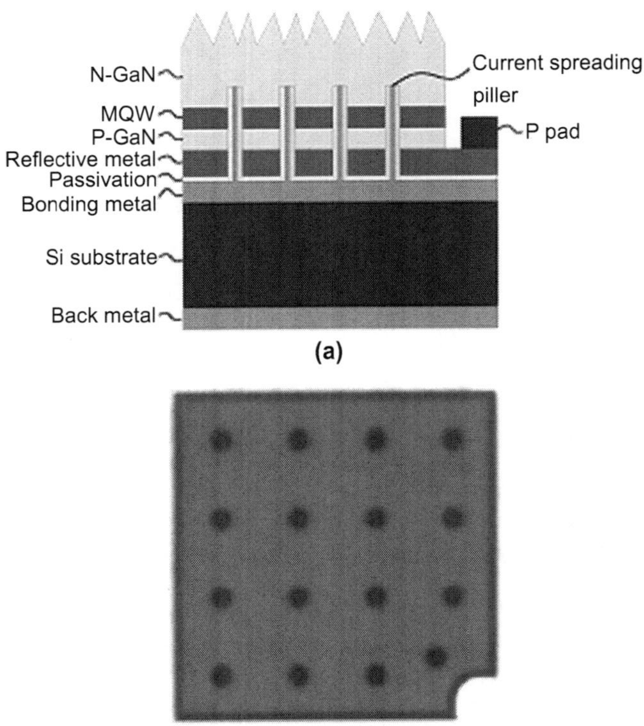

Fig. 5.28 (a) The cross-sectional schematic diagram. (b) Plane view microscope image of thin-film flip LED chips (Zhang et al. 2013).

In Zhang's report (Zhang et al. 2013), the chip is fabricated in a size of 1.1 × 1.1 mm^2, then is encapsulated with silicone. Besides, a white LED lamp packaged with standard YAG phosphor is also prepared. The performance of the blue chip encapsulated by silicone is measured by integral sphere tester, which reveals light output power of 546 mW and external quantum efficiency of 50.3% at injection current of 350 mA. When the injection current increases to 700 mA, the light output power and external quantum efficiency reach 965 mW and 44.3%, respectively. As for the white LED lamp, at injection current of 350 mA, the photometric light output reach 120.1 lm with the corresponding color temperature of 5700 K. These values indicate that the performance of LEDs on Si substrates are comparable with that of LEDs on sapphire.

4. SUMMARY AND PROSPECTS

So far, significant achievement of GaN-based LEDs on Si substrates has been obtained. However, there is still a performance gap between LEDs on sapphire and on Si. Generally, the growth of LEDs on Si is facing a great challenge in controlling the stress and obtaining high crystalline quality of GaN films to meet the requirement of devices. On the one hand, it is hard to achieve uniform LEDs properties including emitting wavelength and voltage of LED devices due to the complicated control in stress. On the other hand, the poorer crystalline quality of GaN films severely reduces the IQE of LEDs. To address these problems, except for optimizing the buffer structure and growth process, the patterned substrate technology also needs to be further studied. In addition, lights can be absorbed by Si substrate, which reduces the LEE of LEDs dramatically. Therefore, Si substrate should be removed and its corresponding technologies also need to be well studied. Additionally, the higher performance of LEDs achieved on SiC or sapphire proposes higher benchmark for GaN-based LEDs on Si substrates.

Even so, performance of LEDs on Si has obtained a great development and many more technologies involved in conquering these challenges are being developed. We do believe that, after hard work, the issues for preparation of LEDs on Si substrates will be achieved and LEDs on Si substrates may replace the traditional LEDs in the future due to their special advantages.

5. ACKNOWLEDGEMENTS

This work is supported by National Science Fund for Excellent Young Scholars of China (No. 51422203), National Natural Science Foundation of China (Nos. 51572091 and 51372001), Distinguished Young Scientist Foundation of Guangdong Scientific Committee (No. S2013050013882), Key Project in Science and Technology of Guangdong Province (Nos. 2014B010119001 and 2011A080801018) and Strategic Special Funds for LEDs of Guangdong Province (Nos. 2011A081301010, 2011A081301012 and 2012A080302002).

6. REFERENCES

Abe, Y., H. Fujimori, A. Watanabe, N. Ohmori, J. Komiyama, S. Suzuki, et al. 2012. Defect propagation from 3C-SiC intermediate layers to III-Nitride epilayers. Jpn. J. Appl. Phys. 51: 35603.

Ando, T., A. B. Fowler and F. Stern. 1982. Electronic properties of two-dimensional systems. Rev. Mod. Phys. 54: 437-672.

Arkun, F. E., R. Dargis, R. Smith, D. Williams, A. Clark and M. Lebby. 2012. Integrated high reflectivity silicon substrates for GaN LEDs. Phys. Status Solidi C 9: 814-817.

Arslan, E., Ö. Duygulu, A. A. Kaya, A. Teke, S. Özcelik and E. Ozbay. 2009. The electrical, optical and structural properties of GaN epitaxial layers grown on Si (111) substrate with SiN_x interlayers. Superlattice. Microst. 46: 846-857.

Bak, S. J., D. H. Mun, K. C. Jung, J. H. Park, H. J. Bae, I. W. Lee, et al. 2013. Effect of Al pre-deposition on AlN buffer layer and GaN film grown on Si (111) substrate by MOCVD. Electron. Mater. Lett. 9: 367-370.

Blake, Adam H., Derek Caselli, Christopher Durot, Jason Mueller, Eduardo Parra, Joseph Gilgen, et al. 2012. InGaN/GaN multiple-quantum-well light-emitting diodes grown on Si (111) substrates with ZrB_2 (0001) buffer layers. J. Appl. Phys. 111: 33107.

Butte, R., E. Feltin, J. Dorsaz, G. Christmann, J. F. Carlin, N. Grandjean, et al. 2005. Recent progress in the growth of highly reflective nitride-based distributed Bragg reflectors and their use in microcavities. Japn. J. Appl. Phys. 44: 7207-7216.

Cao, J., S. Li, G. Fan, Y. Zhang, S. Zheng, Y. Yin, et al. 2010. The influence of the Al pre-deposition on the properties of AlN buffer layer and GaN layer grown on Si (111) substrate. J. Cryst. Growth 312: 2044-2048.

Chen, P., R. Zhang, Z. M. Zhao, D. J. Xi, B. Shen, Z. Z. Chen, et al. 2001. Growth of high quality GaN layers with AlN buffer on Si (111) substrates. J. Cryst. Growth 225: 150-154.

Cheng, K., M. Leys, S. Degroote, B. Van Daele, S. Boeykens, J. Derluyn, et al. 2006. Flat GaN epitaxial layers grown on Si (111) by metalorganic vapor phase epitaxy using step-graded AlGaN intermediate layers. J. Electron. Mater. 35: 592-598.

Cheng, K., M. Leys, S. Degroote, M. Germain and G. Borghs. 2008. High quality GaN grown on silicon (111) using a Si_xN_y interlayer by metal-organic vapor phase epitaxy. Appl. Phys. Lett. 92: 192111.

Cherns, D., S. J. Henley and F. A. Ponce. 2001. Edge and screw dislocations as non-radiative centers in InGaN/GaN quantum well luminescence. Appl. Phys. Lett. 78: 2691-2693.

Clames, J. Y., Ch. Giesen, T. Meyer and M. Heuken. 2006. MOCVD of InGaN-based light emitting structures on silicon substrates with strain optimized buffer layers using long Al pre-deposition. Phys. Status Solidi C 3: 2191-2194.

Cong, G. W., Y. Lu, W. Q. Peng, X. L. Liu, X. H. Wang and Z. G. Wang. 2005. Design of the low-temperature AlN interlayer for GaN grown on Si (111) substrate. J. Cryst. Growth 276: 381-388.

Contreras, O., F. A. Ponce, J. Christen, A. Dadgar and A. Krost. 2002. Dislocation annihilation by silicon delta-doping in GaN epitaxy on Si. Appl. Phys. Lett. 81: 4712-4714.

Dadgar, A., A. Alam, T. Riemann, J. Blasing, A. Diez, M. Poschenrieder, et al. 2001. Crack-free InGaN/GaN light emitters on Si (111). Phys. Status Solidi A 188: 155-158.

Dadgar, A., M. Poschenrieder, O. Contreras, J. Christen, K. Fehse, J. Blasing, et al. 2002. Bright, crack-free InGaN/GaN light emitters on Si (111). Phys. Status Solidi A 192: 308-313.

Dadgar, A., A. Strittmatter, J. Blasing, M. Poschenrieder, O. Contreras, P. Veit, et al. 2003. Metalorganic chemical vapor phase epitaxy of gallium-nitride on silicon. Phys. Status Solidi C 0: 1583-1606.

Dadgar, A., C. Hums, A. Diez, J. Blaesing and A. Krost. 2006. Growth of blue GaN LED structures on 150-mm Si (111). J. Cryst. Growth 297: 279-282.

Dadgar, A., T. Hempel, J. Blaesing, O. Schulz, S. Fritze, J. Christen, et al. 2011a. Improving GaN-on-silicon properties for GaN device epitaxy. Phys. Status Solidi C 8: 1503-1508.

Dadgar, A., J. Blaesing, A. Diez and A. Krost. 2011b. Crack-free, highly conducting GaN layers on Si substrates by Ge doping. Appl. Phys. Express 4: 01101.

Datta, R., M. J. Kappers, M. E. Vickers, J. S. Barnard and C. J. Humphreys. 2004. Growth and characterisation of GaN with reduced dislocation density. Superlattice. Microst. 36: 393-401.

Drechsel, P. and H. Riechert. 2011. Strain controlled growth of crack-free GaN with low defect density on silicon (111) substrate. J. Cryst. Growth 315: 211-215.

Fang, H., M. Katagiri, H. Miyake, K. Hiramatsu, H. Oku, H. Asamura, et al. 2014. Crack-free GaN grown by using mask-less epitaxial lateral overgrowth on Si substrate with thin SiC intermediate layer. Phys. Status Solidi A 211: 744-747.

Feng, Y., H. Wei, S. Yang, H. Zhang, S. Kong, G. Zhao, et al. 2014. Significant quality improvement of GaN on Si (111) upon formation of an AlN defective layer. CrystEngComm 16: 7525.

Fenwick, W. E., N. Li, T. Xu, A. Melton, S. Wang, H. Yu, et al. 2009a. MOCVD growth of GaN on Si (111) substrates using an ALD-grown Al_2O_3 interlayer. J. Cryst. Growth 311: 4306-4310.

Fenwick, W. E., N. Li, M. Jamil, T. Xu, A. Melton, S. Wang, et al. 2009b. Development of new substrate technologies for GaN LEDs: atomic layer deposition transition layers on silicon and ZnO. Proc. SPIE 7231: 723119.

Fleurence, A., W. Zhang, C. Hubault and Y. Yamada-Takamura. 2013. Mechanisms of parasitic crystallites formation in ZrB_2 (0001) buffer layer grown on Si (111). Appl. Surf. Sci. 284: 432-437.

Fritze, S., A. Dadgar, H. Witte, M. Bugler, A. Rohrbeck, J. Blasing, et al. 2012. High Si and Ge n-type doping of GaN doping-limits and impact on stress. Appl. Phys. Lett. 100: 122104.

Fu, Y. K., D. A. Gulino and R. Higgins. 2000. Residual stress in GaN epilayers grown on silicon substrates. J. Vac. Sci. Technol. 18: 965-967.

Guha, S. and N. A. Bojarczuk. 1998. Ultraviolet and violet GaN light emitting diodes on silicon. Appl. Phys. Lett. 72: 415-417.

Haeberlen, M., D. Zhu, C. McAleese, M. J. Kappers and C. J. Humphreys. 2010. Dislocation reduction in MOVPE grown GaN layers on (111) Si using SiN_x and AlGaN layers. J. Phys.: Conf. Ser. 209: 012017.

Hikosaka, T., H. Yoshida, N. Sugiyama and S. Nunoue. 2014. Reduction of threading dislocation by recoating GaN island surface with SiN for high-efficiency GaN-on-Si-based LED. Phys. Status Solid C 11: 617-620.

Honda, Y., Y. Kuroiwa, M. Yamaguchi and N. Sawaki. 2002. Growth of a GaN crystal free from cracks on a (111) Si substrate by selective MOVPE. J. Cryst. Growth 242: 77-81.

Horng, R. H., B. R. Wu, C. H. Tien, S. L. Ou, M. H. Yang, H. C. Kuo, et al. 2014. Performance of GaN-based light-emitting diodes fabricated using GaN epilayers grown on silicon substrates. Opt. Express 22: A179-A187.

Huang, J. Y., Z. Z. Ye, L. Wang, J. Yuan, B. H. Zhao and H. M. Lu. 2002. Comparison of GaN epitaxial films on silicon nitride buffer and Si (111). Solid State Electron. 46: 1231-1234.

Ishikawa, H., K. Yamamoto, T. Egawa, T. Soga, T. Jimbo and M. Umeno. 1998. Thermal stability of GaN on (111) Si substrate. J. Cryst. Growth 189: 178-182.

Ishikawa, H., T. Jimbo and T. Egawa. 2008. GaInN light emitting diodes with AlInN/GaN distributed Bragg reflector on Si. Phys. Status Solidi C 5: 2086-2088.

Jiang, F., C. Xiong, W. Fang, L. Wang, C. Mo and H. Liu. 2007. The characteristics of GaN-based blue LED on Si substrate. J. Lumin. 122-123: 185-187.

Kappers, M. J., R. Datta, R. A. Oliver, F. Rayment, M. E. Vickers and C. J. Humphreys. 2007. Threading dislocation reduction in (0001) GaN thin films using SiN_x interlayers. J. Cryst. Growth 300: 70-74.

Kim, D. K. 2007. Effect of AlN buffer thickness on stress relaxation in GaN layer on Si (111). Solid State Electron. 51: 1005-1008.

Kim, M. H., Y. G. Do, H. C. Kang, D. Y. Noh and S. J. Park. 2001. Effects of step-graded $Al_xGa_{1-x}N$ interlayer on properties of GaN grown on Si (111) using ultrahigh vacuum chemical vapor deposition. Appl. Phys. Lett. 79: 2713-2715.

Komiyama, J., Y. Abe, S. Suzuki, H. Nakanishi and A. Koukitu. 2009. MOVPE of AlN-free hexagonal GaN/cubic SiC/Si hetero-structures for vertical devices. J. Cryst. Growth 311: 2840-2843.

Krost, A. and A. Dadgar. 2002. GaN-based optoelectronics on silicon substrates. Mat. Sci. Eng. B-Solid 93: 77-84.

Krost, A., A. Dadgar, G. Strassburger and R. Clos. 2003. GaN-based epitaxy on silicon: stress measurements. Phys. Status Solidi A 200: 26-35.

Kukushkin, S. A., A. V. Osipov, V. N. Bessolov, B. K. Medvedev, V. K. Nevolin and K. A. Tcarik. 2008. Substrates for epitaxy of gallium nitride: new materials and techniques. Rev. Adv. Mater. Sci. 17: 1-32.

Kung, P., A. Saxler, X. Zhang, D. Walker, T. C. Wang, I. Ferguson, et al. 1995. High quality AlN and GaN epilayers grown on (001) sapphire (100) and (111) silicon substrates. Appl. Phys. Lett. 66: 2958-2960.

Lahreche, H., P. Vennegues, O. Tottereau, M. Laugt, P. Lorenzini, M. Leroux, et al. 2000. Optimisation of AlN and GaN growth by metalorganic vapour-phase epitaxy (MOVPE) on Si (111). J. Cryst. Growth 217: 13-25.

Lau, K. M., B. Zhang, H. Liang, Y. Wang, Z. Feng and K. N. Wei. 2007. High-performance III-nitride blue LEDs grown and fabricated on patterned Si substrates. J. Cryst. Growth 298: 725-730.

Lau, K. M., K. M. Wong, X. Zou and P. Chen. 2011. Performance improvement of GaN-based light-emitting diodes grown on patterned Si substrate transferred to copper. Opt. Express 19: A956-A961.

Laubsch, A., M. Sabathil, J. Baur, M. Peter and B. Hahn. 2010. High-power and high-efficiency InGaN-based light emitters. IEEE T. Electron Dev. 57: 79-87.

Lee, S. J., G. H. Bak, S. R. Jeon, S. H. Lee, S. M. Kim, S. H. Jung, et al. 2008. Epitaxial growth of crack-free GaN on patterned Si(111) substrate. Jpn. J. Appl. Phys. 47: 3070-3073.

Lei, T. and T. D. Moustakas. 1992. A comparative study of GaN epitaxy on Si(001) and Si(111) substrates. Wide Band Gap Semiconductors Symposium. USA pp. 433-439.

Lin, Y., S. Zhou, W. Wang, W. Yang, H. Qian, H. Wang, et al. 2015. Performance improvement of GaN-based light-emitting diodes grown on Si (111) substrates by controlling the reactor pressure for the GaN nucleation layer growth. J. Mater. Chem. C 3: 1484-1490.

Liu, J., J. Zhang, Q. Mao, X. Wu and F. Jiang. 2013. Effects of AlN interlayer on growth of GaN-based LED on patterned silicon substrate. CrystEngComm 15: 3372.

Liu, L. and J. H. Edgar. 2002. Substrates for gallium nitride epitaxy. Mat. Sci. Eng. R 37: 61-127.

Liu, R., F. A. Ponce, A. Dadgar and A. Krost. 2003. Atomic arrangement at the AlN/Si(111) interface. Appl. Phys. Lett. 83: 860-862.

Lu, Y., X. Liu, X. Wang, D. C. Lu, D. Li, X. Han, et al. 2004. Influence of the growth temperature of the high-temperature AlN buffer on the properties of GaN grown on Si (111) substrate. J. Cryst. Growth 263: 4-11.

Lumbantoruan, F., Y. Y. Wong, Y. H. Wu, W. C. Huang, N. M. Shrestra, T. T. Luong, et al. 2014. Investigation of TMAl preflow to the properties of AlN and GaN film grown on Si(111) by MOCVD. IEEE-ICSE2014 Proc., Malaysia: 20-23.

Luo, W., X. Wang, L. Guo, H. Xiao, C. Wang, J. Ran, et al. 2008. Influence of AlN buffer layer thickness on the properties of GaN epilayer on Si(111) by MOCVD. Microelectron. J. 39: 1710-1713.

Marchand, H., L. Zhao, N. Zhang, B. Moran, R. Coffie, U. K. Mishra, et al. 2001. Metalorganic chemical vapor deposition of GaN on Si (111): Stress control and application to field-effect transistors. J. Appl. Phys. 89: 7846-7851.

Markurt, T., L. Lymperakis, J. Neugebauer, P. Drechsel, P. Stauss, T. Schulz, et al. 2013. Blocking growth by an electrically active subsurface layer: the effect of Si as an antisurfactant in the growth of GaN. Phys. Rev. Lett. 110: 036103.

Monemar, B. and B. E. Sernelius. 2007. Defect related issues in the "current roll-off" in InGaN based light emitting diodes. Appl. Phys. Lett. 91: 181103.

Moram, M. A., T. B. Joyce, P. R. Chalker, Z. H. Barber and C. J. Humphreys. 2006. Microstructure of epitaxial scandium nitride films grown on silicon. Appl. Surf. Sci. 252: 8385-8387.

Ng, T. B., D. A. Ewoldt, D. A. Shepherd and M. J. Loboda. 2015. Reflectance analysis on the MOCVD growth of AlN on Si (111) by the virtual interface model. Phys. Status Solidi C 12: 385-388.

Ni, X. F., L. P. Zhu, Z. Z. Ye, Z. Zhao, H. P. Tang, W. Hong, et al. 2005. Growth and characterization of GaN films on Si (111) substrate using high-temperature AlN buffer layer. Surf. Coat. Tech. 198: 350-353.

Nishimura, S., H. Hanamoto, K. Terashima and S. Matsumoto. 2002a. Growth of GaN on Si (100) substrates using BP as a buffer layer-selective epitaxial growth. Mat. Sci. Eng. B-Solid 93: 135-138.

Nishimura, S., S. Matsumoto and K. Terashima. 2002b. Growth of GaN on Si substrates-roles of BP thin layer. Opt. Mater. 19: 223-228.

Nishimura, S., S. Matsumoto and K. Terashima. 2004. GaN on Si substrates for LED and LD applications. Phys. Status Solidi C 1: 238-241.

Norenberg, C., M. A. Moram and P. J. Dobson. 2006. Surface structures of scandium silicides grown on Si (111) studied by STM, AFM and electron diffraction. Surf. Sci. 600: 4126-4131.

Pan, X., M. Wei, C. B. Yang, H. L. Xiao, C. M. Wang and X. L. Wang. 2011. Growth of GaN film on Si (111) substrate using AlN sandwich structure as buffer. J. Cryst. Growth 318: 464-467.

Radtke, G., M. Couillard, G. A. Botton, D. Zhu and C. J. Humphreys. 2010. Scanning transmission electron microscopy investigation of the Si (111)/AlN interface grown by metalorganic vapor phase epitaxy. Appl. Phys. Lett. 97: 251901.

Radtke, G., M. Couillard, G. A. Botton, D. Zhu and C. J. Humphreys. 2012. Structure and chemistry of the Si (111)/AlN interface. Appl. Phys. Lett. 100: 011910.

Raghavan, S. and J. M. Redwing. 2004. In situ stress measurements during the MOCVD growth of AlN buffer layers on (111) Si substrates. J. Cryst. Growth 261: 294-300.

Riemann, T., T. Hempel, J. Christen, P. Veit, R. Clos, A. Dadgar, et al. 2006. Optical and structural microanalysis of GaN grown on SiN submonolayers. J. Appl. Phys. 99: 123518.

Roucka, R., Y. J. Ana, A. V. G. Chizmeshya, V. R. D'Costa, J. Tolle, J. Menendez, et al. 2008. Structural and optical properties of ZrB_2 and $Hf_xZr_{1-x}B_2$ films grown by vicinal surface epitaxy on Si (111) substrates. Solid State Electron. 52: 1687-1690.
Shi, Z., X. Li, G. Zhu, Z. Wang, P. Grunberg, H. Zhu, et al. 2014. Characteristics of GaN-based LED fabricated on a GaN-on-silicon platform. Appl. Phys. Express 7: 82102-82104.
Shur, M. S. and A. Zukauskas. 2005. Solid-state lighting: toward superior illumination. P. IEEE 93: 1691-1703.
Smeeton, T. M., M. J. Kappers, J. S. Barnard, M. E. Vickers and C. J. Humphreys. 2003. Electron-beam-induced strain within InGaN quantum wells: false indium "cluster" detection in the transmission electron microscope. Appl. Phys. Lett. 83: 5419-5421.
Sun, Y. J., T. J. Yu, Z. Z. Chen, X. N. Kang, S. L. Qi, M. G. Li, et al. 2008. Properties of GaN-based light-emitting diode thin film chips fabricated by laser lift-off and transferred to Cu. Semicond. Sci. Tech. 23: 125022.
Sun, Y. J., T. J. Yu, C. Y. Jia, Z. Z. Chen, P. F. Tian, X. N. Kang, et al. 2010. GaN-Based thin film vertical structure light emitting diodes fabricated by a modified laser lift-off process and transferred to Cu. Chinese Phys. Lett. 27: 127303.
Swanepoel, R. 1983. Determination of the thickness and optical constants of amorphous silicon. Journal of Physics E (Scientific Instruments) 16: 1214-1222.
Tanaka, S., M. Takeuchi and Y. Aoyagi. 2000. Anti-surfactant in III-nitride epitaxy - Quantum dot formation and dislocation termination. Jpn. J. Appl. Phys. 39: L831-L834.
Tsao, J. Y., M. E. Coltrin, M. H. Crawford and J. A. Simmons. 2010. Solid-state lighting: an integrated human factors, technology and economic perspective. P. IEEE 98: 1162-1179.
Ulrici, W. 2004. Hydrogen-impurity complexes in III-V semiconductors. Rep. Prog. Phys. 67: 2233.
Wang, T. Y., S. L. Ou, R. H. Horng and D. S. Wuu. 2014a. Improved GaN-on-Si epitaxial quality by incorporating various Si_xN_y interlayer structures. J. Cryst. Growth 399: 27-32.
Wang, T. Y., S. L. Ou, R. H. Horng and D. S. Wuu. 2014b. Growth evolution of Si_xN_y on the GaN underlayer and its effects on GaN-on-Si (111) heteroepitaxial quality. CrystEngComm 16: 5724-5731.
Wang, W., W. Yang, H. Wang and G. Li. 2014c. Epitaxial growth of GaN films on unconventional oxide substrates. J. Mater. Chem. C 2: 9342-9358.
Wang, W., W. Yang, Z. Liu, H. Wang, L. Wen and G. Li. 2015. Interfacial reaction control and its mechanism of AlN epitaxial films grown on Si (111) substrates by pulsed laser deposition. Sci. Rep. 5: 11480.
Xiang, R. F., Y. Y. Fang, J. N. Dai, L. Zhang, C. Y. Su, Z. H. Wu, et al. 2011. High quality GaN epilayers grown on Si (111) with thin nonlinearly composition-graded $Al_xGa_{1-x}N$ interlayers via metal-organic chemical vapor deposition. J. Alloy. Compd. 509: 2227-2231.
Zamir, S., B. Meyler, J. Salzman, F. Wu and Y. Golan. 2002. Enhanced photoluminescence from GaN grown by lateral confined epitaxy. J. Appl. Phys. 91: 1191-1197.
Zang, K. Y., Y. D. Wang, L. S. Wang, S. Y. Chow and S. J. Chua. 2007. Defect reduction by periodic SiN_x interlayers in gallium nitride grown on Si (111). J. Appl. Phys. 101: 093502.
Zhang, B., T. Egawa, H. Ishikawa, Y. Liu and T. Jimbo. 2005a. Thin-film InGaN multiple-quantum-well light-emitting diodes transferred from Si (111) substrate onto copper carrier by selective lift-off. Appl. Phys. Lett. 86: 71111-71113.
Zhang, J. C., D. S. Jiang, Q. Sun, J. F. Wang, Y. T. Wang, J. P. Liu, et al. 2005b. Influence of dislocations on photoluminescence of InGaN/GaN multiple quantum wells. Appl. Phys. Lett. 87: 71908.
Zhang, S., B. Feng, Q. Sun and H. Zhao. 2013. Preparation of GaN-on-Si based thin-film flip-chip LED. J. Semicond. 34: 053006.

Zhou, W., M. Tao, L. Chen and H. Yang. 2007. Microstructured surface design for omnidirectional antireflection coatings on solar cells. J. Appl. Phys. 102: 103101-103105.

Zhu, D., C. McAleese, M. Haberlen, C. Salcianu, T. Thrush, M. Kappers, et al. 2010. InGaN/GaN LEDs grown on Si (111): dependence of device performance on threading dislocation density and emission wavelength. Phys. Status Solidi C 7: 2168-2170.

Zhu, D, C. McAleese, M. Haeberlen, C. Salcianu, T. Thrush, M. Kappers, et al. 2011. Efficiency measurement of GaN-based quantum well and light-emitting diode structures grown on silicon substrates. J. Appl. Phys. 109: 14502.

Zhu, D. and C. J. Humphreys. 2013. Low-cost high-efficiency GaN LED on large-area Si substrate. CS MANTECH Conference New Orleans. Louisiana. USA 269.

Zhu, D., D. J. Wallis and C. J. Humphreys. 2013. Prospects of III-nitride optoelectronics grown on Si. Rep. Prog. Phys. 76: 106501.

6

Understanding the Role of CVD Nanodiamond Thin Films in Solar Energy Conversion

M. A. Fraga,[1,*] L. A. A. Rodríguez,[2,a]
R. S. Pessoa[3,b] and V. J. Trava-Airoldi[1,c]

ABSTRACT

The use of emerging semiconductor materials is fundamental to achieving competitive solar energy conversion systems with low cost and high energy conversion efficiency. A variety of nanostructured thin films has been investigated for application in devices which convert sunlight directly into electricity, particularly those based on photovoltaic or thermionic effect. The challenge of using materials in thin film form is that the bulk properties can change when the material is constrained in size. Understanding the thin films properties and how these properties are affected by synthesis conditions is critical to the successful implementation of solar energy conversion devices. It is imperative that the development of highly efficient conversion devices depends on the proper material selection. Diamond thin films exhibit potentially beneficial characteristics to be employed in photovoltaic and thermionic devices. They possess a range of superior electronic, optical, mechanical and chemical properties to outclass competing wide-bandgap materials. The combination of these properties offers engineering solutions that can shift performance to new levels or enable completely new approaches to challenging problems in different applications. This chapter presents an overview on the synthesis and properties of CVD diamond films and on their applications in solar energy conversion. The basic issues and application challenges are discussed.

[1] Associate Laboratory of Sensors and Materials, National Institute for Space, Research, São José dos Campos-SP, 12227-010, Brazil.
[2] Federal University of São Paulo-ICT, São José dos Campos-SP, 12231-280, Brazil.
[3] Nanotechnology and Plasmas Processes Laboratory, Universidade do Vale do, Paraíba, São José dos Campos, SP, 12244-000, Brazil.
[a] E-mail: laardilar88@gmail.com
[b] E-mail: rspessoa@univap.br
[c] E-mail: vladimir.airoldi@inpe.br
* Corresponding author: mafraga@ita.br

Keywords: Diamond, thin films, solar energy, photovoltaic devices, thermionic devices.

1. INTRODUCTION

Regardless of the type of solar energy conversion technology, there are a number of challenges related to improving its efficiency, whilst reducing costs, to generate electricity at competitive levels with electricity obtained from burning fossil fuels. These challenges are being addressed by the research community mainly through the use of novel materials for solar energy conversion.

There is a definite relationship between the rise of new materials and the development of products based on new technologies. As with other applications, the most established material platform for solar energy conversion determines and limits the state of the technology at the time. Figure 6.1 shows the three generations of development of photovoltaic (PV) cell technologies that are classified depending on the basic material used and the level of commercial maturity.

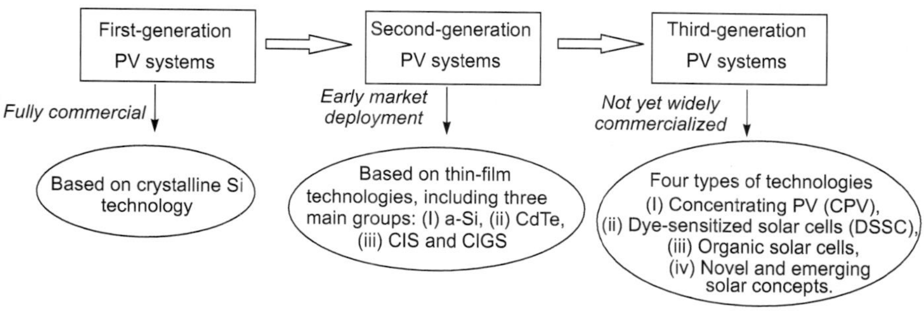

Fig. 6.1 Classification of PV cell technologies.

Photovoltaics (PV) is the direct conversion of sunlight into electricity through a semiconductor p-n junction which absorbs light and generates/separates charge carriers. It is potentially the most widely applicable renewable energy technology, but it remains expensive relative to other technologies as can be seen in Fig. 6.2.

Solar cell technology based on crystalline silicon bulk has been established. Si solar cells with 15-20% efficiencies are commercially available (Saga 2010). However, they are still expensive. Two main approaches have been used in order to overcome this problem: (i) discover innovative methods for making crystalline silicon at lower cost and (ii) employ other materials and/or produce new solar cell designs.

Thin-film PV cells have emerged as an alternative to conventional crystalline Si cells because they have relatively good efficiency and reasonable cost. The concept behind thin-film PV technology is reducing the cell cost by reducing the amount of raw material (use less than 1% of silicon compared to Si wafer-based PV cells). Initially, amorphous silicon thin film-based solar cells led the way. They were already being produced for almost 40 years. Nevertheless, other semiconductors thin films such as Copper Indium Gallium Selenide (CIGS), gallium arsenide (GaAs) and cadmium telluride (CdTe) have gained significant attention. In addition, multi-junction (multi-layer thin films), light-absorbing dyes and organic/polymer solar cells have also been developed.

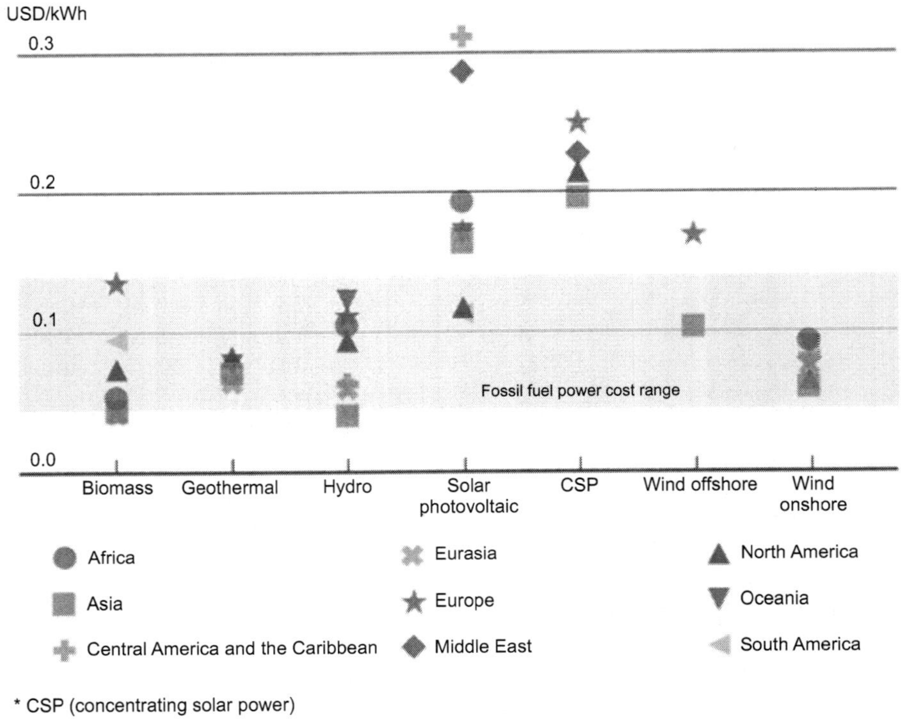

Fig. 6.2 Weighted average cost of electricity by region for utility-scale renewable technologies, compared with fossil fuel power generation costs, 2013/2014 (Source: International Renewable Agency (IRENA) Cost Database, 2014).

Thin-film PV technology has allowed the fabrication of solar cells on different kind of substrates which opens new ideas for solar cell designs besides enabling new applications. For example, crystalline Si thin films grown on glass have been employed in the development of solar cells (called CSG solar devices) that have been recognized as a balance between the low-cost of thin film and the high efficiency of Si bulk.

Another interesting thin-film PV technology is based on the deposition of multi-layer thin films and is called tandem cell. A promising solar cell configuration is combining thin layers of nanocrystalline silicon (nc-Si) and amorphous Si (a-Si). The a-Si thin film has a higher bandgap (1.7 eV) than the crystalline Si (1.1 eV), i.e. it absorbs the visible part of the solar spectrum more strongly than the infrared portion of the spectrum. So, in this cell solar, a-Si absorbs the visible light and leaves the infrared part of the spectrum for the nc-Si. This type of solar cell has exhibited efficiencies above those based on Si wafers (Kim and Fthenakis 2011).

Following the approach of using wide bandgap materials in multilayer thin-film solar cells, different wide-bandgap semiconductors, such as silicon carbide

(SiC), diamond and oxides (TiO_2, SnO_2, ZnO, etc.) have been investigated for this application. Efforts have been devoted to for the growth of nanostructured wide bandgap semiconductor films. It has been shown that the use of nanostructured films in solar energy conversion devices provides various advantages related to final costs (significant reduction in material usage) and higher efficiency. The development of technologies based on nanostructures is considered fundamental to advance the state of the art in PV cells (Candelaria et al. 2012).

Studies on the sensitization of wide bandgap semiconductor films to visible light have been intensively reported in the literature. This process is the basis to form Dye-Sensitized Solar Cell (DSSC) also sometimes referred to as Dye Sensitized Cells (DSC) or Gratzel cell. Among the photoactive materials investigated for use in DSSC, doped nanocrystalline diamond films are considered promising candidates. Diamond films, as other stable carbon forms, have shown high potential to be employed in solar energy conversion systems. These films produced by CVD (Chemical Vapor Deposition) are wide bandgap semiconductors with unique optical, electrical and mechanical properties.

Diamond films are classified in function of the crystal size into four categories: Single Crystalline Diamond (SCD), Micro-Crystalline Diamond (MCD), Nano-Crystalline Diamond (NCD) and Ultra-Nano-Crystalline Diamond (UNCD). In general, these materials possess high hardness, excellent thermal conductivity, favorable electrical properties, chemical inertness and optical transparency over a wide spectral range. Furthermore, they have very low and even Negative Electron Affinity (NEA) characteristics that can be achieved by hydrogenation of their surfaces or by doping. N-type doped diamond films present a reduced effective work function which lead to efficient thermo emission. For this, it is necessary to control the bandgap, electron affinity, doping concentration and the morphology.

Thermionic Electron Emission (TEE) is a phenomenon utilized for direct energy conversion of heat to electricity. In recent years, diamond based thermionic energy conversion devices have been reported. They have shown significant emission at relatively low temperatures (less than 600°C) (Kataoka et al. 2010).

To better understand the role of CVD nanodiamond thin films in photovoltaic and thermionic devices, this chapter addresses the following topics:

- Overview on the synthesis and properties of diamond films with emphasis on CVD techniques.
- *In situ* doping of CVD diamond films.
- Current stage of diamond film technology for photovoltaic and thermionic energy conversion.
- The use of nanodiamond films in solar energy conversion and the steps of development of devices based them.

2. SYNTHESIS AND PROPERTIES OF DIAMOND

Diamond has a lot of physical properties that make it very attractive such as the hardest material known, the highest thermal conductivity at room temperature, excellent optical transparency in a wide wavelength range and inertia to most chemical attacks. Table 6.1 summarizes some properties of diamond.

Table 6.1 Some properties of diamond (adapted from Fraga et al. 2015).

Properties of diamond	Description
Structural	Cubic structure 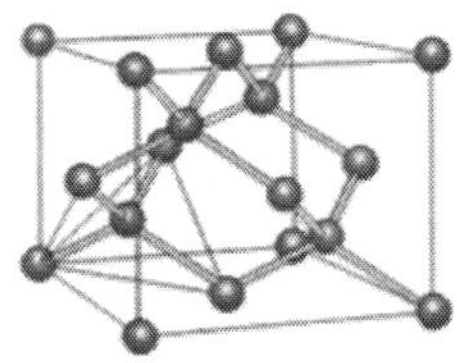
Physical	• High melting point (~4000°C). • Low thermal expansion coefficient at room temperature (1×10^{-6} K^{-1}). • High thermal conductivity at room temperature (2000 $Wm^{-1}K^{-1}$). • Low dielectric constant (5.7).
Electronic	• Wide bandgap (5.45 eV, indirect gap). • High electrical resistivity (10^{13} – 10^{16} Ω·cm). • High saturated electron velocity (2.7×10^7 cm/s). • High electric breakdown field (10 MV/cm).
Mechanical	• High hardness (~90 GPa) and wear resistance. • High Young's modulus (~1050 GPa). • High strength/ tensile (> 1.2 GPa). • Low compressibility (8.3×10^{-3} m^2N^{-1}).
Optical	• High refractive index (2.41). • Optical transparency (UV to far IR). • Highly resistant to damage from irradiation.
Chemical	• High resistance to corrosion. • Chemically and biologically inert.

The density of the diamond is larger than that of graphite, so it is natural to think that a way to produce diamond is converting graphite. However, graphite is the thermodynamically most stable carbon form. This makes the process very difficult due to the need of high pressure and high temperature. The graphite should be compressed and heated to over 2000 K in the presence of a suitable metal catalyst and wait until the diamond crystallizes. The disadvantage of this method is the production of the diamond in the form of single crystals ranging in size from nanometers to millimeters. This motivated the development of diamond synthesis processes based on Chemical Vapor Deposition (CVD).

The number of publications about the synthesis and properties of CVD nano diamond films is elevated. High quality nanodiamond films exhibit most of the properties of the natural diamond, as can be noted in Table 6.2. It is impossible to review all the interesting papers about this subject in one book chapter. Here we will describe a brief historical overview on the motivations behind of the first attempts to synthesize diamond and on the development of deposition processes.

Table 6.2 Some properties of nanodiamond films at room temperature (adapted from Alshekhli 2013).

Property	Value
Wide hardness range (GPa)	10-80
High Young's modulus (GPa)	500-1000
Coefficient friction	0.05-0.5
Wear rate coefficient (mm^3)	10^{-8}-10^{-6}
Roughness (nm)	10-30
High refractive index	2.33-2.44
High electrical resistivity (Ω·cm)	$\sim 10^{11}$
Wide thermal conductivity (W/mK)	20-1000

2.1 Brief Historical Overview

After the Second World War, the companies Eliktriska Allmänna Svenska Aktiebolaget (ASEA) and General Electric Corporation (GE) reported the synthesis of diamond by heating carbon under very high pressure. At that time, the process developed by GE for making diamonds at high pressures and high temperatures became the most industrially used (Bundy et al. 1955) until the rise of CVD techniques.

At the end of the 1950s, there was a great interest in the synthesis of diamond at much lower pressures. The reason was that it would be more advantageous from the economic standpoint and equipment (May 2000). The first experiments involving reduced pressures were carried out by thermal decomposition of carbon-containing gases. The main problem observed was the low growth rate of diamond due to co-deposition of graphite.

In their experiments, Angus (2014) observed that a larger amount of hydrogen gas, during deposition process, increases the diamond growth rate. Atomic hydrogen acts as a selective etchant because it removes the graphite but not the diamond. This allowed the development of a deposition process based on two steps: first, it produced a layer with a small percentage (less than 1%) of diamond bonds and subsequently, a higher percentage of graphically bonded material was removed by a selective hydrogen reduction process at a high temperature and pressure. The results evidenced the importance of high concentration of hydrogen to successful synthesis of diamond. This was also observed by Fedoseev et al. (1978), who showed that the diamond could be grown on non-diamond substrates.

In the 1970s, it was reported that a process used a hot tungsten filament to generate atomic hydrogen. In the 1980s, Japanese researchers published a series of papers with different methods to synthesize diamond using: hot filament (Matsumoto et al. 1982), microwave discharge (Kamo et al. 1983), electron assisted CVD (Sawabe and Inuzuka 1986), plasma CVD (Suzuki et al. 1987) and oxyacetylene torch (Murakawa et al. 1989). Thus during the early 1990s, the synthesis of the diamond by CVD became a well-established and understood field.

2.2 Single and Polycrystalline CVD Diamond Thin Films

In the overall CVD process for diamond growth, the gases are mixed in the reaction chamber before reaching the substrate surface. Once the gases are mixed, they go through an activation zone that provides them energy, heating the gas species to few thousands of Kelvin and causing the dissociation of molecular species into reactive radicals and atoms. Figure 6.3 shows the schematic representation of the hot-filament CVD process. Under the activation zone reactive fragments are mixed and chemically react until they strike the substrate surface where they may adsorb and react and desorbs return into the gas phase or spread near the substrate surface until they find a reaction site. The atomic hydrogen is critical for the diamond growth and one way to produce it is by thermal decomposition of H_2 on the surface of a heated filament (Fig. 6.3). Another way is by electron impact dissociation in plasma.

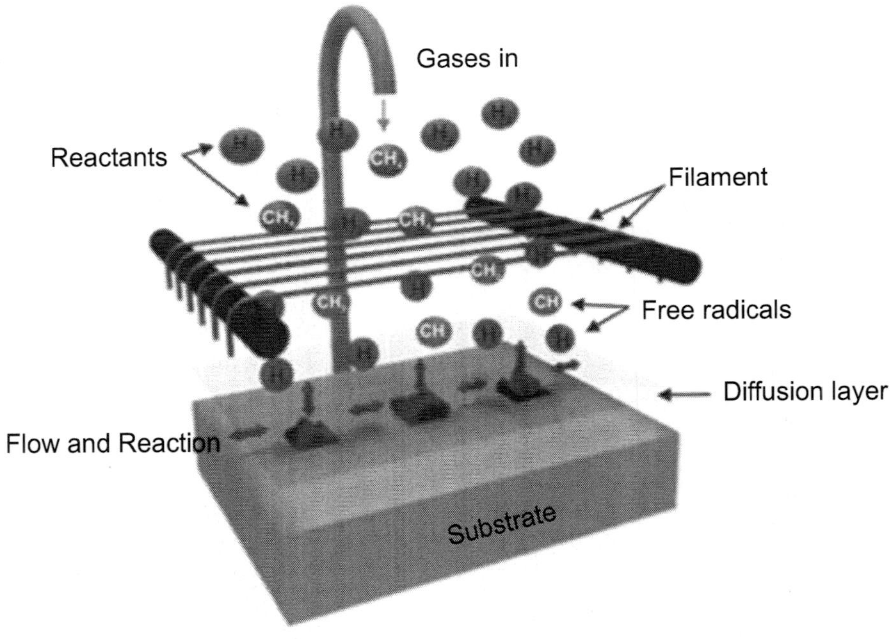

Fig. 6.3 Hot-filament chemical vapor deposition (HFCVD) diamond growth process.

The CVD diamond growth mechanism depends on the chemistry of the gas phase, reactor type and the nature of the diamond crystal used as a substrate (Godwin and Butler 1997). As shown in a simplified form in Fig. 6.4, the diamond growth can be considered as gradual addition of carbon atoms to the existing diamond lattice catalyzed by the presence of atomic hydrogen in excess (May 2000).

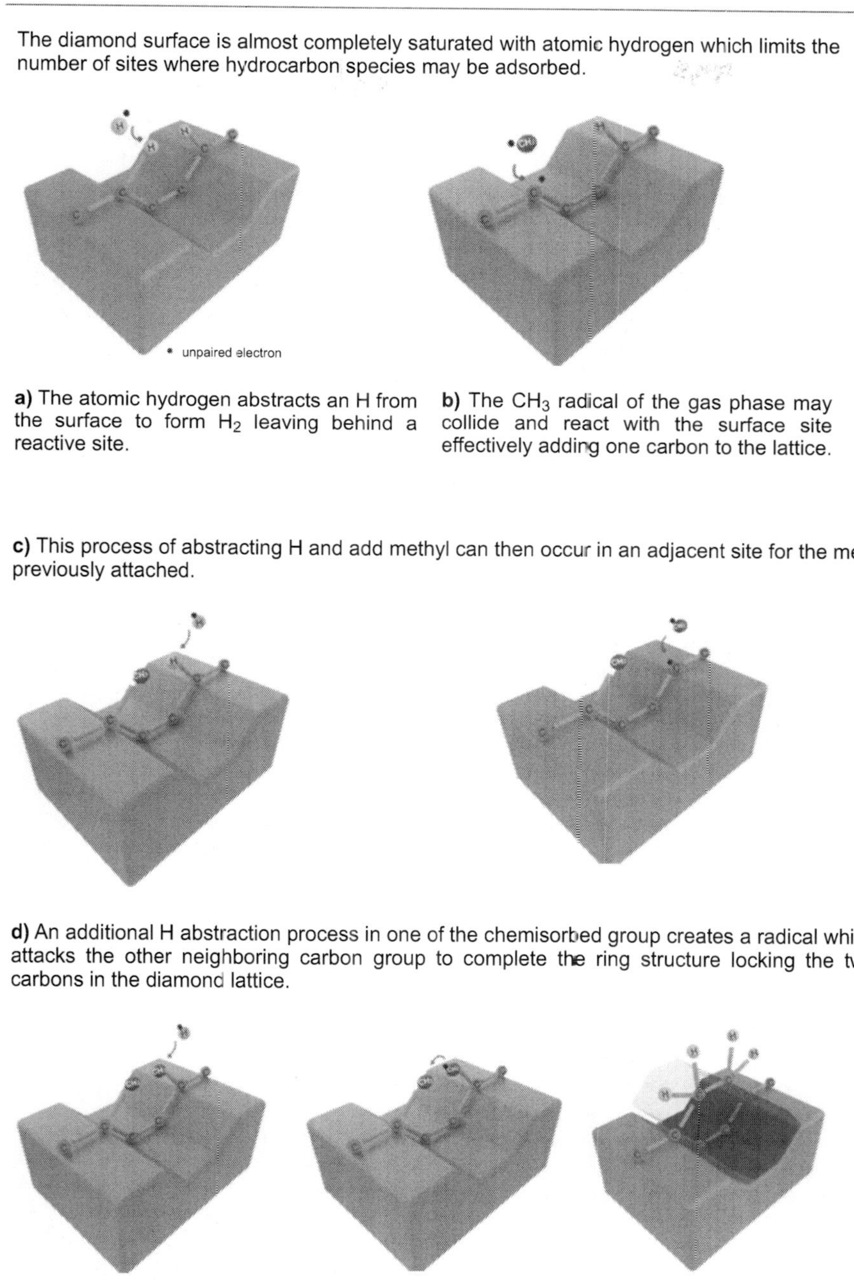

Fig. 6.4 Schematic representation of CVD diamond growth.

Several CVD techniques have been employed to produce diamond. The main difference among them is the type of energy that activates the hydrogen-rich gas phase. The hot filament method called HFCVD (Fig. 6.5a) is the oldest and most widely used to grow diamond at low pressures (Matsumoto et al. 1982). The HFCVD system uses a vacuum pump which pumps a deposition chamber where the precursor gases (mixture of methane and hydrogen) are introduced at a controlled flow rate and pressure. It remains close to 50 Torr while the substrate, that is located a few millimeters (5-10 mm) from the filament, is brought to a temperature close to 850°C. The filament is constituted of refractory metal such as tungsten, tantalum or rhenium and it is electrically heated to 2200°C and then reacts with carbon forming a metal carbide (May 2000). The HFCVD technique is cheap, easy to operate and allows the growth of polycrystalline diamond films at rates compatible with industrial applications. One disadvantage is that the filament material may contaminate the diamond film. This contamination is not an obstacle to use the diamond film in mechanical applications but it can limit its use in electronic devices.

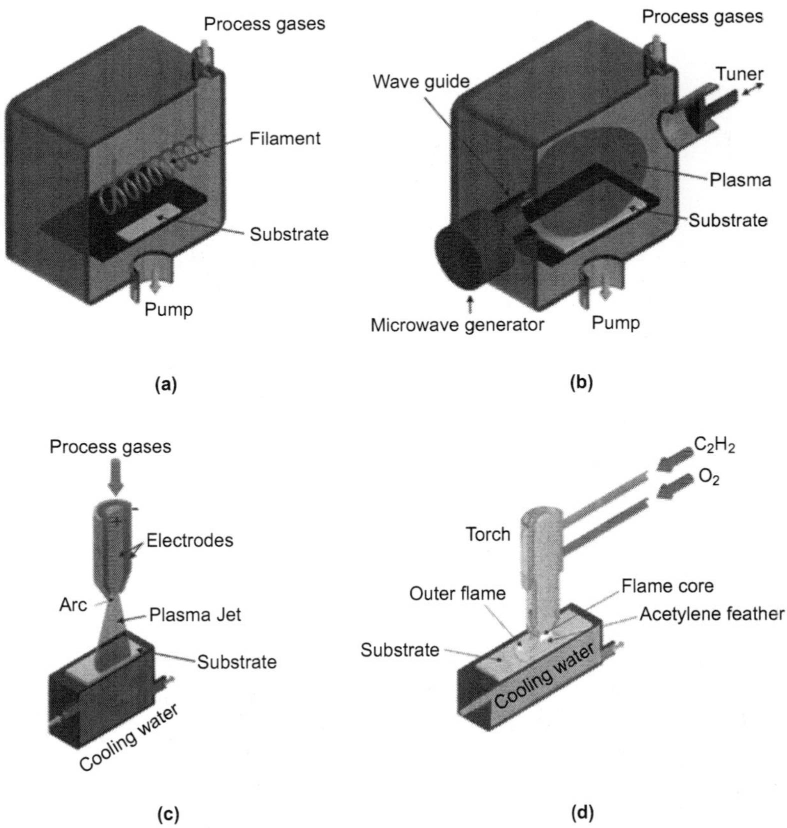

Fig. 6.5 Examples of CVD techniques used to grow diamond: (a) Hot-filament. (b) Microwave assisted. (c) Plasma jet. (d) Oxyacetylene torch.

Some modifications have been incorporated to improve the purity and growth rate of HFCVD diamond film. One is to use in combination with HFCVD, a positive bias voltage applied to the substrate and a negative to the filament resulting in a substrate electron bombardment. Collisions of the neutral gas species with high energy electrons results in a species dissociation. This method is called electron assisted HFCVD (Sawabe and Inuzuka 1986). Another method uses a DC plasma (Suzuki et al. 1987) in which the bias voltage is sufficiently high to start a stable plasma discharge. The H_2 and hydrocarbon decomposition is greatly improved leading to an increase in growth rate. From there, other types of plasma started to be used, as for example, 2.45 GHz Microwave Plasmas and 13.56 MHz RF plasma CVD.

In microwave assisted chemical vapor deposition (MWCVD) configuration, the substrate is placed on substrate holder in a supported tubular reactor with a waveguide directing the microwaves generated by a magnetron (Kamo et al. 1983). The microwaves are guided into the tubular reactor and its energy is transmitted to the gas mixture through collisions and consequently heating, dissociation of gas molecules occur and formation of active species for diamond growth on the substrate immersed in the microwave plasma (Fig. 6.5b). Considering that the use of a metallic filament is not required, this system allows the growth of diamond films with increased purity so it is chosen to grow films for electronic applications.

The fact is that there are a significant number of ions during the discharge, the substrate allows it to be biased with a negative potential during the first minutes of the process and accelerates the ions containing carbon to the substrate allowing them to be implanted into the surface and form a carbon-rich layer which causes an increase in the nucleation rate and a greater film mismatch to the substrate lattice allowing growth of films with preferred orientation or mono crystalline film, this technique is called Bias-Enhanced Nucleation (BEN) (Yugo et al. 1990).

Some other techniques such as plasma jet or oxyacetylene torch were developed as alternatives to MWCVD and HFCVD techniques. In the plasma jet technique (Fig. 6.5c), the gas at high flow rates (liters per minute) passes through a high power electric discharge and forms a jet of ionised particles, atoms and radicals which expands in a secondary chamber to strike the substrate at high speed. The technique most widely used is the DC arc jet where plasma begins with DC arc between the electrodes (Kurihara et al. 1991). Other emerging technologies include fewer electrodes discharges as RF inductively coupled and plasma jet with microwave sources (Kurihara et al. 1991). The main feature of these plasma jet techniques are high growth rates achieved even three times higher than HFCVD and MWCVD (Ohtake and Yoshikawa 1992). However, the deposit area on the substrate is limited to the area where the jet arrives. The cooling of the substrate is a big problem because thermal shock and rapid contraction delaminate the film.

The oxyacetylene torch (Murakawa et al. 1989) system is a very cheap method and can be operated at atmospheric pressure without the need for a vacuum chamber and equipment because combustion is used to produce the plasma (Fig. 6.5d). In this system, the torch is operated with acetylene flow is slightly greater than the oxygen flow creating a region where the flame has many species and radicals. Table 6.3 summarizes the main CVD methods used to grow diamond.

Table 6.3 Main CVD methods used to grow diamond (adapted from Bachmann 1994).

Method	Description
Thermal CVD	Thermal decomposition
	Chemical transport reaction
	Hot filament technique
	Oxyacetylene torch
	Halogen assisted CVD
DC plasma CVD	Low pressure DC plasma
	Medium pressure DC plasma
	Hollow cathode discharge
	DC arc plasma and plasma jet
RF plasma CVD	Low pressure RF glow discharge
	Thermal RF plasma CVD
Microwave plasma CVD	0.9-2.45 GHz
	ECR 2.45 GHz
	8.2 GHz

In terms of its structure, the diamond can be mono or polycrystalline. Polycrystalline CVD diamond has a large number of crystallites or grains coalescing which can be of various degrees of perfection, shape and size depending on growth conditions (Muchnikov et al. 2015). In addition, they may have different structures depending on grain size (Barbosa et al. 2009). The microcrystalline diamond (MCD) films have grains of micron size, from one to hundreds of micrometers of carbon atoms with sp^3 bonds of high purity in the crystalline structure (Fig. 6.6a).

(a) (b)

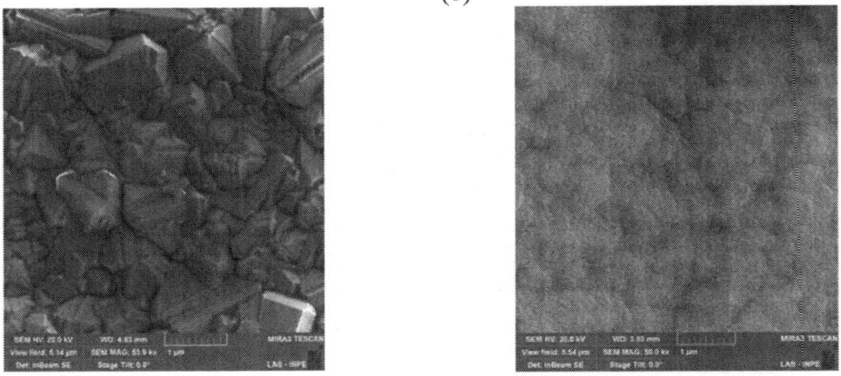

Fig. 6.6 Surface topography of diamond obtained using FEG-SEM: (a) MCD (2.0% CH_4 concentration). (b) NCD (6.0% CH_4 concentration) by HFCVD (Fraga 2015).

The nanocrystalline diamond (NCD) films are those with grain size lower than 500 nm (Fig. 6.6b); whereas the ultrananocrystalline (UNCD) films have fine grains with 3-5 nm (Fig. 6.7). Typically, polycrystalline diamond films with different structures are obtained varying only the methane content in CVD process. For example, Dumpala et al. (2014) produced MCD and NCD coatings by HFCVD varying the methane concentration from 2 to 4%.

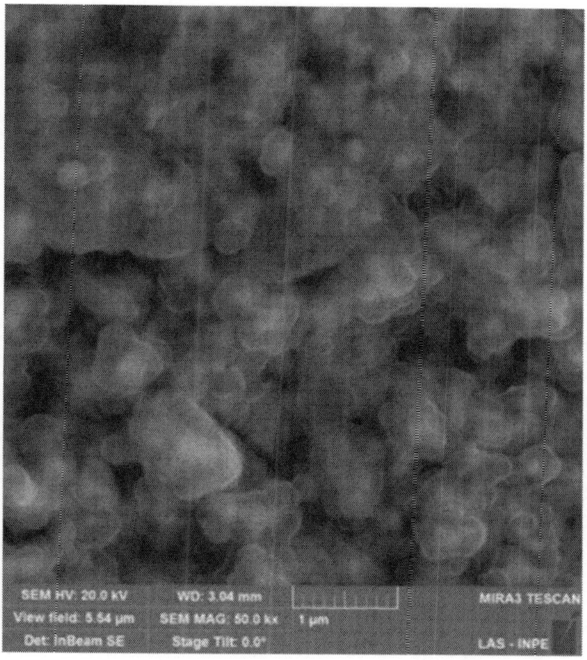

Fig. 6.7 FEG-SEM image of the UNCD (9.0% CH_4 concentration) film grown by HFCVD (Fraga 2015).

Polycrystalline diamond films have been extensively studied for mechanical applications, particularly as coating of hard metal cutting tools. Nanocrystalline diamond films due their lower surface roughness are very useful for machining applications because they allow very smooth finished parts (Almeida et al. 2008). They have also better wear resistance than microcrystalline films (Hu et al. 2008). In this type of mechanical applications, it has been shown that the use of multilayer films nanocrystalline and microcrystalline remarkably improves the behavior of diamond films (Chen and Sun 2013). The boundaries between the crystals of the polycrystalline diamond films are imperfect due to misalignment of adjacent grains and the presence of impurities or high concentration of crystal defects or non-diamond carbon (Clausing 1997).

The oxidation rate of polycrystalline diamond is expected to be much higher than for the single crystalline phase because of the rapid oxidation at grain boundaries. Moreover, the etching rates for MCD diamond can be 100 times higher than for single crystal {100} faces (Clausing 1997). On the other hand, Sato and Kamo (1989) reported that a polycrystalline <100> textured diamond film with relatively large columnar crystals may resist the oxidation.

Grain boundaries and impurities can also affect other properties that are dependent on the crystal perfection such as optical transparency (Findeling-Dufour and Gicquel 1997), carrier mobility in electronic materials (Clausing 1997) and thermal conductivity that is affected by the interaction of phonons with the grain boundary (Graebner et al. 1992).

CVD monocrystalline diamond (or Single Crystalline Diamond, SCD) has a high purity feature that makes it resistant to degradation and thus with optimal mechanical, optical and thermal performance (Linares and Doering 1999). Besides of its wide bandgap, it exhibits large values of carrier mobility (4500 cm^2/Vs (electron) and 3800 cm^2/Vs (hole)) and high thermal conductivity (Yamada et al. 2015). SCD films produced in a microwave plasma CVD reactor using 4% CH_4 at two different pressures (145 and 210 Torr) were investigated by Muchnikov et al. (2010). It was observed that the pressure increase resulted in a growth rate increase from 3.5 to 5 μm/h. However, this did not result in a worse SCD surface quality as can be seen in Fig. 6.8 (Muchnikov et al. 2010).

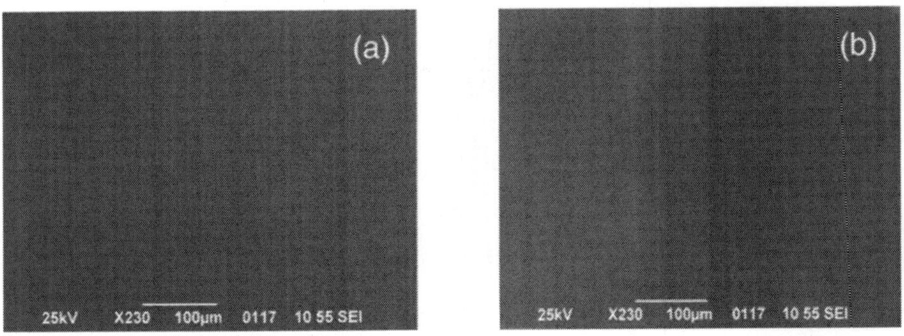

Fig. 6.8 SEM images of SCD surfaces: (a) 145 Torr. (b) 210 Torr after the deposition with MWPD amounted to 200 W/cm^3 in a Microwave Plasma CVD reactor. From (Muchnikov et al. 2010).

Monocrystalline diamond is considered a promising candidate for high-performance electronic devices (Chen et al. 2012). Some of its features has been confirmed by SCD-based devices such as diodes and transistors that have shown a fast response and stable performance at high temperatures (Umezawa et al. 2013). Nowadays, SCD tool plates and optical windows are commercially available. However, in contrast to other established semiconductor material technologies, such as silicon, surface processing of SCD often present challenges related to doping concentration, material purity or extended defect densities and material processing. Further developments in four main areas are important to increase the range of technological applications of SCD materials, namely: size, purity, surface quality and crystal perfection. Such developments have the potential to enable new technologies in fields such as quantum computing, electronics or laser physics (Friel et al. 2009). In the field of solar cells, interesting properties have been found for diamond polycrystalline films diffused and implanted with B and Li (Popovici et al. 1997). These films have been used in photodiodes and photoconductive structures with promising results for UV detection (Polyakov et al. 2005).

Hydrogen plays an important role in significantly modifying the electrical properties of diamond: increasing its conductivity and inducing a negative electron affinity. These modifications are favorable to thermionic emission that makes the hydrogenated diamond an interesting cathode material for low temperature thermionic applications (Paxton et al. 2014). Diamond photocathodes have been demonstrated to be more stable than conventional materials such as cesium iodide (CsI). Furthermore, efficient thermoemitters have been proposed using an emitter surface made by CVD diamond whose work function can be engineered as a function of the hydrogen content (Cicala et al. 2014).

Diamond has small or negative electron affinity that can provide the possibility of a small effective work function (Suzuki et al. 2009) so it is also promising in the field of high efficiency electron emitters. The electron field emission properties of different diamond structures have been reported. A high stability of these properties has been observed. In addition, a turn-on field as low as 0.2-0.3 V/μm for n-type polycrystalline diamond has been observed (Okano et al. 1997).

Several reports have described the excellent thermionic emission properties of n-type diamond. It was found at temperatures below 900 K, a work function of 1.5-1.9 eV and a saturation current density of 0.1-10 A/cm^2 for nitrogen-doped diamond grown on molybdenum substrates (Koeck and Nemanich 2006), an effective work function of 2.4 eV at 550°C for homoepitaxial doped diamond films (Kataoka et al. 2010) and a work function 0.9 eV with thermionic emission current stable up to 765°C for polycrystalline phosphorus-doped diamond films on metal substrates (Koeck et al. 2009).

2.3 Doping

The performance of solar energy devices depends strongly on the properties of their active materials. Diamond acts as a semiconductor when one of the carbon atoms is replaced with a dopant atom. Compared to conventional semiconductor materials, diamond exhibits outstanding properties which make it an alternative material for different emerging applications. However, as other wide bandgap materials, efficient doping of diamond is a challenge and its control is necessary to fully exploit the potential of this material.

It has been known that improvements in the conductivity of diamond films are an important issue for its wide use in solar cells. According to Kalish (1999), the applications of diamond thin films in devices can be categorized into two groups depending on the requirements: (i) using highly conductive diamond and (ii) using the semiconducting properties of diamond. Diamond films have been doped during CVD growth process or by ion implantation. *In situ* doping during growth is most used because ion implantation in diamond is complicated. Furthermore, *in situ* process is cheaper.

P-type conduction of diamond is realized mainly by boron doping. Boron, an acceptor forms states 0.35 eV above the valence band maximum, can be implanted successfully and it can also be readily incorporated in CVD diamond growth into the polycrystalline or single crystal material (Mainwood 2006, Koeck et al. 2009). As a boron source, the diborane (B_2H_6) or trimethylborane ($B(CH_3)_3$) is added to deposition gas mixture. At low boron doping levels, the diamond acts as an extrinsic semiconductor whereas at high levels acts as a semimetal. Boron-doped nanocrystalline diamond layers have been considered promising to replace p-NiO as photocathode material (Krysova et al. 2015). Another application of boron-doped diamond film is as a substrate for growth of CdTe layer, which forms a p-diamond/n-CdTe PV cell based on the concept of an inverted p-n heterojunction solar cell (Huth et al. 2001).

N-type doping is recognized as beneficial for electron emission. On the other hand, in contrast to p-type doping that is well established; effective n-type doping is most difficult and not still fully understood. The main reason for this difficulty in forming n-type semiconducting diamond is the passivation of the donors by the presence of hydrogen or crystalline imperfection in the doped film. In order to confirm the n-type conduction of diamond films, Hall measurements have been the most used quantitative method. A problem observed in the polycrystalline diamond films is that the grain boundary may influence the conductivity, which adds ambiguity to the results. N-type conductivity of diamond has been investigated especially using nitrogen or phosphorus (Fig. 6.9). The incorporation of nitrogen atoms in substitutional sites of diamond form a deep donor level with an activation energy 1.6-1.7 eV thus no electrical conduction can be observed at room temperature (Farrer 1969).

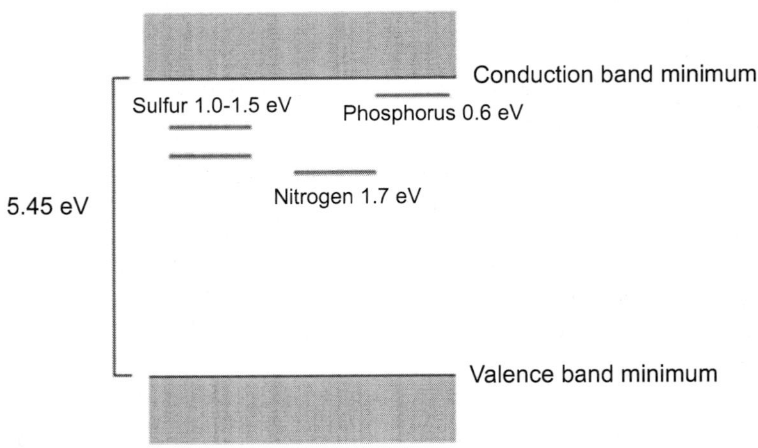

Fig. 6.9 Donor levels of the impurities used to obtain n-type diamond.

Nitrogen-doped diamond has been produced using N_2 or NH_3 as doping precursors in CVD growth. It has been observed that the addition of nitrogen based gas in CVD process increases the growth rate besides influencing the crystal quality and the morphology of diamond films. Phosphorus has been shown as alternative dopant (charge carrier activation energy 0.6 eV). The introduction of phosphorous during the chemical vapor deposition of diamond is made by introducing phosphine (PH_3) gas. In this process, large amounts of hydrogen atoms are incorporated in the diamond film, which inhibits the electrical activation. Two factors have been limiting for successful controlled doping of phosphorus into the diamond lattice: (i) the phosphorus atom is larger than carbon which results in some defects and structural displacement during the phosphorus incorporation (lattice mismatch and covalent bond length) and (ii) the phosphorus incorporation is more favorable on (111) surface compared to (100) surface (Chen et al. 2001). Phosphorus doping into single crystalline diamond (111) during microwave plasma-enhanced CVD (MPCVD) growth has been used in highquality p-i-n junction and UV light emission of around 240 nm was demonstrated even at room temperature (Kato et al. 2007). Lithium ion implantation is another process that can be used to transform intrinsic diamond into an n-type semiconductor. However, the high dose of Li ions implanted causes a graphitization in diamond crystals resulting in a problem for practical applications (Kraft 2007).

Co-doping of diamond, with nitrogen-boron or boron-sulfur, has also been reported. However, although the co-doping of boron-sulfur has shown to exhibit n-type conductivity, no significant reduction in the work function has been observed (Kalish 1999). In the literature, there are some review articles which give a detailed overview on the doping of diamond films by different methods and their different applications (Kalish 1999, Kraft 2007).

3. APPLICATIONS OF NANODIAMOND THIN FILMS IN SOLAR ENERGY CONVERSION

One of the most important green energy sources is solar. It is estimated that the energy irradiated from the sun that reaches the earth's surface is 10,000 times more than the global population currently consumes in one year (Graetzel 2001). However, despite its potential, solar energy still contributes little to the world's total energy production (Maçaira et al. 2013).

There have been many researches in order to enable broad and effective use of solar energy. Major research challenges associated to the use of this energy form are addressed to: (i) high cost, when compared with fuels like oil and coal and (ii) difficulty of efficient sunlight to energy conversion or storage. Overcoming these challenges depends mainly on the improvement in the methods used for harvesting solar energy.

The most common methods used to convert solar energy into chemical and/or electrical energy are photovoltaics, photoelectrochemical and artificial photosynthesis. Photovoltaic devices based in p-n junction are recognized as the most efficient method of harvesting solar energy due to their good conversion efficiency (laboratory scale devices can achieve as high as 25% sun to power conversion efficiency – PCE (Maçaira et al. 2013, Abdin et al. 2013). Nevertheless, these devices exhibit two important limitations. First, they produce only electricity, which makes storage difficult. There are numerous devices and systems for storing electricity, however during their charging and discharging processes significant losses may occur besides these devices have short operating lifetimes. Second, although many technological advances have been made to make them cheaper, the cost of materials, fabrication and installation of the photovoltaic modules are still high (Abdin et al. 2013).

In recent years, the area of photovoltaic devices is facing new challenges. The advantages of developing devices based on nanostructured materials and conductive polymers have been highlighted. For example, these materials are relatively inexpensive to manufacture and they can be used on flexible substrates and molded or painted according to household devices or architecture for decoration applications (Abdin et al. 2013, Sanders 2002). With these new materials, it is possible to completely leaving the classical solid-state junction device, replacing the phase in contact with the semiconductor by an electrolyte (gel, liquid or solid organic) forming a photoelectrochemical (PEC) structure also called photoelectrochemical cell (Gusenbauer 2006). These PEC devices have shown themselves as an alternative method to photovoltaic devices because they allow the photocatalytic conversion of solar energy not only in electricity, but also into chemical energy, for example hydrogen (Abdin et al. 2013, Sanders 2002, Gusenbauer 2006).

3.1 Photovoltaic Energy Conversion: The Use of Diamond Films in Solar Cells

The research of solar cells has been mainly concentrated in the development of new cell concepts or in the improvement of the materials of the electrodes that constituted the solar cells. Although the use of diamond thin films is still reduced in this field, some studies point to the use of polycrystalline diamond as a replacement for the silicon solar cells currently used in space applications. Some characteristics of diamond material can be pointed out for this application (Salisbury 2015, Prelas et al. 1993, Popovici et al. 1995):

- It can withstand the high levels of radiation typical of the space environment. By contrast, the performance of silicon cells degrades by about 50% after 10 years in orbit.
- It can operate at high temperatures. As a result, they can be used with low-weight inflatable solar collectors resulting in an energy system that produces more electricity per pound, a critical factor in space applications.
- It has a potential conversion efficiency of 50% as compared to 10-15% for silicon solar cells (considering mainly the ultraviolet radiation conversion into electricity).

Concerning the development of Dye-Sensitized Solar Cell (DSSC), the choice of photoelectrodes with thermal stability and good photoactivity is fundamental. The common photoelectrodes for DSSCs is TiO_2 as photoanode and Pt as photocathode. However, Pt represents the 'photoelectrochemical silent' cathodes, which means that they serve solely for the dark electron-transfer towards the electrolyte mediator. Among the alternative photocathode materials, the B-doped nanocrystalline diamond (BDD) can be considered a promising replacement of p-type semiconductors such as p-NiO (Krysova et al. 2015). Diamond films made by CVD are attractive due to their excellent chemical and electrochemical stability, optical transparency and favorable electrical properties. The electrochemical inertness of BDD is beneficial in view of the corrosive nature of certain electrolyte solutions using, e.g., the I_3^-/I^- redox couple as the mediator, although the large charge-transfer resistance of BDD might be an issue. BDD has a high hole diffusion coefficient (2–30 cm^2/s), which compares auspiciously to the values obtained for p-NiO (4×10^{-8} cm^2/s). BDD is better optically transparent than p-NiO, approaching the optical quality of Indium-Tin Oxide (ITO) at certain levels of doping (Krysova et al. 2015).

Another research area in DSSC is the sensitization of wide bandgap semiconductors to visible light, which is intensively studied in the area of n-doped electrode. However, for the case of p-type electrode few studies were performed. Recent work published by Krysova et al. (2015) demonstrated a novel simple and versatile synthetic strategy for the surface modification of boron-doped diamond. In a two-step procedure, polyethyleneimine is adsorbed on the hydrogenated diamond surface and subsequently modified with a model light-harvesting donor-p-bridge-acceptor molecule (coded P1). It was observed that the sensitized diamond exhibits stable cathodic photocurrents under visible-light illumination in aqueous electrolyte solution with dimethyl viologen serving as an electron mediator (Krysova et al. 2015).

Kovalenko et al. (2015) demonstrated the possibility of replacing indium–tin oxide (ITO) with heavily BDD (when doped with a boron level of approximately $(3-5) \times 10^{20}$ cm^{-3}, diamond becomes metallically conductive). For this, polymer-fullerene bulk heterojunction solar cells were fabricated to test the potential of BDD application in photovoltaic devices. It was concluded in this initial research that the efficiency of BDD-based devices was lower compared to those using ITO-based architecture. An issue pointed to improve the efficiency of the aforementioned device is concerning the B/C ratio and layer thickness of BDD thin film (Kovalenko et al. 2015).

Finally, the use of BDD for artificial photosynthesis application (water splitting) was recently proposed by researchers from Fraunhhofer Institute (Yang and Nebel 2011). They mounted a diamond/NiO_x p-type DSSC using transparent, insulating diamond as a substrate, conductive, boron-doped diamond as an electrode and $Ni(OH)_2$ as a semiconducting transducer layer. In such a p-type DSSC, diamond acts as hole acceptor, donating electrons through the intermediate transducer layer into the ground state of the light absorbing dye molecules. Integrating diamond into such a model system requires the deposition of a p-type inter-layer ($Ni(OH)_2$) on the surface of the diamond. Initial tests with I_3^-/I^- electrolyte indicated photochemical sensitivity of the structure (Yang and Nebel 2011).

3.2 Thermionic Energy Conversion: The Use of Diamond as an Electron Emitter

The main advantage of thermionic energy conversion technology is the relatively simple and low cost material requirements compared to photovoltaics and potential high conversion efficiency achievable when waste heat recovery is included (Cryan et al. 2015). Thermionic emission is the promotion of electrons to the vacuum from a hot surface of a conducting material. Thermionic energy conversion is a relatively unexplored technology for the efficient conversion of thermal energy directly to electrical energy. In recent years, thermionic emission properties of different cathode materials and its potential uses in solar energy conversion have been much discussed. There is a demand by cathode materials with low work function values (<2 eV) for low temperature thermionic electron emission, which is a key phenomenon for waste heat recovery applications (Sherehiy 2014). N-type diamond, doped with nitrogen, phosphorus and sulfur, meets this requirement.

3.2.1 Thermionic Energy Conversion

The thermionic emission occurs when thermally excited electrons are emitted from the surface of a material which is then used to drive a load. In Fig. 6.10 the schematic representation of a thermionic converter (TEC) is shown. Its structure is very simple and consists of: two electrodes, an emitter and a collector, which are enclosed in an evacuated container (Paxton 2011). The electrodes are connected by electrical leads. The emitter is thermally connected to a heat source whereas the collector is thermally attached to a heat sink. This converter type is called vacuum TEC.

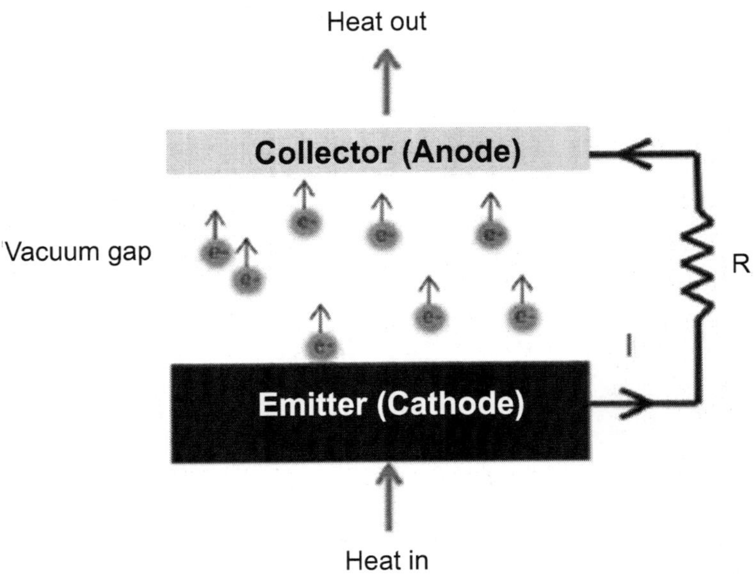

Fig. 6.10 Schematic view of the thermionic emission converter (adapted from Paxton 2011).

The vacuum TEC acts as a heat engine: the heat is taken in by the emitter, ejected by the collector and useful electrical work is done in the load. Its operating principle can be described as follows: the emitter is heated and emits electrons thermionically into vacuum. These electrons traverse the space between electrodes, strike the collector and are absorbed. Then, the electrons pass outside the collector through the lead and into an external load where electric work is done. Subsequently, the electrons return to the emitter via the lead and thus complete the circuit (Smith 2007). Emission current density increases exponentially with the temperature of the electron emitter leading to a highly efficient conversion device at elevated temperatures. It has been reported that a TEC can achieve high efficiency (>30%).

Two factors influence the electron transport across a TEC which can result in the reduction of performance of the device: (i) back emission that is the thermionic electron emission from the collector to the emitter and (ii) the negative space charge due to electrons emitted into the evacuated space between electrodes, a net negative charge develops. This negative space charge repels some of the emitted electrons, screening the slower ones and limiting the output current of the device. An ideal TEC model assumes that back emission and space charge can be ignored (Hatsopoulous and Gyftopoulos 1973). It has been determined that the space between electrodes of a vacuum TEC must be around 1 μm in order to avoid the negative space charge. Another approach is the use of a negative electron affinity emitter such as diamond in order to mitigate the negative space charge effect providing a new way forward in vacuum thermionics (Smith 2007).

In solar applications, the absorber, which is the side of the cathode exposed to radiation, is an important element of the thermionic conversion stage. It must have good adhesion and lattice compatibility with the thermionic emitter layer deposited on the opposite surface not exposed to the radiation. It also needs to be electrically conductive to allow passage of the thermionic current without significant Ohmic loss. Other requirements for the absorber include: high solar absorptivity, thermal and mechanical stability at high operation temperatures and high thermal conductivity to feed thermal energy to the thermionic emitter side without incurring a large temperature gradient (Selvakumar and Barshilia 2012).

As explained earlier, in a thermionic converter, the emitter is at elevated temperature whereas the collector, located at some distance from emitter, is kept at a much lower temperature. Therefore, the efficient electron emission to the vacuum depends on the temperature and on the work function of the materials used for the collector and emitter. The work function of the emitter should be larger than that of the collector. The optimal values of the work function are less than or equal to 2 eV for the emitter and around 1 eV for the collector. Moreover, the emitter should exhibit stability in the temperature range from 700 to 1000 K (Sherehiy 2014).

Initially, thermionic energy conversion devices were based on electron emission from standard metal surfaces. Later the electron emission properties of advanced materials started to be investigated in order to allow the thermionic energy conversion at lower temperatures. Currently, the following cathode materials have been employed as an emitter: coated cathodes (BaO, SaO, CaO), boride cathodes (LaB_6 and the whole family of the hexaboride materials), thoriated tungsten and porous tungsten impregnated with Ba based multi component oxide (Sherehiy 2014). Diamond due to its Negative Electron Affinity (NEA) has been shown as a potential candidate to replace these materials.

3.2.2 *Diamond as a Material for Thermionic Devices*

When discussing the potential of a material as an emitter in a thermionic converter, its electronic properties are the main parameter. Among all the semiconductor materials, diamond possesses a unique feature that is the presence of the true negative electron affinity (NEA). This allows a condition such that the vacuum level of the surface falls below the conduction band minimum.

Several diamond surface orientations with H-termination have exhibited NEA, which can be associated to lowering of electron affinity of diamond by electrostatic effect of the surface termination by hydrogen (Maier et al. 2001). The undoped diamond films have exhibited work function value of order of 3.3 eV (Uppireddi et al. 2009). The specific electronic properties of diamond have also allowed that it can have its work function lowered by two processes: first, through surface modification inducing dipole moments, affecting the electron affinity and second by doping with appropriate dopants causing a change in the position of the Fermi level (Field 1979).

Recent studies on thermionic emission from the n-type doped diamond have shown that it can be possible to tune work function of 0.9-2.5 eV which would be of significant importance to improve thermionic energy conversion. It has been reported that phosphorus-doped nanocrystalline diamond films grown on W foil have work function values from 1.0 to 1.33 eV (Sherehiy 2014). In addition, Koeck et al. (2009) observed a work function value as low as 0.9 eV for the phosphorus-doped diamond film grown on different metal substrates.

Thermionic emission from nitrogen-doped nanocrystalline (NCD) and ultra nanocrystalline diamond (UNCD) films have resulted in observed work functions as low as 1.3 eV (Suzuki et al. 2009). These results show that the n-type conductivity of diamond is required to achieve high emission currents from the thermionic diamond energy converters. Robinson et al. (2006) observed that uniformity of emission is an important aspect in the application of thermionic emission materials in practical devices. They detected variations in the apparent work function across the surface of boron-doped diamond films using scanning Kelvin-probe force microscopy. Paxton et al. (2014) investigated polycrystalline chemical vapor deposited diamond films by monitoring the isothermal thermionic emission current behavior. They observed that the desorption of hydrogen (deuterium) from diamond to be more complex than originally thought. A detailed understanding of the emission degradation mechanism will prove crucial for any future implementation attempts with hydrogenated diamond as a low-temperature (< 1000°C) for thermionic emitter for applications such as energy conversion.

Researchers from University of Bristol have developed a nanodiamond-based solar energy converter in order demonstrate that it exhibits performance at much lower temperatures than those quoted for conventional metal-based converters. Figure 6.11 shows the schematic representation of the solar energy converter proposed. As can be observed, nanodiamond films are used on an emitter and collector. The researchers used alithiated diamond surface and they concluded that it is excellent for thermionic devices. However, the reduction of the resistance of the doped diamond is still a challenge to produce an economically viable and high efficiency thermionic converter.

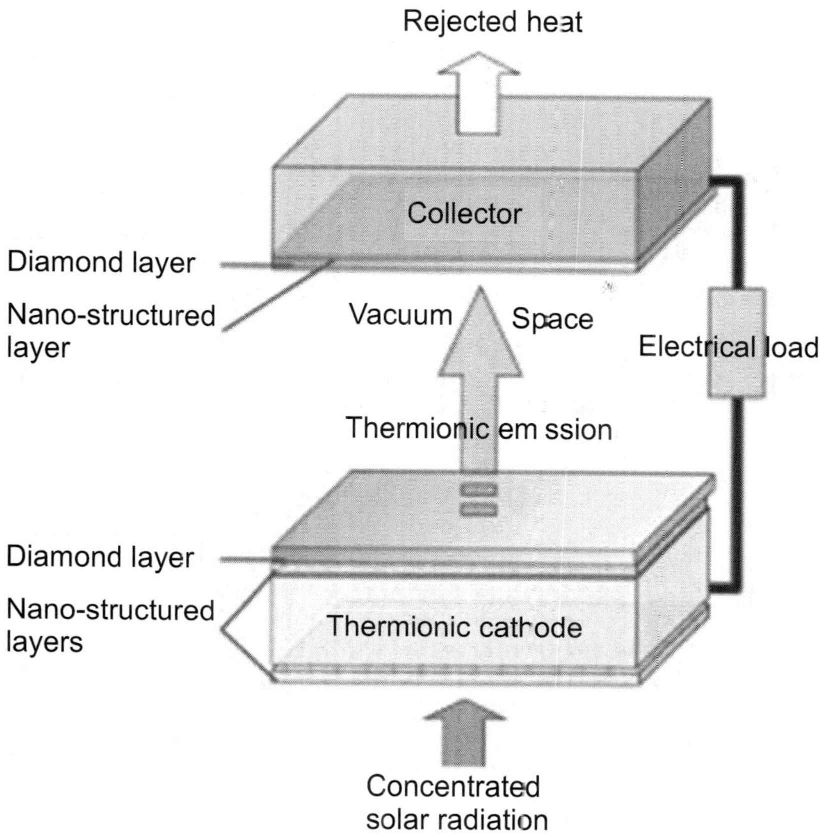

Fig. 6.11 Schematic representation of a thermionic solar energy converter using nanodiamond layers (Source: Project completion summary – Thermionic diamond for concentrated solar energy conversion, University of Bristol).

4. FINAL REMARKS

This chapter presented the potential of diamond films as materials for solar energy conversion systems. It were reviewed the main CVD methods used to synthesize diamond. Furthermore, the effect of growth conditions on the different properties of diamond films was discussed. Efficient doping of diamond is crucial to produce a surface with a reduced effective work function. The evolution and the current stage of the photovoltaic and thermionic devices based on CVD nanodiamond films promise the development of more efficient solar energy conversion devices.

5. ACKNOWLEDGEMENTS

This research was supported by the São Paulo Research Foundation FAPESP (process number 14/18139-8) and CAPES.

6. REFERENCES

Abdin, Z., M. A. Alim, R. Saidur, M. R. Islam, W. Rashmi, S. Mekhilef, et al. 2013. Solar energy harvesting with the application of nanotechnology, Renew. Sust. Energ. Rev. 26: 837-852.

Almeida, F. A., J. Sacramento, F. J. Oliveira and R. F. Silva. 2008. Micro- and nano-crystalline CVD diamond coated tools in the turning of EDM graphite. Surf. Coat. Technol. 203(3-4): 271-276.

Alshekhli, O. 2013. Pulsed electron deposition and characterization of nanocrystalline diamond thin films, PhD Thesis, Laurentian University, Ontario.

Angus, J. C. 2014. Diamond synthesis by chemical vapor deposition: the early years. Diamond Relat. Mater. 49: 77-86.

Bachmann, P. K. 1994. Microwave plasma CVD and related techniques for low pressure diamond synthesis. pp. 31-53. *In*: A. Lettington and J. W. Steeds (eds.). Thin Film Diamond. Chapman and Hall, London.

Barbosa, D. C., F. A. Almeida, R. F. Silva, N. G. Ferreira, V. J. Trava-Airoldi and E. J. Corat. 2009. Influence of substrate temperature on formation of ultrananocrystalline diamond films deposited by HFCVD argon-rich gas mixture. Diamond Relat. Mater. 18(10): 1283-1288.

Bundy, F. P., H. T. Hall, H. M. Strong and R. H. Wentorf. 1955. Man-made diamonds. Nature 176: 51-55.

Candelaria, S.L., Y. Shao, W. Zhou, X. Li, J. Xiao, J. Zhang, et al. 2012. Nanostructured carbon for energy storage and conversion. Nano Energy 1: 195-220.

Chen, C. F., C. L. Tsai and C. L. Lin. 2001. Electronic properties of phosphorus-doped triode-type diamond field emission arrays. Mater. Chem. Phys. 72(2): 210-213.

Chen, G. C., B. Li, Z. Q. Yan and F. X. Lu. 2012. Single crystalline diamond grown by repositioning substrate in DC arcjet plasma enhanced chemical vapor deposition. Diamond Relat. Mater. 21: 83-87.

Chen, N. C. and F. H. Sun. 2013. Cutting performance of multilayer diamond coated silicon nitride inserts in machining aluminum-silicon alloy. Transactions of Nonferrous Metals Society of China (English Edition) 23(7): 1985-1992.

Cicala, G., V. Magaletti, A. Valentini, M. A. Nitti, A. Bellucci and D. M. Trucchi. 2014. Photo- and thermionic emission of MWPECVD nanocrystalline diamond films. Appl. Surf. Sci. 320: 798-803.

Clausing, R. E. 1997. Diamond morphology. pp. 19-47. *In*: M. A. Prelas, G. Popovici and L. K. Bigelow (eds.). Handbook of Industrial Diamonds and Diamond Films. Taylor & Francis, New York, USA.

Cryan, M. J., S. Nunez-Sanchez, N. Ahmad, C. Wan, J. Stokes, M. Z. Othman, et al. 2015. Diamond based solar thermionic energy converters. 11th World Conference on Thermophotovoltaic Generation of Electricity (TPV-11). Hamburg, Germany.

Dumpala, R., N. Kumar, C. R. Kumaran, S. Dash, B. Ramamoorthy and M. S. R. Rao. 2014. Adhesion characteristics of nano- and micro-crystalline diamond coatings: Raman stress mapping of the scratch tracks. Diamond and Relat. Mater. 44: 71-77.

Farrer, R. G. 1969. On the substitutional nitrogen donor in diamond. Solid State Commun. 7(9): 685-688.

Fedoseev, D. V, K. S. Uspenskaya, V. P. Varnin and S. P. Vnukov. 1978. Effect of hydrogen on diamond growth from a gaseous phase. Bulletin of the Academy of Sciences of the USSR, Division of Chemical Science 27(6): 1088-1091.

Field, J. E. 1979. The Properties of Diamond. Academic Press, London, UK.

Findeling-Dufour, C. and A. Gicquel. 1997. Study for fabricating large area diamond single-crystal layers. Thin Solid Films 308-309: 178-185.

Fraga, M. A. 2015. Growth of CVD diamond by HFCVD. Research project report presented to National Institute for Space Research (INPE).

Fraga, M. A., M. Bosi and M. Negri. 2015. Silicon carbide in microsystem technology – thin film versus bulk material. pp. 1-30. *In*: S. E. Saddow and F. La Via (eds.). Advanced Silicon Carbide Devices and Processing. InTech, Rijeka, Croatia.

Friel, I., S. L. Clewes, H. K. Dhillon, N. Perkins, D. J. Twitchen and G. A. Scarsbrook. 2009. Control of surface and bulk crystalline quality in single crystal diamond grown by chemical vapour deposition. Diamond Relat. Mater. 18(5-8): 808-815.

Godwin, D. and J. Butler. 1997. Theory of diamond CVD. pp. 527-581. *In*: M. A. Prelas, G. Popovici and L. K. Bigelow (eds.). Handbook of Industrial Diamonds and Diamond Films. Taylor & Francis, New York, USA.

Graebner, J. E., S. Jin, G. W. Kammlott, J. A. Herb and C. F. Gardinier. 1992. Unusually high thermal conductivity in diamond films. Appl. Phys. Lett. 60(13): 1576-1578.

Graetzel, M. 2001. Photoelectrochemical cells. Nature 414: 338-344.

Gusenbauer, A. 2006. Photoelectrochemical Cells based on Photoactive Polymers. PhD Thesis, Johannes Kepler Universitat, Linz.

Hatsopoulous, G. N. and E. P. Gyftopoulos. 1973. Thermionic Energy Conversion, Volume 1. MIT Press, Cambridge, Massachusetts.

Hu, J., Y. K. Chou and R. G. Thompson. 2008. Nanocrystalline diamond coating tools for machining high-strength Al alloys. Int. J. Refract. Met. Hard Mater. 26(3): 135-144.

Huth, P. V., J. E. Butler, W. Jaegermann and R. Tenne. 2001. New measurements on the diamond/CdTe solar cell. Proceedings of the 13th Workshop on Quantum Solar Energy Conversion Kirchberg in Tirol, Österreich.

IRENA Report, "Renewable power generation costs in 2014," January 2015.

Kalish, R. 1999. Doping of diamond. Carbon 37: 781-785.

Kamo, M., Y. Sato, S. Matsumoto and N. Setaka. 1983. Diamond synthesis from gas phase in microwave plasma. J. Cryst. Growth 62(3): 642-644.

Kataoka, M., C. Zhu, F. A. M. Koeck and R. J. Nemanich. 2010. Thermionic electron emission from nitrogen-doped homoepitaxial diamond. Diamond Relat. Mater. 19(2-3): 110-113.

Kato, H., T. Makino, S. Yamasaki and H. Okushi. 2007. n-type Diamond Growth by Phosphorus Doping. MRS Proceedings, 1039, 1039-P05-01.

Kim, H. C. and V. M. Fthenakis. 2011. Comparative life-cycle energy payback analysis of multi-junction a-SiGe and nanocrystalline/a-Si modules. Prog. Photovolt. Res. Appl. 19: 228-239.

Koeck, F. A. M. and R. J. Nemanich. 2006. Emission characterization from nitrogen-doped diamond with respect to energy conversion. Diamond Relat. Mater. 15(2-3): 217-220.

Koeck, F. A. M., R. J. Nemanich, A. Lazea and K. Haenen. 2009. Thermionic electron emission from low work-function phosphorus doped diamond films. Diamond Relat. Mater. 18(5-8): 789-791.

Kovalenko, A., P. Ashcheulov, A. Guerrero, P. Heinrichová, L. Fekete, M. Vala, et al. 2015. Diamond-based electrodes for organic photovoltaic devices, Sol. Energy Mater. Sol. Cells 134: 73-79.

Kraft, A. 2007. Doped diamond: a compact review on a new, versatile electrode material. Int. J. Electrochem. Sci. 2: 355-385.

Krysova, H., Z. Vlckova-Zivcova, J. Barton, V. Petrak, M. Nesladek, P. Cigler, et al. 2015. Visible-light sensitization of boron-doped nanocrystalline diamond through non-covalent surface modification. Phys. Chem. Chem. Phys. 17: 1165-1172.

Kurihara, K., K. Sasaki and M. Kawarada. 1991. Diamond synthesis By Dc plasma jet CVD. Mater. Manuf. Processes 6(2): 241-256.

Linares, R. and P. Doering. 1999. Properties of large single crystal diamond. Diamond Relat. Mater. 8(2-5): 909-915.

Maçaira, J., L. Andrade and A. Mendes. 2013. Review on nanostructured photoelectrodes for next generation dye-sensitized solar cells. Renew. Sustain. Energy Rev. 27: 334-349.

Maier, F., J. Ristein and L. Ley. 2001. Electron affinity of plasma-hydrogenated and chemically oxidized diamond (100) surfaces. Phys. Rev. B 64: 165411.

Mainwood, A. 2006. Theoretical modelling of dopants in diamond. J. Mater. Sci. Mater. Electron. 17: 453-458.

Matsumoto, S., Y. Sato, M. Tsutsumi and N. Setaka. 1982. Growth of diamond particles from methane-hydrogen gas. J. Mater. Sci. 17(11): 3106-3112.

May, P. W. 2000. Diamond thin films: a 21st-century material. Philos. Trans. R. Soc. London, Ser. A 358(1766): 473-495.

Muchnikov, A. B., A. L. Vikharev, A. M. Gorbachev, D. B. Radishev, V. D. Blank and S. A. Terentiev. 2010. Homoepitaxial single crystal diamond growth at different gas pressures and MPACVD reactor configurations. Diamond Relat. Mater. 19(5-6): 432-436.

Muchnikov, A. B., A. L. Vikharev, D. B. Radishev, V. A. Isaev, O. A. Ivanov and M. A. Gorbachev. 2015. A wafer of combined single-crystalline and polycrystalline CVD diamond. Mater. Lett. 139: 1-3.

Murakawa, M., S. Takeuchi and Y. Hirose. 1989. Diamond coating of a long substrate by use of a combustion flame. Surf. Coat. Technol. 39: 235-240.

Ohtake, N. and M. Yoshikawa. 1992. Nucleation effects and characteristics of diamond film grown by arc discharge plasma jet chemical vapor deposition. Thin Solid Films 212(1-2): 112-121.

Okano, K., T. Yamada, H. Ishihara, S. Koizumi and J. Itoh. 1997. Electron emission from nitrogen-doped pyramidal-shape diamond and its battery operation. Appl. Phys. Lett. 70(16): 2201-2203.

Paxton, W. F. 2011. Characterization of the thermal electron emission properties of boron-doped polycrystalline diamond films for use in energy conversion. Master Thesis. Vanderbilt University, Nashville, Tennessee, USA.

Paxton, W. F., M. M. Brooks, M. Howell, N. Tolk, W. P. Kang and J. L. Davidson. 2014. Role of deuterium desorption kinetics on the thermionic emission properties of polycrystalline diamond films with respect to kinetic isotope effects. J. Appl. Phys. 115: 234904.

Polyakov, V. I., I. Rukovishnikov, N. M. Rossukanyi, V. G. Ralchenko, F. Spaziani and G. Conte. 2005. Photoconductive and photovoltaic properties of CVD diamond films. Diamond Relat. Mater. 14(3-7): 594-597.

Popovici, G., M. A. Prelas, S. Khasawinah, T. Sung, V. I. Polyakov, P. I., Perov, et al. 1995. Diamond Photovoltaics: characterization of CVD diamond film-based heterostructures for light to electricity conversion, Applications of Diamond Films and Related Materials: Third International Conference, 1995.

Popovici, G., A. Melnikov, V. V. Varichenko, T. Sung, M. A. Prelas, R. G. Wilson, et al. 1997. Diamond ultraviolet photovoltaic cell obtained by lithium and boron doping. J. Appl. Phys. 81: 2429-2431.

Prelas, M. A., E. J. Charlson, E. M. Charlson, J. M. Meese, G. Popovici and G. Stacy. 1993. Diamond photovoltaic cell as a first-wall material and energy conversion system for inertial confinement fusion. Laser and Particle Beams 11(1): 65-79.

Project completion summary: Thermionic diamond for concentrated solar energy conversion, University of Bristol. http://www.eon.com/content/dam/eon-com/ueber-uns/innovation/Diamond Converter_Ext_PCS_RES_12452_130124.pdf, accessed in: 08/10/2015.

Robinson, V. S., Y. Show, G. M. Swain, R. G. Reifenberger and T. S. Fisher. 2006. Thermionic emission from surface-terminated nanocrystalline diamond. Diamond Relat. Mater. 15: 1601-1608.

Saga, T. 2010. Advances in crystalline silicon solar cell technology for industrial mass production. NPG Asia Mater. 2: 96-102.

Salisbury, D. F., webpage: http://exploration.vanderbilt.edu/news/news_diamond.htm, access in August 20, 2015.

Sanders, B., Cheap, Plastic Solar Cells May be on the Horizon, UC Berkeley Campus News. 28 March 2002. http://www.berkeley.edu/news/media/releases/2002/03/28_solar.html. Accessed in 17/08/2015.

Sato, Y. and M. Kamo. 1989. Texture and some properties of vapor deposited diamond films, Surf. Coat. Technol. 40: 183-198.

Sawabe, A. and T. Inuzuka. 1986. Growth of diamond thin films by electron-assisted chemical vapour deposition and their characterization. Thin Solid Films 137(1): 89-99.

Selvakumar, N. and H. C. Barshilia. 2012. Review of physical vapor deposited (PVD) spectrally selective coatings for mid- and high-temperature solar thermal applications. Sol Energy Mater. Sol. Cells 98: 1-23.

Sherehiy, A. 2014. Thermionic Emission Properties of Novel Carbon Nanostructures. Ph.D. Thesis, University of Louisville, Louisville, Kentucky, USA.

Smith, J. R. 2007. Thermionic Energy Conversion and Particle Detection Using Diamond and Diamond-Like Carbon Surfaces. Ph.D. Thesis, North Carolina State University, Raleigh, North Carolina, USA.

Suzuki, K., A. Sawabe, H. Yasuda and T. Inuzuka. 1987. Growth of diamond thin films by dc plasma chemical vapor deposition. Appl. Phys. Lett. 50(12): 728-729.

Suzuki, M., T. Ono, N. Sakuma and T. Sakai. 2009. Low-temperature thermionic emission from nitrogen-doped nanocrystalline diamond films on n-type Si grown by MPCVD. Diamond Relat. Mater. 18(10): 1274-1277.

Umezawa, H., Y. Kato and S. Shikata. 2013. 1 Ω on-resistance diamond vertical-schottky barrier diode operated at 250°C. Appl. Phys. Express 6: 011302, 1-5.

Uppireddi, K., T. L. Westover, T. S. Fisher, B. R. Weiner and G. Morell. 2009. Thermionic emission energy distribution from nanocrystalline diamond films for direct thermal-electrical energy conversion applications. J. Appl. Phys. 106: 043716.

Yamada, H., A. Chayahara, Y. Mokuno and S. I. Shikata. 2015. Numerical microwave plasma discharge study for the growth of large single-crystal diamond. Diamond Relat. Mater. 54: 9-14.

Yang, N. and C. E. Nebel. Annual Report, Fraunhhofer Institute, 2011.

Yugo, S., T. Kimura and T. Muto. 1990. Effects of electric field on the growth of diamond by microwave plasma CVD. Vacuum 41(4-6): 1364-1367.

7

A Discussion on the Use of Metal-Containing Diamond-Like Carbon (Me-DLC) Films as Selective Solar Absorber Coatings

M. A. Fraga,[1,*] G. Leal,[2,a] M. Massi[2,b] and V. J. Trava-Airoldi[1,c]

ABSTRACT

The wide use of sunlight for energy generation still faces challenges related to high cost and low efficiency. The successful expansion of the capacity for solar energy production depends on technological advances in several fields, particularly materials science and engineering. The key mechanisms of solar energy conversion are dominated by the intrinsic properties of the active materials that constitute the photovoltaic or solar thermal collector systems. An intensive research on the synthesis and properties of materials for solar energy conversion, especially nanomaterials, has been carried out in order to build solar systems with lower cost and higher efficiency. Carbon-based nanomaterials have been recognized as promising candidates due to their outstanding physicochemical properties. Among them, Metal-containing Diamond-Like Carbon (Me-DLC) coatings have been investigated to enhance absorption of solar radiation and also enhance heat transfer to working fluid in solar collectors. In this chapter, the focus is on the synthesis, properties and solar absorber coating applications of Me-DLC films. The fundamental mechanisms related to selective solar absorbers are described. The main synthesis methods of Me-DLC films are reviewed. Furthermore, the material properties of Me-DLC films are discussed as a function of types of transition metals, such as Cr, W, Ti, Pt, etc. which are incorporated into the DLC film. The

[1] Associate Laboratory of Sensors and Materials, National Institute for Space, Research, São José dos Campos-SP, 12227-010, Brazil.

[2] Federal University of São Paulo-ICT, São José dos Campos-SP, 12231-280, Brazil.

[a] E-mail: gabriela.leal@unifesp.br

[b] E-mail: massi.marcos@unifesp.br

[c] E-mail: vladimir.airoldi@inpe.br

* Corresponding author: mafraga@ita.br

promising results reported in literature using Me-DLC films as selective solar absorber coatings are also discussed.

Keywords: Diamond-like carbon, transition metal doping, cermet coatings, selective solar absorbers.

1. INTRODUCTION

The crescent demand for energy has driven the development of new materials, processes and mechanisms with the aim of increasing energy production based on the renewable resources. The use of such resources is a convenient choice with respect to conventional fuels and their environmental impacts (Kirchain and Alonso 2011). Solar energy conversion is recognized as the most valuable source of sustainable and renewable energy. However, the widespread use of solar energy conversion systems still faces some challenges, especially related to efficiency and cost of production.

The two main methods currently used for solar energy conversion are illustrated in Fig. 7.1, namely: (i) photovoltaic cell which converts solar radiation into electricity and (ii) thermal solar device (known as solar collector) which converts the solar energy into heat energy by absorbing it.

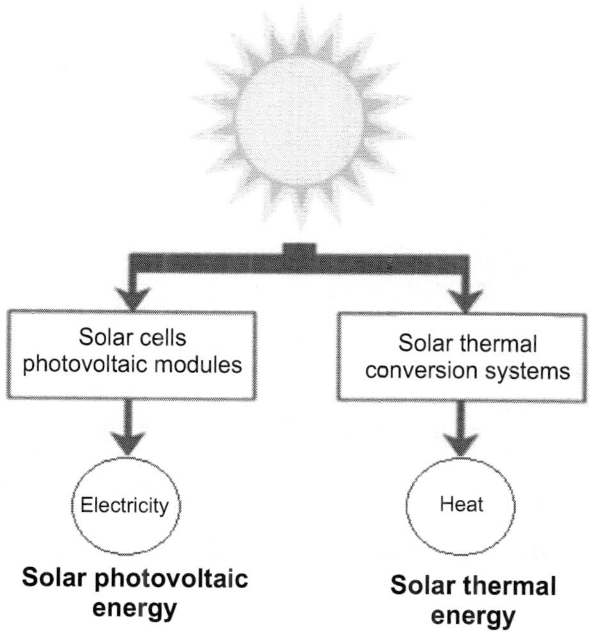

Fig. 7.1 Methods used for solar energy conversion.

In recent years, intense research and development activities have been devoted to fabrication of photovoltaic and solar collector systems with lower cost and higher efficiency. The application of nanoscale materials is a promising way to meet both challenges (Oelhafen and Schuler 2005). Nanotechnology has been recognized as an outstanding option to make possible the development of more economically viable materials for solar energy conversion by improving their efficiency while reducing their cost and size. Nanostructured materials are defined as objects that have at least one dimension in the range of 1-100 nm. They can be categorized by the number of nanoscale dimensions into four groups (Haick 2013).

- 0-D nanostructures are materials in which each spatial dimension has from 1 to 100 nm. Examples: nanoparticles and quantum dots.
- 1-D nanostructures are materials with a characteristic diameter between 1 and 100 nm and a length that could be much greater. Examples: nanotubes, nanowires, quantum wires and nanorods.
- 2-D nanostructures are nanotextured surfaces with a thickness between 1 and 100 nm, while the other two dimensions are much greater. Examples: thin films, planar quantum wells and superlattices.
- 3-D nanostructures are bulky materials with all dimensions above 100 nm. Examples: bulk nanocrystalline films and nanocomposites.

There is a pressing challenge to produce and store renewable energy. In this context, different types of nanomaterials have been investigated for the development of alternative clean and sustainable energy technologies. Much of the discussion is centered on the use of carbon-based nanostructures, such as diamond, graphite, graphene, carbon nanotubes and amorphous carbon (including diamond-like carbon - DLC). They are considered very promising candidates due to their electrical, thermal and optical properties. Hence a great deal of research has been conducted to employ them in solar energy conversion systems.

DLC films have emerged as favorable coating materials, with different applications in solar systems, because their optical and mechanical properties such as bandgap, refractive index and hardness can be tailored to satisfy the requirements of the intended application (Candelaria et al. 2012). A potential application of these coatings is as selective solar absorbers.

The absorber coating plays the most significant role in solar thermal collectors: it absorbs and converts solar radiation into heat. The function of the collector system is to heat a fluid passing through the conduit by converting incident solar radiation into thermal energy with minimum heat losses. Figure 7.2 shows a cross-sectional schematic representation of a simple flat plate solar collector used for domestic hot water application (Tesfamichael 2000). These collectors are not efficient because of the losses caused by convection and conduction. The fluid temperature can achieve 400°C if evacuated tubes are used in the collectors, reducing these losses (Kalogirou 2004).

There are already high-performing selective absorbers developed but some of them have problems with long-term durability, moisture resistance, adhesion, scratch resistance and more costly production techniques (Bostrom 2006). In order to achieve a highly efficient solar collector, it is imperative to use spectrally selective absorber coating with chemical, physical and optical properties to be able to ensure good performances in terms of energy efficiency, temporal stability and durability at the operating temperatures (Pratesi et al. 2014).

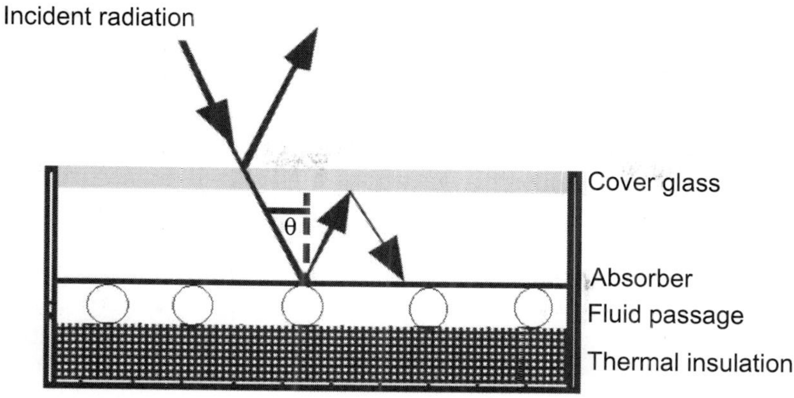

Fig. 7.2 Cross-sectional view of a commonly used flat plate solar collector (adapted from Tesfamichael 2000).

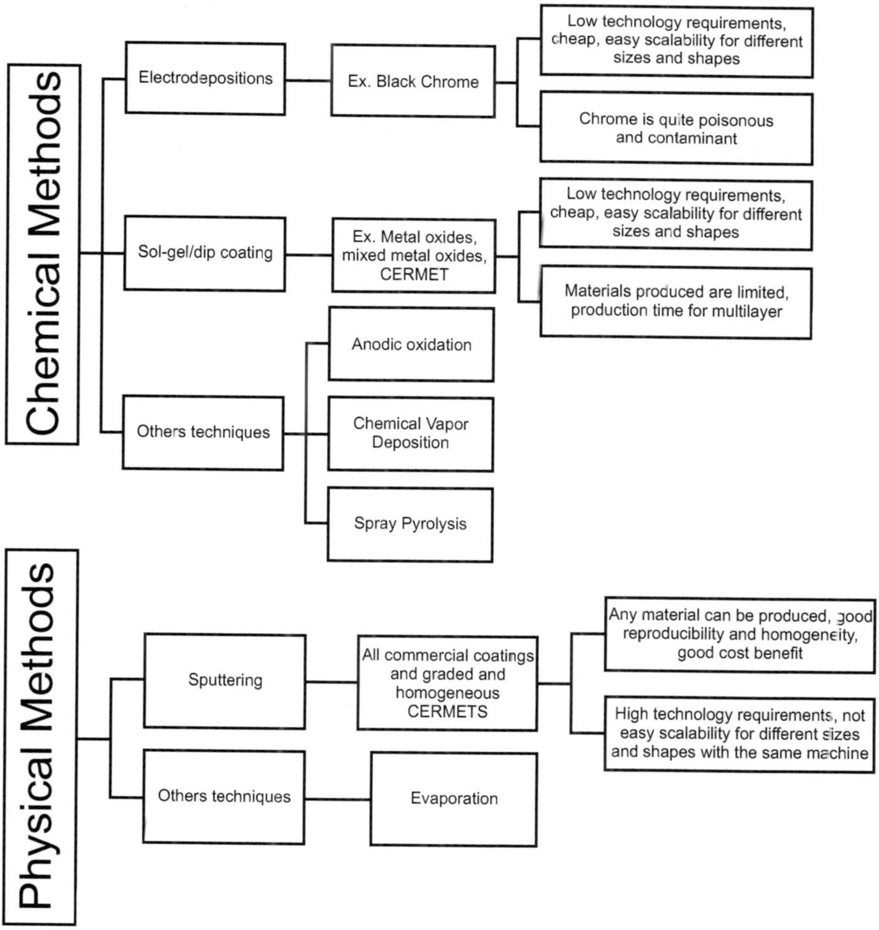

Fig. 7.3 Methods of production of selective absorber coatings.

The selective solar absorber can be designed using different type of materials such as: (i) semiconductors with suitable bandgaps, (ii) metal-dielectric nanocomposites (called cermets), which are black for solar wavelengths but transparent for the heat and (iii) roughened metallic surfaces, in which higher absorptivity solar is obtained by multiple absorption-reflection of the light inside the surface grooves (Nejati 2008).

A large number of deposition techniques have been used for the production of different selective absorber coatings, mainly wet chemical and Physical Vapor Deposition (PVD) processes (Fig. 7.3).

PVD is the most promising technology because wet chemical methods are not environmentally friendly (Selvakumar and Barshilia 2012). In addition, PVD coatings exhibit higher chemical and thermal stability. Among the PVD methods, magnetron sputtering techniques have been widely used for large-area deposition of selective solar coatings. As outlined in Fig. 7.3, besides sputtering, electrodeposition and sol-gel/dip coating techniques are commercial methods currently employed to produce absorber coatings.

Relatively thick coatings of the order of 1 μm are required for solar absorbers. Thus, the induced stress and cracking are challenges (Williams et al. 2014). The large residual stress of DLC films limits the maximum film thickness because it reduces the adhesion to substrate causing delamination and cracking of the film. For metal-containing DLC (Me-DLC, also known as metal-doped DLC) film, the adhesion is significantly improved whereas the hardness and the tribological properties are conserved (Dai et al. 2007).

Me-DLC films are nanostructured materials containing clusters with dimensions on the nanometer length scale. These films are cermet materials produced by introduction of metal (like chromium, tungsten, platinum, titanium among other transition metals) into DLC (a-C or a-C:H) matrix during deposition process (Sánchez-López and Fernández 2008). Thereby their properties depend on the metal type used and on the metal content incorporated in the film. It is possible to optimize the absorption coefficient of a Me-DLC film by varying the metal content. This makes them potential candidates for use as selective solar absorber coatings.

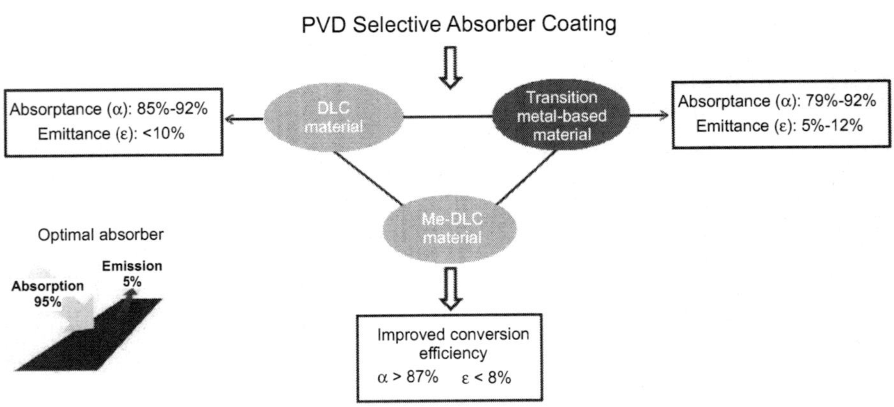

Fig. 7.4 Absorptance and emittance of PVD solar selective coatings: DLC, transition metal-based and Me-DLC materials.

A selective absorber coating considered optimal for reaching high efficiency should have high absorption (>95%) and low emittance (<5%). Figure 7.4 shows the absorptance and emittance for DLC, transition metal-based and Me-DLC materials produced by PVD techniques.

A service lifetime of more than 25 years is expected for Me-DLC coatings due to the absence of grain boundaries which might be susceptible to corrosion. On the other hand, the oxidation of the metal particles could decrease the absorber lifetime (Tinchev et al. 2010).

In this chapter, Me-DLC film technology and its use in selective solar absorbers is addressed as follows. Next different types of selective solar absorber coatings are described. Then an overview on the synthesis and properties of Me-DLC films are provided. The effects of the metal type on the properties of Me-DLC film are discussed. Later the Me-DLC solar selective coatings, using different transition metals, are described and their performances are compared.

2. SELECTIVE SOLAR ABSORBER COATINGS

The solar absorber is a key component in the solar thermal system, so its performance significantly affects the efficiency of all systems. The most basic structure of the absorber consists of a thin solar radiation absorbing layer over a metal surface, mainly aluminum and copper. The working principle is to convert solar radiation into heat, which is then passed through the metal surface and into the working fluids of the system (Jain 2014).

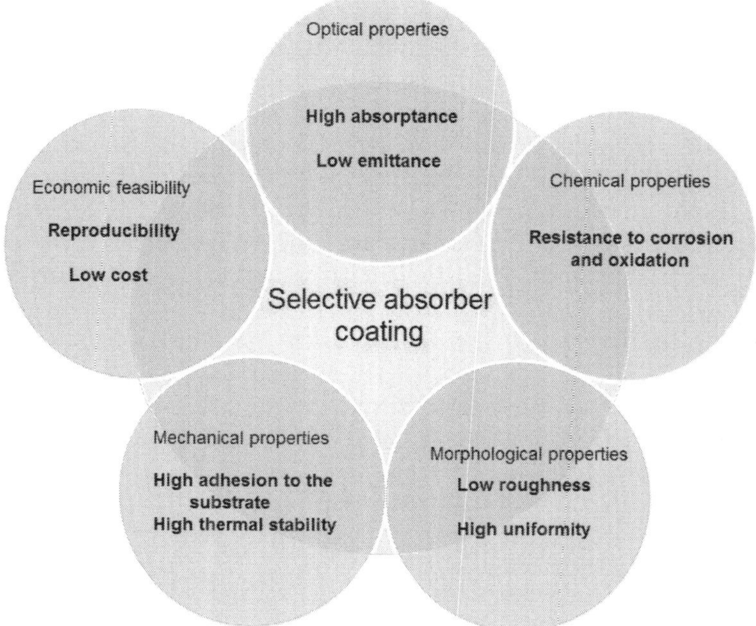

Fig. 7.5 Main requirements for selective absorber coatings.

A good absorber is expected to be: (i) selective (high absorptivity and low emissivity) and (ii) thermally stable to operate properly solar energy harvesting at elevated temperatures over time. Figure 7.5 summarizes the required properties of a selective solar absorber surface. Along with the suitable material properties of the coatings, their production process should have low cost and high repeatability (Ladgaonkar and Patil 2014).

An effective selective absorber coating should have a high solar absorptance (α) in the solar radiation range (0.3-2.0 μm) and a low emittance (ε) in the heat irradiation (2.0-50 μm) over the operating temperature of an absorber plate (Lee 2007). The ideal is that a spectrally selective absorber has zero reflectance in the visible region and high reflectance in the infrared region (Selvakumar and Barshilia 2012).

Long term thermal stability is also an important parameter for spectrally selective coatings used in collectors. For estimating the acceptable service lifetime of an absorber coating, it is necessary to calculate the Performance Criterion (PC) function that is defined by:

$$PC = -\Delta\alpha + 0.25\Delta\varepsilon \times 0.05 \quad (1)$$

where $\Delta\alpha$ is the change of solar absorptance and $\Delta\varepsilon$ is the emittance on the solar fraction. The International Energy Agency-Solar Heating and Cooling program recommends an initial thermal stability test at 250°C for 200 hours followed by an adhesion test. After this test, the absorber coating is qualified without further tests if the PC $\leq$ 0.015 and surface adhesion is 0.15 MPa (Selvakumar and Barshilia 2012).

In nature, there are no materials with ideal solar selective properties (i.e. that absorb all solar radiation without emitting). Metals have low solar absorptance and low emittance whereas semiconductors and metal oxides have moderate absorptance and high thermal emittance. However, reports show that some metal compounds and semiconductors after their compositions have been modified greatly to achieve the ideal characteristics. These materials are known as intrinsic selective absorbers. Besides intrinsic absorption, the selective coatings can be based on other different optical absorption mechanisms including light trapping, particulate coatings, semiconductor-metallic layers, multilayer films and quantum size effects (Jain 2014).

Spectrally selective coatings have been designed based on six main concepts: (i) intrinsic absorber, (ii) metal-semiconductor tandem (also known as absorber-reflector tandem), (iii) multilayer absorber, (iv) metal-dielectric composite (also known as cermet), (v) textured surface and (vi) solar-transmitting coating on a blackbody absorber. Table 7.1 describes these concepts. Of them, cermet layer deposited on a highly infrared-reflective metal is the most flexible method to achieve a good selectivity (Nejati 2008). Furthermore, the performance of cermet is the highest of the simple structures.

Table 7.1 Types of selective absorber coatings.

Type of selective solar absorber	Selective coating material	Some literature examples	Schematic design
Intrinsic	Intrinsic selective properties are found in transition metals and semiconductors, but both need to be greatly modified to serve as an intrinsic absorber. Drawback: There are no ideal intrinsic materials.	Transition metal such as tungsten and semiconductor materials such as MoO_3-doped Mo, Si doped with B and SnO_2.	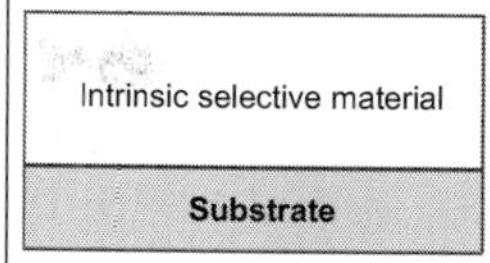
Semiconductor metal-tandem	Semiconductors with band-gaps from about ~0.5 eV (2.5 μm) to 1.26 eV (1.0 μm) absorb short-wavelength radiation and the underlying metal provides low emittance to give the desired spectral selectivity to semiconductor-metal tandems. Drawback: The semiconductor coatings have high refractive indices. Thereby to obtain high solar absorptance, it is necessary using an anti-reflection layer.	Si (1.1 eV), Ge (0.7 eV) and PbS (0.4 eV).	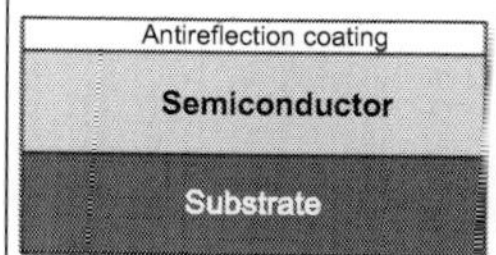
Multilayer	The selective effect is because the multiple reflectance passes through the bottom dielectric layer and is independent of the selectivity of the dielectric. A thin semitransparent reflective layer, typically a metal, separates two quarter-wave dielectric layers. Drawbacks: This absorber type is very sensitive to the thickness of layers. Another problem is the multilayer deterioration at elevated temperature due to inter-diffusion between the layers.	Several multilayer absorbers using different metals (e.g., Mo, Ag, Cu, Ni) and dielectric layers (e.g., Al_2O_3, SiO_2, CeO_2, ZnS). A typical example of a multilayer absorber is Al_2O_3/Mo/Al_2O_3.	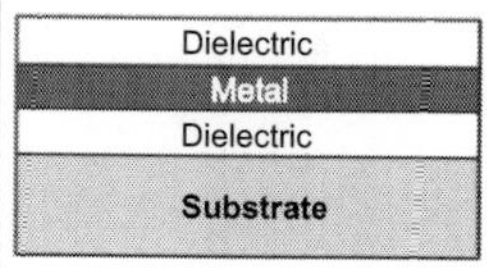
Cermet (Metal-dielectric composites)	The highly absorbing metal-dielectric composite or cermet, consists of fine metal particles in a dielectric or ceramic matrix or a porous oxide impregnated with metal. Drawback: Oxidation of metallic particles in cermet.	Nickel pigmented anodic aluminum oxide and sputtered nickel/nickel oxide.	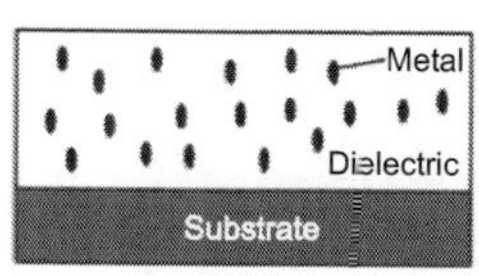

Table 7.1 Contd.

Type of selective solar absorber	Selective coating material	Some literature examples	Schematic design
Textured surface	Textured surfaces are known as optical trapping surfaces. The rough surface absorbs solar energy by trapping the light through geometric effects of multiple reflection and absorption. Drawback: It is not much used because thermal emittance is too high and durability is poorer.	A dendrite structure of Re, W and Ni made by CVD and textured Cu, Ni and stainless steel made by sputter etching.	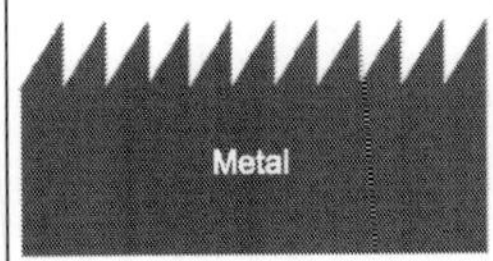
Solar-transmitting coating/ blackbody-like	The selective solar-transmitting coating can be a highly doped semiconductor over an absorber with a proven long-term durability. Drawback: Control of doping is crucial for enhancing selectivity.	SnO_2:F, SnO_2:Sb, In_2SO_3:Sn, ZnO:Al.	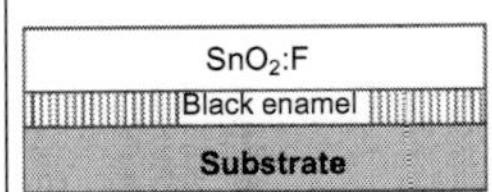

Generally, the selective absorbers based on cermet coatings are integrated with an Anti-Reflection layer (AR) and/or an IR reflective layer (either an intrinsically IR-reflecting substrate or an IR-reflective coating on a substrate). Figure 7.6 shows the single and double cermet absorber structures. Each layer has its own function: AR coating reduces the solar reflection off the surface; the cermet provides selective absorption, the IR reflector (typically metals with low intrinsic emissivity) reduces radiation losses and the substrate is usually metal (to conduct heat well) or glass (for lower cost) (Cao et al. 2014).

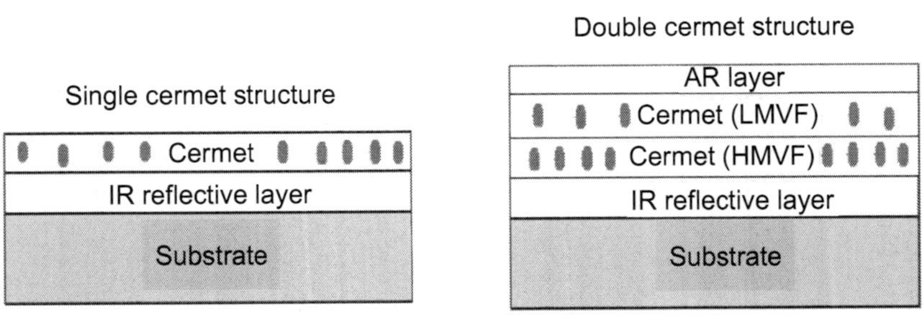

Fig. 7.6 Cermet absorber structures: single and double with AR layer.

In double cermet structures, the two cermet layers have different metal volume fractions: one with High Metal Volume Fraction (HMVF) and another with Lower Metal Volume Fraction (LMVF). The cermet layers are located between an IR reflective layer and an AR layer. This configuration has photo-thermal efficiency higher than the single cermet structure (Nejati 2008).

It is necessary to optimize metal volume fraction, in LMVF and HMVF cermet layers, to obtain the best optical properties. Table 7.2 shows some examples of cermet structures with their absorptance and emittance. The main cermet coatings that are commercially available include Pt-Al_2O_3, Ni-Al_2O_3, Mo-Al_2O_3, W-Al_2O_3, Mo-SiO_2, Ni-SiO_2, etc.

Table 7.2 Comparison between the characteristics of single and double cermet structures.

Structure	Cermet	Deposition technique	α	ε	Ref.
Single cermet	Ni-Al_2O_3	RF-Sputtering	0.94	0.07	Sathiaraj et al. 1990
	Ni-Al_2O_3	Electroplating	0.85	0.08	Salmi et al. 2000
	Ag-Al_2O_3	Evaporation	0.7	0.1	Bogaerts and Lampert 1983
	Au-MgO	RF-Sputtering	0.93	0.09	Fan 1981
	Cr-Cr_2O_3	RF-Sputtering	0.92	0.08	Fan 1981
Double cermet	Al-N	DC-Sputtering	0.96	0.08	Zhang et al. 1999
	W-AlN	DC-Sputtering	0.92	0.08	Zhang 2000
	Mo-AlN	DC-Sputtering	0.94	0.11	Zhang 2000
	SS-AlN	DC-Sputtering	0.95	0.05	Lee 2001

Metal-containing DLC coatings are a typical example of cermet material, in which metal nanoparticles are embedded in a dielectric matrix. They are recognized as promising for the fabrication of selective solar absorbers coatings based on this concept because of their ability to control their optical properties by varying the metal content in the deposition process and thus optimize their absorption coefficients (Tinchev et al. 2010).

Selective absorber coatings based on a-C:H/Cr films are already commercially produced for solar hot water applications as will be shown in later. Next properties of Me-DLC films and their synthesis processes, which are important steps towards to designing high-performance coatings for solar thermal energy conversion, will be discussed.

3. SYNTHESIS AND PROPERTIES OF Me-DLC FILMS

A great variety of methods have been developed to deposit Me-DLC films with a wide range of properties to meet the demands of different applications. Figure 7.7 shows the most used techniques to grow Me-DLC thin films. Besides traditional PVD methods, such as magnetron sputtering and cathodic arc plasma, dual plasma deposition, combining PVD and PECVD (Plasma Enhanced Chemical Vapor Deposition), have been used. In general, the deposition techniques can also be used to produce hydrogenated DLC films by changing the carbon source from pure carbon to hydrocarbon gases. Each technique has its own advantages and disadvantages in terms of deposition rate, film quality, uniformity and cost.

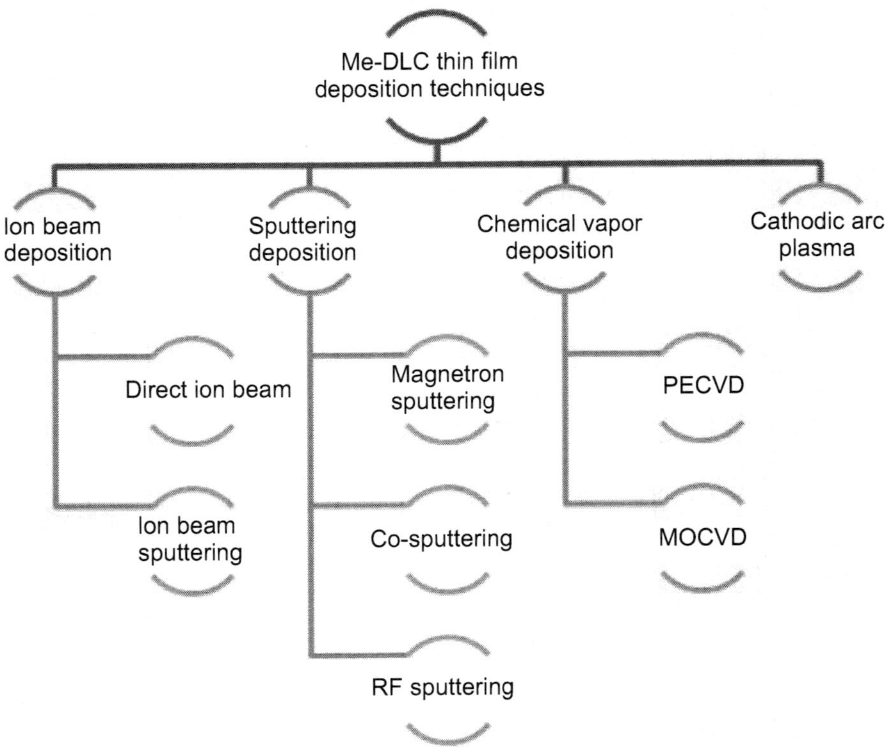

Fig. 7.7 Main techniques used to deposit Me-DLC films.

The choice of the most suitable technique depends on the desired properties of the Me-DLC coating. Furthermore, the metal type incorporated into DLC film is also an important parameter. For example, hydrogenated W-DLC and Mo-DLC coatings exhibit high thermal stability (up to 500°C) avoiding oxidation of the film, whereas Ti-DLC exhibits better tribological properties.

Independent of the metal used, the high mechanical strength and enhancing optical absorption are typical characteristics of metal-containing DLC films when compared to pure DLC.

Here the fundamental properties of Me-DLC and DLC films are described and compared. In addition, the basic principles of the main techniques employed to deposit Me-DLC films are presented with the focus in PVD processes.

3.1 Properties of DLC and Me-DLC Films

According to the International Union of Pure and Applied Chemistry (IUPAC) definition "Diamond-like carbon (DLC) films are hard amorphous films with a significant fraction of sp^3 hybridized carbon atoms and which can contain a significant amount of hydrogen. At the atomic level, amorphous DLC presently refers to the group of carbon materials with strong chemical bonding composed of a mixture of sp^2 and sp^3 arrangements of atoms incorporated into an amorphous structure. Depending on the deposition conditions, the amorphous carbon material may contain diamond crystallites. These materials are not called diamond unless a full three-dimensional crystalline lattice of diamond is proven."

A common definition of DLC films is that they are metastable amorphous carbon materials (a-C, ta-C, a-C:H and ta-C:H) that display some of the typical properties of diamond. Table 7.3 compares the properties of different forms of amorphous carbon with diamond and graphite ones.

DLC films, hydrogenated or non-hydrogenated, are usually applied as coatings to other materials that benefit from their outstanding properties similar to diamond with the advantage of being much cheaper to produce. In contrast to the production of diamond films, DLC films can be deposited at room temperatures (Stoica 2012).

Table 7.3 Comparison of the properties of amorphous carbons with those of the diamond and graphite (adapted from Robertson 2002).

	sp^3 (%)	H (%)	Density ($g \cdot cm^{-3}$)	Gap (eV)	Hardness (GPa)
Diamond	100	0	3.515	55	100
Graphite	0	0	2.267	0	0.2-10
ta-C	80-88	0	3.1	2.5	80
ta-C:H	70	30	2.4	2.0-2.5	50
a-C:H	40-60	30-50	1.2-2.2	1.1-4	<10-20
Sputtered C	5	0	2.2	0.5	3-6
Glassy C	0	0	1.3-1.55	1.6	3

Ferrari and Robertson (2000) proposed the model, shown in Fig. 7.8, which represents the DLCs and other carbon films in a ternary phase diagram. This diagram is considered a comprehensive study of the structural and chemical properties of these films. The regions of the DLC films are identified and based on the fraction of sp^3 bonds and hydrogen content. The films are classified ranging from hydrogenated amorphous carbons (or a-C:H) to tetrahedral amorphous carbon (or ta-C) (Robertson 2002).

Doping of DLC films have been investigated in order to overcome two limitations of pure DLC. First, the DLC coating thickness is limited by the buildup of residual stress, which can lead to delamination failure as film thicknesses increase. The second factor is that at relatively low temperatures (250°C) the properties of the DLC begin to degrade (i.e. DLC converts to graphite). By 400°C the graphitisation process is rapid, which limits the maximum service temperature of DLC to around 300°C (Hainsworth and Uhure 2007).

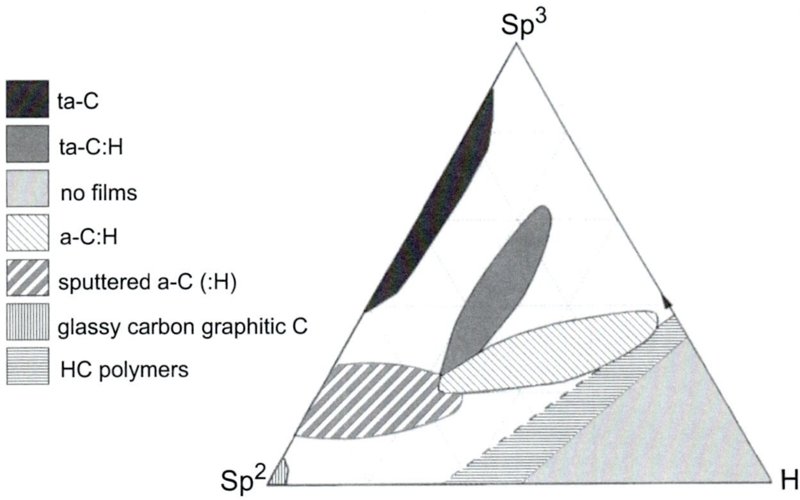

Fig. 7.8 Ternary phase diagram of bonding in amorphous carbon (adapted from Robertson 2002).

Hence the motivations to carry out doping are: reducing the internal stress, increasing the adhesion strength and increasing the maximum service temperature of the DLC film. The dopants can be categorized into two groups: (i) light elements such as nitrogen, silicon, oxygen, fluorine and deuterium and (ii) metals such as titanium, copper, tungsten and chromium.

Structural investigations of metal-doped DLC (Me-DLC) materials have shown that the metallic component does not distribute homogeneously in the DLC matrix but forms small metal carbide clusters with nanometer sizes (Fig. 7.9), which leads to a unique combination of properties, e.g., good adhesion with the substrate, high hardness, low friction coefficient, high thermal stability and high electrical conductivity (Li et al. 2015).

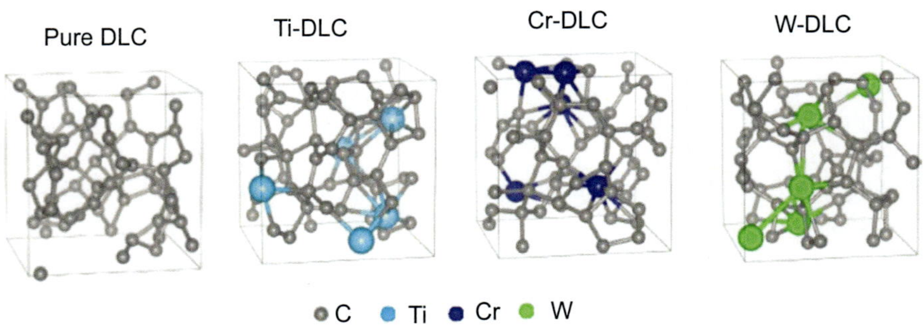

Fig. 7.9 Configurations for DLC and for the (Ti, Cr or W)-DLC films with a metal concentration of 7.81 at.% (adapted from Li et al. 2015).

Figure 7.10 shows a plot which relates the percentage of sp^3 bonding and hydrogen content for the various forms of carbon, including a-C:H:Me, a-C:Me and a-C:H:X (X = Si, O, N, F). This figure shows the two extremes, diamond and graphite and every combination of bonding in between. Their properties range from graphite-like to diamond-like to polymer-like, depending on the deposition conditions. Hardness definitely increases with increased sp^3/sp^2 ratio (Martin 2011).

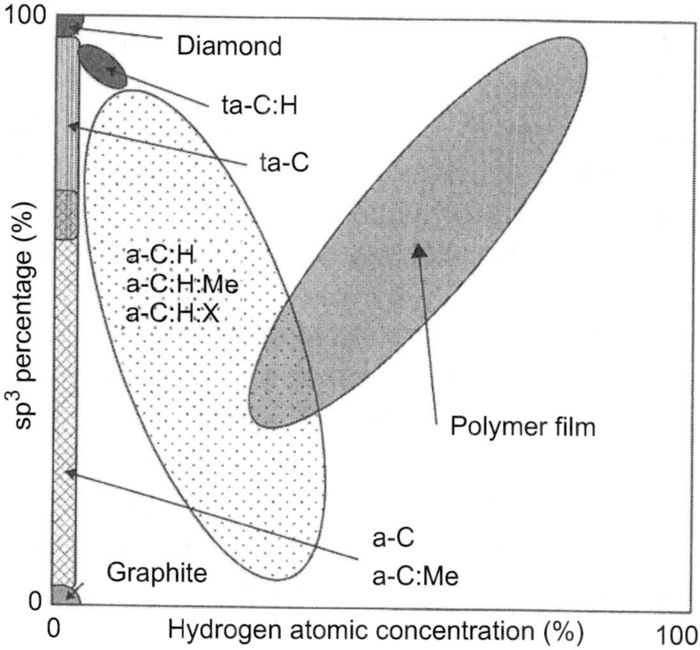

Fig. 7.10 Percent sp^3 bonding for thin film carbon allotropes (adapted from Martin 2011).

Me-DLC films exhibit improved hardness and elastic modulus and can give a combination of hardness and ductility. The hardness is tailored by the coating deposition process and can range from 500 to 2500 VHN (Hainsworth and Uhure 2007). Table 7.4 summarizes the main Me-DLC properties in comparison with pure DLC.

Table 7.4 Main metal-doped DLC material properties in comparison with pure DLC.

–Lower compressive stress.
–Higher hardness/Young's modulus ratio.
–Improved adhesion (Cr, Ti, W).
–Better toughness.
–Improved biocompatibility (Ti).
–Improved electric conductivity.
–Better wear resistance in certain cases for Cu, Ti, Cr and W (for metal content < 30-40%).
–Improved tribological performance with certain lubricants (Ti, W).

Hydrogenated Me-DLC coatings produced by PVD processes are commercially available. One in particular, the W-DLC coating has been shown as a chemical inert film with high elasticity (Young's modulus range from 100-120 GPa), good wear resistance and residual compression with a residual stress of ~900 MPa (Hainsworth and Uhure 2007). A Cr-adhesion layer with a W-DLC coating on a steel substrate showed interface toughness superior to that of pure DLC. The advantage of this is that failure occurs within the DLC itself rather than at the coating-substrate interface (Wang et al. 1998).

Regarding the optical properties, the refractive indexes of Me-DLC coatings with small metal contents are similar and do not depend on the metal type. For example, Pt-DLC (1.5 at.% Pt), Cu-DLC (1.5 at.% Pt) and W-DLC (3 at.% W) coatings have the following refractive indexes, respectively: 1.75, 1.69 and 1.8 (Rutkuniene et al. 2013). On the other hand, the refractive index depends on metal content. An example: the refractive index value of W-DLC film (3 at.% W) is 1.8 whereas the of W-DLC (10 at.% W) is around of 2.2. This possibility of varying the optical constants of W-DLC film to optimal values by adjusting the film composition is an advantage (Gampp et al. 1994).

The small metal content into Me-DLC films affects the emittance: DLC containing a small fraction of metallic elements shows a significant increase in the whole range of measurement, while the transmittance in the IR range is considerably decreased so the coating is almost opaque. In the case of Ti-DLC coatings, it was observed that they behave almost as a black body, where the transmittance just goes to zero (Sánchez-López and Fernández 2008).

3.2 An Overview on The Main Techniques Used to Deposit Me-DLC Films

Metal doping has been developed to create a two dimensional array of nanocrystalline metal clusters and metallic carbides within the amorphous carbon matrix and thus improves DLC films by enhancing adhesion, thermal stability and toughness. First reports on the synthesis and properties of Me-DLC were published in late 1980s. Klages and Memming (1989) discussed the properties of various types of metals (Au, Cu, Ag, Cr, Ti, Ta and W) containing hydrogenated amorphous carbon films. Dimigen and Klages (1991) investigated Me-DLC films deposited by sputtering of a metal target in argon + hydrocarbon plasma. They observed a granular structure in the films consisting of nanocrystalline metals or carbides particles in a DLC matrix.

Figure 7.11 shows the year in which each DLC type was first deposited by a certain technique. Nowadays, there are several methods employed for the Me-DLC films, which provide a good flexibility in tailoring their properties according to specific needs or potential applications. It is important to understand the interrelationship of coating properties, synthesis process and desired application. Since the deposition conditions influence the material properties (coating thickness, composition, microstructure, optical, mechanical and electrical characteristics) whereas the coating properties determine its potential application.

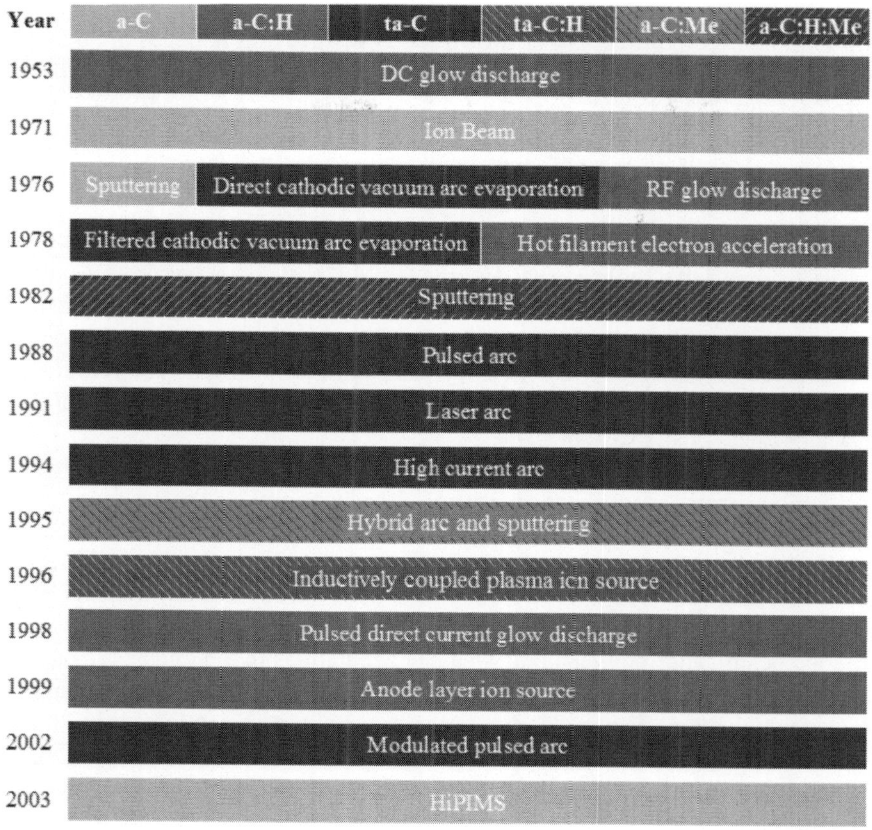

Fig. 7.11 Types of DLC thin films with a historical overview (adapted from Martin 2010).

Table 7.5 Deposition process characteristics (adapted from Martin 2010).

	Ion plating	Sputtering	Chemical vapor deposition	Cathodic arc
Mechanism of production on deposited species	Thermal energy	Momentum transfer	Chemical reaction	Thermal energy
Deposition rate	Can be very high – up to 250000 Å/min	Low, excepted to metal or co-sputtering	Moderate – 200 to 25000 Å/min	Can be very high
Deposition species	Atoms and ions	Atoms and ions	Atoms	Ions
Complex shaped objects coverage	Good, but non-uniform thickness	Good, non-uniform and uniform thickness	Good	Good, but non-uniform thickness
Coverage of small blind holes	Poor	Poor	Limited	Poor
Energy of deposited species	Can be high ~1 to 100 eV	Can be high ~1 to 100 eV	Can be high with plasma assisted CVD	Can be high

Although the majority of the techniques for metal insertion in DLC films use *in situ* doping, hybrid processes, on which the DLC thin film is deposited by one technique and the metal doping is realized by other technique, have also been reported (Stoica 2012). Here we describe four deposition techniques: ion beam deposition (a special case of ion plating), sputtering, CVD and cathodic arc. Table 7.5 compares some characteristics of these techniques.

3.2.1 Ion Beam Deposition

The first diamond-like carbon was deposited in 1971 by Aisenberg and Chabot (1971) by ion deposition. In this technique, the carbon ions are produced by plasma sputtering of a graphite cathode at the ion source and electrostatically accelerated towards the target (Robertson 2002). They observed that, with this configuration, the kinetic energy supplies the energy necessary to the nucleation and growth of the thin film (Aisenberg and Chabot 1971).

Since then, DLC thin films have been widely produced by ion beam deposition. This technique allows an independent control of the ion energy and of the ion current density (Gago et al. 2006). Ion beam deposition can be considered an ion implantation where bombarding ions penetrate to subsurface layers and bond with target atoms to grow high quality films. The carbon source used in ion beam deposition can be a solid carbon target or a hydrocarbon gas (Toyoda and Matsui 2014).

The ion beam deposition techniques gained adaptations in the ion source or ion gun and in the interaction between the gas molecules and substrate atoms with the ions (Gago et al. 2006). For the incorporation of metal atoms into DLC thin films, the hybrid technique is commonly used. Dai and Wang (2011a, b) deposited Al-DLC thin films using a direct ion beam deposition system with a linear ion source to produce DLC thin films with acetylene gas while magnetron sputtering system was used for doping the thin film with aluminum. The direct ion beam deposition uses a reactive gas that is introduced in the ion source where they are ionized to produce reactive ions. When the ions collide with the surface, they are absorbed and react to form the thin film (Gago et al. 2006). In Fig. 7.12a, a schematic representation of a direct ion beam deposition system is shown. Ion sources operate efficiently in an ion energy range of 100-1000 eV. This technique is widely used to deposit hydrogenated amorphous carbon; however the deposition of free hydrogen amorphous carbon is not possible (Gago et al. 2006).

Besides direct ion deposition, other deposition techniques have been used to produce non-hydrogenated DLC films including ion assisted evaporation, pulsed laser deposition, Mass-Selected Ion Beam Deposition (MSIBD), sputtering, filtered cathodic arc deposition and arc discharge. Such techniques can also be used to produce hydrogenated DLC films by changing the carbon source from pure carbon to hydrocarbon gases.

Bharathy et al. (2010, 2012) used a hybrid ion beam system to enhance the mechanical properties of DLC thin films with tungsten nanoparticles. They have used a direct ion beam deposition with an argon and methane mixture to produce the DLC thin films and an ion beam sputtering deposition system using argon as ion beam gas and tungsten target for doping the thin film. In ion beam sputtering technique, an ion beam with high energy is used to sputter a target material enabling the deposition from the sputtered atoms on the substrate surface. In Fig. 7.12b the schematic representation of an ion beam sputtering deposition system can be seen. The advantages of this technique are the lower pressures, cleaner vacuum atmosphere and flexibility in the selection of the target geometry, morphology and thermal or electrical characteristics. On the other hand, the ion beam sputtering has a relatively low deposition rate (Gago et al. 2006).

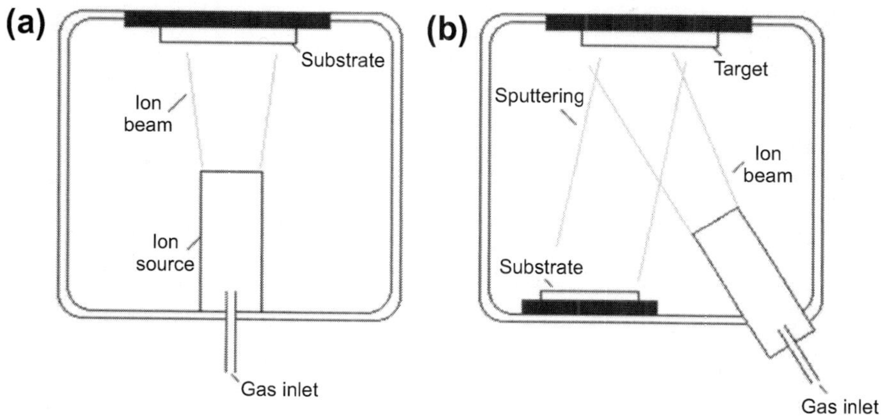

Fig. 7.12 Schematic representation of ion beam deposition systems: (a) Direct ion beam. (b) Ion beam sputtering.

3.2.2 Sputtering

Sputtering is the most used technique to deposit DLC thin films (Robertson 2002). In this process, the target is placed in the cathode and the substrate in the anode. They are placed at a fixed distance (target-substrate distance) inside a vacuum chamber with an inert gas flow. When the discharge voltage is applied, the inert gas is ionized, forming the plasma. The ions are accelerated in direction of the target sputtering and the ejected atoms form the thin film (Bräuer 2014).

Different configurations can be used to improve the sputtering process, as can be observed in Fig. 7.13. Sonoda and co-workers (2009) used a DC magnetron co-sputtering to deposit Ti/C nanocomposite DLC thin films. In this technique, a magnetron cathode is used to enhance the ions formation, resulting in an increase in the deposition rate. They used two cathodes, one to the carbon target and the other to titanium target to deposit Ti-DLC thin films. Besides Me-DLC films, the reactive sputtering is widely used to deposit oxides, nitrides or carbides formed by the addition of corresponding reactive gas (Bräuer 2014).

Cui et al. (2012) deposited Ti-DLC thin films in a DC reactive magnetron sputtering system using a mixture of methane and argon flow for the formation of DLC thin films while a Ti target was used for doping.

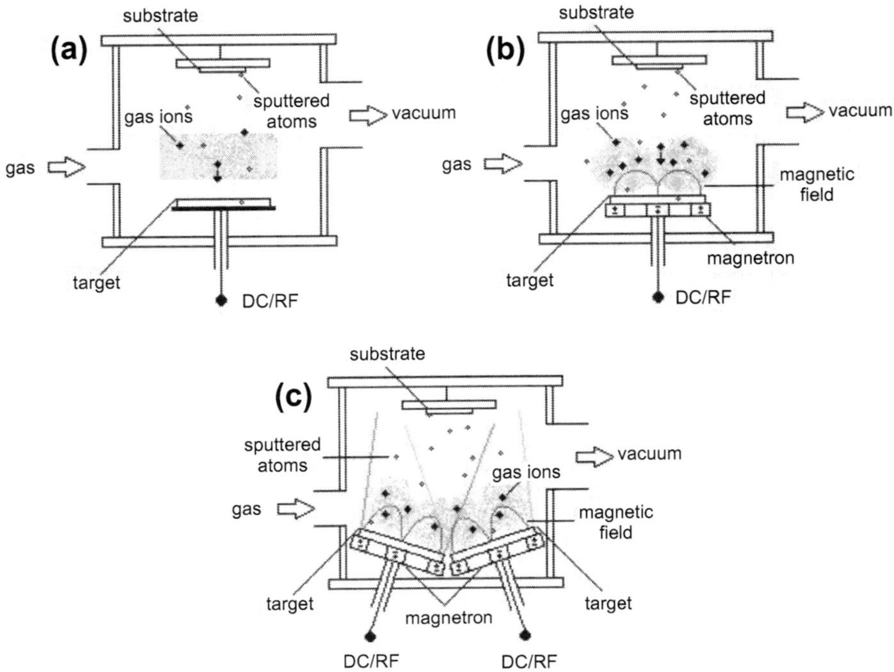

Fig. 7.13 Different configurations of sputtering: (a) Conventional sputtering. (b) Magnetron sputtering. (c) Co-sputtering.

Harding (1976) prepared metal carbide films on copper by reactive sputtering of pure metals (Cr, Fe, Mo, W, Ni and Ta) or metal mixtures (e.g., stainless steel) in a gas mixture containing 1-2% methane in argon. The seven sputtered metal carbide selective surfaces produced exhibited high absorptance values (75 to 90%) and low emittance values (2 to 5%).

In a subsequent paper, the same author also demonstrated that the deposition of a a-C:H on top of the sputtered chromium carbide absorber layer increased the absorptance by 2 with 1% increase in the emittance (Harding 1980).

3.2.3 Chemical Vapor Deposition

In Chemical Vapor Deposition (CVD), the thin film is deposited on a heated substrate due to a gas chemical reaction that occurs on or near the substrate (Carlsson and Martin 2010). The schematic of CVD process is showed in Fig. 7.14a. CVD techniques allow controlling sp^3 percentage in DLC films only modifying process parameters such as gas flow, substrate bias, discharge and power. The disadvantage of CVD is that the chemical reactions occur at high temperature (more than 600°C); however the use of plasma can offset this problem (Pierson 1999).

Zhao and co-workers (2010) deposited Ti-DLC thin films using Plasma Enhanced Chemical Vapor Deposition technique (PECVD, also called plasma assisted/activated chemical vapor deposition) with an RF frequency to deposit DLC and an unbalanced magnetron sputtering to dope with Ti. The PECVD can be isothermal (called arc plasma) that uses approximately 1 MHz of frequency or non-isothermal (called glow-discharge) that uses a frequency between 3.45 MHz and 2.45 GHz [normally using microwave (MW) or radio frequency (RF)]. In order to deposit Me-DLC thin films are more common than the use of MW-PECVD and RF-PECVD, as shown in Figs. 14b and c, respectively (Pierson 1999).

A Metal Organic Chemical Vapor Deposition (MOCVD) can be used to deposit Me-DLC thin films using a lower temperature (Yang 2014). Luithardt and Benndorf (1997) used MOCVD with a capacitively coupled RF plasma to deposit Fe-DLC thin films. This technique uses an organometallic precursor to deposit thin films (Kern and Schuegraf 2002). A metal organic is a compound in which an atom of one element is bound to carbon atoms and normally is used a metal of IIA, IIB, IIIB, IVB, VB and VIB groups (Pierson 1999). In Fig. 7.14d a representation of the system is shown.

Gampp et al. (1994) noticed that with the suitable choice of the deposition parameters, in a combined CVD-r.f. sputtering process, the composition, optical constants and morphology of a-C:H/W and a-C:H/Cr coatings can be tailored in such a way as to render them suitable for the application as solar selective absorber coatings in flat plate collectors.

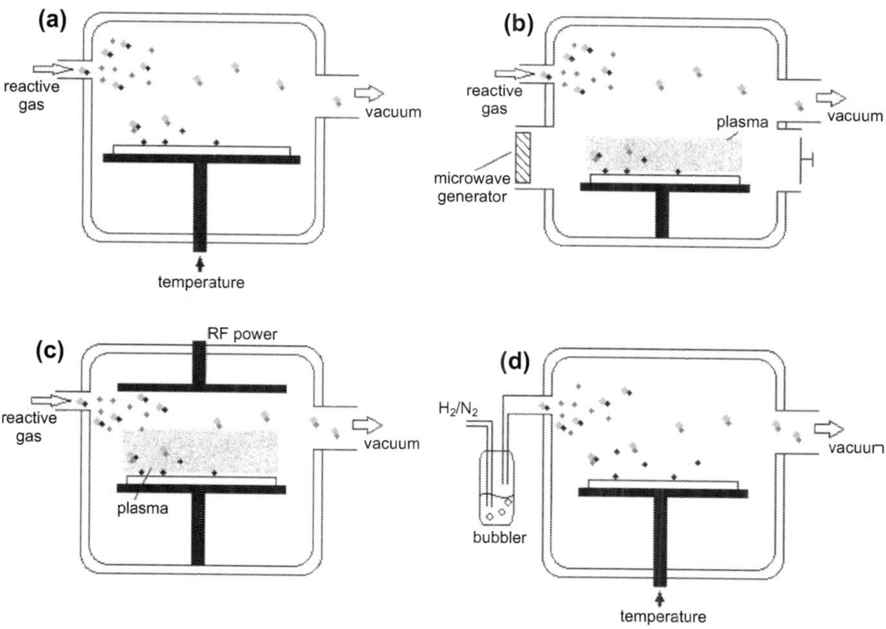

Fig. 7.14 Schematic representation of CVD process: (a) Thermal CVD. (b) MW-PECVD. (c) RF-PECVD. (d) MOCVD.

3.2.4 Cathodic Arc Deposition

Cathodic Arc Deposition (CAD also called Cathodic Arc Vacuum Deposition CAVD) is not a usual technique, but its history began along with the history of electricity and the ability to generate electric current (Robertson 2002, Koskinen 2014). When the cathodic arc discharge is applied on the electrode plasma is formed, so the solid electrode and the ionized species condense on the substrate forming the thin film and the process do not required an ambient gas because the plasma species have originated on the solid cathode (Anders 2010, Koskinen 2014). The CAD schematics can be seen in Fig. 7.15a.

The arc can be continuous or pulsed. For the pulsed mode, a capacitor bank is used to strike the arc. With this technique, it is possible to miniaturize sources and have lower average power (Robertson 2002, Koskinen 2014). In Fig. 7.15b, the schematic representations of pulsed cathodic arc deposition system is shown.

The disadvantage of cathodic arc deposition is the possibility of deposition of macroparticles on the substrate. This problem can be solved using a filter, which prevents macroparticles reaching the substrate. Sansongsiri et al. (2008) used a pulsed dual cathode filtered cathodic arc vacuum deposition to produce Mo-DLC thin films. In this system, the plasma pass along a toroidal magnetic filter duct, which improve the technique by using highly ionized plasma resulting in a high deposition rate. The schematic of a filtered cathodic arc deposition is shown in Fig. 7.15c.

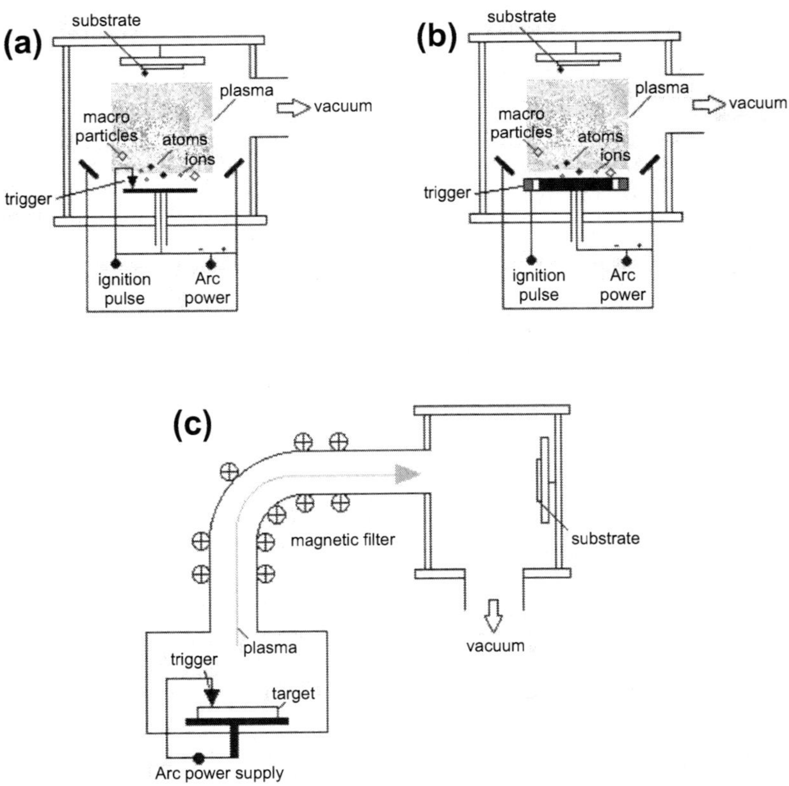

Fig. 7.15 Schematic representation of: (a) Cathodic arc deposition. (b) Pulsed cathodic arc deposition. (c) filtered cathodic arc deposition.

Yin et al. (1996) discussed the possibility of applying the cathodic arc deposition technology for the production of solar thermal selective surfaces. One of the structures produced was an a-C:H/a-C:H-SS cermet/SS selective surface. The acetylene flow rate was increased by a few steps from near the Stainless Steel (SS) layer to the top layer of almost pure a-C:H material. A low reflectance in the solar radiation range was observed, which is apparently due to the absorption by the a-C:H-SS cermet layer. This result indicated that it is possible to apply the cathodic arc deposition technique to metal-dielectric cermet deposition in commercial production (Yin et al. 1996).

4. EFFECTS OF METAL TYPE ON THE PROPERTIES OF Me-DLC FILMS

DLC thin films have excellent properties like high hardness, high wear resistance and low friction coefficient (Weber et al. 2006). However, they exhibit high residual stress (Qiang et al. 2015). According to many authors, the main benefits of incorporating metals into DLC coating are the reduction of compressive internal stress and enhancement of adhesion (Zhao et al. 2010, Qiang et al. 2015). In order achieve such benefits the incorporation of metal should be limited to a maximum 30-40 at.%. Above this value it produced an increment in hard carbide and metallic phases resulting in increased abrasive and adhesive wear combined with higher friction coefficients (Sánchez-López and Fernández 2008).

Me-DLC thin films consist of metal or metal carbide clusters in a DLC matrix, depending on the type of metal and its concentration in the thin film. The transition metal of group IV to VI (Ti, Cr, W and others) bond with carbon forming metal carbide phases, while other metals (Fe, Al, Ni and others) are relatively inert to carbon (Bewilogua et al. 2000, Sonoda et al. 2009, Dai and Wang 2011a). These metal clusters or metal carbide clusters in DLC matrix can significantly modify the thin film structure, affecting the DLC properties (Dai and Wang 2011b).

Broitman and Hultman (2014), using the keywords 'diamond-like' and 'carbon films', performed a research in Google Scholar database to determine the number of scientific papers published on DLC thin films. Using the same mechanism and adding the keyword 'metal doped', we performed a research which resulted in 460 papers published on Me-DLC thin films from 1970 to June 2015. The number of papers for each metal is summarized in Fig. 7.16.

4.1 Deposition Rate and Chemical Composition

The deposition rate of a thin film depends mainly on the deposition method used. As previously shown, among the thin film deposition techniques described, the ion beam deposition and cathodic arc deposition have a higher deposition rate (100 – 250,000 Å/min), while CVD has a moderated deposition rate (200 – 25,000 Å/min) and sputtering has the slowest deposition rate (25 – 10,000 Å/min). However, the deposition rate is also affected by deposition conditions, including pressure, gas flow, temperature, delivery power and mainly thin film material (Martin 2010).

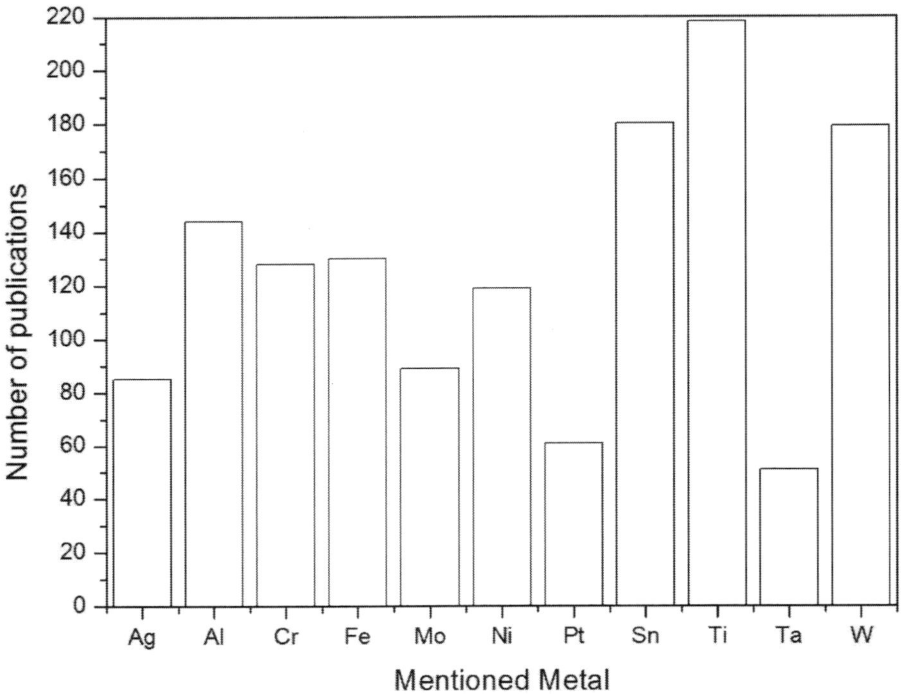

Fig. 7.16 Number of scientific papers published that mentioned metal doping of DLC film.

In the sputtering process, high deposition rates (~10,000 Å/min) can be obtained only for target materials of high thermal conductivity, as for example some metals like copper or through the use of magnetron that can significantly increase the deposition rate (Ohring 2002, Depla et al. 2010). In Fig. 7.17, it can be seen that the sputtering yield of carbon and some metals used to deposit Me-DLC thin films as a function of ion energy using argon as the projectile (Grill 1999). When a reactive magnetron sputtering is used, the metals sputtering yield can significantly decrease due to the compounds formation on the target (Ohring 2002). The use of RF instead DC power supply can also decrease the deposition rate (Depla et al. 2010).

In the CVD technique, the deposition rate can be increased when assisted by plasma (PECVD) and controlling some process parameters, such as electrode, reactant type, reactant flow and RF power (Pierson 1999). The deposition rate using cathodic arc process also depends on the process parameters, mainly the power applied. The DC cathodic arc technique exhibits high deposition rate, while the pulsed cathodic arc is lower (Gago et al. 2006).

Me-DLC thin films chemical composition depends on the metal type, due to its deposition rate and consequently, on deposition technique and parameters. According to Bewilogua et al. (2000), low content of metal at DLC matrix (atomic ratios of Me/C up to approximately 0.3) is capable of modifying DLC properties as the compressive stress. However, many authors also study the properties of Me-DLC films with high metal content.

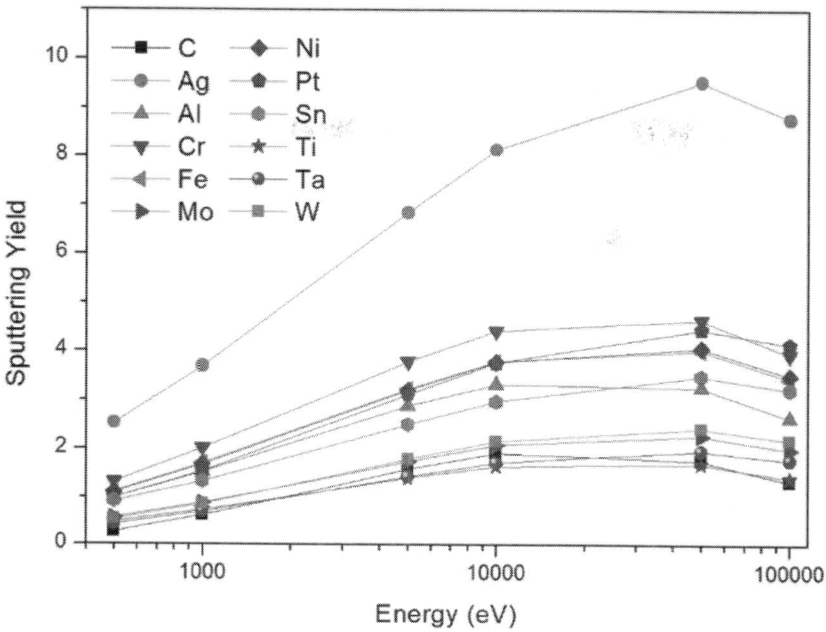

Fig. 7.17 Sputtering yield of carbon and some metals used to deposit Me-DLC thin films as function of argon ion energy (adapted from Matsunami et al. 1984).

4.2 Structure and Morphology

Some authors believe that the properties of Me-DLC thin films are modified not only by the simple addition of dopants, but also by changes that metals can induce in the DLC structure (Grill 1999, Silva and Carey 2003, Schultes et al. 2006). This due to the fact that metal impurities are not arranged as atomic impurities, but as metal clusters or metal carbide cluster. In addition, it has been observed that the metal insertion in DLC matrix can favor the formation of sp^2 bonding, inducing the DLC graphitization (Schultes et al. 2006).

Gayathri et al. (2015) deposited Ag, Ni and Ti-DLC thin films by pulsed laser deposition technique with 20 at.% of metal concentration. Using Raman spectroscopy, they observed a third peak at 1200 cm^{-1} in Me-DLC, absent on DLC, attributed to distorted C-C sp^3 stretching vibration. In addition, they also observed a shift in G peak position towards lower wavenumber to Me-DLC compared with DLC thin films that they attributed to the increase in cluster size of aromatic ring and to the reduction in internal compressive stress. The I_D/I_G ratio also varied with the metal insertion in the thin film. The Ag-DLC thin films presented the higher I_D/I_G ratio (0.74), indicating that silver induces the formation of unstrained sp^2 large clusters, followed by Ti-DLC (0.62), DLC (0.54) and Ni-DLC (0.44). The Ni-DLC lower I_D/I_G ratio indicates smaller sp^2 cluster size and higher fraction of sp^3 bonding that some authors attributed to the good lattice math with diamond (Gayathri et al. 2015). Table 7.6 compares these results with those obtained for Me-DLC films deposited by PECVD.

Takeno et al. (2008) investigated Me-DLC (Me = Co, Mo and W) thin films deposited by PECVD and co-sputtering with different metal contents. Using Raman spectroscopy, they observed a variation in I_D/I_G ratio with metal concentration. For the Co-DLC thin film, the I_D/I_G varies from approximately 0.15 to more than 0.4 with metal concentration whereas for Mo-DLC varies from approximately 0.12 to almost 0.5. However, W-DLC thin films I_D/I_G presented a constant value (approximately 0.2).

Table 7.6 Raman parameters for DLC and Me-DLC films.

Sample	Me content (%)	Deposition process	G peak position (cm^{-1})	I_D/I_G	Ref.
DLC	–	PECVD	1628	0.6	Rutkuniene et al. 2013
Pt-DLC	1.5	PECVD	1553	0.6	
Cu-DLC	1.5	PECVD	1578	–	
DLC	–	PLD	1558	0.54	Gayathri et al. 2015
Ag-DLC	20	PLD	1555	0.74	
Ti-DLC	20	PLD	1536	0.62	
Ni-DLC	20	PLD	1526	0.44	

The morphology of Me-DLC films has been explained as metal or metal carbides clusters on a carbon matrix. Schiffmann et al. (1999) investigated Me-DLC (Me = Au, Pt, Fe and W) thin films produced by PVD + CVD hybrid process. Using transmission electron and scanning tunneling microscopy, they observed that:

(i) The particles size and center of mass distance rise with the increasing metal content.
(ii) The minimum cluster size seems to be independent of the metal type, while cluster sizes spread at higher metal contents.
(iii) The Me-DLC thin films that form carbide (W and Fe) form smaller clusters than those films with noble metals (Au, Pt).
(iv) Metals with higher melting point, in general, presented smaller particles size and center of mass distance.

4.3 Mechanical Properties

The mechanical properties of DLC can vary according to its type. For example, the hardness of DLC materials varies from <10 GPa (a-C:H) to 80 GPa (ta-C) (Robertson 2014). This variation in hardness as well as in other mechanical characteristics has also been observed for Me-DLC thin films related to metal type and its content.

Bharathy et al. (2010) observed that doping DLC thin films with metallic elements can modify mechanical and tribological properties, such as reducing internal stress, enhance adhesion strength, improve wear resistance and reduce friction coefficient among other properties. In addition, other studies have reported that these properties depend not only on metal type, but also on deposition technique and dopant concentration.

Table 7.7 Mechanical and tribological properties of Me-DLC as function of metal type, metal concentration and deposition technique.

Material	Me at%	Deposition technique	H (GPa)	E (GPa)	HU (GPa)	FC	S (GPa)	Ref*
DLC	0	PECVD	10.1	83.4	–	–	–	1
DLC	0	PECVD	28	260	11.5	0.2	–	2
DLC	0	H IBD	25.6	180	–	–	–	3
DLC	0	OM CVD	24	–	–	0.15	–	4
DLC	0	H IBD	17.5	10.0	–	0.17	0.5	5
DLC	0	H IBD	–	–	–	0.207	–	6
Al-DLC	0.68	H IBD	18	26	–	0.08	1.6	7
Al-DLC	6.93	H IBD	12	21	–	0.04	0.7	7
Al-DLC	11.04	H IBD	11	20	–	0.03	0.7	7
Al-DLC	17.6	H IBD	8	17	–	0.025	0.2	7
Ag-DLC	4.3	H IBD	–	–	–	0.114	–	6
Cr-DLC	0.27	H IBD	12.5	14.5	–	0.22	0.85	5
Nb-DLC	–	R MS	–	–	–	0.2	–	8
Ti-DLC	4.6	OM CVD	17	–	–	0.23	–	4
Ti-DLC	10.8	OM CVD	14	–	–	0.23	–	4
Ti-DLC	7.2	OM CVD	12	–	–	0.22	–	4
Ti-DLC	0.4	OM CVD	8	–	–	0.18	–	4
Ti-DLC	18	U MS	14	130	6	0.2	–	2
Ti-DLC	–	R MS	–	–	–	0.25	–	8
Ti-DLC	0.3	MF MS	9.5	12.9	–	0.03	0.35	9
W/Ti-DLC	0.3/2.6	MF MS	12.8	14.9	–	0.02	0.38	9
W/Ti-DLC	0.3/10.4	MF MS	8.7	11.1	–	0.06	0.62	9
W/Ti-DLC	0.3/34.5	MF MS	6.0	9.1	–	0.08	0.96	9
W-DLC	13.7	H IBD	18	160	–	–	–	3
W-DLC	–	R DC MS	–	–	16-23	0.1-0.2	–	10
W-DLC	–	R MS	–	–	–	0.3	–	8

PE – Plasma Enhanced, OM – Organometallic, CVD – Chemical Vapor Deposition, H – Hybrid, IBD – Ion Beam Deposition, U – Unbalanced, R – Reactive, MF – Medium Frequency, DC – Direct Current, MS – Magnetron Sputtering.

*[1](Zhao et al. 2010), [2](Michler et al. 1998), [3](Bharathy et al. 2010), [4](Peters and Nastasi 2003), [5](Dai and Wang 2011b), [6](Yu et al. 2013), [7](Dai and Wang 2011a), [8](Bewilogua et al. 2000), [9](Qiang et al. 2015), [10](Weber et al. 2006).

Park et al. (2012) investigated W, Mo and Cr-DLC thin films deposited by RF magnetron sputtering. The films presented a hardness of approximately 14.5 GPa, 15 GPa and 18 GPa for Cr-DLC, Mo-DLC and W-DLC, respectively. They also observed that the film hardness depends not only on the sp^3 bonding fraction in DLC, but also on the metal carbide nanocrystalline sp^2 clusters. The friction coefficient of the thin films varied from about 0.5 to 1.0. The lowest value was for W-DLC while Cr-DLC presented the highest value. They concluded that the low friction coefficient of W-DLC thin film is due to structural transition from amorphous carbon matrix to W doped nanocrystalline carbon structure.

Table 7.7 lists the DLC material type, deposition technique and metal concentration and its tribological and mechanical properties: hardness (H), elastic modulus (E), universal hardness (HU), friction coefficient (FC) and residual stress (S).

5. THE USE OF Me-DLC FILMS AS SELECTIVE ABSORBER COATINGS

Estimating the service lifetime of a selective absorber coating is a challenge because during the aging of solar collectors two different types of degradation mechanisms may occur: (i) due to inherent instability of the coating and (ii) due to interactions between the coating and metallic substrate.

The performance characterization of Me-DLC films as selective solar absorber coatings have been an active research field since the paper by Gampp et al. published on 1994. In their studies, the deterioration mechanisms of sputtered a-C:H/W and a-C:W/Cr films deposited on different substrates (Fig. 7.18) were investigated by accelerated aging tests in order to evaluate the film stability itself at high temperatures in the air. As can be observed, the a-C:H/Cr coating was deposited only on a conducting substrate. This is because the substrate bias voltage cannot be applied with dc power supply on an isolating glass.

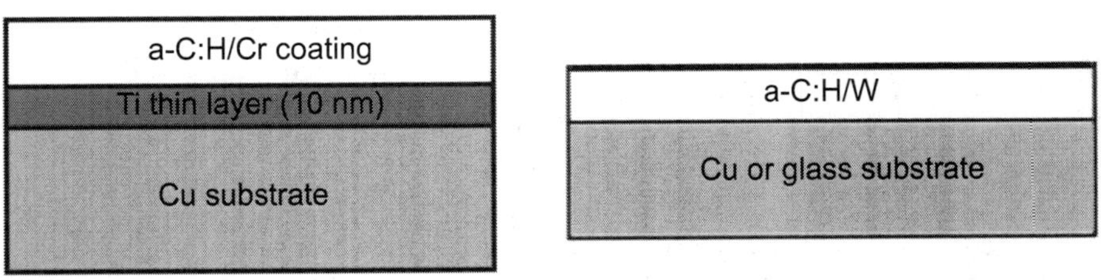

Fig. 7.18 Structures investigated by Gampp et al. (1994).

The samples were submitted to accelerated aging cycles. After every time step the reflectance and in case of the a-C:H/W coating on glass, transmission spectra were recorded. These experiments at elevated temperatures (200 to 300°C) revealed the good film stability of a-C:H/Cr coating and its potential for use as solar absorber (Gampp et al. 1994).

As the results obtained for the a-C:H/Cr absorber coatings were better than that observed for a-C:H-W, the same group of researchers investigated in more detail the degradation of these coatings, deposited by combined PECVD and sputtering process, at elevated temperatures in the air. Two different configurations were studied, as shown in Fig. 7.19.

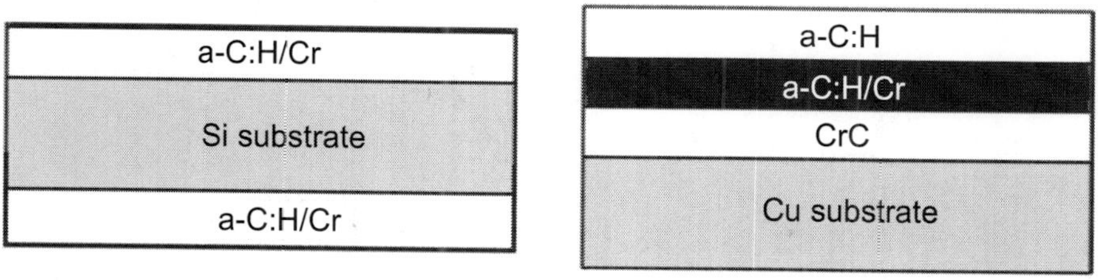

Fig. 7.19 Configurations used for solar absorbers based on a-C:H/Cr coating.

In the first configuration, they used silicon substrates with low electrical resistivity, between 10 and 50 Ω·cm, to transmit the DC substrate voltage during the coating deposition process and high enough for transmission measurements in the near infrared, above about 1200 nm. The silicon substrate allows the exclusion of the unwanted aging mechanisms caused by the combination with the copper substrate. In the second configuration, a-C:H/Cr coating (15 at.% Cr) was deposited on a CrC layer which ensures the adhesion and suppresses the diffusion of copper through the coating. Above a-C:H/Cr coating, an anti-reflective layer of pure a-C:H was deposited. The test results showed that a minimum service lifetime of 25 years multilayered can be expected for a-C:H/Cr absorber under standard conditions. Table 7.8 compares this absorber with others shown in the literature.

Table 7.8 Comparison among PVD absorber coating based on different materials.

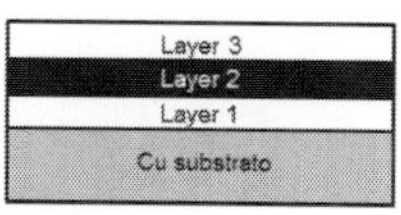	**Gampp et al. 1998**	**Barshilia et al. 2009**	**Barshilia et al. 2008**	**Selvakumar et al. 2010**
Layer 3	a-C:H	Al_2O_3	Si_3N_4	HfOx
Layer 2	a-C:H/Cr	Al	$TiAlN_xO_y$	Mo
Layer 1	CrC	Al_2O_3	TiAlN	HfO_2
α	91-95%	90%	96%	90-92%
ε	8-5%	6%	7%	7-9%

Since 2004, the company IKARUS COATINGS has been commercializing PVD a-C:H/Cr absorber coatings for solar hot water applications (Oelhafe and Schuler 2005). These coatings exhibit a solar absorptance around of 0.92 and emittance of 0.08 at 100°C and thermal stability above 250°C in vacuum. In air at temperature of 220°C, the coating failed after 280 hours.

Despite good results obtained for a-C:H/Cr absorber coatings, chromium is an environmentally harmful heavy metal and therefore should be avoided (Tinchev et al. 2010).

In the last decade, promising applications of Ti-DLC films as hard, wear resistant and biocompatible coatings were reported. The outstanding results obtained for these materials especially related to mechanical and tribological properties have motivated studies on their use as selective solar absorber coatings.

Schuler et al. (2000) investigated a-C:H/Ti based multilayered solar absorber coatings (Fig. 7.20). For these coatings, a solar absorptance of 0.87 and emittance of 0.06 at 100°C was achieved. In the air, the coating lifetime should exceed 150 hours at 220°C.

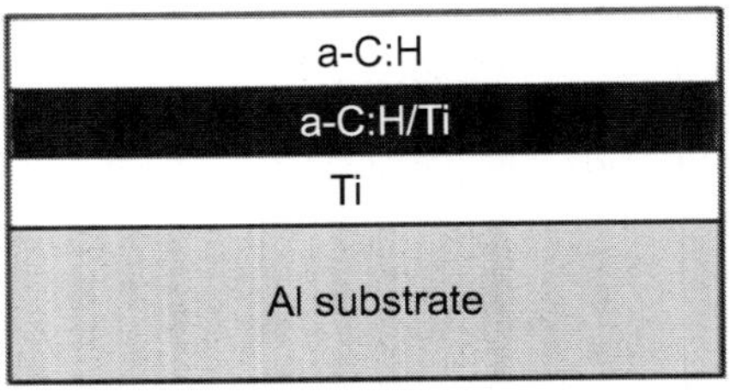

Fig. 7.20 Configuration used for solar absorber based on a-C:H/Ti coating.

Recently, Lan et al. (2014) investigated a double-cermet structured, using two a-C:H/Pt thin films with different Pt content deposited by co-sputtering, as selective solar absorbers. Figure 7.21 shows the structure developed on Cu substrate and using a-C:H film as an anti-reflective (AR) layer. The structure exhibited an absorptance of 0.91 when tested without the AR layer and of 0.96 with the addition of this layer. After annealing at 400°C, the absorptance remains the same which demonstrates the potential double-layered a-C:H/Pt coatings for high temperature solar absorbers.

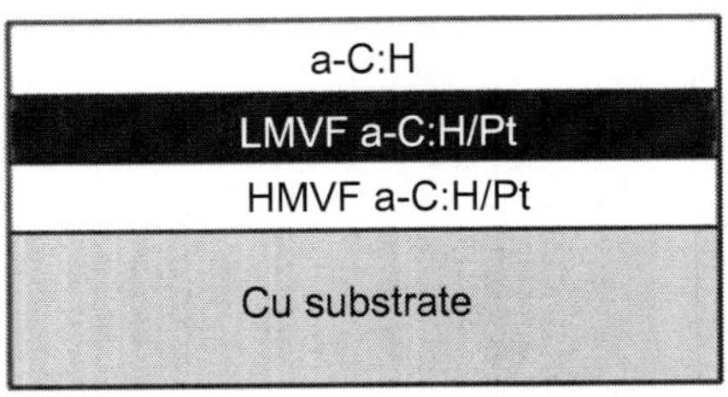

Fig. 7.21 Configuration used for solar absorber based on double a-C:H/Pt coating.

From this literature review it is revealed that Me-DLC films can be considered an alternative to the traditional absorber coatings. The perspectives with regard to their use as selective solar absorbers are as follows:

- Double cermet structures based on Me-DLC (LMVF and HMVF) display the greatest potential for the development of high-temperature solar selective absorbers. Given that single cermet structures, like DLC-Cr, have exhibited degradation at elevated temperatures in the air due to oxidation of the chromium carbide.

- An important field of research on Me-DLC films is understanding fully the mechanisms of graphitization and oxidation of the a-C:H matrix, that are dominant for the degradation of the optical performance of Me-DLC absorber coatings, in order to reduce them.
- A high optical selectivity has been achieved for a-C:H/Ti based multilayered solar absorber coatings, which become this configuration as an alternative to the commercially available double cermet based on Me-DLC coatings.

6. FINAL REMARKS

Me-DLC films were described with focus on their use as selective solar absorbers coatings. Different deposition processes, especially PVD techniques, have been employed to deposit Me-DLC films for these applications. The suitable choice of the metal type and its content allows that the properties of Me-DLC films can be optimized. Up to now Me-DLC films with Ti, Cr, W and Pt have been investigated as selective absorbers using different configurations. In general, they exhibit high absorptance, low emittance and thermal stability. PVD a-C:H/Cr coatings are commercially available.

7. ACKNOWLEDGEMENTS

This research was supported by the São Paulo Research Foundation – FAPESP (processes number 14/18139-8 and 13/17045-7) and by CAPES/PVNS.

8. REFERENCES

Aisenberg, S. O. L. and R. Chabot. 1971. Ion-beam deposition of thin film of diamond-like carbon. J. Appl. Phys. 42(7): 2953-2958.

Anders, A. 2010. Unfiltered and filtered cathodic arc deposition. pp. 467-531. *In*: Martin, P. M. (ed.). Handbook of Deposition Technologies for Films and Coatings Third Edition. Science, Applications and Technology. Elsevier Inc., Burlington, MA, USA.

Barshilia, H. C., N. Selvakumar, K. S. Rajam, D. V. Sridhara Rao and K. Muraleedharan. 2008. Deposition and characterization of TiAlN/TiAlCN/Si3N4 tandem absorbers prepared using reactive direct current magnetron sputtering. Thin Solid Films 516(18): 6071-6078.

Barshilia, H. C., N. Selvakumar, G. Vignesh, K. S. Rajam and A. Biswas. 2009. Optical properties and thermal stability of pulsed-sputter-deposited $Al_xO_y/Al/Al_xO_y$ multilayer absorber coatings. Sol. Energ. Mat. Sol. Cells 93: 315-323.

Bewilogua, K., C. V. Cooper, C. Specht, J. Schröder, R. Wittorf and M. Grischke. 2000. Effect of target material on deposition and properties of metal-containing DLC (Me-DLC) coatings. Surf. Coat. Technol. 127(2-3): 223-231.

Bharathy, P. V., D. Nataraj, P. K. Chu, H. Wang, Q. Yang, M. S. R. N. Kiran, et al. 2010. Effect of titanium incorporation on the structural, mechanical and biocompatible properties of DLC thin films prepared by reactive-biased target ion beam deposition method. Appl. Surf. Sci. 257(1): 143-150.

Bharathy, P. V., Q. Yang, M. S. R. N. Kiran, J. Rha, D. Nataraj and D. Mangalaraj. 2012. Reactive biased target ion beam deposited W-DLC nanocomposite thin films – Microstructure and its mechanical properties. Diamond Relat. Mat. 23: 34-43.

Bogaerts, W. F. and C. M. Lampert. 1983. Materials for photothermal solar energy conversion. J. Mater. Sci. 18(10): 2847-2875.

Bostrom, T. 2006. Solution-Chemically Derived Spectrally Selective Solar Absorbers with System Perspectives on Solar Heating. Ph.D. Thesis, Uppsala University, Uppsala, Sweden.

Bräuer, G. 2014. Magnetron sputtering. pp. 57-73. *In*: Hashmi, M. S. J., G. F. Batalha, C. J. Van Tyne and B. S. Yilbas (eds.). Comprehensive Materials Processing Vol. 4 – Films and Coatings: Technology and Recent Development. Elsevier, Waltham, MA, USA.

Broitman, E. and L. Hultman. 2014. Advanced carbon-based coatings. pp. 389-412. *In*: Hashmi, M. S. J., G. F. Batalha, C. J. Van Tyne and B. S. Yilbas (eds.). Comprehensive Materials Processing Vol. 4 – Films and Coatings: Technology and Recent Development. Elsevier, Waltham, MA, USA.

Candelaria, S. L., Y. Shao, W. Zhou, X. Li, J. Xiao, J. Zhang, et al. 2012. Nanostructured carbon for energy storage and conversion. Nano Energy 1: 195-220.

Cao, F., K. McEnaney, G. Chen and Z. Ren. 2014. A review of cermet-based spectrally selective solar absorbers. Energy Environ. Sci. 7: 1615-1627.

Carlsson, J.-O. and P. M. Martin. 2010. Chemical vapor deposition. pp. 314-363 *In*: Martin, P.M. (ed.). Handbook of Deposition Technologies for Films and Coatings Third Edition. Science, Applications and Technology. Elsevier Inc., Burlington, MA, USA.

Cui, J., L. Qiang, B. Zhang, X. Ling, T. Yang and J. Zhang. 2012. Mechanical and tribological properties of Ti-DLC films with different Ti content by magnetron sputtering technique. Appl. Surf. Sci. 258(12): 5025-5030.

Dai, M.-J., K.-S. Zhou, S.-S. Lin, H.-J. Hou, X.-G. Zhu, H.-W. Li, et al. 2007. A study on metal-doped diamond-like carbon film synthesized by ion source and sputtering technique. Plasma Processes Polym. 4: S215-S219.

Dai, W. and A. Wang. 2011a. Deposition and properties of Al-containing diamond-like carbon films by a hybrid ion beam sources. J. Alloys Compd. 509(13): 4626-4631.

Dai, W. and A. Wang. 2011b. Synthesis, characterization and properties of the DLC films with low Cr concentration doping by a hybrid linear ion beam system. Surf. Coat. Technol. 205 (8-9): 2882-2886.

Depla, D., S. Mahieu and J. E. Greene. 2010. Sputter deposition process. pp. 253-296. *In*: Martin, P. M. (ed.). Handbook of Deposition Technologies for Films and Coatings Third Edition. Science, Applications and Technology. Elsevier Inc., Burlington, MA, USA.

Dimigen, H. and C. P. Klages. 1991. Microstructure and wear behavior of metal-containing diamond-like coatings. Surf. Coat. Technol. 49: 543-547.

Fan, J. C. C. 1981. Sputtered films for wavelength-selective applications. Thin Solid Films 80(1-3): 125-136.

Ferrari, A. C. and J. Robertson. 2000. Interpretation of Raman spectra of disordered and amorphous carbon. Phys. Rev. B. 61: 14095.

Gago, R., I. Jiménez and J. M. Albella. 2006. Thin film growth by ion-beam-assisted deposition techniques. pp. 345-381. *In*: Pauleau, Y. (ed.). Materials Surface Processing by Directed Energy Techniques First Edition. Elsevier, San Diego, CA, USA.

Gampp, R., P. Gantenbein, Y. Kuster, P. Reimann, R. Steiner and P. Oelhafen. 1994. Characterization of a-C:H/W and a-C:H/Cr solar selective absorber coatings. Proc. Spie. 2255: 92-106.

Gampp, R., P. Oelhafen, P. Gantenbein, S. Brunold and U. Frei. 1998. Accelerated aging tests of chromium containing amorphous hydrogenated carbon coatings for solar collectors. Sol. Energy Mater. Sol. Cells 54: 369-377.

Gayathri, S., N. Kumar, R. Krishnan, T. R. Ravindran, S. Amirthapandian, S. Dash, et al. 2015. Influence of transition metal doping on the tribological properties of pulsed laser deposited DLC films. Ceram. Int. 41(1): 1797-1805.

Grill, A. 1999. Electrical and optical properties of diamond-like carbon. Thin Solid Films 355: 189-193.

Haick, H. 2013. Nanotechnology and Nanosensors. Israel Institute of Technology, Haifa.

Hainsworth, S. V. and N. J. Uhure. 2007. Diamond like carbon coatings for tribology: production techniques, characterization methods and applications. Int. Mater. Rev. 52: 153-174.

Harding, G. L. 1976. Sputtered metal carbide solar-selective absorbing surfaces. J. Vac. Sci. Technol. 13: 1070-1072.

Harding, G. L. 1980. Absorptance and emittance of metal carbide selective surfaces sputter deposited onto glass tubes. Sol. Energy Mater. 2: 469-481.

Jain, A. 2014. Simulating and Characterizing Carbon Based Solar Thermal Absorbers. Master's Thesis in Physics. The Arctic University of Norway, Tromsø, Norway.

Kalogirou, S. A. 2004. Solar thermal collectors and applications. Prog. Energy Combust. Sci. 30: 231-295.

Kern, W. and K. K. Schulgraf. 2002. Deposition technologies and applications: introduction and overview. pp. 11-48. *In*: Seshan, K. (ed.). Handbook of Thin-Film Deposition Processes and Techniques Second Edition. Principles, Methods, Equipment and Applications. Noyes Publications, Norwich, NY, USA.

Kirchain, R. and E. Alonso. 2011. Materials availability and recycling. pp. 549-564. *In*: D. S. Ginley and D. Cahen (eds.). Fundamentals of Materials for Energy and Environmental Sustainability. Cambridge University Press.

Klages, C. P. and R. Memming. 1989. Microstructure and physical properties of metal-containing hydrogenated carbon films. Mater. Sci. Forum. 52: 609-644.

Koskinen, J. 2014. Cathodic-arc and thermal-evaporation deposition. pp. 03-55. *In*: Hashmi, M. S. J., G. F. Batalha, C. J. Van Tyne and B. S. Yilbas (eds.). Comprehensive Materials Processing Vol. 4 – Films and Coatings: Technology and Recent Development. Elsevier, Waltham, MA, USA.

Ladgaonkar, P. S. and A. M. Patil. 2014. Development and performance evaluation of selective coating of absorber tube for parabolic trough collector. IJAIEM 3: 169-174.

Lan, Y., S. Brahma, Y. H. Tzeng and J. Ting. 2014. Platinum containing amorphous hydrogenated carbon (a-C:H/Pt) thin films as selective solar absorbers. Appl. Surf. Sci. 316: 398-404.

Lee, B. 2001. The progress and prospect of middle/high temperature evacuated tubular solar collector. Renew. Energ. 24(3-4): 539-544.

Lee, K. D. 2007. Preparation and characterization of black chrome solar selective coatings. J. Korean Phys. Soc. 51: 135-144.

Li, X., P. Ke and A. Wang. 2015. Probing the stress reduction mechanism of diamond-like carbon films by incorporating Ti, Cr or W carbide-forming metals: ab initio molecular dynamics simulation. J. Phys. Chem. C 119: 6086-6093.

Luithardt, W. and C. Benndorf. 1997. Single source deposition of Mo-C:H films using metal-organic precursors. Diamond Relat. Mater. 6: 533-536.

Martin, P. M. 2010. Deposition technologies: an overview. pp. 01-31. *In*: Martin, P. M. (ed.). Handbook of Deposition Technologies for Films and Coatings Third Edition. Science, Applications and Technology. Elsevier Inc., Burlington, MA, USA.

Martin, P. M. 2011. Superhard Thin Films. Vacuum and Coating Technology http://www.vtc-mag.com/articles/PMM-WEB072012/.

Matsunami, N., Y. Yamamura, Y. Itikawa, N. Itoh, Y. Kazumata, S. Miyagawa, et al. 1984. Energy dependence of the ion-induced sputtering yields of monatomic solids. At. Data Nucl. Data Tables 31: 1-80.

Michler, T., M. Grischke, I. Traus, K. Bewilogua and H. Dimigen. 1998. DLC Films deposited by bipolar pulsed DC PACVD. Diamond Relat. Mater. 7(2-5): 459-462.

Nejati, M. 2008. Cermet Based Solar Selective Absorbers: Further Selectivity Improvement and Developing New Fabrication Technique. Ph.D. Thesis, Universität des Saarlandes, Saarbrücken, Germany.

Oelhafen, P. and A. Schuler. 2005. Nanostructured materials for solar energy conversion. Solar Energy 79: 110-121.

Ohring, M. 2002. Materials Science of Thin Films, Second Edition. Deposition and Structure. Academic Press, San Diego.

Park, Y. S., T.-H. Jung, D.-G. Lim, Y. Park, H. Kim and W. S. Choi. 2012. Tribological properties of metal doped a-C film by RF magnetron sputtering method. Mater. Res. Bull. 47(10): 2784-2787.

Peters, A. M. and M. Nastasi. 2003. Titanium-doped hydrogenated DLC coatings deposited by a novel OMCVD-PIIP technique. Surf. Coat. Technol. 167(1): 11-15.

Pierson, H. O. 1999. Handbook of Chemical Vapor Deposition (CVD) Second Edition – Principles, Technology and Applications. Noyes Publications, Park Ridge, NJ, USA.

Pratesi, S., E. Sani and M. De Lucia. 2014. Optical and Structural Characterization of Nickel Coatings for Solar Collector Receivers. Int. J. Photoenergy. Article ID 834128, 7 pages, 2014.

Qiang, L., K. Gao, L. Zhang, J. Wang, B. Zhang and J. Zhang. 2015. Further improving the mechanical and tribological properties of low content Ti-doped DLC film by W incorporating. Appl. Surf. Sci. 353: 522-529.

Robertson, J. 2002. Diamond-like amorphous carbon. Mater. Sci. Eng., R 37: 129-281.

Robertson, J. 2014. Diamond-like carbon films, properties and applications. pp. 101-139. *In*: Sarin, V. K. and C. E. Nebel (eds.). Comprehensive Hard Materials Vol. 3 – Super Hard Materials. Elsevier, Waltham, MA, USA.

Rutkuniene, Z., A. Grigonis and L. Vigricaite. 2013. The influence of metal impurities to a-C:H films. Przeglad Elektrotechniczny 89: 288-290.

Sánchez-López, J. C. and A. Fernandez. 2008. Doping and alloying effects on DLC coatings. pp. 311-338. *In*: C. Donnet and A. Edermir (eds.). Tribology of Diamond-like Carbon Films. Springer, New York, USA.

Sansongsiri, S., A. Anders and B. Yotsombat. 2008. Electrical properties of a-C:Mo films produced by dual-cathode filtered cathodic arc plasma deposition. Diamond Relat. Mater. 17(12): 2080-2083.

Salmi, J., J-P. Bonino and R. S. Bes. 2000. Nickel pigmented anodized aluminium as solar selective absorbers. J. Mater. Sci. 35(6): 1347-1351.

Sathiaraj, T. S., R. Thangaraj, H. Al Sharbaty, M. Bhatnagar and O. P. Agnihotri. 1990. Ni-Al_2O_3 selective cermet coatings for photothermal conversion up to 500°C. Thin Solid Films 190(2): 241-254.

Schiffmann, K. I., M. Fryda, G. Goerigk, R. Lauer, P. Hinze and A. Bulack. 1999. Sizes and distances of metal clusters in Au-, Pt-, W- and Fe-containing diamond-like carbon hard coatings: a comparative study by small angle X-ray scattering, wide angle X-ray diffraction, transmission electron microscopy and scanning tunneling microscopy. Thin Solid Films. 347(1-2): 60-71.

Schuler, A., J. Geng, P. Oelhafen, S. Brunold, P. Gantenbein and U. Frei. 2000. Application of titanium containing amorphous hydrogenated carbon films (a-C:H/Ti) as optical selective solar absorber coatings. Sol. Energy Mater. Sol. Cells 60: 295-307.

Schultes, G., P. Frey, D. Goettel and O. Freitag-Weber. 2006. Strain sensitivity of nickel-containing amorphous hydrogenated carbon (Ni:a-C:H) thin films prepared by r.f. sputtering using substrate bias conditions. Diamond Relat. Mater. 15(1): 80-89.

Selvakumar, N., H. C. Barshilia, K. S. Rajam and A. Biswas. 2010. Structure, optical properties and thermal stability of pulsed sputter deposited high temperature $HfOx/Mo/HfO_2$ solar selective absorbers. Sol. Energy Mater. Sol. Cells 94: 1412-1420.

Selvakumar, N. and H. C. Barshilia. 2012. Review of physical vapor deposited (PVD) spectrally selective coatings for mid- and high-temperature solar thermal applications. Sol. Energy Mat. Sol. Cells 98: 1-23.

Silva, S. R. P. and J. D. Carey. 2003. Enhancing the electrical conduction in amorphous carbon and prospects for device applications. Diamond Relat. Mater. 12(2): 151-158.

Sonoda, T., S. Nakao and M. Ikeyama. 2009. Deposition of Ti/C nano-composite DLC films by magnetron DC sputtering with dual targets. Vacuum 84(5): 666-668.

Stoica, A. 2012. Preparation of nanostructured carbon films by Plasma Enhanced Chemical Vapor Deposition. Ph.D. Thesis, Masaryk University, Brno, Czech Republic.

Takeno, T., Y. Hoshi, H. Miki and T. Takagi. 2008. Activation energy in metal-containing DLC films with various metals of various concentrations. Diamond Relat. Mater. 17(7-10): 1669-1673.

Tesfamichael, T. 2000. Characterization of Selective Solar Absorbers Experimental and Theoretical Modeling. Ph.D. Thesis, Uppsala University, Uppsala, Sweden.

Tinchev, S. S., P. I. Nikolova and Y. T. Dyulgerska. 2010. Thermal solar absorber made of diamond-like carbon thin films. JPCS 223: 012017.

Toyoda, N. and S. Matsui. 2014. Ion beam deposition: Recent developments. pp. 187-200. *In*: Hashmi, M. S. J., G. F. Batalha, C. J. Van Tyne and B. S. Yilbas (eds.). Comprehensive Materials Processing Vol. 4 – Films and Coatings: Technology and Recent Development. Elsevier, Waltham, MA, USA.

Vetter, J. 2014. Surface & Coatings Technology 60 years of DLC coatings: Historical highlights and technical review of cathodic arc processes to synthesize various DLC types and their evolution for industrial applications. Surf. Coat. Technol. 257: 213-240.

Wang, J. S., Y. Sugimura, A. G. Evans and W. K. Tredway. 1998. The mechanical performance of DLC films on steel substrates. Thin Solid Films 325: 163-174.

Weber, M., K. Bewilogua, H. Thomsen and R. Wittorf. 2006. Influence of different interlayers and bias voltage on the properties of a-C:H and a-C:H:Me coatings prepared by reactive d.c. magnetron sputtering. Surf. Coat. Technol. 201(3-4): 1576-1582.

Williams, B. A., E. S. Aydil and L. F. Francis. 2014. Processing of Solar Absorber Coatings from Nanocrystal Dispersions. Proc. 17th International Coating Science and Technology Symposium, San Diego, CA, USA.

Yang, F. H. 2014. Modern metal-organic chemical vapor deposition (MOCVD) reactors and growing nitride-based materials. pp. 27-81. *In*: Huang, J., H.-C, Kuo and S.-C Shen (eds.). Nitride semiconductor light- emitting diodes (LEDs) – Materials, technologies and applications. Woodhead Publishing. Philadelphia, PA, USA.

Yin, Y., D. R. McKenzie and W. D. McFall. 1996. Cathodic arc deposition of solar thermal selective surfaces. Sol. Energy Mater. Sol. Cells 44: 69-78.

Yu, X., Y. Qin, C. B. Wang, Y. Q. Yang and X. C. Ma. 2013. Effects of nanocrystalline silver incorporation on sliding tribological properties of Ag-containing diamond-like carbon films in multi-ion beam assisted deposition. Vacuum 89(1): 82-85.

Zhang, Q.-C., K. Zhao, B.-C. Zhang, L.-F. Wang, Z.-L. Shen, D.-Q. Lu, et al. 1999. High performance Al–N cermet solar coatings deposited by a cylindrical direct current magnetron sputter coater. J. Vac. Sci. Technol. A 17: 2885.

Zhang, Q.-Z. 2000. Recent progress in high-temperature solar selective coatings. Sol. Energy Mater. and Sol. Cells 62(1-2): 63-74.

Zhao, F., H. Li, L. Ji, Y. Wang, H. Zhou and J. Chen. 2010. Ti-DLC films with superior friction performance. Diamond Relat. Mater. 19(4): 342-349.

8

ZnO Thin Films: The Most Potential Semiconductor Material as Buffer Layers in Thin Film Solar Cells

Nowshad Amin,[1,*] Jamilah Husna[2,a] and Mohammad Mezbaul Alam[3,b]

ABSTRACT

Zinc Oxide (ZnO) is a prominent photovoltaic material used as either buffer or window layers in Copper-Indium-Gallium-Sulfide (CIGS) or Cadmium-Telluride (CdTe) thin film solar cells. In this chapter, ZnO will be explored from the material point of view to the thin film growth and characterization, leading to application to thin film solar cell devices. Here, zinc oxide thin films were deposited by Radio Frequency (RF) magnetron sputtering at optimized conditions. Then, the effects of annealing on the structural, optical and electrical properties of the ZnO films have been discussed. Chronological investigation will show how the post deposition annealing possesses a positive effect in terms of film quality, roughness and crystalline property that can be utilized in the CIGS or CdTe thin film solar cells as window or buffer layers. Finally, initial implementation as solar cell buffer layers has been demonstrated through CIGS and CdTe thin film solar cells.

Keywords: II-VI compounds, ZnO thin films, sputtering growth, solar cells, buffer layers.

[1] Department of Electrical, Electronic and Systems Engineering, Faculty of Engineering and Built Environment, The National University of Malaysia, 43600 Bangi, Selangor, Malaysia.

[2] Institute of Microengineering and Nanoelectronics (IMEN), The National University of Malaysia, 43600 Bangi, Selangor, Malaysia.

[3] Advanced Materials Research Chair, Chemistry Department, College of Sciences, King Saud University, Riyadh 11451, Saudi Arabia.

[a] E-mail: miila_jv@yahoo.com

[b] E-mail: alamohammad@ksu.edu.sa

* Corresponding author: nowshad@ukm.edu.my

1. INTRODUCTION

Over the past century, researchers worldwide have made rapid and meaningful advances in the arena of material sciences, especially in semiconductor physics and applications. Almost all the research works relating to the technological development in the recent past have been explored on the fabrication of electronic devices like solar cells, transistors and integrated circuits etc. which are mainly based on silicon (Si) technology. However, being an indirect bandgap semiconductor, silicon is not suitable for injection or excitation by lasers and quite unsuitable for application in light emitting applications. Moreover, the narrow bandgap of 1.1 eV prevents visible light emission and enhances recombination (Low et al. 2008). Furthermore, due to low carrier mobility, it is unsuitable in high speed optoelectronic devices (Gunawan et al. 2008). Therefore, due to the performance limitations of some of the silicon based devices as well as the high cost factor researchers have been compelled to look into other semiconductor materials, especially II-VI compound semiconductors.

2. II-VI COMPOUND SEMICONDUCTORS

A group of compounds of elements that belong to II-VI group of periodic table indicate excellent semiconducting properties. Bandgap of these compound semiconductors have attracted greater interest attributed to their potential applications in many areas, such as UV sensors, Light Emitting Diodes (LED), high speed high power-electronic devices and spintronic devices (Look 2001, Service 1997, Özgür et al. 2005). The most potential representatives of this group are Zinc Selenide (ZnSe), Zinc Telluride (ZnTe), Cadmium Sulfide (CdS) and Cadmium-Telluride (CdTe) etc. These materials become attractive as some of them have wider bandgap greater than 1.5 eV, whereas some make matching in visible spectrum of sunlight, such as CdTe with its bandgap of 1.5 eV. This property indicates the light emission in the visible region of solar spectrum, especially in the blue to green region. However, device fabrication using II-VI compound semiconductors presents challenges to researchers due to the stoichiometric composition to adjust the n and p conductivities of the same compound that is necessary for device fabrication. Transparent Conducting Oxide (TCO) semiconductors such as Indium Tin Oxide (ITO), fluorine doped SnO_2 and aluminum doped ZnO are constantly receiving attention, owing to their low cost of fabrication, chemical robustness and high thermal conductance (Minami 2005). These transparent n-type materials with resistivity of 10^{-4}-10^{-5} ohm-cm can be used as transparent electrodes in thin film transistors, organic light-emitting diodes and solar cells.

2.1 Zinc Oxide (ZnO) as a Potential Candidate

Zinc oxide is a group II-VI compound semiconductor whose iconicity shows in between covalent and ionic semiconductors. Over the past decades, there has been a revival and subsequent rapid expansion in the research on ZnO as a semiconductor. Zinc oxide has been well recognized as one of the most promising oxide semiconductor materials owing to its excellent optical, electrical and acoustic properties. Zinc oxide has a direct bandgap of 3.36 eV with a high exciton binding energy (60 meV), in the near-UV spectral region and a large free-exciton binding energy, so that excitonic emission processes can persist at or even above room temperature. The advantages of ZnO in comparison to other materials are its low price, non-toxicity and easy manufacturability, thus making it a highly potential material for industrial applications.

2.1.1 ZnO Thin Films

During the recent decades, ZnO thin films, one of the important representatives of transparent II-IV oxide materials, are considered as a choice of material for application in optoelectronic industry due to their wide direct bandgap of 3.37 eV at room temperature and this bandgap can be changed via alloying with Cadmium Oxide (CdO) (Vijayalakshmi et al. 2008), Magnesium Oxide (MgO) (Lorenz et al. 2003) etc. The earlier works on optoelectronic and the piezoelectric properties of ZnO was started by Bell Labs a decade before the 1960s. However, this early interest in ZnO electronics seemed to have lost the attention of the R&D world at least by the 1980s, simply due to the development of the computer industry. Hence researchers thought that working in non silicon materials was a waste of time. The growing applicability of ZnO in electronics is finding new dimensions for the semiconductor industries. ZnO and its ternary alloys have the potential to compete with other materials which governs the optoelectronic industries as it is inexpensive, relatively abundant, chemically stable, easy to prepare and non-toxic. However, the fabrication of ZnO based optical devices suffer from the lack of reproducible, low resistive, high conductivity p-type ZnO thin films owing to the presence of native defects. Fabrication of stable p-type ZnO is a challenge to researchers because the electronic and optical properties of ZnO are very sensitive to minute concentrations of dopants, impurities and to microscopic perturbations of the lattice (Queisser and Haller 1998).

2.1.2 Crystal Structure of ZnO

Most of the group II-VI binary compound semiconductors crystallize in either cubic zincblende or hexagonal wurtzite structure, as shown in Fig. 8.1, where each anion is surrounded by four cations at the corners of a tetrahedron and vice versa. This tetrahedral coordination is the characteristic of sp^3 covalent bonding (Özgür et al. 2005). ZnO has the ionicity between covalent and ionic semiconductor. The crystal structures shared by ZnO are wurtzite, zinc blende and rarely observed rocksalt structure. At ambient conditions, the thermodynamically stable phase is wurtzite. On cubic substrates zinc blende structure may be grown and at relatively high pressure rocksalt structure can be obtained. The crystalline structure of ZnO is explained by a number of alternating planes consisting of tetrahedrally coordinated O^{2-} and Zn^{2+} ions which are stacked alternately along the *c*-axis (Wang 2004). ZnO also has a considerable ionic character that tends to increase the bandgap beyond the one that is expected from the covalent bonding.

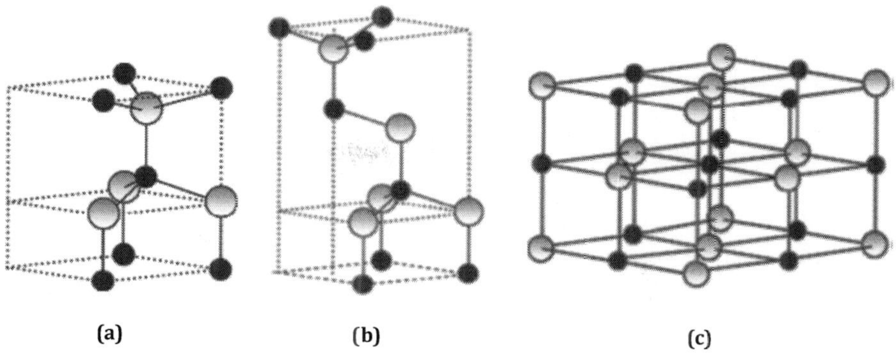

Fig. 8.1 Various ZnO crystal structures: (a) Hexagonal wurtzite structure. (b) Cubic zinc blende structure. (c) Cubic rocksalt structure (shaded gray and black spheres represent and O atoms, respectively (Özgür et al. 2005).

2.1.3 Energy Band Structure of ZnO

To use ZnO with its potential utility proper understanding of the electronic band structure is very important. Several theoretical approaches like greens function method (Rössler 1969), Local Density Approximation (LDA) (Vogel et al. 1995), GW approximation (Oshikiri and Aryasetiawan 1999) and first principles (Xu and Ching 1993) have been used to calculate band structures of ZnO. A number of experimental methods such as X-ray or UV reflection/absorption or emission techniques, photoelectron spectroscopy (PBS) (Langer and Vesely 1970), Angle-Resolved Photoelectron Spectroscopy (ARPES) (Ozawa et al. 2005), Low Energy Electron Diffraction (LEED), Empirical Pseudopotential Method (EPM) also have been employed by researchers to determine the band structure of this material. Oxygen 2p, 2s states and zinc 3d states mainly construct the valence band of ZnO. In the valance bands of ZnO a lot of low curvature bands are found which are the characteristics of low mobilization and localization of electrons in the d-orbitals as the d-orbitals of Zn have high energy. The conduction bands mainly occur due to the Zn 4s states.

For three different phases of ZnO the valance band spectrum is different near the Γ point. Both wurtzite and zinc blende structure of ZnO have direct bandgap but the electronic band structure differs near Γ point with the gap value 0.97 eV (Zhang et al. 2007). From Fig. 8.2 it can be noticed that when phase transition from wurtzite to rocksalt takes place by the application of pressure, the bandgap changes from direct to indirect in nature. This change occurs due to the symmetrical dependence of the interaction between oxygen p-states and zinc d-states.

2.1.4 Effect of Native Defects and Various Dopants

ZnO is a good host for incorporating foreign elements as Zn loses two valence electrons to O when Zn combines with O. Thus due to loss of an outer O shell the size of Zn changes from 1.33 Å to 0.74 Å, but the O atom increases in size due to addition of an outer shell from 0.64 Å to 1.4 Å. The wide difference in size between the zinc and oxygen atoms leaves relatively large open spaces for the incorporation of foreign atoms. The study of doping in ZnO is very important because of the fact that some of its properties can be noticeably changed by introducing dopants and most of the doping materials (Cd, Mn, Mg, Co, Al, P, N, Sn, etc.) that are used to dope ZnO to have different applications are also easily available.

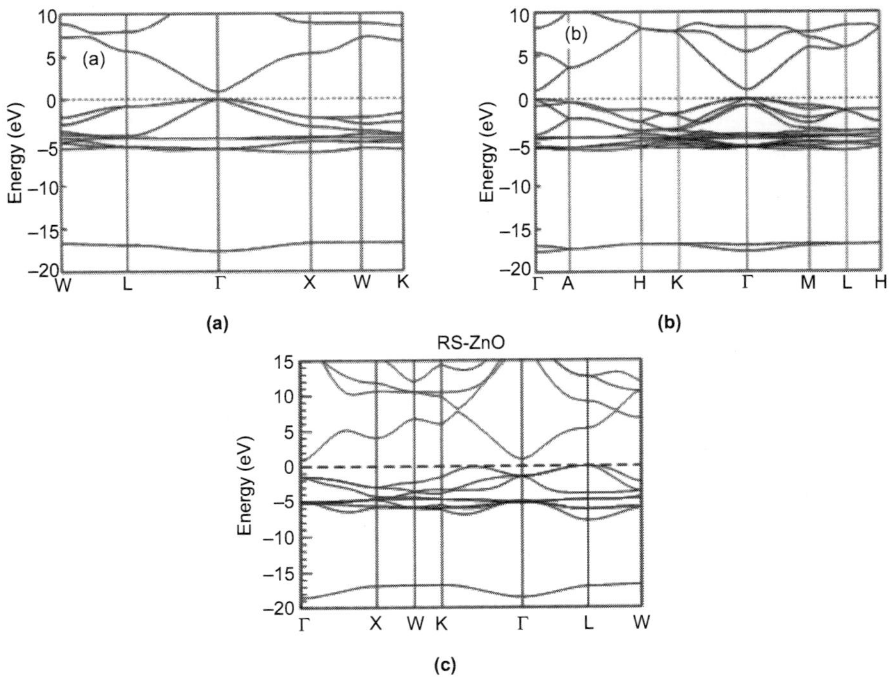

Fig. 8.2 Band structures of ZnO in different forms: (a) Zinc blende structure. (b) Wurtzite structure (Zhang et al. 2007). (c) Rocksalt structure (Dixit et al. 2010).

2.1.4.1 Native Defects of ZnO

For the successful application of a semiconductor proper understanding of the behavior of native defects is essential. The different native defects are created in the ZnO crystal lattice owing to (i) Oxygen vacancy (V_O) (ii) Zn interstitials (Zn_i) and (iii) Zn vacancy (V_{Zn}) (Look et al. 2003).

(i) **Oxygen vacancy (V_O):** Oxygen vacancy is the most common defect existing in ZnO crystal lattice. It has the least formation energy among other defects and behaves as a deep level donor so that it cannot contribute to the n-type conductivity of ZnO. However they can compensate p-type doping.

(ii) **Zn interstitials:** Zinc interstitial atom (Zn_i) could occupy at the octahedral or tetrahedral sites in the ZnO wurzite crystal structure. However, Zn interstitials will be more stable at the octahedral sites compared to that at the tetrahedral sites where the geometric constraints are more severe (Janotti and Van de Walle 2006, Janotti and Van de Walle 2007).

Zn_i produces a shallow donor level below the conduction band which is responsible for n-type conductivity in ZnO. However, Zn_i will be present in very low concentrations in ZnO crystal owing to the high formation energies under n-type conditions and hence Zn interstitials are unlikely to be responsible for unintentional n-type conductivity. Moreover, they are fast diffusers and hence unlikely to be stable as isolated point defects.

(iii) **Zinc vacancy:** Zinc vacancies (V_{Zn}) are deep acceptors and have low energies under n-type conditions. It is suggested that they can occur as the dominant compensating acceptor in n-type ZnO (Tuómisto et al. 2003). It may be mentioned that zinc vacancies are a possible source of the often-observed green luminescence in ZnO (Reynolds et al. 1997, Reynolds et al. 1998).

2.1.4.2 Acceptor Impurities

The elements for the acceptor dopants in ZnO are group-I elements (Li and Na) and group-V elements (N, P, As, Sb). The substitution of group-I acceptors have more shallow energy levels than the group-V acceptors (Park et al. 2002, Yamamoto et al. 2000, Lee and Chang 2004). If Fermi level is near around valence band ($E_F = E_v$), the group-I elements like Li and Na have a tendency to incorporate at interstitial site (Park et al. 2002, Yamamoto et al. 2000, Lee and Chang 2004) which are donors and cause the compensation of the substitutional group-I element acceptor states. The group-V elements can be incorporated at Zn-sites but due to the large size mismatch this causes self compensation. Acceptor like impurities generally has much higher ionization and formation energy (Park et al. 2002).

2.1.4.2.1 p-type Doping in ZnO

The role of crystal defects in ZnO is a subject of considerable interest as good p-type conductivity requires an understanding of the interactions between acceptors and crystal defects. The p-type conductivity in ZnO by doping with group-I elements is not possible owing to the compensation of electrons. Theoretically N in O site or Li in Zn site is the best candidate for p-type dopants as the atomic size mismatch among these three elements with the host atom is small. However, N is not able to act as p-type doping in ZnO due to low solubility, low rate of substitution to the O site, compensation by the doping induced defects and passivation by hydrogen.

2.1.4.3 Donor Impurities

The group-III impurities like depleted B, Al, Ga and In act as donor impurities in ZnO. When these elements are substituted in the Zn site, they can act as shallow donors in ZnO (Meyer et al. 2004). The extra valence electron of these impurities is loosely bound and occupies effective-mass states near the Valance Band Maximum (VBM) at low temperatures. As the temperature rises this extra electron is excited to the conduction band and is free to move. Donor-like defects are also formed by substitution of oxygen by group VII elements (Meyer et al. 2004). Highly conducting ZnO with carrier concentration of 10^{20} cm^{-3} can be achieved by doping with Al or Ga. Substitution of Al in Zn site predict shallow level 120 meV below the conduction band minimum and this defect has low formation energy in Zn rich condition (Zhang et al. 2001). Hydrogen, which is difficult to remove from the crystal growth environment, is an excellent shallow donor and contributes to the free electron concentration of n-type ZnO (Van de Walle 2001).

2.2 Important Properties and Device Applications of ZnO

Nowadays there are many exciting applications of ZnO thin films made into various materials and products, where this material is a highly potential candidate for industrial applications. The bandgap of ZnO is 3.44 eV at low temperatures and 3.37 eV at room temperature. This enables applications in optoelectronics in the blue/UV region, including light-emitting diodes, laser diodes and photodetectors. Optically pumped lasing has been reported in ZnO platelets, thin films (solar cells), clusters consisting of ZnO nanocrystals and ZnO nanowires.

The free-exciton binding energy in ZnO is 60 meV. This large exciton binding energy indicates that efficient excitonic emission in ZnO can persist at room temperature and higher. Since the oscillator strength of excitons is typically much larger than that of direct electron–hole transitions in direct gap semiconductors, the large exciton binding energy makes ZnO a promising material for optical devices that are based on excitonic effects.

In piezoelectric materials, an applied voltage generates a deformation in the crystal and vice versa. These materials are generally used as sensors, transducers and actuators. The low symmetry of the wurtzite crystal structure combined with a large electromechanical coupling in ZnO gives rise to strong piezoelectric and pyroelectric properties (Janotti and Van de Walle 2009).

As compared to many materials with which ZnO competes, it is inexpensive, relatively abundant, chemically stable, easy to prepare and nontoxic. Since the Bronze Age, ZnO has been used domestically as an alloying agent to make bronze. This oxide has applications in various materials and products such as plastics, ceramics, glass, cement, rubber, lubricants, white paints, ointments, adhesive, sealants, pigments, foods, batteries, ferrites, fire retardants, porcelain enamels and rubber manufacture etc. but there are few exciting applications of ZnO in optoelectronics. Some important properties and related device application of ZnO are discussed below.

2.2.1 Wide Direct Bandgap of ZnO

The light generation by electron-hole recombination in ZnO is high due to the wide direct bandgap (3.4 eV) where the radiative life time is short. This material has applications in light-emitting diodes, laser diodes and photo-detectors which operate in the blue or UV region (Look 2001, Ozgur et al. 2005). ZnO based optically pumped laser (Bagnall et al. 1997) and p-n homojunctions (Mandalapu et al. 2006) have also been reported.

2.2.2 Application in Solar Cell

ZnO could play an important role in the solar cell where it is regularly used as a back electrode. ZnO may also be used for electrode material in organic PV or dye sensitive cells. Due to its wide direct bandgap and transparency in the visible region it can be used as window material in thin film solar cells (Ghosh et al. 2009).

2.2.3 *Transparent Electronics: The Role of ZnO*

Transparent electronics is an upcoming technology for next generation of optoelectronic devices which satisfies the requirements of invisible objects. Through this technology the wind-shield of a car can be transformed into a TV screen or a road map at the touch of a button. The transparent window of a house can be transformed into a video screen using this interesting technology. Transparent oxide semiconductors provide space for the utilization in this new technology as they have high conductivity with high visual transparency. Transparent Thin Film Transistors (TFTs) can be fabricated using these materials because they have the unique feature of optical transparency in the visible region and controllable electrical conductivity. In 2003 researchers at the Oregon State University first developed transparent electronic circuit using transparent ZnO substrate (Hoffman et al. 2003). Among other different new compound materials that can provide transparent electronic circuit, the use of ZnO is advantageous because it is a wide bandgap material (3.37 eV) with high transparency in the visible region of the solar spectra and is less light sensitive and electrical conductivity can be modified with appropriate doping or post annealing. Another advantage of using ZnO deals with the fact that it is possible to be grown at room temperature which is advantageous for the fabrication of electronic devices, where the response speed is important (Hoffman et al. 2003).

2.2.3.1 Transparent Conducting Oxide: Al Doped ZnO (AZO)

In many display applications Indium Tin Oxide (ITO) is used as a transparent conductor due to its combination of environmental stability, relatively low electrical resistivity and high transparency. Stoichiometric ZnO is insulating and it can be made conductive by doping with hetero atoms or by introducing O vacancies (Lee et al. 2004). To enhance the conductivity of ZnO the most effective dopants are Al, Ga and F. The most popular replacement for ITO is Al doped ZnO (AZO) where it could substantially lower costs of transparent conductors as well as, non-toxic and high stability under hydrogen plasma (Tang and Cameron 1994, Tsuchiya et al. 1994).

Applications of this AZO include the front transparent window owing to its good optical transmission in the visible wavelength region (400-700 nm), compared with Sn doped indium oxide (ITO) films and contact on thin-film amorphous silicon and CuInGaSe (CIGS) solar cells. AZO is also used in gas sensing and Surface Acoustic Wave (SAW) devices and UV emitters.

2.3 Thin Film Solar Cells

Thin film solar cells consist of layers of different materials in thin film form. In general, the solar cell consists of substrates to hold the thin film layers, Transparent Conducting Oxide (TCO) to draw the current to the outer circuit, buffer layer (p or n-type) to create the junction with the absorber layer with minimum absorption losses and to drive the generated carriers to the electrode, absorber layer (i or p-type) for generating carriers on absorbing light with minimal transmission or reflection losses and metal contact layer as the bottom electrode. Each material has different physical and chemical properties and affects the overall performance of the device. A critical understanding of behavior of these individual components is essential for designing a device. Also important are the interfaces between different layers, since each layer has a different crystal structure, microstructure, lattice constant, electron affinity/work function, thermal expansion coefficient, diffusion coefficient, chemical affinity, mobility, mechanical adhesion etc. the interfaces can cause stress, defect and interface states, surface recombination centers, inter diffusion and chemical changes with attendant electro-optical changes (Mathew 2010).

2.3.1 *Application of ZnO in Thin Film Solar Cells*

At present ZnO has received growing attention as a potential semiconductor in thin films solar cell; commonly used as a front contact or window material in thin film solar cells. It has also become a material of interest for silicon thin film solar cell application and may be used for electrode material in organic PV or dye sensitive cells. Currently ZnO is one of the prominent materials in solar cell industries used as a window layer, transparent conducting oxide (TCO) and buffer layer.

2.3.2 *ZnO Buffer Layer for Solar Cells*

In the heterojunction solar cells absorber, the buffer layers are made from two different materials and this introduces a fair possibility of using materials having different bandgaps. A buffer layer with a wider bandgap could enhance the amount of light reaching the junction which is the main bottleneck in homojunction solar cells. Yet another advantage of heterojunction solar cells (p-n junction between different semiconductors) compared to homojunctions is that recombination in the wide bandgap semiconductor is quite low. On the other hand, the risk from interface recombination is higher for heterojunction due to defects and imperfections at the junction. Buffer layer, in combination with other window layers, can minimize the interface recombination losses and help to attain large band-bending.

2.4 The Role of Buffer Layer in Thin Film Solar Cells

The main function of a buffer layer in a heterojunction cell is to form a junction with the absorber layer while admitting a maximum amount of light to the junction region and absorber layer (Mccandless and Hegedus 1991). In addition, this layer should have minimal absorption losses and should be capable of driving out the photo generated carriers with minimum recombination losses and transporting the photo generated carriers to the outer circuit with minimal electrical resistance. For high optical throughput with minimal resistive loss the bandgap of the window layer should be as high as possible and the layer should be as thin as possible to maintain low series resistance. It is also important that any potential 'spike' in the conduction band at the heterojunction is minimized for optimal minority carrier transport. Lattice mismatch (and consequent effects) at the junction is important for consideration for epitaxial or highly oriented layers. In the case of microcrystalline layers, mismatch varies spatially and thus the complicated effect, if any, averages out. For a good buffer layer, some requirements could be achieved by prospective materials, such as large energy bandgap for high optical transmission in the visible region, optimal band discontinuities, optimal doping density and less lattice mismatch (Kushiya 2004).

Efficiency gain of the thin film solar cells greatly depends upon the quality and thickness of the buffer layer. The standard CIS solar cell needs optimized thickness of buffer layer between the absorber layer and the transparent front contact layer to improve its efficiency. It drives out the photo generated carriers with minimal losses while coupling light to the junction with minimum absorption losses, yielding a highly efficient solar cell. Thin film heterojunction solar cells provide more light towards the junction as it has a wide bandgap buffer layer in contrast with optimal low bandgap absorber layer. This provides the most reliable way of increasing the efficiency of the cell. The beneficial effects of the buffer layer ranges from modifying the absorber surface chemistry to protecting the sensitive interface during the subsequent window deposition (Greer 2006). Favorable properties of the interface are suggested to be related to the match between lattice parameters.

The current understanding is that candidates for buffer material should hold wider bandgap for limited light absorption. The process for deposition should have a capability to passivate the surface states of the absorber layer and should provide an alignment of the conduction band with the absorber to yield better efficiency. Binary sulfides, oxides and oxy-sulfides are frequently investigated as potential Cd-free buffer layers. Buffer layers also enhance the durability of the cell as the absorber is usually sensitive and needs protection during the impact of ions while sputtering the intrinsic and doped TCO layers. Furthermore, the absorber film becomes better suited for air exposure when protected by a buffer layer (Mathew 2010).

2.4.1 Alternative Buffer Layers

The candidates for alternative buffer material should have four common properties, which are as mentioned below:

1. The material should be n-type in order to form a p-n junction with the absorber layer.
2. The bandgap should be wide for limited light absorption.
3. The process for deposition should be low cost and suitable for wide area deposition.
4. In addition the technique should have the capability to passivate the surface states of the absorber layer.

As is well known, CdS is commonly used as a buffer layer in order to obtain better efficiency of thin film solar cells. In order to achieve environmental safety and manufacturing issue, it is desirable to replace CdS with an alternative buffer material (Kushiya 2004). Nowadays, films based on ZnS, ZnSe, ZnO, (Zn, Mg)O, $In(OH)_3$, In_2S_3, In_2Se_3 and InZnSe are deposited on differently processed absorbers and tested as an alternative to the traditional CdS buffer by different groups.

2.4.2 Techniques for ZnO Thin Film Deposition

The physical properties of the films strongly depend on the structure, morphology and the impurity. Because of this reason, the process of deposition plays an important role in controlling properties of ZnO thin films. In the last decade, different techniques were used to grow a ZnO buffer layer. Generally, ZnO thin films are usually prepared by several methods such as Molecular Beam Epitaxy (MBE), Chemical Vapor Deposition (CVD), Metal Organic Chemical Vapor Deposition (MOCVD), sol-gel processing, spray pyrolysis, spin coating and Radio Frequency (RF) magnetron sputtering (Gardeniers et al. 1998).

2.4.3 Sputtering Method

In this study the sputtering technique for the growth of ZnO buffer layer was employed. The sputtering technique is a process of physical vapor deposition for depositing thin films; it means ejecting material from a target and depositing it on a substrate where the target is the source material. A schematic representation is shown in Fig. 8.3. Sputtering starts when a negative charge is applied to the target material causing a plasma or glow discharge. Positive charged gas ions generated in the plasma region are attracted to the negatively biased target plate at a very high speed. This collision creates a momentum transfer and ejects atomic size particles from the target. These particles are deposited as a thin film onto the surface of the substrates. Because of the low substrate temperatures used, sputtering is an ideal method to deposit contact metals for thin film transistors. This technique is usually used to fabricate thin film sensors, photovoltaic thin films (solar cells), metal cantilevers and interconnects.

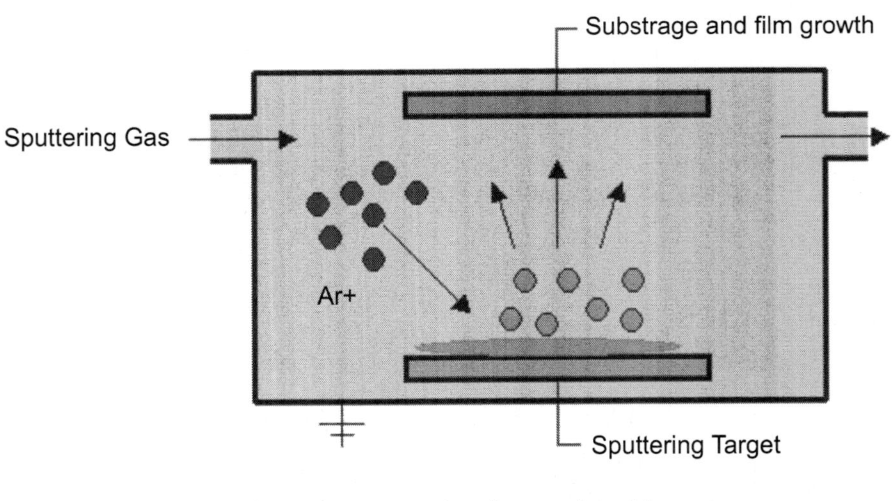

Fig. 8.3 Schematic representation of sputter deposition system.

Magnetron sputtering can be done either in DC or RF modes. DC sputtering is done with conducting materials. If the target is a non-conducting material, the positive charge will build up on the material and it will stop sputtering. RF sputtering can be done both conducting and non-conducting materials. Magnets are used to increase the percentage of electrons that take part in ionization of events and thereby increase the probability of electrons striking the argon atoms, increasing the length of the electron path and hence increasing the ionization efficiency significantly. The properties of sputtered ZnO thin films are known to depend on deposition parameters such as RF power, pressure, substrate temperature and ambient atmosphere (Gardeniers et al. 1998).

The sputtering technique has many advantages compared to other technique depositions, such as it can be carried out at low temperature, low cost with uniform properties, high purity, the films can be controlled easily and on the other hand, better controlled films will be obtained from the sputtering process. The sputtering technique is capable of growing high-resistive thin films on large area substrates. In addition, the thickness of the film grown is easy to control using this technique and thus recognized as a competitive alternative for other methods used for ZnO thin-films deposition. However, the disadvantage of this technique is that the growth rate is low.

3. DEPOSITION AND CHARACTERIZATION OF ZnO BY RF MAGNETRON SPUTTERING TECHNIQUE

Here the methodology that was adopted for deposition of intrinsic zinc oxide (i-ZnO) buffer layer is presented. The substrate cleaning procedure was shown at the beginning of this chapter. Then, the deposition process of i-ZnO buffer layers will be conducted. On the other hand, the annealing treatment that will be performed at different temperatures in order to obtain good properties for i-ZnO thin films will also be discussed. The structural characterizations of ZnO buffer layers thin film will be done using atomic force microscope (AFM) (NTEGRA Prima), scanning electron microscopy (SEM) (Philips, XL30), Field Emission Scanning Electron Microscopy (FE-SEM) and X-ray diffraction (XRD) (D8ADVANCE, BRUKERAX8). UV-Vis measurement is carried out to obtain the optical properties of ZnO buffer layers. The electrical properties of ZnO buffer layers will be examined using Ecopia HMS-3000 Hall Measurement System. The various thicknesses of ZnO buffer layer that were used in this work are in the range of 50-200 nm and the films will be annealed at temperatures of 250°C, 350°C and 400°C for 30 minutes. The details of deposition processes of ZnO thin films as flow chart are shown in Fig. 8.4.

3.1 Deposition of ZnO Thin Films By RF Sputtering

The intrinsic zinc oxide thin films layer were prepared using pure ZnO as target by RF magnetron sputtering and substrate temperature was set at room temperature (27°C). The sputtering chamber was evacuated at 1.0×10^{-5} Torr; then, prior to sputtering the target was employed to pre-sputtering for about 5 minutes. The working pressure was kept stable at 1.0×10^{-2} Torr during the film deposition, with the argon flow of 10 sccm and the sputtering power was maintained at 40 W.

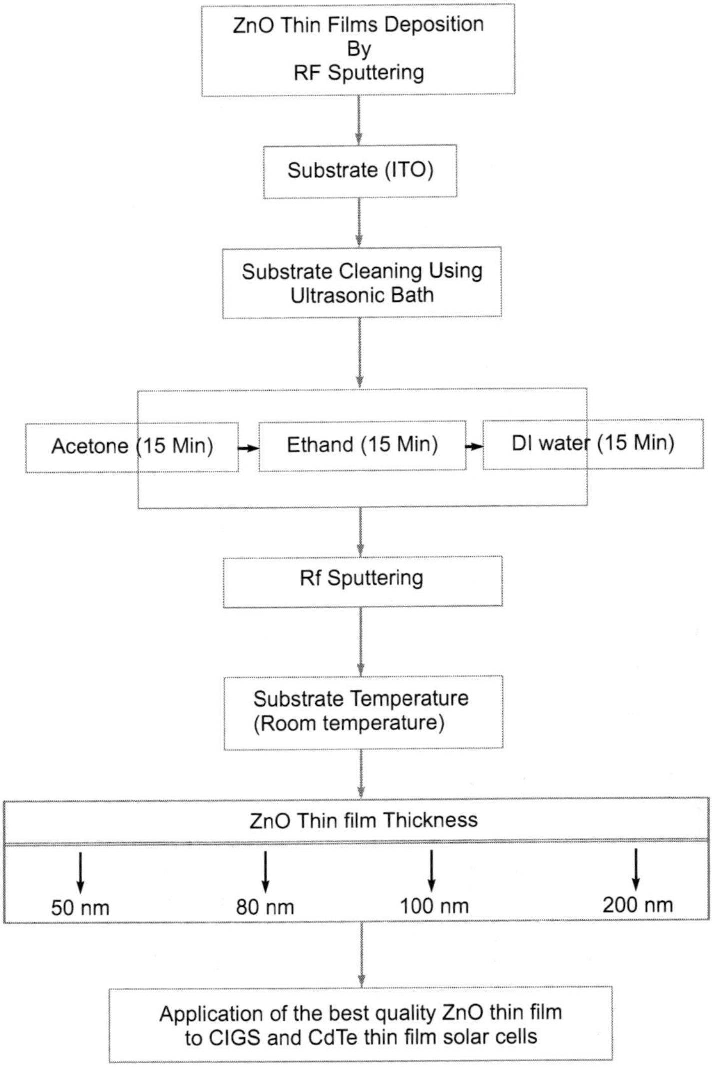

Fig. 8.4 Flow chart of ZnO thin films deposition by RF sputtering.

In a sputtering system, the thickness of film can be varied by changing the sputtering gas pressure, the sputtering time and the applied voltage. ZnO has been fabricated by altering the sputtering time in order to get layers with different thickness, where the thickness varied between 50-200 nm. Thickness depends on the total gas pressure during sputtering, besides the plasma power. One way to avoid oxygen vacancies and to possibly increase the film's resistivity, would be to sputter i-ZnO in an oxygen containing environment. The sputtering machine that was used for the deposition of ZnO thin films is shown in Fig. 8.5.

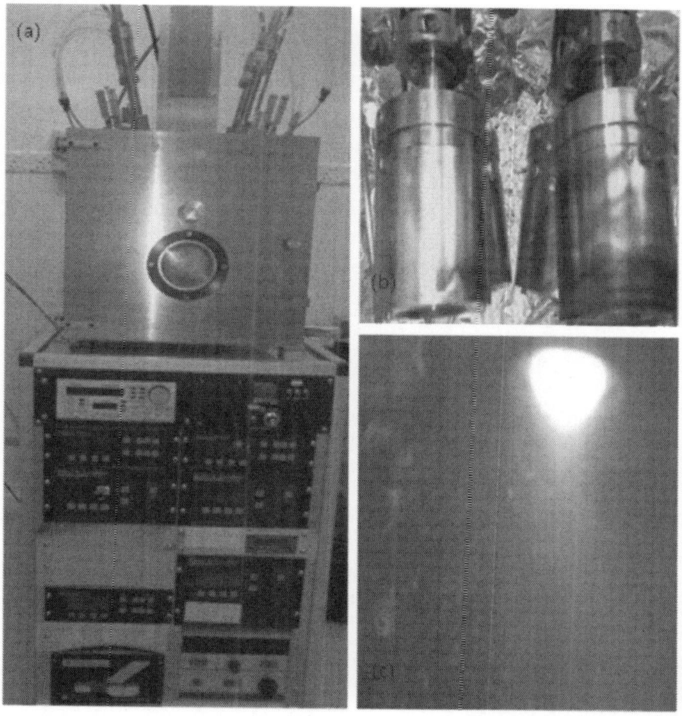

Fig. 8.5 (a) Sputtering chamber. (b) Sputtering guns. (c) Plasma observed during ZnO deposition.

The influence of RF sputtering power, pressure in chamber, substrate temperature and deposition rate on the properties of ZnO thin films was reported where at lower substrate temperature and with lower deposition rate perpendicularly oriented ZnO were obtained (Gardeniers et al. 1998). The parameters of the sputtering deposition process are summarized in Table 8.1.

Table 8.1 Deposition Parameters by RF sputtering Technique.

Variable of process	Range/Value
Substrate temperature	Room temperature (25°C)
Base pressure	2×10^{-5} Torr
Working pressure	1.0×10^{-2} Torr
RF power	40 watt
Gas flow	10 sccm
Growth time	20-100 minute

In this study, the effect of post annealing treatment on the optical properties of the deposited ZnO thin films is also executed. The first step which is the annealing process was started in the vacuum, where the base pressure was kept at range 50-80 Torr and the gas flow at 13 sccm. Furthermore, the working pressure was kept stable in the range values of 350 to 400 Torr. When the temperature reached the value heater was set, the time for annealing was started at that time. The temperature of annealing treatment was performed at 250°C, 350°C and 400°C in nitrogen-oxygen mixed gas ambient. Less than 1% oxygen was used in the mixed gas to prevent further oxidation of the films. Hence, the influence of post deposition annealing on the structural and optical properties include the electrical properties of ZnO thin films were then subjected to various investigations. Figure 8.6 shows the instrument that was used to anneal the films. The identification of annealed samples is described in Table 8.2. The detailed information of i-ZnO annealing treatment parameters are summarized as shown in Table 8.3. Moreover, sample identification of ZnO annealed samples with details are shown in Table 8.4.

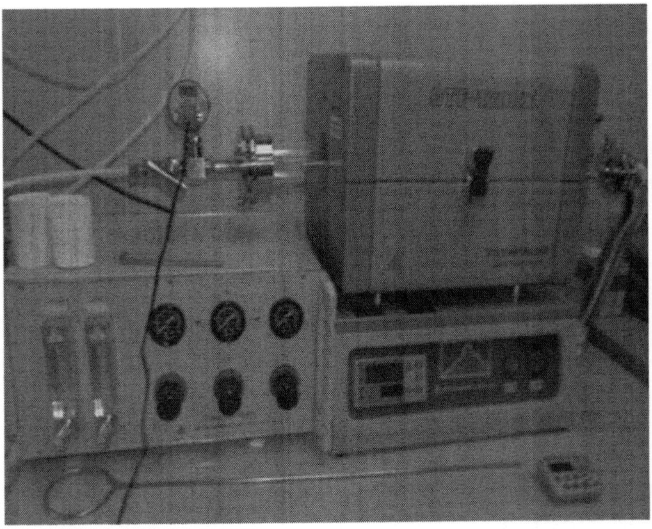

Fig. 8.6 Annealing chamber for thermal annealing at controlled environment.

Table 8.2 Identification of ZnO thin films as-deposited sample.

Sample ID	Thickness (nm)	Substrate temp. (°C)	Sputtering time (min)
1	50	25	42
2	80	25	58
3	100	25	82
4	200	25	160

Table 8.3 Annealing treatment parameters.

Variable of process	Range/Value
Base pressure	50-80 Torr
Working Pressure	350-400 Torr
Gas Flow	13 SCCM
Time	30 min
Temperature	250-400°C

Table 8.4 Identification of Annealed ZnO samples.

Sample ID	Thickness (nm)	Substrate temp. (°C)	Annealing temp. (°C)	Annealing time (min)
1	50	25	250-400	30
2	80	25	250-400	30
3	100	25	250-400	30
4	200	25	250-400	30

3.2 Application of i-ZnO Film in CIGS and CdTe Solar Cells

The best quality i-ZnO films, in terms of physical, optical and electrical properties, as found from the above experiments are applied to CIGS and CdTe thin film solar cells. The configurations of both solar cells are as shown in Fig. 8.7.

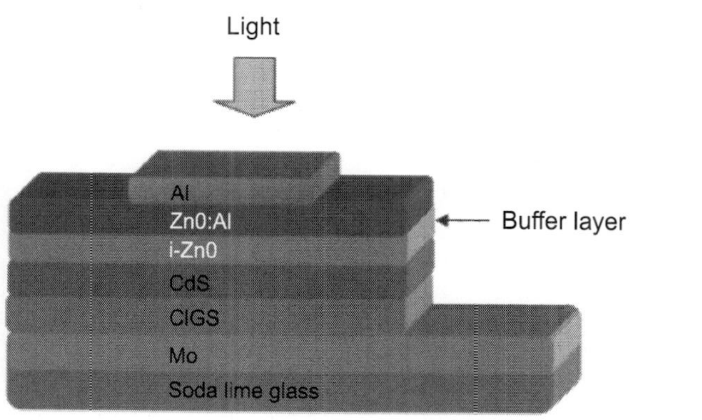

(a) CIGS

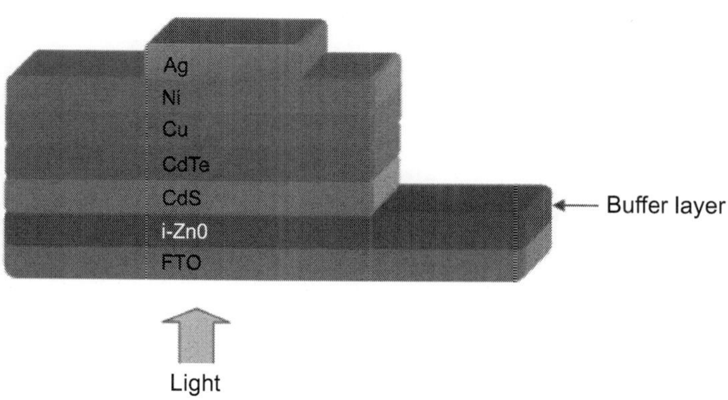

(b) CdTe

Fig. 8.7 Structure of CIGS and CdTe solar cells with ZnO Buffer Layer.

4. PROPERTIES OF PHYSICAL VAPOR DEPOSITED ZnO

4.1 Structural Analysis by XRD

X-ray Diffraction (XRD) measurement has been carried out to investigate the crystalline property of the ZnO thin films and the results are shown in Fig. 8.8. ZnO films were polycrystalline with (1 0 1), (1 0 2) and (0 0 2) oriented crystallites of hexagonal wurtzite structures having the prominent alignment along (0 0 2) as shown in Fig. 8.6a-e. The as-deposited ZnO film as shown in Fig. 8.6a has (1 0 1), (1 0 2) and (0 0 2) peaks related to ZnO, with some unknown ones possibly from the ITO beneath. The films treated at annealing temperature of 400°C and above show the dominance of (0 0 2) over others. Assuming a homogeneous strain across the films, the crystallite size may be estimated from the Full-Width at Half-Maximum (FWHM) of (0 0 2) diffraction peak using Scherer's Formula (Song et al. 2002) as shown in Eq. (1).

$$D = \frac{0.9\lambda}{B\cos\theta} \tag{1}$$

Here λ, θ and B are the X-ray wavelength, Bragg diffraction angle and FWHM of the ZnO (0 0 2) diffraction peak, respectively (Fang et al. 2005). As can be found in Fig. 8.9, it is quite obvious that the grain size increases after annealing treatment, which is supposed to be due to recrystallization of the film.

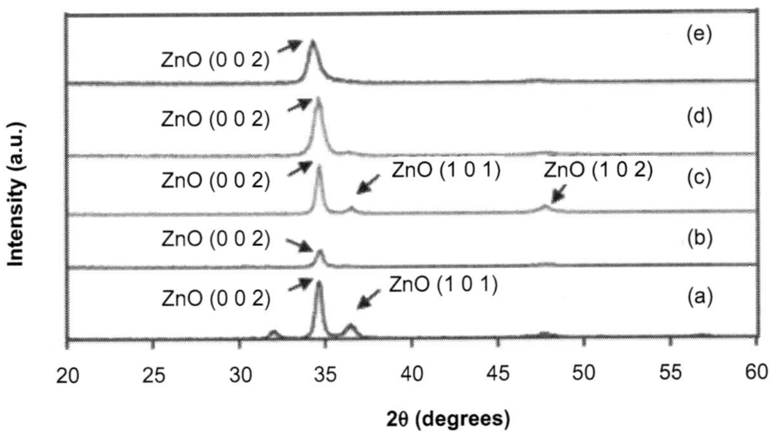

Fig. 8.8 XRD patterns of ZnO thin films: (a) as-deposited and annealed samples at (b) 250°C. (c) 350°C. (d) 400°C. (e) 450°C in nitrogen-oxygen ambient.

Scanning Electron Microscope (SEM) of the films reveal the presence of hexagonal shape of the particles and EDX (Energy Dispersive X-ray spectroscopy) measurements confirm the removal of impurities as a result of the annealing treatment (Elilarassi and Chandrasekaran 2010). EDX measurement was performed associated with the SEM inspection, to determine the atomic ratio of Zn to O. As can be observed, when the films are annealed at 250°C and 350°C, atomic percentile, as shown in Table 8.5, value of Zn gets larger compared to as deposited films, however at higher annealing temperature (e.g., 400°C and 450°C) Zn content reduces making it Zn poor film.

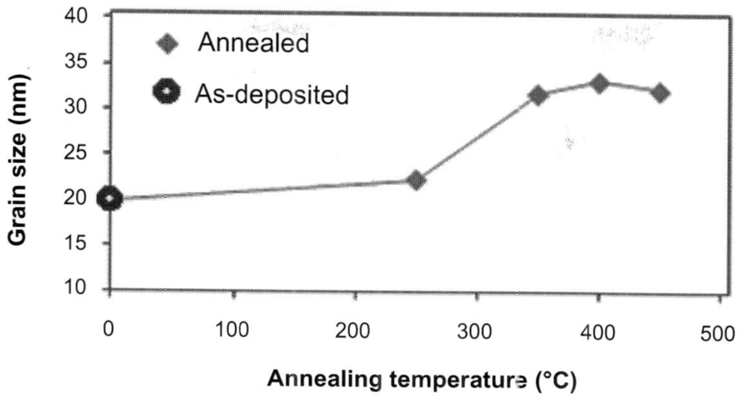

Fig. 8.9 Effect of thermal annealing (in nitrogen-oxygen ambient) on the grain size of ZnO thin films.

Table 8.5 Percentile ratio of Zinc and Oxygen in ZnO thin films (as-deposited and annealed).

Element (Zn:O)	Atomic % of Zn	Atomic % of O
As Deposited	37.9	62.1
250°C	38.7	61.3
350°C	41.6	58.4
400°C	36.0	64.0
450°C	35.6	64.4

The surface topography and growth morphology (surface and cross-section images in the same image) of the ZnO thin films were investigated by Atomic Force Microscopy (AFM) and Field Emission Scanning Electron Microscopy (FESEM). It is evident that the surface morphology changes on the annealing process. As can be seen from the FESEM images in Fig. 8.10, annealing treatment changed the surface of films more homogeneously with better coverage and reduction of the porosity of the films (pinhole free films).

Figure 8.11 shows the surface topography and the roughness of the samples that were measured using AFM. It shows that the Root Mean Square (RMS) values of surface roughness of these films were between 6.83 to 14.90 nm indicating better smoothness of the films. The surface roughness reaches the maximum value of 14.90 nm for the sample annealed at 400°C. However, it was also observed that annealing treatment had little or almost no effect on the surface roughness.

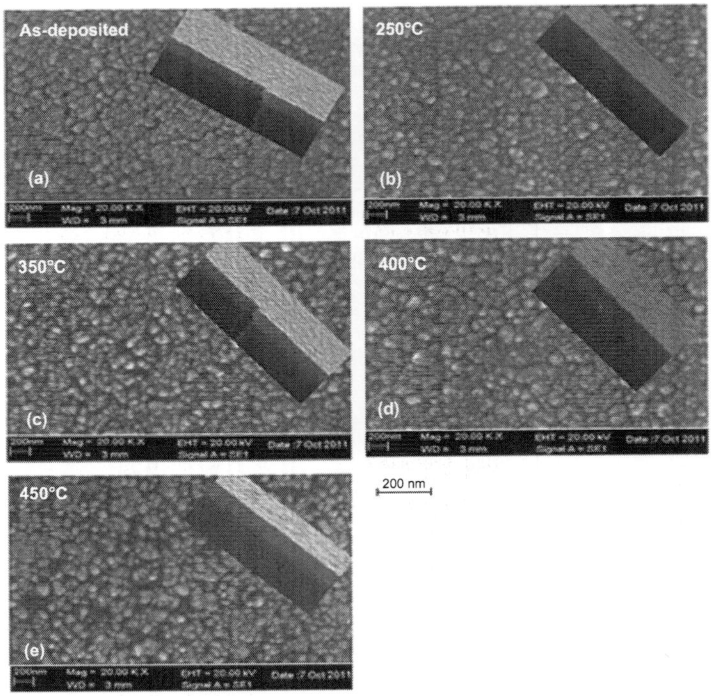

Fig. 8.10 FESEM images of ZnO thin films: (a) As deposited and annealed samples at (b) 250°C. (c) 350°C. (d) 400°C. (e) 450°C in nitrogen-oxygen ambient.

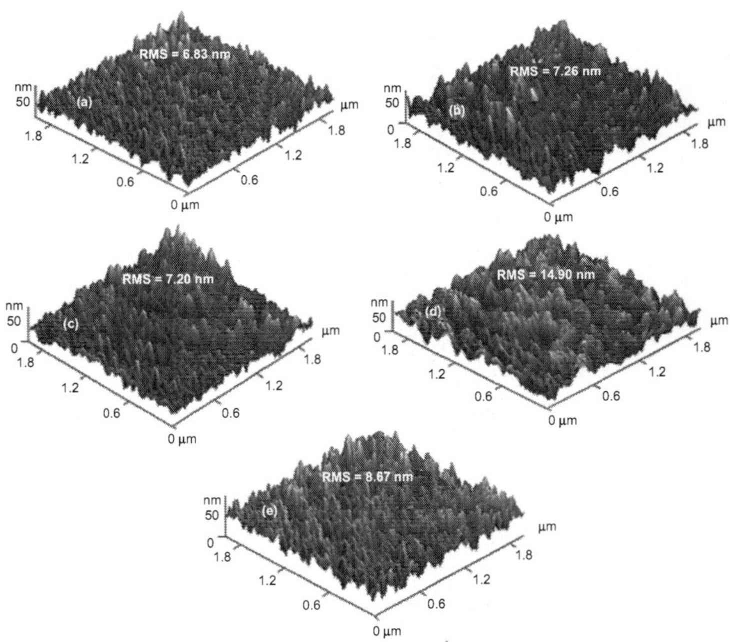

Fig. 8.11 AFM images of ZnO thin films: (a) As deposited and annealed samples at (b) 250°C. (c) 350°C. (d) 400°C. (e) 450°C in nitrogen-oxygen ambient.

One of the vital parameters for the application of ZnO thin films is to have higher optical transmittance (Flickyngerova et al. 2008). The optical transmittance of the film was measured by UV-Visible spectrometer (Perkin Elmer instrument lambda 35), in the range from 350 to 950 nm. Moreover, the optical transmission spectra of the as deposited and annealed ZnO thin films at different temperatures demonstrate good optical transmittance (over 85%) in the visible and near infrared as shown in Fig. 8.12.

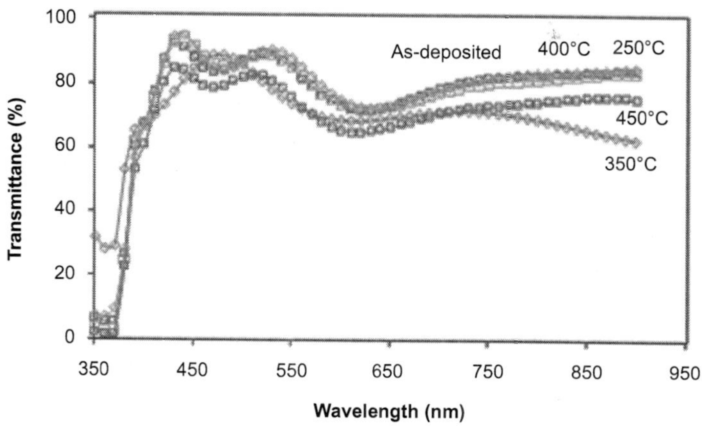

Fig. 8.12 Optical transmittance spectra of ZnO thin films, as deposited and annealed samples at various temperatures.

The variation of absorption coefficient, α with respect to photon energy (hv) was found to comply with the relation: $\alpha hv = A(hv\text{-}E_g)^{1/2}$ for the allowed direct transition where A is the edge width parameter and E_g is the optical bandgap. The optical bandgap values are obtained by extrapolating the linear portion of the plots of $(\alpha hv)^2$ versus hv to $\alpha = 0$ (Chaabouni et al. 2004). As shown in Fig. 8.13, the optical bandgap of ZnO films initially shifted to blue (3.12 to 3.23 eV) as annealed at 400°C and a red shift (3.23 to 3.12 eV) was observed in the annealing temperature range of 250 to 350°C.

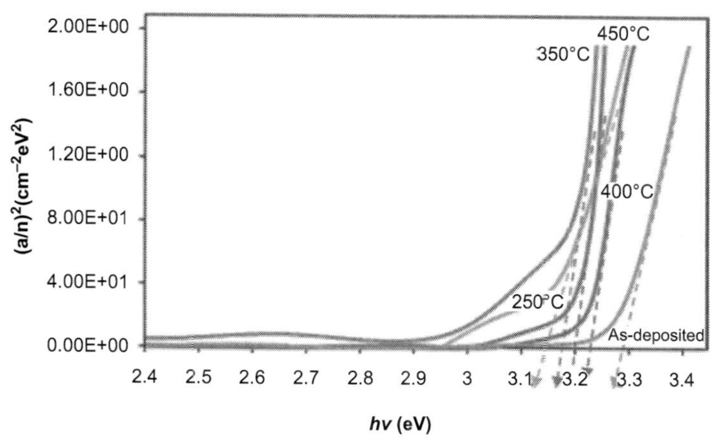

Fig. 8.13 Derivation of ZnO thin film energy bandgap for as deposited and annealed samples at different temperature.

4.2 Electrical Property Analysis

Hall Effect measurement was carried out to determine the electrical properties of ZnO thin films, in order to evaluate the resistivity value of films. Recently, some groups (Konagai and Kushiya 2004, Mazzamuto et al. 2008) have been using zinc oxide (ZnO) material as a buffer layer into CdTe and CIGS cells with resistivity of 10^3 Ω-cm. As shown in Fig. 8.14 and Table 8.6, the results show that the resistivity values of deposited films with different thickness are in the range of 1.96×10^4 to 4.69×10^4 Ω-cm. It is possible to obtain ZnO films with a wide range of resistivities, from 10^{-4} to 10^9 Ω-cm by the RF sputtering technique, due to intrinsic and extrinsic defects (Ondo-Ndong et al. 2003). Moreover, it is shown that the resistivity decreases when the thickness films are increased from 50 nm to 200 nm, however for any given thickness resistivity increases with annealing temperature as can be seen in Figures 8.15 to 8.18. Some references that are mentioned above suggested that the recommended resistivity for buffer layer must be at least around 10^3 Ω-cm or more and the resistivity results for deposited samples of ZnO films that were obtained from this study attained 4.69×10^4 Ω-cm.

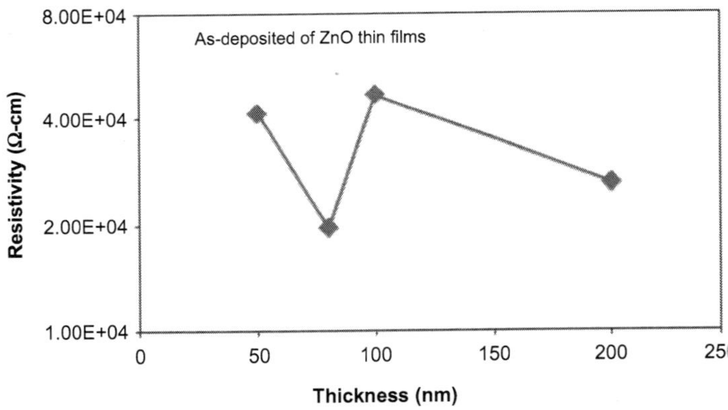

Fig. 8.14 Resistivity of as deposited ZnO thin film with different thickness.

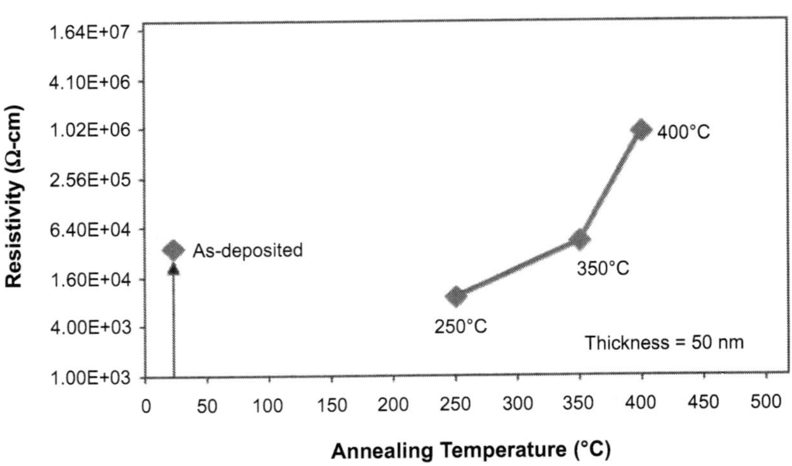

Fig. 8.15 Resistivity of annealed ZnO thin film at thickness of 50 nm.

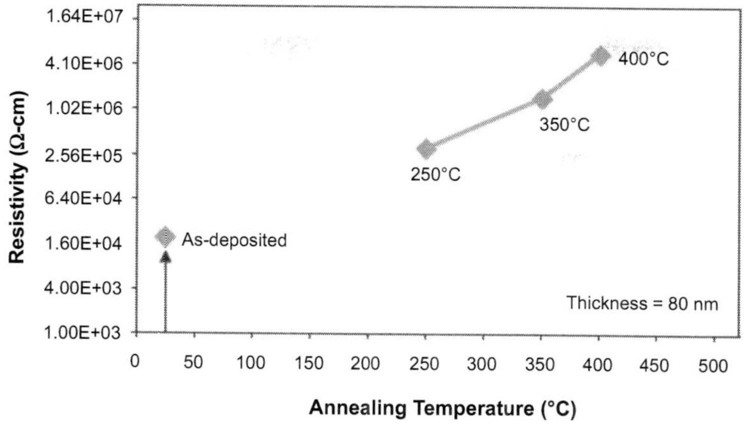

Fig. 8.16 Resistivity of annealed ZnO thin films at thickness of 80 nm.

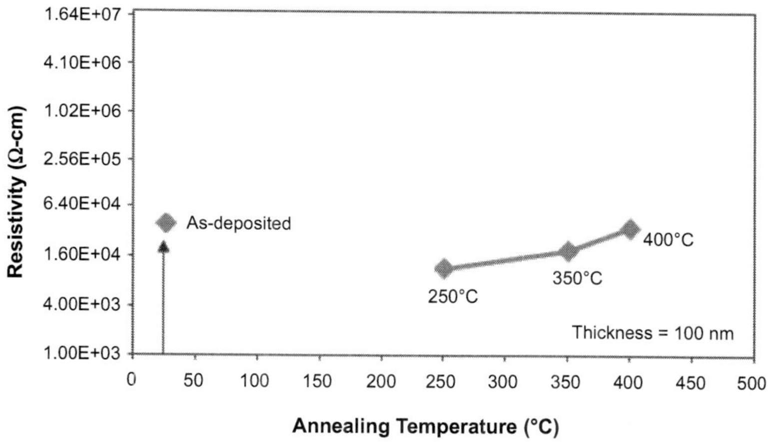

Fig. 8.17 Resistivity of annealed ZnO thin films at thickness of 100 nm.

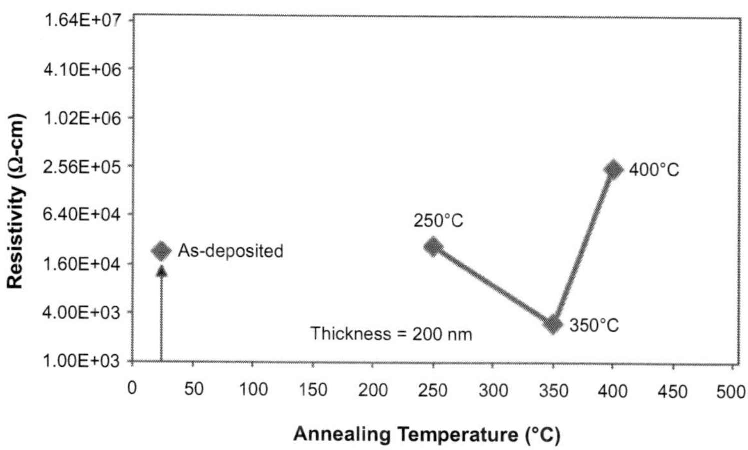

Fig. 8.18 Resistivity of annealed ZnO thin films at thickness 200 nm.

Table 8.6 Resistivity of ZnO as deposited sample.

Samples ID	Thickness (nm)	Resistivity (Ω-cm)
As deposited ZnO_1	50	4.13×10^4
As deposited ZnO_2	80	4.96×10^4
As deposited ZnO_3	100	4.69×10^4
As deposited ZnO_4	200	2.62×10^4

The influence of post-annealing temperatures on electrical properties of ZnO buffer layer was investigated using Hall measurements. As shown in Table 8.7, the films resistivity increases as the anneal temperature increases from 250 to 400°C. The resistivity values for annealed samples were in the order of 2.74×10^3 to 5.89×10^6 Ω-cm, respectively, where the maximum value is obtained at annealing temperature of 400°C and thickness of 80 nm. On average, the results in this study show that the resistivity values that are obtained are high and suitable as a buffer layer for solar cells.

Table 8.7 Resistivity of ZnO thin films as annealed sample at different annealing temperature.

Sample ID	Thickness (nm)	Resistivity (Ω-cm)		
		Annealing temperature		
		250°C	350°C	400°C
ZnO1	50	8.97×10^3	4.31×10^4	9.20×10^5
ZnO2	80	3.16×10^5	1.50×10^6	5.89×10^6
ZnO3	100	1.16×10^4	1.93×10^4	3.72×10^4
ZnO4	200	2.74×10^3	3.01×10^4	2.53×10^5

5. CIGS AND CDTE THIN FILM SOLAR CELLS WITH ZnO THIN FILM BUFFER LAYER

CIGS layers were prepared on Mo-coated soda-lime glass substrates by RF sputtering and the structure of complete cells of CIGS in this study is Mo/CIGS/CdS/i-ZnO/ZnOAl/Al. For the results of I-V characterization for CIGS solar cells with ZnO thin films as the buffer layer as shown in the Fig. 8.19, where a conversion efficiency of 5.64% (V_{oc}: 0.59 V, J_{sc}: 14.5 mA/cm^2, *FF*: 66%) was achieved as an initial attempt. Furthermore, CdTe layers were prepared by RF sputtering with the cell structure of FTO/ZnO/CdS/CdTe/Cu/Ni/Ag. The details results of I-V characterization of complete cells of CdTe solar cells with ZnO thin films as buffer layer as shown in the Fig. 8.20, where a conversion efficiency of 4.29% (V_{oc}: 0.48V, J_{sc}: 12.4 mA/cm^2, *FF*: 72%) was achieved as an initial test cell. A number of optimization of layers and their interfaces are required to achieve higher latent or potential conversion efficiency.

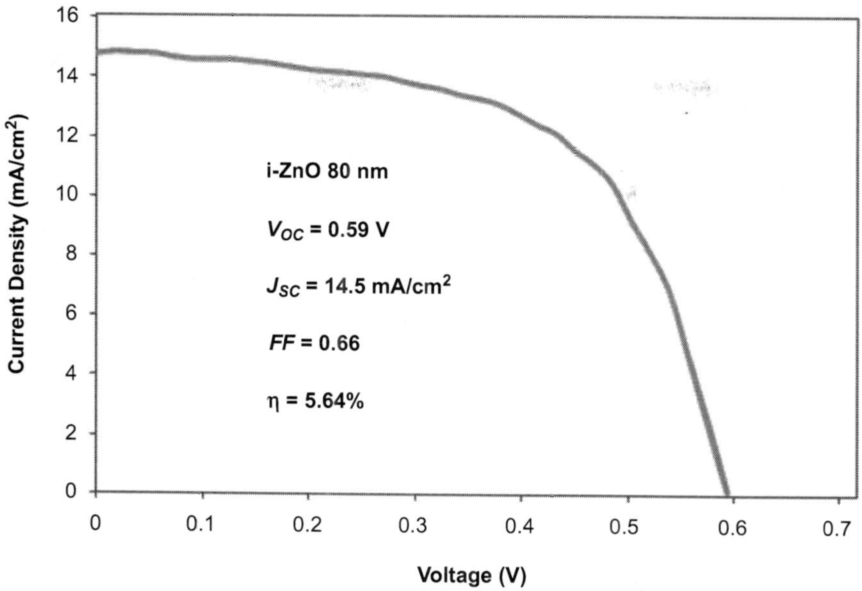

Fig. 8.19 Current density–voltage characteristics for CIGS solar cells fabricated with i-ZnO buffer layer.

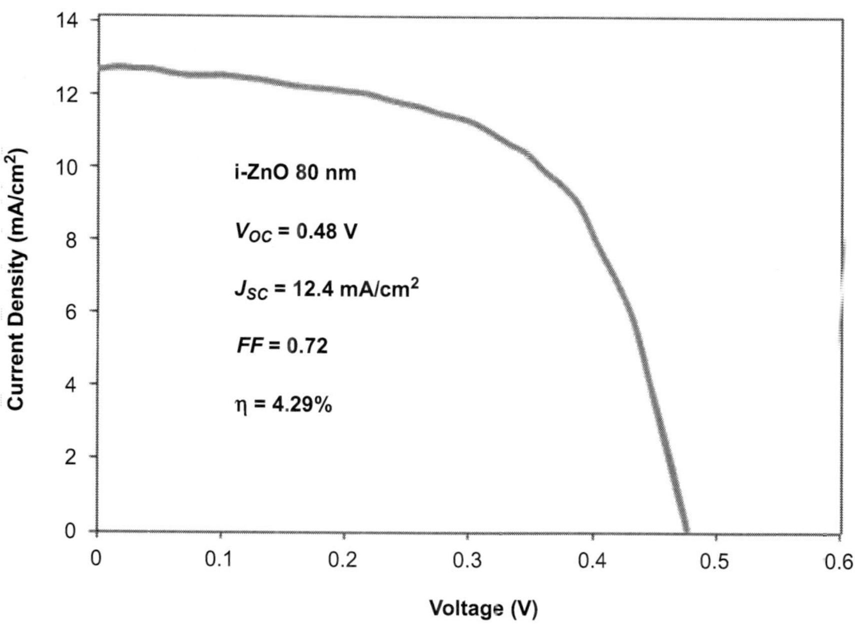

Fig. 8.20 Current density–voltage characteristics for CdTe solar cells fabricated with i-ZnO buffer layer.

Nonetheless, the above results only show the possibility of inserting ZnO thin films as buffer layers in both categories of thin film solar cells. Conversion efficiency in both cases can be achieved higher through rigorous optimization of interfacial defects between layers by careful selection of growth parameters and post-deposition treatments such as annealing. Therefore, the study only opens the door of extracting the benefit of ZnO thin films in potential thin film solar cells.

6. CONCLUSIONS

Intrinsic Zinc Oxide (i-ZnO) films are regarded as one of the most potential buffer layer materials in thin film photovoltaics. In this study, ZnO films are deposited on top of ITO coated glass substrates by RF magnetron sputtering technique. After annealing treatment the film surface becomes quite smooth and uniform compared to the samples without annealing. Moreover, the annealing treatment improves the morphology of ZnO film as obvious from AFM and SEM images. Additionally, the films show optical transmission over 85% in the visible wavelength range. XRD results show that the films were polycrystalline with the preferential orientations along (0 0 2), (1 0 1) and (1 0 2) direction, but when the annealing temperature is increased above 400°C, the films are shown (0 0 2) to be more dominant. It is found that the ZnO thin films with post annealing treatment exhibit good structural and optical properties with a smoother surface. Based on the results, it can be concluded that the properties of the ZnO thin films are suitable for solar cell application, especially for the buffer layer in the CIGS and CdTe solar cells, tuning the bandgap to suitable values upon annealing. The resistivity results are as mentioned below: 4.13×10^4 Ω-cm at thickness 50 nm, 4.96×10^4 Ω-cm at 80 nm thicknesses, 4.69×10^4 Ω-cm 100 nm thickness and 2.62×10^4 Ω-cm at thickness 200 nm as found from Hall Effect Measurement. In the annealing treatment, the resistivity films become higher compared to as deposited films. Moreover, the highest resistivity for i-ZnO thin films was achieved when the annealing temperature was at 400°C, which has the value of 5.89×10^6 Ω-cm to be more dominant than others. In conclusions, 80 nm thickness and 400°C annealing temperature are the best parameters for electrical properties of i-ZnO thin films that were obtained later. In conclusions, overall results on intrinsic ZnO thin films that have been obtained from this work prove the possibility to solar cell application, in particular as buffer layers for both CIGS and CdTe solar cells. The films with thickness of 80 nm and 400°C annealing temperature are the best parameters for electrical and optical properties of i-ZnO thin films. The optimum ZnO film has been adapted to both CIGS and CdTe thin film solar cells. Initially, a conversion efficiency of 5.6% was achieved for CIGS thin films solar cells and conversion efficiency of 4.29% was achieved for CdTe thin films solar cells. However, it is expected to have full potential once interfacial and other process optimization will be executed in both thin film solar cell structures as has already been done by many groups.

7. SCOPES OF RESEARCH ON ZnO AS BUFFER LAYER IN THIN FILM SOLAR CELLS

In future, the intrinsic zinc Oxide (i-ZnO) thin films that were developed can further be improved in the quality of the properties of the films used in the large area of photovoltaic application. Special attention should be made on using the best parameters that will be used in this technique (sputtering) to deposit the buffer layer films. The optimization to produce the layers can also be done to maximize the results. On the other hand, this can give a new insight for the failure mechanisms observed for highly textured buffer layer. Further improvements of the quality of annealed films of buffer layer can be investigated at the researcher's laboratory using more variation of temperature of annealing. The most important point is that, some requirements that have to be looked into is that the material should be n-type in order to form a p-n junction with the absorber layer. Moreover, there should be a large bandgap for limited light absorption. The process for deposition should be low cost and suitable for wide area deposition. In addition, the technique should be capable to passivate the surface states of the absorber layer.

8. ACKNOWLEDGEMENTS

The authors would like to express heartfelt gratitude to the predecessors in ZnO research for decades. The authors also wish to express their thanks to people in both The National University of Malaysia, Malaysia and King Saud University, Saudi Arabia for research facilities where experimental studies have been conducted and analyzed. Last but not least, Prof. Biswajit Ghosh and his student Dr. Madhumita Das from Jadavpur University, India are also acknowledged for their introductory textual help to make this chapter a readable one.

9. REFERENCES

Bagnall, D. M., Y. F. Chen, Z. Zhu, T. Yao, S. Koyama, M. Y. Shen, et al. 1997. Optically pumped lasing of ZnO at room temperature. Appl. Phys. Lett. 70: 2230.

Chaabouni, F., M. Abaab and B. Rezig. 2004. Effect of the substrate temperature on the properties of ZnO films grown by RF magnetron sputtering. Mat. Sci. and Eng. B 109: 236-240.

Dixit, H., R. Saniz, D. Lamoen and B. Partoens. 2010. The quasiparticle band structure of zincblende and rocksalt ZnO. J. Phys. Condens. Matter. 22: 125505.

Elilarassi, R. and G. Chandrasekaran. 2010. Effect of annealing on structural and optical properties of zinc oxide films. Mater. Chem. Phys. 121: 378-384.

Fang, Z. B., Z. J. Yan, Y. S. Tan, X. Q. Liu, Y. Y. Wang, W. Lee, et al. 2005. Influence of post-annealing treatment on the structure properties of ZnO films. Appl. Surf. Sci. 241: 303-308.

Flickyngerova, S., A. Rehakova, V. Tvarozek and I. Novotny. 2008. Sputtered of ZnO:Al thin films for application in photovoltaic solar cells. Advanc. in Elec. and Elec. Eng. 7: 382-384.

Gardeniers, J. G. E., Z. M. Rittersma and G. J. Burger. 1998. Preferred orientation and piezoelectricity in sputtered ZnO films. J. Appl. Phys. 83: 7844-7854.

Ghosh, B., M. Das, P. Banerjee and S. Das. 2009. Fabrication of the SnS/ZnO heterojunction for PV applications using electrodeposited ZnO films. Semicond. Sci. Technol. 24: 025024.

Greer, J. 2006. Large-area commercial pulsed laser deposition. pp. 191-213. *In*: R. Eason (ed.). Pulsed Laser Deposition of Thin Films: Applications-Led Growth of Functional Materials. Wiley Interscience, Hoboken.

Gunawan, O., L. Sekaric, A. Majumdar, M. Rooks, J. Appenzeller, J. W. Sleight, et al. 2008. Measurement of carrier mobility in silicon nanowires. IBM Research Division, T. J. Watson Research Center, Yorktown Heights, New York 10598, Nano Lett. 8(6): 1566-1571.

Hoffman, R. L., B. J. Norris and J. F. Wager. 2003. ZnO-based transparent thin-film transistors. Appl. Phys. Lett. 82: 733.

Janotti, A. and C. G. Van de Walle. 2006. New insights into the role of native point defects in ZnO. J. Cryst. Growth 287: 58-65.

Janotti, A. and C. G. Van de Walle. 2007. Native point defects in ZnO. Phys. Rev. B 75: 165202.

Janotti, A. and C. G. Van de Walle. 2009. Fundamentals of zinc oxide as a semiconductor. Reports on Progress in Physics 72(12): 126501.

Konagai, M. and K. Kushiya. 2004. Development of Cu(InGa)Se_2 thin-film solar cells. Thin Film Solar Cells. Springer Series in Photonics 13: 183-210.

Kushiya, K. 2004. Development of Cu(InGa)Se_2-based thin-film PV modules with a Zn (O, S, OH) x buffer layer. Solar Energy 77: 717-724.

Langer, D. W. and C. J. Vesely. 1970. Electronic core levels of zinc chalcogenides. Phys. Rev. B 2: 4885-4892.

Lee, E. C. and K. J. Chang. 2004. Possible p-type doping with group-I element in ZnO. Phys. Rev. B 70: 115210.

Lee, W., M.-C. Jeong and J.-M. Myoung. 2004. Catalyst-free growth of ZnO nanowires by metal-organic chemical vapour deposition (MOCVD) and thermal evaporation. Acta Materialia 52: 3949-3957.

Look, D. C. 2001. Recent advances in ZnO materials and devices. Mater. Sci. Eng. B 80(1-3): 383-387.

Look, D. C., C. Coskun, B. Chaflin and C. G. Farlow. 2003. Electrical and optical properties of defects and impurities in ZnO. Physica. B 32: 340-342.

Lorenz, M., E. Kaidashev, H. V. Wenckstern, V. Riede, C. Bundesmann, D. Spemann, et al. 2003. Optical and electrical properties of epitaxial (Mg, Cd) x Zn 1− x O, ZnO and ZnO:(Ga, Al) thin films on c-plane sapphire grown by pulsed laser deposition. Solid State Electron. 47: 2205-2209.

Low, J. J., M. L. Kreider, D. P. Pulsifier, A. S. Jones and T. H. Gilani. 2008. Band gap energy in silicon. Am. J. Undergrad. Res. 7: 27-32.

Mandalapu, L. J., Z. Yang, F. X. Xiu, D. T. Zhao and J. L. Liu. 2006. Homojunction photodiodes based on Sb-doped p-type ZnO for ultraviolet detection. Appl. Phys. Lett. 88: 092103.

Mathew, M. 2010. Buffer Layer for Thin Film Heterojunction. Thesis Chapter. Thin Film Photovoltaic Division, Department of Physics, Cochin University of Science and Technology, India.

Mazzamuto, S., L. Vaillant, A. Bosio, N. Romeo, N. Armani and G. Salviati. 2008. A study of the CdTe treatment with a Freon gas such as CHF_2Cl. Thin Solid Films 516: 7079-7083.

MaCandless, B. E. and S. S. Hegedus. 1991. Influence of CdS Window layers on thin film CdS/CdTe solar cell performance. Photovoltaic Specialists Conf.: 967-972.

Meyer, B. K., H. Alves, D. M. Hofmann, W. Kriegseis, D. Forster, F. Bertram, et al. 2004. Bound exciton and donor–acceptor pair recombinations in ZnO. Phys. Status Solidi 241: 231-260.

Minami, T. 2005. Transparent conducting oxide semiconductors for transparent electrodes. Semicond. Sci. Technol. 20: S35-S44.

Ondo-Ndong, R., F. Pascal-Delannoy, A. Boyer, A. Giani and A. Foucaran. 2003. Structural properties of zinc oxide thin films prepared by r.f. magnetron sputtering. Mater. Sci. Eng. B 97: 68-73.

Oshikiri, M. and F. Aryasetiawan. 1999. Band gaps and quasiparticle energy calculations on ZnO, ZnS and ZnSe in the zinc-blende structure by the GW approximation. Phys. Rev. B 60: 10754-10757.

Ozawa, K., K. Sawada, Y. Shirotori and K. Edamoto. 2005. Angle-resolved photoemission study of the valence band structure of ZnO(1010). J. Phys. Condens. Matter. 17: 1271-1278.

Özgür, Ü., Ya. I. Alivov, C. Liu, A. Teke, M. A. Reshchikov, S. Doğan, et al. 2005. A comprehensive review of ZnO materials and devices J. Appl. Phys. 98: 041301.

Park, C. H., S. Zhang and S. H. Wei. 2002. Origin of p-type doping difficulty in ZnO: the impurity perspective. Phys. Rev. B 66: 073202.

Queisser, H. J. and E. E. Haller. 1998. Defects in semiconductors: some fatal, some vital. Science 281: 945-950.

Reynolds, D. C., D. C. Look, B. Jogai and H. Morkoc. 1997. Similarities in the bandedge and deep-centre photoluminescence mechanisms of ZnO and GaN. Solid State Commun. 101: 643.

Reynolds, D. C., D. C. Look, B. Jogai, J. E. Van Nostrand, R. Jones and J. Jenny. 1998 Source of the yellow luminescence band in GaN grown by gas-source molecular beam epitaxy and the green luminescence band in single crystal ZnO. Solid State Commun. 106: 701-704.

Rössler, U. 1969. Energy bands of hexagonal II-VI semiconductors. Phys. Rev. 184: 733-738.

Service, R. F. 1997. Will UV lasers beat the blues? Science 276: 895.

Song, D., J. Zhao, A. Wang, P. Widenborg, W. Chin and A. Aberle. 2001. 8% efficient ZnO/Si heterojunction solar cells prepared by magnetron sputtering. 17th European PV Conference, Munich.

Tang, W. and D. C. Cameron. 1994. Aluminum-doped zinc oxide transparent conductors deposited by the sol-gel process. Thin Solid Films 238: 83-87.

Tsuchiya, T., T. Emoto and T. Sei. 1994. Preparation and properties of transparent conductive thin films by the sol-gel process. J. Non-Cryst. Solids 178: 327-332.

Tuomisto, F., V. Ranki and K. Saarinen. 2003. Evidence of the Zn vacancy acting as the dominant acceptor in n-type ZnO. Phys. Rev. Lett. 91: 205502.

Van de Walle, C. G. 2001. Defect analysis and engineering in ZnO. Physica. B Condens. Matter. 899: 308-310.

Vijayalakshmi, S., S. Venkataraj and R. Jayavel. 2008. Characterization of cadmium doped zinc oxide (Cd : ZnO) thin films prepared by spray pyrolysis method. J. Phys. D: Appl. Phys. 41: 245403.

Vogel, D., P. Krüger and J. Pollmann. 1995. Ab initio electronic-structure calculations for II-VI semiconductors using self-interaction-corrected pseudopotentials. Phys. Rev. B 52: R14316-R14319.

Wang, Z. L. 2004. Zinc oxide nanostructures: growth, properties and applications. J. Phys. Condens. Matter. 16: R829-R858.

Xu, Y. N. and W. Y. Ching. 1993. Electronic, optical and structural properties of some wurtzite crystals. Phys. Rev. B 48: 4335-4351.

Yamamoto, T. and H. Katayama-Yoshida. 2000. Unipolarity of ZnO with a wideband gap and its solution using codoping method. J. Cryst. Growth 214/215: 522.

Zhang, S. B., S.-H. Wei and A. Zunger. 2001. Intrinsic n-type versus p-type doping asymmetry and the defect physics of ZnO. Phys. Rev. B 63: 075205.

Zhang, X-.Y., Z-.W. Chen, Y.-P. Qi, Y. Feng, L. Zhao, L. Qi, et al. 2007. Ab initio comparative study of zincblende and wurtzite ZnO. Chin. Phys. Lett. 24(4): 1032-1034.

9

Formation of Nanoporous α-Fe_2O_3 Thin Film as Photoanode by Anodic Oxidation on Iron

Monna Rozana,[1,a] Atsunori Matsuda,[2] Go Kawamura,[2] Wai Kian Tan[2] and Zainovia Lockman[1,*]

ABSTRACT

Sunlight can be transformed into useful energy via photovoltaic effect or by photolysis of water. The latter can be done by using a photoelectrochemical (PEC) cell with hematite (α-Fe_2O_3) as a photoanode for oxygen evolution and platinum electrode for hydrogen generation. α-Fe_2O_3 is a low bandgap material and thus electron-hole pairs can be generated when the oxide is illuminated with sunlight. Nonetheless, α-Fe_2O_3 displays low water oxidation efficiencies thus the overall solar to hydrogen conversion is often very small. The limitations of α-Fe_2O_3 can be due mostly to the carrier dynamics, which can be overcome by nanostructuring approach i.e. forming thin film comprising of nanostructure. One process that can be adopted to form nanostructured α-Fe_2O_3 film is anodic oxidation of iron. Anodic oxidation is chosen to obtain α-Fe_2O_3 thin film with self-ordered pores in nanoscale. The process is simple, cost-effective and rather efficient in the formation of thin film with such nanostructure. However, despite the success in surface nanostructuring, the anodic film is often hydrated and amorphous, hence requiring post-annealing treatment. Annealing results in the formation of internal oxide layer which resembles the multi layered scale oxide when iron is thermally oxidized. These internal layers, comprising of oxide of various phases, may influence the electron transport properties, lowering the overall efficiency of the PEC cell. In this chapter, a review on the recent progress in nanoporous α-Fe_2O_3 film formation by

[1] Green Electronics Nanomaterials Group, Science and Engineering of Nanomaterials Team, School of Materials and Mineral Resources, Universiti Sains Malaysia, Penang 14300, Malaysia.

[2] Department of Electrical and Electronic Engineering, Faculty of Engineering, Toyohashi University of Technology, Aichi 441-8580, Japan.

[a] E-mail: narozana@gmail.com

* Corresponding author: zainovia@usm.my

anodic process of iron and the photocurrent performance of the formed oxide will be presented, highlighting on efforts made to improve on the PEC cell efficiency.

Keywords: α-Fe_2O_3, nanoporous film, anodic oxidation, photoanode.

1. INTRODUCTION

This chapter presents literature review on thin α-Fe_2O_3 film comprising of self-ordered nanopores formed by anodic oxidation of iron for an application as electrode in a photoelectrochemical (PEC) cell. PEC cells can have various different functions but here we shall dwell on the use of it to split water to produce hydrogen or oxygen gas. The production of hydrogen is seen as a promising technology for solving the energy crisis and depletion of resources at this moment. Since water itself does not absorb significant radiation within the solar spectrum, a light absorbing specie (termed photoconverter) must be used to transduce the radiant energy to chemical (or electrical) energy in the form of Electron-Hole Pairs (EHPs) which are responsible to drive the reaction. A PEC cell is composed of such photoconvertor: semiconductor electrode (photoanode) as illustrated in Fig. 9.1a. When illuminated with photon of energy larger than its bandgap, the photoanode will generate EHPs. Because of this the chemical equilibrium in the PEC cell is now disturbed, a photovoltage that gives rise to a photocurrent will be developed. The photocurrent can be measured and the circuit of the electrons is completed by electron transfer processes over the cathode|electrolyte interfaces. An oxidation reaction in the electrolyte takes place at the anode and simultaneously a reduction takes place at the cathode. As mentioned, PEC cell is now being routinely used to split water and from an electron transfer to H^+ in the electrolyte, hydrogen gas can be produced in the cathode and evolution of oxygen can be observed at the anode (Fig. 9.1b).

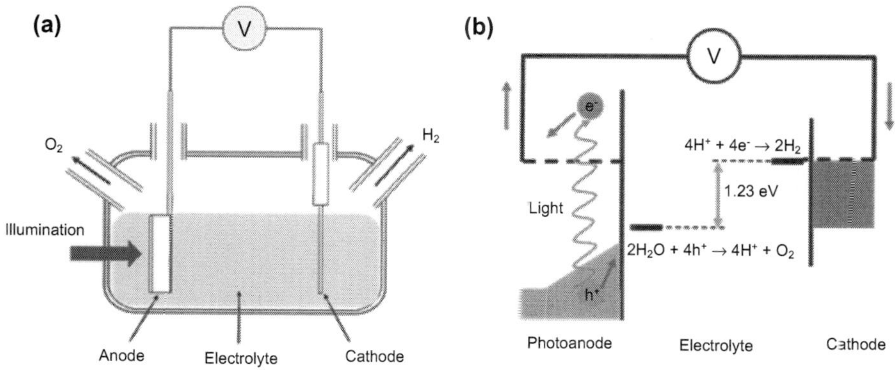

Fig. 9.1 (a) A typical PEC cell. (b) Diagram showing hydrogen gas production at the cathode.

Despite the rather simple description on how water can be transformed to hydrogen (or oxygen) under illumination of light, the transformation process is far from straightforward, especially if the conversion is done under sunlight. Typically small gap n-type semiconductors are chosen as the photoconvertor material but not all n-type semiconductors are photoactive or are stable in the electrolyte used. Among all semiconductors materials, oxide semiconductors are known to be the most stable i.e. they do not corrode in electrolyte under illumination but most oxides to date are wide gap semiconductors. Thus only a small fraction of sunlight is used to produce the EHPs and they have low photoconversion efficiency.

The most important figure of merit for a semiconductor photoanode is the photoconversion efficiency for water splitting, which is defined as the ratio of the chemical potential energy stored in the form of hydrogen molecules to the incident radiative energy (Murphy et al. 2006). There are several acceptable methods utilized in describing the efficiency of a PEC cell: Incident Photon to Current Efficiencies (IPCE) at an electrode potential given relative to a certain counter electrode like Reversible Hydrogen Electrode (RHE) or Solar-to-Hydrogen (STH) efficiency i.e. measurement of hydrogen formation. A comprehensive work by Murphy et al. (2006) outlined all of the common and applicable methods to determine if indeed a photoelectrode is efficient or not. Whilst both have advantages and own limitations, at the moment IPCE is said to be more accurate and has been applied in determining efficiency of a PEC cell. Comparison, by simply looking at the photocurrent can also be done, but again to a lesser accuracy as photocurrent generated is very sensitive to the illumination source.

2. KEY REQUIREMENTS OF A PHOTOANODE

A free energy change for the conversion of one molecule of water to H_2 and ½ O_2 under standard conditions is $\Delta G = 237.2$ kJ mol^{-1} (Chen et al. 2012). This corresponds to ΔE^o of 1.23 V per electron transferred. A photoanode material must be able to absorb radiant light with photon energies of >1.23 eV then to convert the energy to hydrogen and oxygen (Walter et al. 2010). Therefore photoanode material must have conduction band-edge energy and valence band-edge energy that straddles the electrochemical potential for hydrogen and oxygen evolution. An efficient photoanode should also have a characteristic of having high absorption coefficient, which implies that most of the incident photons will be absorbed in the material and transformed to EHPs. Both photoactive n-type and p-type semiconductors can be used to produce reasonable photocurrent, provided the semiconductor possess valence band energies more positive than the water oxidation potential and conduction band energies sufficiently more negative for proton reduction. Wide energy gap n-type semiconductors often straddle the potential for O_2/H_2O redox couple and hence can be selected as photoanode material that can derive the water photoelectrolysis for hydrogen or oxygen evolution.

To ensure all EHPs play the role in driving the process, they must not be allowed to recombine and ought to be separated as fast as possible. Carrier dynamics in the photoanode material is indeed an important aspect in designing and selecting a material for a high efficiency photoanode. In another words, the carriers must have a long life until they can travel to the surface of the semiconductor to be transferred to the electrolyte. Not many materials have the ability to sustain their photogenerated EHPs and most EHPs will recombine within picosecond of them being generated.

Due to these losses, energy required for photoelectrolysis at a photoanode is normally higher (1.6-2.4 eV) than the said 1.23 V (Grimes et al. 2008). It is ideal to separate the EHPs right after the process of their generation. The process of separation is explained next.

When a n-type semiconductor photoanode is brought into contact with liquid, electrons transfer at the junction will produce an electrode with excess positive charge, arising from the ionized dopant atoms in the semiconductor and the solution will be in an excess of negative charge (Krol 2012a). The positive charge in the photoanode will be spread out over the depletion width in the semiconductor whereas a Hemholtz layer of the negative charge region will be spread out in the solution. Electric field will be produced within this region and EHP separation process may happen here. The field strength developed can be high enough to then direct the free holes to be transferred to the solution.

Typically the depletion region can be on the order of hundreds of nanometers thus EHPs formed within this region will be able to be separated and transferred (Krol 2012b). The width and potential developed within the depletion region are dependent on dopant levels in the photoanode. Once holes are transferred out, electrons are set to move to the back contact and will be measured and collected at the cathode. If the energy is sufficiently high for the formation of H_2, bubbles of H_2 will be seen at the cathode. Holes transfer process is also an important characteristic that needs to be understood in selecting a photoanode material. First, it is related to the lifetime of the photogenerated carriers as mentioned previously and then on the process of carriers separation. Kinetics plays a role here in determining, which process would occur at a faster rate; for example if the rate of photogenerated carriers formation is slower than the rate of recombination then the material will not have enough EHPs. Ideally, a material has to produce as much EHPs as possible and within picoseconds; the EHPs ought to be separated. The separation occurs at the depletion region hence a wide enough region with high built in potential must be designed. Once the EHPs are separated, the conductivity of the carriers becomes an important metric. The movement of electrons to the back contact for example must be set to be as smooth as possible with minimum hindrance that can induce scattering.

Indeed, a material with high conductivity is preferred to promote the transfer of electrons from the oxides layer onto the back contact (Gan et al. 2014) and yield high efficiency PEC. To achieve this, mobility and number of electrons in the material have to be as high as possible. During the electrons transport from the surface to the photoanode to the back contact, electrons will undergo various scattering processes. Ideally anode system free from scattering centers is preferable. Purity, phases, crystallinity, ionized impurities, grain boundaries and other factors which may influence the mobility of the electrons must be taken into consideration. As for an efficient holes transfer at the semiconductor|liquid junction, a material which has long hole diffusion length is preferred as one would prefer to have most of the holes formed to be transferred out to the electrolyte for oxygen formation. In certain materials, short diffusion length of holes may reduce the efficiency of the photoanode for oxygen formation hence many of the holes are recombined with free electrons before they can be successfully transferred out to the electrolyte at the junction. The oxidative power of holes is also an important consideration and it must, as mentioned be adequate for oxygen evolution.

Photoanode must also be able to withstand harsh conditions; obviously it cannot photocorrode under illumination in the liquid it is immersed in and must not undergo any process on its surface that may degrade it when polarized under radiation. Moreover, in view of process scalability, the material must not be too expensive, should be easily scalable and by itself not possess any harmful effect to the environment.

To date, there is no single material screened for a photoanode that has all the properties listed above. Among all the materials, TiO_2 is perhaps the one which fulfill most of the requirements. TiO_2 is in fact the most explored photoanode material since the seminal work reported by Fujishima and Honda (1972) despite the wide bandgap the material possesses. A photoanode material should have a small bandgap to utilize a significant portion of visible light from the sun for EHPs formation. Quest in finding the ideal semiconductor material is ongoing by various methods including *ab initio* calculations as well as extensive photoelectrochemical studies on doped materials have been published (Torres et al. 2004, Tilley et al. 2010). Among several oxide semiconductors proposed, α-Fe_2O_3 is seen as an interesting candidate since it is a material with bandgap of ~2.2 eV hence EHPs can be produced when α-Fe_2O_3 is irradiated with sunlight. Moreover, α-Fe_2O_3 is an extrinsic semiconductor with n-type character could have originated from the interstitial Fe^{2+} which act as donors. Nonetheless, α-Fe_2O_3 has its own limitation as shown by the typical characteristics its possesses as shown in Table 9.1. The life of EHPs formed in the oxide is reported to be very short whereby 70% of EHPs recombine within 8 ps (Cherepy et al. 1998). On the other hand, the absorption coefficient of photon by α-Fe_2O_3 is reported to be in the order 10^3 cm^{-1} (Galuza et al. 1998). The absorption length in the oxide is 118 nm at photon wavelength 550 nm (Balberg and Pinch 1978). Morin (1951) reported that pure sintered crystalline hematite has very low electrical conductivity and the conductivity of the oxide appears to be rather low generally due to the low electron mobility of the oxide. The holes diffusion length has been estimated to be 2-4 nm (Kennedy and Frese 1978). Despite all these limitations, there have been many works on this oxide as a photoelectrode.

Early works on the use of α-Fe_2O_3 as photoanode were focused on the assessment of the oxide to perform photoassisted electrolysis of water. The first work on this subject was by Hardee and Bard (1976) whereby they reported on photoassisted electrolysis of water on α-Fe_2O_3 at wavelengths longer than 400 nm. Following this work, more data on photocurrent measurement of the oxide irradiated with light of wavelength from 400 to 700 nm were published for example by Kennedy and Frese (1978). Assessment on the photostability of the oxide was also performed in the early days; most work reached a consensus by stating that α-Fe_2O_3 is stable under irradiation in a pH range from 4 to 14. Parallel to this, doping work was also being reported perhaps in an effort to increase the conductivity of the oxide. Ti, Nb and Si have been successfully doped in α-Fe_2O_3 giving decent photocurrent values and IPCE efficiency (Leygraf et al. 1982, Cesar et al. 2006, Cesar et al. 2009, Glasscock et al. 2007). Generally the photocurrent performance of pure α-Fe_2O_3 photoanode have been reported to be much lower than its maximum calculated photocurrent density of 12.6 mA cm^{-2} under the AM 1.5 G solar spectrum (100 mW cm^{-2}) (Tilley et al. 2010, Murphy et al. 2006). However when dopants are used, increase in carrier density and reduction in charge recombination may improve electrical conduction of α-Fe_2O_3 (Iandolo et al. 2015, Satsangi et al. 2010).

Table 9.1 Characteristics of α-Fe_2O_3.

Characteristics	Values	References
EHP generation	560-640 nm	Sivula 2012, Galuza et al. 1998, Lindgren et al. 2003
EHP recombination	70% of EHPs recombine within 8 ps	Cherepy et al. 1998
Absorption coefficient	10^3 cm^{-1}	Galuza et al. 1998
Absorption length	118 nm at photon wavelength 550 nm	Balberg and Pinch 1978
Conductivity	10^{-14} Ω^{-1} cm^{-1}	Morin 1951
Conduction electron concentration	10^{18} cm^{-3} at 1,000 K	Morin 1954
Electron mobility	10^{-2} $cm^2/V/s$	Morin 1954
Holes diffusion length	2-4 nm	Kennedy and Frese 1978

3. NANOSTRUCTURED α-FE_2O_3 PHOTOANODE

Despite all the extensive work on improving the photochemical properties of α-Fe_2O_3, the oxide remains a poor photoanode and hence not many reports appeared on this material in the 1990s. However the material seemed to emerge again in early 2000s, but with research direction having been shifted on improving it by nanostructuring. A seminal work by Lindgren et al. in 2002 can be regarded as the most imperative in looking at the need of the use of nanostructured α-Fe_2O_3. Sivula et al. (2011) outlined and discussed the origin of the poor performance of α-Fe_2O_3 and concluded that most of the limitations are due to the use of standard planar single crystal or sintered disk electrode geometries. Controlling the electrode morphology by nanostructuring it in various dimensions and structure was then proposed to be the way to further improve the PEC properties of α-Fe_2O_3. The progress of the use of nanostructured α-Fe_2O_3 was reviewed by Satsangi et al. (2010), Wheeler et al. (2012) and Sivula (2012). The most recent review on α-Fe_2O_3 was published in 2015 by Iandolo et al. (2015). The strategies to reduce the high onset potential for oxygen production at the photoanode was reviewed and discussed. The advancement in lowering onset potentials was focused on the way to enhance reaction kinetic, reduction of surface and bulk recombination, increasing photovoltage and shifting the flat band potential. Another comprehensive review is by Zandi and Hamann (2015); in this work they focused on circumventing depletion region recombination in order to improve the efficiency of α-Fe_2O_3 electrodes.

Introduction of vacancies (for example oxygen vacancies) also can help in the improvement of the conductivity of the photoanode. Vacancies can be introduced by doping or by selecting a fabrication process that would induce their formation. Creation of vacancies has many different advantages including intentionally adding electrons and the state level possessed by the vacancy can act as electron traps (Rocket 2008). One way to include oxygen vacancies in α-Fe_2O_3 is by fabricating the oxide in condition with lower pressure of oxygen. Anodic oxidation, due to the concentration gradient of oxygen can also be thought as an effective way in oxygen vacancy creation. In fact more importantly, oxidation produces surface oxide which depend on the anodization parameters that can be easily nanostructured.

Apart from anodic process, nanostructuring of α-Fe_2O_3 photoanode can be achieved by typical thin film formation techniques: spray pyrolysis (Satsangi et al. 2008), atmospheric pressure chemical vapor deposition (Kay et al. 2006, Tilley et al. 2010), chemical vapor deposition (Cesar et al. 2009, Tahir et al. 2009), spin coating (Souza et al. 2009a, 2009b) and thermal oxidation (Hiralal et al. 2011, Vincent et al. 2012). Often porous film is desired as it will have more contact area with the electrolyte. Porous film can be made in the form of ordered or random pores (can be divided into macropores, mesopores, micropores), tubes, wires, rods and others. According to the International Union of Pure and Applied Chemistry (IUPAC), macropores have the pore size larger than 50 nm, mesopores 2-50 nm and micropores have a pore smaller than 2 nm in diameter. Nanoporous material is a subset of a porous material which has pore diameter of 1-100 nm. Anodic oxidation as mentioned can be adopted for nanostructuring purposes. It is worth mentioning that the primary advantage commonly associated with nanostructured photoanode compared to planar system is the decoupling of the directions of light absorption and charge carrier collection. Highly ordered nanostructure, for example nanorods will have high surface area and a small enough diameter which reduces the distance of the minority carries to travel to the surface of the oxide to be transferred to the electrolyte. As the distance for holes to travel is smaller, more of them can be expected to be removed from the oxide before they can recombine. Similarly, in the case of nanoporous oxide, if the pore wall can be tailored to have a thickness of less than the diffusion length of the minority carriers, then they will have to travel only a small distance before being transferred to the electrolyte. More minority carriers are thought to be able to reach the interface before recombination hence more free electrons are expected to give higher photocurrent of the photoanode.

4. ANODIC OXIDATION FOR NANOPOROUS α-Fe_2O_3 THIN FILM FORMATION

Anodic oxidation is an electrochemical process to produce a thin oxide film on a metallic substrate. If the substrate used is iron, then anodic film on iron would be likely iron oxide but not necessarily in a form of pure α-Fe_2O_3. One interesting feature of anodic oxidation process is the ability for the process to produce self-ordered nanostructures, growing perpendicular to the substrate as illustrated in Fig. 9.2, control over the wall thickness as well as the depth of the porous structure.

Scanning Electron Microscope (SEM) image of the surface anodized iron is shown in Fig. 9.2 as well. Based on ISO definition (ISO/TS 80004-4:2011), nanoporous material is solid material with nanopores, while the nanopore is a cavity with at least one dimension in the nanoscale. From this figure, one can conclude that nanoporous oxide is formed on iron after oxidation as typically the pore diameter is 50 nm and pore wall of << 10 nm. The pores are elongated in depth (or length) and can reach several microns depending on the oxidation time and voltage applied during the anodization process. The following will explain further on how nanopores can be formed by anodic oxidation of iron.

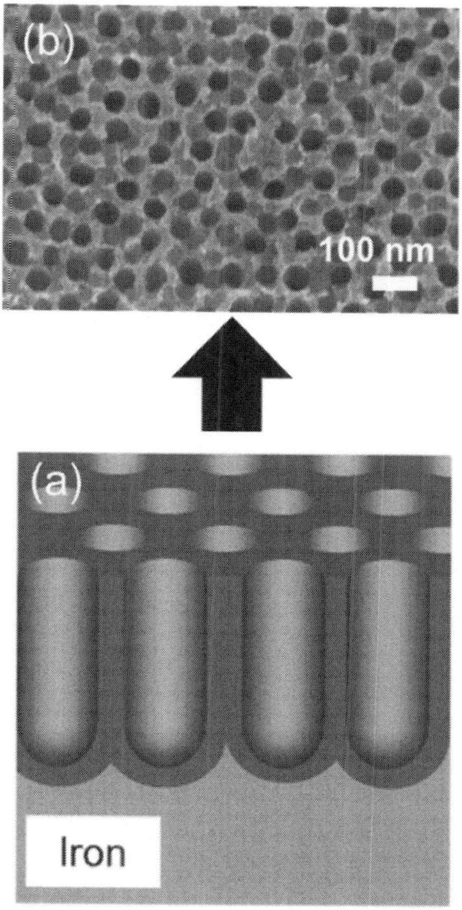

Fig. 9.2 Anodic iron oxide in form of nanopores: (a) illustration. (b) surface SEM morphology

4.1 Introduction to Anodization Process for Nanostructuring

Anodic oxidation of metals has been utilized to protect metal components from corrosion and for decorative purposes for more than 100 years (Aladjem 1973). However, as the field is developing, electrochemical anodic oxidation has become a technique not only for protection or to increase the aesthetic appearance of the metal components but done explicitly to produce a thin film which depending on the metal anodized will have many unique properties leading to various interesting applications. For instance anodizing aluminum resulted in Al_2O_3 film which has raised substantial scientific and technological interest in recent years due to the dielectric property of the oxide which can be used in electrolytic capacitors (Ban et al. 2014). Anodized titanium on the other hand has been widely studied as the TiO_2 film produced, being a photoactive semiconductor that has many applications especially in the field of solar energy utilization (Grimes and Mor 2009, Kowalski et al. 2013, Sauvage et al. 2010). TiO_2 is also a well-known photocatalysts hence anodizing Ti can be used to produce TiO_2 photocatalyzer that can degrade organic pollutants in contaminated air and water (Lee and Park 2013, Sreekantan et al. 2009, Ao and Lee 2005).

As mentioned, anodic oxidation is a process that resulted in the formation of thin oxide film on a metallic substrate. When a Metal (M) is exposed to a sufficiently high anodic voltage in an oxidizing electrolyte, an oxidation reaction will take place. Generally the reaction can be written as Eq. (1):

$$M \rightarrow M^{n+} + ne \tag{1}$$

If the metal ions, M^{n+} are solvatized in the electrolyte, no anodic film will form and the metal will continuously dissolve. Depending on the rate of dissolution, the metal components can be completely corroded or only a thin layer will be removed as a typical process in electropolishing of metal. If the electrolyte is comprised of species that can react with the solvated ions, then a metal oxide can form Eq. (2).

$$M^{n+} + O^{2-} \rightarrow MO \tag{2}$$

Now the properties of the Metal Oxides (MO) will determine the fate of the underlying metal. If the MO is easily dissolved then the metal will not be protected and it will degrade. If the MO does not dissolve, a barrier layer is formed and such a compact film may protect the underlying metal from further corrosion. Normally this can be achieved at relatively low anodizing voltage and by an appropriate choice of electrolyte. A competition between solvatization and oxide formation can also be observed whereby the anodic film will dissolve but at the same time oxidized to form MO. Therefore, depending on the electrolyte used, the surface oxide formed on anodized metal can either be compact or porous. Porous film can be further classified as random pores and ordered pores or pores formed by loosely bounded small particulates on the metal surface forming a 3 dimensional network of interconnected particles.

Ordered pores can be in a form of discreet pores, characterized by elongated holes forming perpendicular to the underlying metal substrate. The holes can be of micron of depth but the diameter is normally less than 100 nm hence they are termed nanopores as described earlier. An anodic oxide film with such morphology is called nanoporous oxide. Anodizing aluminium in acidic electrolyte is a well-known example whereby the resulting anodic Al_2O_3 consists of such nanopores (Sulka 2008). Moreover, not only are the nanopores elongated, forming channels of equal diameter and length, the channels are actually highly densely packed in hexagonal arrangements. The first work reporting on this observation was by Masuda and Fukuda (1995) whereby they reported that under specific experimental conditions pores growth is highly organized. Applications of such structure are endless and one of the indirect uses of it is as a template to produce nanostructured material. The use of nanoporous Al_2O_3 or termed Anodic Aluminium Oxide (AAO) as a template for synthesis nanomaterials was first reported by Uosaki et al. (1990). Following these pioneering works, many reports have appeared on the use of AAO as a template material. The pores are filled with the desired material; metal, polymer or semiconductors, the AAO template will be etched away to yield to the 1-dimensional material like nanorods or nanotubes.

Anodic process to yield self-organized pores on other metals such as W (Ng et al. 2015), Nb (Stojadinović et al. 2015), Ti (de Tacconi et al. 2006), Zr (Berger et al. 2008) and Fe (Cheng et al. 2012b, Kim et al. 2004) and semiconductors like Si (Kim et al. 2004) and InP (Sato et al. 2006) have received more attention. In the early works, focus was given to produce anodic oxide with similar morphology: close-packed cells in a hexagonal arrangement with nanopores at their centers. Optimization of the process to have control over pore diameter and length was done. Anodic oxidation voltage has a strong influence the diameter and the length of the nanopores formed.

When the morphology of the AAO is inspected further, the pore cell is found to consist of a scalloped shaped bottom. This is readily observed especially when the oxide layer is removed from the metallic substrate; exploring this allows for the proposal of several mechanistic models that describe the occurrence of the self-organized AAO.

The principles and mechanisms of formation have then been adopted to explain the development of similar structures on other metals as mentioned earlier. Among the metals (and semiconductors) stated, titanium has been the most studied considering the anodic layer (i.e. TiO_2) possesses many excellent properties for applications in environment remedy and energy production as stated earlier (Mor et al. 2006, Chen and Samuel 2007). The work published by Zwilling et al. (1999) can be regarded as the first report on the formation of self-organized TiO_2 nanostructure by the anodic process. In the said work chromic acid electrolyte containing hydrofluoric acid was used to form what appears to be a nanotubular structure. Despite the poor wall appearance, this work presents the most significant findings such that fluoride ions are needed in the electrolyte to allow the transformation of nanopores to nanotubes. Following this, more reports have emerged on the formation of oxide nanotubes especially TiO_2 with fluoride ions as the main ingredient in the electrolyte that allows for the pores to tubes transformation (Wang and Lin 2009, Wang et al. 2009).

4.2 Anodization of Iron for Nanoporous Formation

Anodic oxidation of iron is not a new topic (Nagayama and Cohen 1962, Nagayama and Cohen 1963). Anodization of iron to produce nanopores and nanotubes is however relatively new (Prakasam et al. 2006, Albu et al. 2009, Mohapatra et al. 2009, Rangaraju et al. 2009, Habazaki et al. 2010, Rangaraju et al. 2010, Zhang et al. 2010, Jagminas et al. 2011, Cheng et al. 2012a, Cheng et al. 2012b, Konno et al. 2012, Lee et al. 2014, Rozana et al. 2014, Schrebler et al. 2014, Momeni et al. 2015, Mushove et al. 2015, Rozana et al. 2015, Wang et al. 2015). There are four main components in the process of electrochemical anodization: (i) metal foil, (ii) platinum, (iii) power supply and (iv) electrolyte as shown in Fig. 9.3a. The electrolyte can consist of fluoride ions, oxygen ions, hydroxide ions and/or water in viscous electrolyte. The positively charged electrode is the anode where the oxidation reaction occurs and the negatively charged electrode is the cathode where reduction reaction is dominating. When the anode is iron metal (and platinum as cathode), the applied voltage generates an electric field across the iron and causes the anodic film in form of a barrier oxide layer to grow as oxygen ions migrate inward under the influence of the electric field and the iron ions eject out from the metal (Fig. 9.3b). Anodic film has a different volume than the parent metal, thus it is expected to be under compressive stress. Following the flow concept for the growth of porous anodic Al_2O_3 (Houser and Hebert 2009), oxide flow is assumed to originate from the plasticity of the barrier oxide layer generated by compressive stresses due to the competition of strong anion adsorption with O_2 incorporation at growing oxide lattice. To minimize stress the oxide behaves viscoplastically and is continually being pushed upwards from the metal|oxide interface forming scalloped structure which will then develop into pores.

The transition from pores to tubes structures can occur by increasing the electrolyte temperature. According to LaTempa et al. (2009), the transformation from nanopores to nanotubes was observed when the electrolyte temperature was increased from 35°C to 55°C. Another method to obtain the nanotubes is by modifying the electrolyte composition. Rozana et al. (2015) modified fluoride-ethylene glycol electrolyte with adding 1 M potassium hydroxide. The addition of KOH was found to be beneficial on the transformation from nanopores to nanotubes.

There are two possible routes that can be used to explain the formation of α-Fe_2O_3 film by anodic process:

1. By reacting Fe with H_2O forming Fe_2O_3 (Mohapatra et al. 2009, LaTempa et al. 2009).
   Iron oxidizes to form a thin layer of Fe_2O_3 on iron metal at the solid-electrolyte interface shown in Eq. (3).

$$2\,Fe + 3\,H_2O \rightarrow Fe_2O_3 + 6\,H^+ + 6\,e^- \qquad (3)$$

   The thin oxide layer is susceptible to breakdown whereby high voltage across the thin oxide may induce significant polarization of the oxide forming small pits on its surface. In the presence of F^-, pits will grow due to chemical etching forming larger pores Eq. (4). As the pores are enlarged, F^- diffusion to the bottom of the pore will induce further dissolution of the oxide|electrolyte interface at this region thinning the barrier layer. Growth occurs inwards as the barrier layer is now thinned very much. It will eventually dissolve and once it dissolves, new oxides will form at the bottom of the pore. The cycle will continue to form inwards growth of the elongated pores.

$$Fe_2O_3 + 12\,F^- + 6\,H^+ \rightarrow 2\,[FeF_6]^{3-} + 3\,H_2O \qquad (4)$$

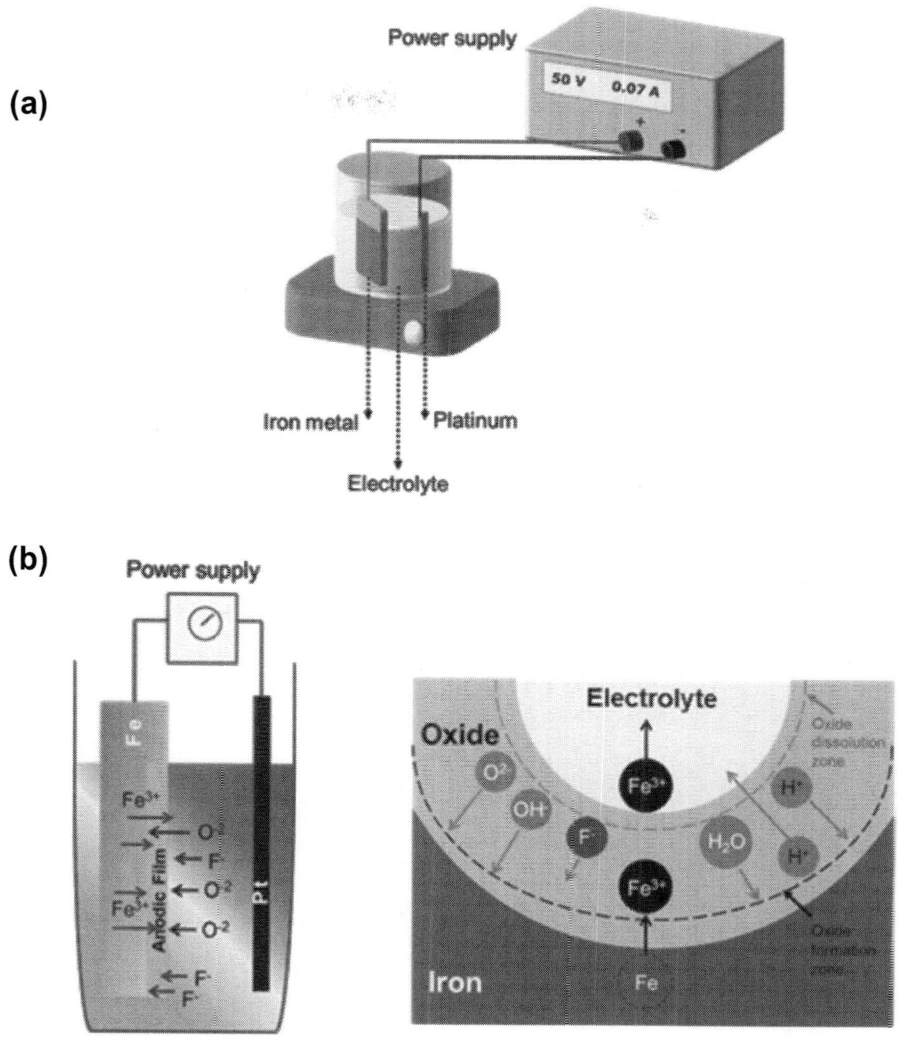

Fig. 9.3 (a) The component of anodic oxidation. (b) The process of anodic oxidation.

2. By reacting Fe with OH^- forming FeOOH (Rozana et al. 2015, Jagminas et al. 2011).

   When iron foil is immersed in an electrolyte, a barrier layer is formed on iron surface (Fig. 9.4a). Over applying a voltage across the initial thin film, a field is developed and a high enough field will induce polarization of the oxide, rupturing it to form small pits which develop into pores. The bottom of the pore as sketched in Fig. 9.4b is now the most active region whereby as oxidation proceeds, more Fe ions are ejected out from Fe surface. Parallel to this, due to the high field effect, inward migration of anions like hydroxide, oxygen and fluoride will also take place. The reaction of the species especially oxygen ions with the iron ions will thicken the oxide layer. At the same time oxide-hydroxide will form as well (Eqs. 5 and 6). The inward migration of F^- ions occurs concurrently with the O^{2-} migration across the oxide due to the high field effect but it is thought that F^- will reside at the interface of the oxide|metal as F^- can travel faster than OH^- or O^{2-} forming a defective fluoride rich layer at this region.

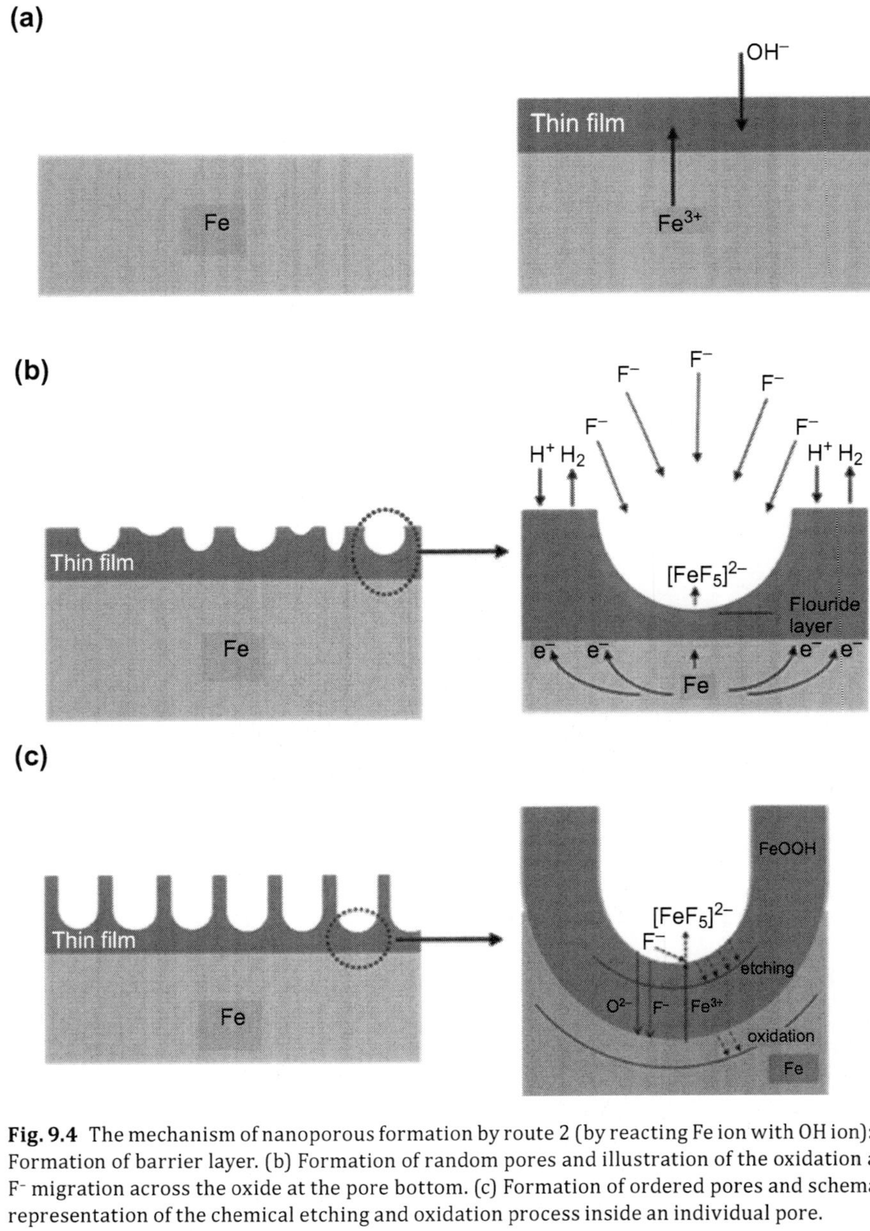

Fig. 9.4 The mechanism of nanoporous formation by route 2 (by reacting Fe ion with OH ion): (a) Formation of barrier layer. (b) Formation of random pores and illustration of the oxidation and F^- migration across the oxide at the pore bottom. (c) Formation of ordered pores and schematic representation of the chemical etching and oxidation process inside an individual pore.

Oxide has different volume than the underlying metal thus it is under compressive stress. To respond to the high stress within itself, the film will be plastically deformed, where the pore walls will be pushed upward. As this happens, the defective fluoride layer will be pushed towards the cell boundaries and accumulate preferentially here. At the same time, chemical dissolution of FeOOH and $Fe(OH)_2$ at the pore bottom due to etching by F^- and H^+ will continuously occur, thinning the oxide as oxidation proceeds (Eqs. 7 and 8). A balance between the chemical dissolution and oxidation at the pore bottom is required as to induce the inwards growth of the pores (Fig. 9.4c), forming elongated pores.

$$2\,Fe + 4\,OH^- \rightarrow 2\,FeOOH + H_2 + 4\,e^- \quad (5)$$

$$Fe + 2\,OH^- - 2\,e^- \rightarrow Fe(OH)_2 \quad (6)$$

$$FeOOH + 5\,F^- + 3\,H^+ \rightarrow [FeF_5]^{2-} + 2\,H_2O \quad (7)$$

$$Fe(OH)_2 + 5\,F^- + 2\,H^+ \rightarrow [FeF_5]^{2-} + 2\,H_2O + e^- \quad (8)$$

As the pores are enlarged and elongated, F^- at the bottom of the pore will induce further dissolution of the oxide|electrolyte interface at this region thinning the barrier layer (Fig. 9.4c). This specie can also be incorporated in the growing oxide as mentioned to form a stable fluoride compound. A typical X-Ray Diffraction (XRD) result confirmed the existence of FeOOH, $Fe(OH)_2$ and $FeF_5{\cdot}H_2O$ in the as-anodized film formed in NH_4F-ethylene glycol with 1 M KOH addition (Rozana et al. 2015). The existence of FeOOH and $Fe(OH)_2$ in the as-prepared film in NH_4F-ethylene glycol with H_2O addition was also confirmed by Mössbauer spectroscopy (Jagminas et al. 2011). It is therefore best to conclude that as-anodized iron needs to be thermally heated so as to remove any hydrated phases, fluorine rich phases and to induce phase formation of α-Fe_2O_3 before it can be used as photoanode.

As mentioned earlier, introduction of vacancies can help in the improvement of the conductivity of the photoanode. Vacancies can be introduced by doping or by selecting a fabrication process that would induce their formation. During the formation of oxide layer by anodic oxidation process (Raja et al. 2005), iron ions dissolve into solution and then cation vacancies are generated. These cation vacancies, because of their negative charge, can migrate along the electric field and reach metal|oxide interface. To maintain electrical neutrality oxygen vacancies could be generated. The presence of these oxygen vacancies may improve the conductivity of α-Fe_2O_3 film formed by anodic oxidation. However too little work been reported on this and is not very conclusive.

5. CURRENT PROGRESS ON ANODIC α-Fe_2O_3 AS PHOTOANODE

α-Fe_2O_3 is the most stable state of iron oxide at ambient conditions. α-Fe_2O_3 is a hexagonal crystal system consisting of iron atoms surrounded by six oxygen atoms. The strong absorption of yellow to UV photons in the visible region and transmission of orange to infra-red photons gives α-Fe_2O_3 the characteristic red color. It is well known that the PEC response of a semiconductor is affected by the intrinsic characteristic of the material. Work on the photoresponse of anodic α-Fe_2O_3 started a decade ago. In fact the first few papers assessing the possibility of using α-Fe_2O_3 as photoanode were published using anodic α-Fe_2O_3. For example Wilhelm et al. (1979) reported on the semiconductor properties of iron oxides electrode and in their work, the properties of anodic α-Fe_2O_3 (with γ-Fe_2O_3 at the surface) at the surface were reported: bandgap of 1.9 eV and donor density of 10^{20} cm^{-3}. Recent comprehensive work on the physicochemical characterization and photoelectrochemical properties of anodic iron oxide films on magnetron-sputtered iron films is reported by Santamaria et al. (2013). Two types of iron oxide films were reported: (i) barrier type and (ii) porous type. From the energy bandgap measurement, the barrier type layer annealed at 450°C has a value of 2.0 eV and lower gap of 1.9 eV is reported for sample annealed at lower temperature. Annealed porous layer at 450°C has an energy gap of 1.97 eV. The photoelectrochemical properties of the anodized sputter-deposited iron will be discussed later. Mushove et al. 2015 however reported an indirect energy gap of 2.4 eV for anodic α-Fe_2O_3 film with nanotubular structure.

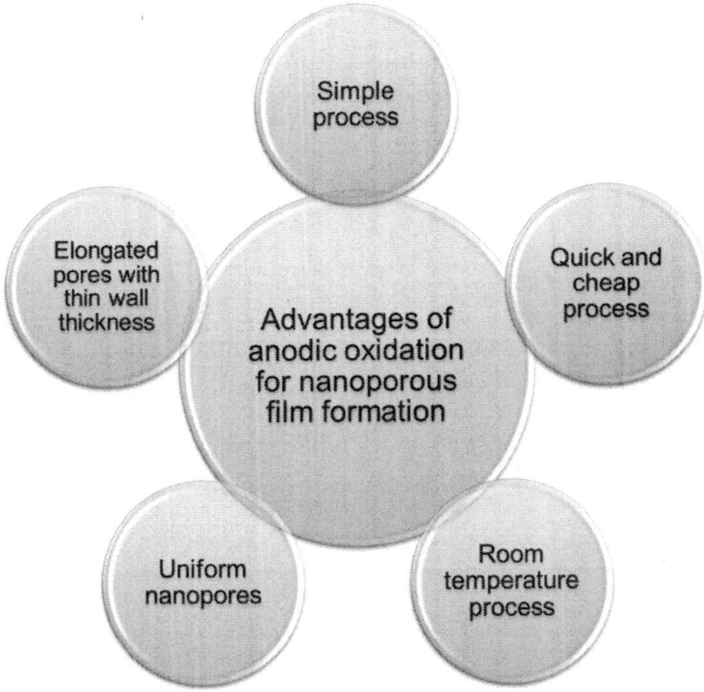

Fig. 9.5 The advantages of anodic oxidation process.

As mentioned, research efforts are now targeting into fabrication of α-Fe_2O_3 film with ordered nanoporous structure. Following the success on the formation of TiO_2 nanotubes by anodization, research progress on anodizing iron for the formation of self-aligned, ordered nanoporous or nanotubes α-Fe_2O_3 has been developed (Prakasam et al. 2006, Mohapatra et al. 2009, Rangaraju et al. 2009, Zhang et al. 2010, Rangaraju et al. 2010, Lee et al. 2014, Rozana et al. 2014, Schrebler et al. 2014, Momeni et al. 2015, Wang et al. 2015, Mushove et al. 2015). The advantages of anodic oxidation are shown in Fig. 9.5. Works on the formation of anodic nanoporous α-Fe_2O_3 for photoanode applications started in 2006 by Prakasam et al. (2006). Prakasam and co-workers proposed potentiostatic anodization of iron in a glycerol-based electrolyte containing fluoride ions as a viable process for nanostructuring the α-Fe_2O_3 film. The iron substrate beneath the α-Fe_2O_3 anodic layer is said to serve as a conducting substrate to transport the photogenerated electrons to the outer circuit.

5.1 Morphology

Typically there are five types of anodic α-Fe_2O_3 reported and used as photoanodes based on the film morphology (Table 9.2): pores (type A), tubes (type B), multi-layer tubes (type C), wave-like tubes (type D) and pores covered by dendritic layer (type E). The typical image of each type is shown in Fig. 9.6. The pore diameter and length of type A is in between 20-250 nm and 0.2-2 μm, respectively. For the tubes (type B), the outer diameter and length are ~70 nm and 4 μm, respectively. Multi-layer tubes (type C), wave-like tubes (type D) has a similar thickness of ~3 μm. A pore covered by dendritic layer of type E has the nano-dendrite arm is typically ~50 nm wide.

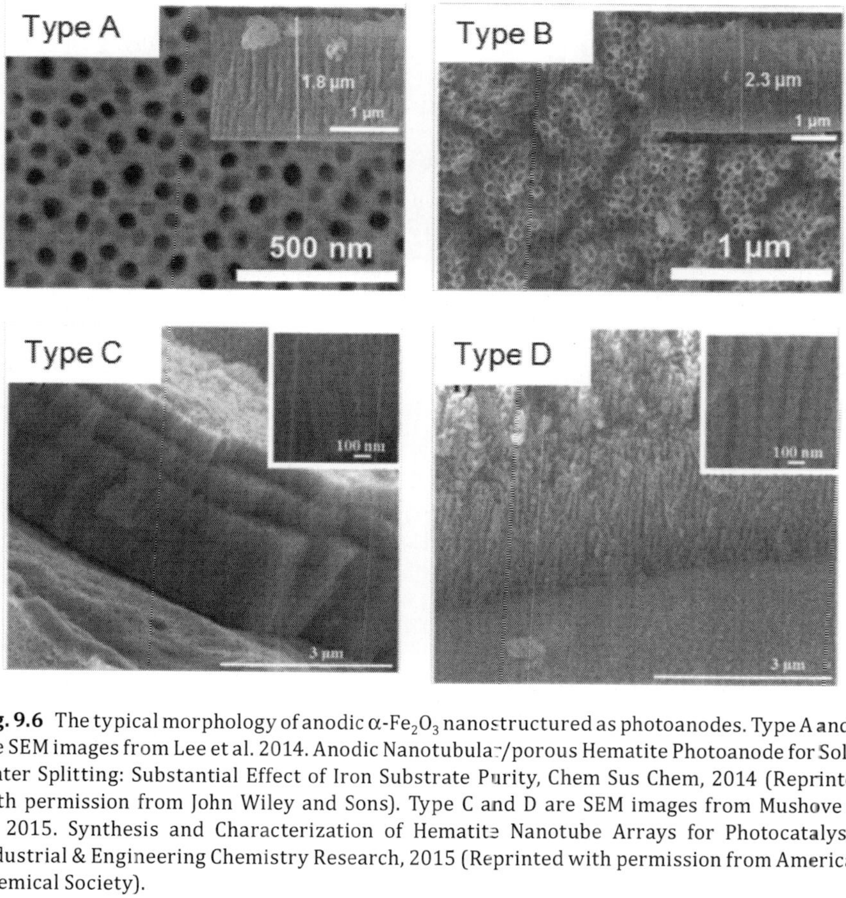

Fig. 9.6 The typical morphology of anodic α-Fe_2O_3 nanostructured as photoanodes. Type A and B are SEM images from Lee et al. 2014. Anodic Nanotubular/porous Hematite Photoanode for Solar Water Splitting: Substantial Effect of Iron Substrate Purity, Chem Sus Chem, 2014 (Reprinted with permission from John Wiley and Sons). Type C and D are SEM images from Mushove et al. 2015. Synthesis and Characterization of Hematite Nanotube Arrays for Photocatalysis, Industrial & Engineering Chemistry Research, 2015 (Reprinted with permission from American Chemical Society).

Table 9.2 Types of morphology of anodic α-Fe_2O_3.

Typical morphology of anodic film on iron (for PEC application)	
Type A	Nanopores
Type B	Nanotubes
Type C	Multi-layer nanotubes
Type D	Wave-like tubes
Type E	Nanopores covered by dendritic layer

In the anodic process, the three most important parameters for α-Fe_2O_3 nanoporous film formation are listed as following and typical values adopted are also shown.

- Electrolyte : 0.3-0.5 wt.% NH_4F + ethylene glycol and > 1 vol% water
- Applied voltage : 40-50 V
- Anodization time : 3 minutes-1 hour

The anodizing electrolyte and applied voltage reported from the current literature for nanostructured anodic film on iron are tabulated in Table 9.3. Anodic film on sputter-deposited iron in ethylene glycol has a similar morphology; nanopores, but the pores are not exactly rounded like the typical Type A film (Santamaria et al. 2013).

Table 9.3 The anodizing electrolyte and applied voltage for anodization of iron to produce nanostructured anodic film on iron.

Type of structure	Electrolyte	Applied voltage (V)	Ref.
A	Glycerol + 0.5 wt.% NH_4F + 1% HF + 0.2% 0.1 M HNO_3 at 10°C	90	(Prakasam et al. 2006)
B	Ethylene glycol + 0.5 wt.% NH_4F + 3 vol% H_2O	50	(Mohapatra et al. 2009)
B	Ethylene glycol + 0.1 M NH_4F + 3 vol% H_2O	50	(Rangaraju et al. 2009)
E	Two step anodization: (1) 90 vol% of ethylene glycol + 0.1 M NH_4F + 3 vol% H_2O and 10 vol% of 0.2 M $Na_5P_3O_{10}$ + 0.05 M NH_4F (2) 0.2 M $Na_5P_3O_{10}$ + 0.05 M NH_4F	Step (1): 20 Step (2): 50	(Rangaraju et al. 2009)
B	Ethylene glycol + 0.3 wt.% NH_4F + 2 vol% H_2O	50	(Zhang et al. 2010)
B	Ethylene glycol + 0.1 M NH_4F + 3 vol% H_2O	50	(Rangaraju et al. 2010)
A (but less ordered)	Ethylene glycol + 0.1 M NH_4F + 1.5 M H_2O	Constant current 50 A m^{-1}	(Santamaria et al. 2013)
A and B	Ethylene glycol + 0.14 M NH_4F + 1.7 M H_2O	50	(Lee et al. 2014)

Table 9.3 Contd.

Type of structure	Electrolyte	Applied voltage (V)	Ref
A	Ethylene glycol + 0.1 M NH_4F + 3 vol% H_2O	50	(Rozana et al. 2014)
A and B	Ethylene glycol + 0.5 wt.% NH_4F + 3 wt.% H_2O	50	(Schrebler et al. 2014)
B	Ethylene glycol + 0.1 M NH_4F + 1 M H_2O at 25°C	40	(Momeni et al. 2015)
A	Ethylene glycol + 0.14 M NH_4F + 1.7 M H_2O at room temperature	50	(Wang et al. 2015)
B	Ethylene glycol + 0.14 M NH_4F + 1.7 M H_2O at 60°C	50	(Wang et al. 2015)
B	Ethylene glycol + 0.37 wt.% NH_4F + 2 wt.% H_2O at 25 ± 2°C	40	(Mushove et al. 2015)
C	Ethylene glycol + 0.37 wt.% NH_4F + 2 wt.% H_2O at 25 ± 2°C	0 to 45 V at rate of 2.5 V/s, then pulsing then voltage between 45 and 20 V.	(Mushove et al. 2015)
D	Ethylene glycol + 0.37 wt.% NH_4F + 2 wt.% H_2O at 25 ± 2°C	Triangular profile with voltages between 30-60 V	(Mushove et al. 2015)

Type A morphology was reported in 2006 and the oxide was produced in glycerol-based electrolyte (Prakasam et al. 2006). Following this, Mohapatra et al. (2009) reported the formation of type B in fluoride-ethylene glycol with water addition and type E by two-step anodization (Rangaraju et al. 2009). Later several researchers used fluoride-ethylene glycol with water addition electrolyte to produce type A and B and the effect of temperature of the electrolyte was studied. The anodization condition was also modified using ultrasonic waves (Schrebler et al. 2014). Still using the similar electrolyte, type C was synthesized by increasing the applied voltage from 0 to 45 V at a rate of 2.5 V/s, then pulsing the voltage between 45 and 20 V, for 72 and 60 seconds, respectively. Type D morphology can be synthesized by applying triangular profile applied voltages between 30-60 V for a given period of time (Mushove et al. 2015). However, as can be seen, only organic electrolyte has been chosen to produce nanoporous iron oxide film (unlike anodic TiO_2 nanotubes formation where by acidic and buffered solutions are also used).

5.2 Phases

The electrochemical behavior of iron has been studied extensively since the beginning of the 19th century, seeing that iron (and its alloys) has always been an imperative ingredient in human civilization. Comprehensive review on the topic of passivation includes works by Macdonald (1999) and Schmuki (2002). Early work on the electrochemistry of iron was done in acid solution, but over the last 30 years studies have been carried out mainly in neutral buffered solution. A major work in neutral solutions was reported by Nagayama and Cohen (1962) using pH 8.4 sodium borate boric acid buffer solution. They considered that in the passive region iron is covered by a thin film of cubic oxide of the γ-Fe_2O_3/Fe_3O_4 type, similar with the film that is formed by the reaction of clean iron with oxygen or dry air. The composition of the passive film on iron depends on the type of electrochemical treatment for forming the film and the nature of the solution in which it is formed. In organic electrolyte, typically the as-anodized iron oxide film is amorphous or consisted of fluoride compounds like FeF_2 and FeOOH and $Fe(OH)_2$. The following described typical phases of the anodic film with morphologies as mentioned in Table 9.3.

Type A. Glancing Angle XRD has been routinely used for oxide film phase investigation. As mentioned the as-anodized nanoporous anodic film is always reported to be amorphous and by annealing at relatively low temperatures of around 300-450°C, the nanoporous film can be crystallized to a mixture of α-Fe_2O_3 and Fe_3O_4 (Prakasam et al. 2006, Santamaria et al. 2013, Rozana et al. 2014, Wang et al. 2015). Along with the amorphous structure of the as-anodized film, it has also been reported that FeF_2 can be observed and the fluoride compound can be removed when the anodic film is annealed for example at temperature of 450°C. In this temperature a mixture of α-Fe_2O_3 and Fe_3O_4 are observed (Rozana et al. 2014). A similar observation is reported by Santamaria et al. (2013) whereby according to the XRD pattern of the as-prepared films, the layers were found to be amorphous. Transmission electron micrographs of the anodic films were presented by Habazaki et al. (2010) showing amorphous layer formation of the as-anodized samples. The amorphous film due to its hydrated nature, poor crystallinity as well as being contaminated by fluoride species is not photoactive; the anodic film has to be annealed.

XRD patterns as a function of annealing temperature have been reported by various authors, for example Santamaria et al. (2013). Typically both α-Fe_2O_3 and Fe_3O_4 are detected for annealed samples. Apart from XRD, compositional information of the anodic film surface can also be analyzed using X-ray photoelectron spectroscopy (Habazaki et al. 2010). Raman spectroscopy on the other hand has been applied in relating the annealing temperatures to the phase formation after the anodic nanoporous film is annealed. According to Santamaria et al. (2013), a high intensity Fe_3O_4 Raman peak can be detected for porous films on anodized sputter-deposited iron annealed at 350-450°C along with the expected α-Fe_2O_3 peaks. There appears to be either phase transformation that occurs when the anodic film is annealed or a new oxide layer formation underneath the hematite layer which may result in the Fe_3O_4 oxide formation. The anodic oxide layer is normally found to be thicker after annealing indicating that thermal oxidation may have occurred to the underlying metal. Oxidation of iron may result in the formation of Fe_3O_4 and/or FeO depending on the annealing temperature. Despite the phase transformation and oxide thickening, fluoride contents seem to be much reduced with annealing especially at temperature $>$400°C. It is however best to anneal nanoporous film at $\leq$600°C as higher temperature would diminish the nanopores and thus form rather compact oxide. While it is interesting to note that phase formation and transformation have occurred after annealing, not a lot has been done to investigate the effect of the multilayered oxide on the photoelectrochemical properties or to try to correlate on the phases formed with the properties.

Type B. The XRD pattern for as-anodized nanoporous film is amorphous and can be converted to α-Fe_2O_3 after annealing for example at 400°C for 1 hour in oxygen (Zhang et al. 2010). When hydrogen was used, α-Fe_2O_3 can still be observed but minor peaks of γ-Fe_2O_3 can also be detected from XRD (Rangaraju et al. 2009). However as the peaks position of γ-Fe_2O_3 is very close to Fe_3O_4, it is difficult to differentiate them. A similar observation was also reported by Mohapatra et al. (2009) whereby the amorphous nanotubular anodic film on iron can be transformed to α-Fe_2O_3 and Fe_3O_4 when annealed at 500°C for 6 hours in oxygen and only α-Fe_2O_3 was detected for sample annealed at 500°C in hydrogen. Reduced partial pressure condition for annealing is concluded preferring the formation of hematite. Little has been reported so far on the effect of annealing conditions to the H_2 generation on the oxide.

Type C and D. Similar to the previous two morphology types, anodic film with type C morphology (multi-layered nanotubes) and Type D (wave-like tubes) are amorphous in their as-anodized condition. Again, it can be converted to α-Fe_2O_3 when annealed at high temperature i.e. 600°C for 2 hours (Mushove et al. 2015).

Type E. Rangaraju et al. (2009) reported that the as-prepared pores covered by dendritic layer can be annealed in the acetylene environment at 550°C for 10 minutes for the formation of predominantly γ-Fe_2O_3 with minor peaks of α-Fe_2O_3. The existence of γ-Fe_2O_3 was further confirmed by High Resolution Transmission Electron Microscopy (HRTEM) result showing the lattice of 0.341 nm. It is not conclusive so far on the effect of γ-Fe_2O_3 on the photocurrent or H_2 generation on anodized iron.

6. THE PHOTOCURRENT PERFORMANCE OF NANOSTRUCTURED α-Fe_2O_3 PHOTOANODES

The current-applied bias behavior is regarded as an indirect way to measure the rate of H_2 generation by water splitting if the Faradaic efficiency for water oxidation is assumed unity. The photocurrent is typically done in 1 M NaOH or 1 M KOH electrolyte under illumination of AM 1.5 with intensity of 100 mW cm^{-2}. The potential values gathered from literature were converted to the potential vs Ag/AgCl here to make comparison easier. Under dark conditions, the electrodes show very small current generation, almost negligible. Anodic α-Fe_2O_3 film will produce photocurrent when it is illuminated with photon of the right wavelength. A Gartner-Butler model can be used in describing semiconductor/liquid junction under illumination and according to this model, nanoporous α-Fe_2O_3 film shows a supralinear dependence of photocurrent with electrode potential (Santamaria et al. 2013). The dependence of photocurrent on potential is described by Santamaria et al. (2013) as due to the transport of injected photocarriers dominated by a trap limited mobility in the localized states of the semiconductor (Poole-Frankel mechanism).

Since the first report on the photocurrent of anodic nanoporous α-Fe_2O_3 in 2006, it is clear that major advances have been made on photocurrent values produced by anodic α-Fe_2O_3 photoanode as summarized in Table 9.4. Type A structure i.e the nanoporous film has the highest photocurrent of 0.51 mA cm^{-2} at 0.6 V vs Ag/AgCl whilst type B gave the highest photocurrent of 1.41 mA cm^{-2} at 0.5 V vs Ag/AgCl. It is perhaps not only the structure which can affect the photocurrent formed; the type B film was also reported to consist of pure α-Fe_2O_3 phase. Photocurrent of anodic film with morphologies of type C and D are surprisingly remarkable. Mushove et al. (2015) produced high photocurrent and IPCE of 14% for wave-like tubes (type D). They argued in term of electrochemical surface area that improves the conversion efficiency. Type D structure possessed higher electrochemically active surface area (17.6 cm^2/cm^2) than type B (9.9 cm^2/cm^2) and type C (10.8 cm^2/cm^2).

Table 9.4 Comparison of photocurrent of nanostructured anodic α-Fe_2O_3 photoanode (of types as listed in Table 9.2).

Type of structure	Photocurrent (mA cm^{-2})	Electrolyte	Source and intensity of light	Ref.
A	0.26 at 0.6 V vs Ag/AgCl	1 M NaOH	AM 1.5 100 mW cm^{-2}	(Prakasam et al. 2006)
A	0.51 at 0.6 V vs Ag/AgCl	1 M NaOH + 0.5 M H_2O_2	AM 1.5 100 mW cm^{-2}	(Prakasam et al. 2006)
A	0.45 at 0.6 V vs Ag/AgCl	1 M NaOH + 0.5 M H_2O_2	AM 1.5 100 mW cm^{-2}	(Rozana et al. 2014)
A	0.40 at 0.5 V vs Ag/AgCl	1 M KOH	AM 1.5 100 mW cm^{-2}	(Wang et al. 2015)
A	0.25 at 0.4 V vs Ag/AgCl	1 M KOH	AM 1.5 100 mW cm^{-2}	(Lee et al. 2014)
B	1.41 at 0.5 V vs Ag/AgCl	1 M KOH	AM 1.5 87 mW cm^{-2}	(Mohapatra et al. 2009)
B	0.8 at 0.5 V vs Ag/AgCl	0.01 Na_2SO_4	AM 1.5 100 mW cm^{-2}	(Zhang et al. 2010)
B	0.40 at 0.5 V vs Ag/AgCl	1 M KOH	AM 1.5 100 mW cm^{-2}	(Rangaraju et al. 2010)
B	0.03 at 0.4 V vs Ag/AgCl	1 M KOH	AM 1.5 100 mW cm^{-2}	(Lee et al. 2014)
B	0.68 at 0.5 V vs Ag/AgCl	1 M NaOH	AM 1.5 with a UV cut-off filter ($\lambda > 420$ nm) 100 mW cm^{-2}	(Momeni et al. 2015)
E	1.8 at 0.5 V vs Ag/AgCl	1 M KOH	The solar simulator passed through AM 1.5 filter	(Rangaraju et al. 2009)
C	2 at 0.41 V vs Ag/AgCl	1 M KOH + 0.5 M H_2O_2	AM 1.5 100 mW cm^{-2}	(Mushove et al. 2015)
D	4 at 0.41 V vs Ag/AgCl	1 M KOH + 0.5 M H_2O_2	AM 1.5 100 mW cm^{-2}	(Mushove et al. 2015)

In order to further improve the photocurrent, the choice of electrolyte is also important. At the moment, most work on the α-Fe_2O_3 film made by anodic process utilizes KOH or NaOH as the electrolyte. Several works as seen in Table 9.4 used alkaline solution but with H_2O_2 added in it. As mentioned, for higher conversion efficiency, holes need to be transferred efficiently to the electrolyte. H_2O_2 serves as hole scavengers. Nyquist plots have been routinely used to investigate charge transfer process between the semiconductor and the electrolyte. Reduction in charge transfer resistance in α-Fe_2O_3 when H_2O_2 is added in the electrolyte may result in higher photoconversion efficiency. Mott–Schottky analysis can be used to measure charge carrier densities in α-Fe_2O_3 film. However, the method is only valid for planar electrode and not nanoporous film.

7. THE CURRENT CHALLENGES OF NANOSTRUCTURED α-Fe_2O_3 BASED PHOTOANODES FORMED BY ANODIC OXIDATION

In general as seen in Table 9.4, there is a major improvement on photocurrent value in nanostructured α-Fe_2O_3 made by the anodic oxidation process. It is also conclusive that the nanoporous (or nanotubular) anodic film can be easily formed by using the most appropriate anodization parameters but the film needs to be annealed before it can be used as a photoanode. However, there are two remaining challenges that need to be faced on applying anodic nanostructured Fe_2O_3 as a photoanode material.

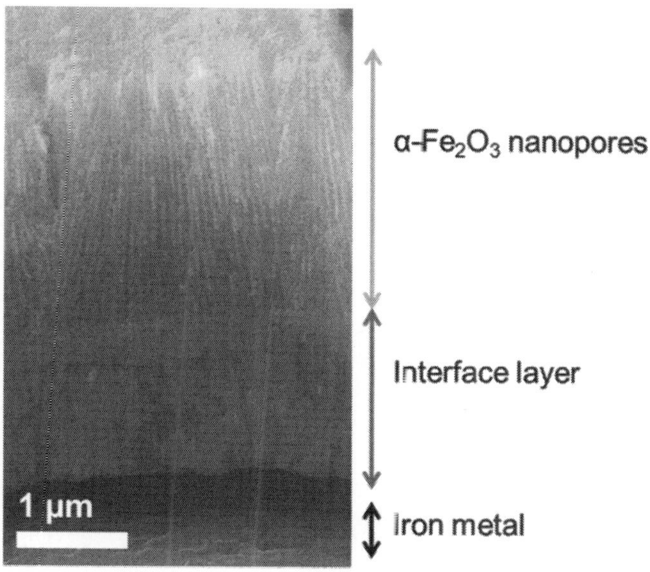

Fig. 9.7 The SEM of cross section of α-Fe_2O_3 nanopores after annealing at 450°C for 2 hours in air.

1. The oxide|metal interface

As mentioned, annealing is needed for the anodic film as to remove undesired hydrated or hydroxylated layer, to remove fluoride rich layer as well as to generally improve (or induce) on the crystallinity of the desired α-Fe_2O_3. But when annealing is done, not only α-Fe_2O_3 will be crystallized; at a high enough temperature and in an oxidizing atmosphere, formation of internal layer which resembles the multi layered scale oxide when iron is thermally oxidized is also possible. The creation of such interfacial layer (Fig. 9.7) hinders the transport of photogenerated electron to the Fe substrate reducing the efficiency of PEC cell. Annealing can be done in reducing gas, such as hydrogen to reduce the interfacial layer formation but under such conditions phase transformation to γ-Fe_2O_3 may result which again is not really desirable for a photoanode. As mentioned, a mixture of α-Fe_2O_3 and Fe_3O_4 is often observed when nanoporous anodic film is annealed in an oxidizing condition. A typical XRD of annealed nanoporous film on iron is shown in Fig. 9.8. A pure α-Fe_2O_3 is desired as Fe_3O_4 is not photoactive but such a phase seems to be unavoidable and is always observed at the interface layer between nanopores and the remaining iron metal (Fig. 9.7). An annealing condition that can reduce this is needed.

2. The wall thickness of the nanopores

One reason for nanostructuring is to produce oxide with nanostructures of less than the hole diffusion length hence holes can be transferred out to the electrolyte before they can recombine with photogenerated electrons. For the case of anodic process, control of the anodization condition can lead to nanopores with wall thickness of less than the diffusion length of the holes. But as mentioned, annealing of anodic film is needed. Along with crystallizing the oxide film, annealing can increase the wall thickness of the pores hampering the effort of utilizing the nanostructuring to improve hematite as a photoanode. A right condition for annealing as to reduce interface oxide formation or hematite transformation to other phases as well as to suppress wall growth is required in order to produce an ideal α-Fe_2O_3 with photocurrent value (12 mA cm^{-2}) and correspondingly high PEC efficiency.

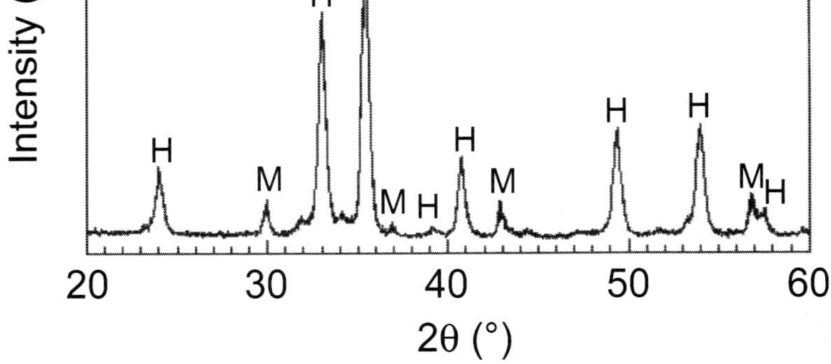

Fig. 9.8 XRD pattern of α-Fe_2O_3 nanopores after annealing at 450°C for 2 hours in air.

8. CONCLUSIONS

A decade of research producing nanostructured α-Fe_2O_3 as photoanode material by anodic oxidation have revealed the obstacle that prevented it to reach the ideal photocurrent value. From 2006 -2015, by modifying the anodic oxidation parameters, from nanopores to wave-like nanotubes, the photocurrent is significantly increased from 0.51 mA cm^{-2} at 0.6 V vs Ag/AgCl to 4 mA cm^{-2} 0.41 V vs Ag/AgCl in H_2O_2 added electrolyte. Even though anodic process results in nanostructuring of the anodic film which may be beneficial in hole transfer process, the as-anodized film is often very hydrated and amorphous and needs to be annealed before use. Often annealing resulted in the formation of other phases than α-Fe_2O_3, for example Fe_3O_4 near the metal|oxide interface. A pure α-Fe_2O_3 is desired as Fe_3O_4 is not photoactive. Furthermore, the growth of the nanoporous boundaries occurs after annealing reducing the effectiveness to remove holes from its surface. While progress on nanoporous film synthesis is constantly advancing and more new findings and important results are reported, many aspects of the anodic oxide need to be explored so as to utilize the oxide in a real device. Continuous effort to solve the obstacles and further fundamental understanding of the PEC properties of anodic nanostructured α-Fe_2O_3 is therefore strongly needed.

9. ACKNOWLEDGMENTS

One BAJA Long Term Research Grant Scheme (LRGS), Ministry of Education Malaysia, Project 2, 304/PBAHAN/6050235 is greatly acknowledged. The authors are grateful for support by the ASEAN University Network/Southeast Asia Engineering Education Development Network (AUN/SEED-Net), Japan International Cooperation Agency (JICA).

10. REFERENCES

Aladjem, A. 1973. Anodic oxidation of titanium and its alloys. J. Mater. Sci. 8(5): 688-704.

Albu, S. P., A. Ghicov and P. Schmuki. 2009. High aspect ratio, self-ordered iron oxide nanopores formed by anodization of Fe in ethylene glycol/NH_4F electrolytes. Phys. Status Solidi RRL 3(2-3): 64-66.

Ao, C. H. and S. C. Lee. 2005. Indoor air purification by photocatalyst TiO_2 immobilized on an activated carbon filter installed in an air cleaner. Chem. Eng. Sci. 60(1): 103-109.

Balberg, I. and H. L. Pinch. 1978. The optical absorption of iron oxides. J. Magn. Magn. Mater. 7(1): 12-15.

Ban, C., Y. He, X. Shao and L. Wang. 2014. Anodizing of etched aluminum foil coated with modified hydrous oxide film for aluminum electrolytic capacitor. J. Mater. Sci.: Mater. Electron. 25(1): 128-133.

Berger, S., F. Jakubka and P. Schmuki. 2008. Formation of hexagonally ordered nanoporous anodic zirconia. Electrochem. Commun. 10(12): 1916-1919.

Cesar, I., A. Kay, J. A. G. Martinez and M. Grätzel. 2006. Translucent thin film Fe_2O_3 photoanodes for efficient water splitting by sunlight: nanostructure-directing effect of Si-doping. J. Am. Chem. Soc. 128(14): 4582-4583.

Cesar, I., K. Sivula, A. Kay, R. Zboril and M. Grätzel. 2009. Influence of feature size, film thickness and silicon doping on the performance of nanostructured hematite photoanodes for solar water splitting. J. Phys. Chem. C 113(2): 772-782.

Chen, H. M., C. K. Chen, R. Liu, L. Zhang, J. Zhang and D. P. Wilkinson. 2012. Nanoarchitecture and material designs for water splitting photoelectrodes. Chem. Soc. Rev. 41(17): 5654-5671.

Chen, X. and S. M. Samuel. 2007. Titanium dioxide nanomaterials: synthesis, properties, modifications and applications. Chem. Rev. 107(7): 2891-2959.

Chen, Z., H. N. Dinh and E. Miller. 2013. Experimental considerations. pp. 17. *In*: Photoelectrochemical Water Splitting. Springer, New York.

Cheng, H., Z. Lu, R. Ma, Y. Dong, H. E. Wang, L. Xi, et al. 2012a. Rugated porous Fe_3O_4 thin films as stable binder-free anode materials for lithium ion batteries. J. Mater. Chem. 22(42): 22692-22698.

Cheng, H., L. Zheng, C. K. Tsang, J. Zhang, H. E. Wang, Y. Dong, et al. 2012b. Electrochemical fabrication and optical properties of periodically structured porous Fe_2O_3 films. Electrochem. Commun. 20: 178-181.

Cherepy, N. J., D. B. Liston, J. A. Lovejoy, H. M. Deng and J. Z. Zhang. 1998. Ultrafast studies of photoexcited electron dynamics in gamma- and alpha-Fe_2O_3 semiconductor nanoparticles. J. Phys. Chem. B 102: 770.

de Tacconi, N. R., C. R. Chenthamarakshan, G. Yogeeswaran, A. Watcharenwong, R. S. de Zoysa, N. A. Basit, et al. 2006. Nanoporous TiO_2 and WO_3 films by anodization of titanium and tungsten substrates: influence of process variables on morphology and photoelectrochemical response. J. Phys. Chem. B 110(50): 25347-25355.

Fujishima, A. and K. Honda. 1972. Electrochemical photolysis of water at a semiconductor electrode. Nature 238(5358): 37-38.

Galuza, A. I., A. B. Beznosov and V. V. Eremenko. 1998. Optical absorption edge in alpha-Fe_2O_3: The exciton-magnon structure. Low Temp. Phys. 24: 726.

Gan, J., X. Lu and Y. Tong. 2014. Towards highly efficient photoanodes: boosting sunlight-driven semiconductor nanomaterials for water oxidation. Nanoscale 6(13): 7142-7164.

Glasscock, J. A., P. R. F. Barnes, I. C. Plumb and N. Savvides. 2007. Enhancement of photoelectrochemical hydrogen production from hematite thin films by the introduction of Ti and Si. J. Phys. Chem. C 111(44): 16477-16488.

Grimes, C. A., O. K. Varghese and S. Ranjan. 2008. Photoelectrolysis. pp. 115. *In*: C. A. Grimes, O. K. Varghese and S. Ranjan (eds.). Light, Water, Hydrogen. Springer, New York.

Grimes, C. A. and G. K. Mor. 2009. TiO_2 Nanotube Arrays: Synthesis, Properties and Applications. Springer Science & Business Media, New York.

Habazaki, H., Y. Konno, Y. Aoki, P. Skeldon and G. E. Thompson. 2010. Galvanostatic growth of nanoporous anodic films on iron in ammonium flouride-ethylene glycol electrolytes with different water contents. J. Phys. Chem. C 114: 18853-18859.

Hardee, K. L. and A. J. Bard. 1976. Semiconductor electrodes: V. The application of chemically vapor deposited iron oxide films to photosensitized electrolysis. J. Electrochem. Soc. 123(7): 1024-1026.

Hiralal, P., S. Saremi-Yarahmadi, B. C. Bayer, H. Wang, S. Hofman, K. G. U. Wijayantha, et al. 2011. Nanostructured hematite photoelectrochemical electrodes prepared by the low temperature thermal oxidation of iron. Sol. Energy Mater. Sol. Cells 95(7): 1819-1825.

Houser, J. E. and K. R. Hebert. 2009. The role of viscous flow of oxide in the growth of self-ordered porous anodic alumina films. Nat. Mater. 8(5): 415-420.

Iandolo, B., B. Wickman, I. Zoric and A. Hellman. 2015. The rise of hematite: origin and strategies to reduce the high onset potential for the oxygen evolution reaction. J. Mater. Chem. A 3(33): 16896-16912.

Jagminas, A., K. Mazeika, N. Bernotas, V. Klimas, A. Selskis and D. Baltrunas. 2011. Compositional and structural characterization of nanoporous produced by iron anodizing in ethylene glycol. Appl. Surf. Sci. 257: 3893-3897.

Kay, A., I. Cesar and M. Grätzel. 2006. New benchmark for water photooxidation by nanostructured α-Fe_2O_3 films. J. Am. Chem. Soc. 128(49): 15714-15721.

Kennedy, J. H. and K. W. Frese. 1978. Photooxidation of water at α-Fe_2O_3 electrodes. J. Electrochem. Soc. 125(5): 709-714.

Kim, D. A., S. I. Im, C. M. Whang, W. S. Cho, Y. C. Yoo, N. H. Cho, et al. 2004. Structural and optical features of nanoporous silicon prepared by electrochemical anodic etching. Appl. Surf. Sci. 230(1-4): 125-130.

Konno, Y., E. Tsuji, P. Skeldon, G. E. Thompson and H. Habazaki. 2012. Factors influencing the growth behaviour of nanoporous anodic films on iron under galvanostatic anodizing. J. Solid State Electrochem. 16: 3887-3896.

Kowalski, D., D. Kim and P. Schmuki. 2013. TiO_2 nanotubes, nanochannels and mesosponge: self-organized formation and applications. Nano Today 8(3): 235-264.

Krol, R. V. D. 2012a. Photoelectrochemical measurements. pp. 69. *In*: R. V. D. Krol and M. Grätzel (eds.). Photoelectrochemical Hydrogen Production. Springer, New York.

Krol, R. V. D. 2012b. Principles of photoelectrochemical cells. pp. 13. *In*: R. V. D. Krol and M. Grätzel (eds.). Photoelectrochemical Hydrogen Production. Springer, New York.

LaTempa, T. J., X. Feng, M. Paulose and C. A. Grimes. 2009. Temperature-dependent growth of self-assembled hematite (α-Fe_2O_3) nanotube arrays: rapid electrochemical synthesis and photoelectrochemical properties. J. Phys. Chem. C 113(36): 16293-16298.

Lee, C. Y., L. Wang, Y. Kado, M. S. Killian and P. Schmuki. 2014. Anodic nanotubular/porous hematite photoanode for solar water splitting: substantial effect of iron substrate purity. ChemSusChem 7(3): 934-940.

Lee, S. and S. Park. 2013. TiO_2 photocatalyst for water treatment applications. J. Ind. Eng. Chem. 19(6): 1761-1769.

Leygraf, C., M. Hendewerk and G. A. Somorjai. 1982. Mg- and Si-doped iron oxides for the photocatalyzed production of hydrogen from water by visible light (2.2 eV $\leq h\nu \leq$ 2.7 eV). J. Catal. 78(2): 341-351.

Lindgren, T., H. Wang, N. Beermann, L. Vayssieres, A. Hagfeldt and S. Lindquist. 2002. Aqueous photoelectrochemistry of hematite nanorod array. Sol. Energy Mater. Sol. Cells 71(2): 231-243.

Lindgren, T., L. Vayssieres, H. Wang and S. E. Lindquist. 2003. Photo-oxidation of water at hematite electrodes. pp. 83-103. *In*: A. I. Kokorin and D. W. Bahnemann (eds.). Chemical Physics of Nanostructured Semiconductors,. VSP International Science Publishers, The Netherlands.

Macdonald, D. D. 1999. Passivity–the key to our metals-based civilization. Pure Appl. Chem. 71(6): 951-978.

Masuda, H. and K. Fukuda. 1995. Ordered metal nanohole arrays made by a two-step replication of honeycomb structures of anodic alumina. Science 268(5216): 1466-1468.

Mohapatra, S. K., S. E. John, S. Banerjee and M. Misra. 2009. Water photooxidation by smooth and ultrathin α-Fe_2O_3 nanotube arrays. Chem. Mat. 21(14): 3048-3055.

Momeni, M. M., Y. Ghayeb and F. Mohammadi. 2015. Solar water splitting for hydrogen production with Fe_2O_3 nanotubes prepared by anodizing method: effect of anodizing time on performance of Fe_2O_3 nanotube arrays. J. Mater. Sci.: Mater. Electron. 26(2): 685-692.

Mor, G. K., O. K. Varghese, M. Paulose, K. Shankar and C. A. Grimes. 2006. A review on highly ordered, vertically oriented TiO_2 nanotube arrays: fabrication, material properties and solar energy applications. Sol. Energy Mater. Sol. Cells 90(14): 2011-2075.

Morin, F. J. 1951. Electrical properties of alpha-Fe_2O_3 and alpha-Fe_2O_3 containing titanium. Phys. Rev. 83: 1005.

Morin, F. J. 1954. Electrical properties of alpha-Fe_2O_3. Phys. Rev. 93: 1195.

Murphy, A. B., P. R. F. Barnes, L. K. Randeniya, I. C. Plumb, I. E. Grey, M. D. Home, et al. 2006. Efficiency of solar water splitting using semiconductor electrodes. Int. J. Hydrog. Energy 31(14): 1999-2017.

Mushove, T., T. M. Breault and L. T. Thompson. 2015. Synthesis and characterization of hematite nanotube arrays for photocatalysis. Ind. Eng. Chem. Res. 54(16): 4285-4292.

Nagayama, M. and M. Cohen. 1962. The anodic oxidation of iron in a neutral solution I. The nature and composition of the passive film. J. Electrochem. Soc. 109(9): 781-790.

Nagayama, M. and M. Cohen. 1963. The anodic oxidation of iron in a neutral solution II. Effect of ferrous Ion and pH on the behavior of passive iron. J. Electrochem. Soc. 110(6): 670-680.

Ng, C., K. A. Razak and Z. Lockman. 2015. Effect of annealing temperature on anodized nanoporous WO_3. J. Porous Mat. 22(2): 537-544.

Piazza, S., M. Santamaria, C. Sunseri and F. D. Quarto. 2003. Recent advances in photocurrent spectroscopy of passive films. Electrochimica acta 48(9): 1105-1114.

Prakasam, H. E., O. K. Varghese, M. Paulose, G. K. Mor and C. A. Grimes. 2006. Synthesis and photoelectrochemical properties of nanoporous iron (III) oxide by potentiostatic anodization. Nanotechnology 17: 4285-4291.

Raja, K. S., M. Misra and K. Paramguru. 2005. Formation of self-ordered nano-tubular structure of anodic oxide layer on titanium. Electrochimica Acta 51(1): 154-165.

Rangaraju, R. R., A. Panday, K. S. Raja and M. Misra. 2009. Nanostructured anodic iron oxide film as photoanode for water oxidation. J. Phys. D: Appl. Phys. 42(13): 135303.

Rangaraju, R. R., K. S. Raja, A. Panday and M. Misra. 2010. An investigation on room temperature synthesis of vertically oriented arrays of iron oxide nanotubes by anodization of iron. Electrochimica Acta 55(3): 785-793.

Rocket, A. 2008. The Materials Science of Semiconductors. Springer, New York.

Rozana, M., M. A. Azhar, D. M. Anwar, G. Kawamura, K. A. Razak, A. Matsuda, et al. 2014. Effect of applied voltage on the formation of self-organized iron oxide nanoporous film in organic electrolyte via anodic oxidation process and their photocurrent performance. Adv. Mater. Res. 1024: 99-101.

Rozana, M., K. A. Razak, G. Kawamura, A. Matsuda and Z. Lockman. 2015. Formation of aligned iron oxide nanopores as Cr adsorbent material. Adv. Mater. Res. 1087: 460-464.

Santamaria, S., S. Terracina, Y. Konno, H. Habazaki and F. Di Quarto. 2013. Physicochemical characterization and photoelectrochemical analysis of iron oxide films. J. Solid State Electrochem. 17: 3005-3014.

Sato, T., T. Fujino and H. Hasegawa. 2006. Self-assembled formation of uniform InP nanopore arrays by electrochemical anodization in HCl based electrolyte. Appl. Surf. Sci. 252(15): 5457-5461.

Satsangi, V. R., S. Kumari, A. P. Singh, R. Shrivastav and S. Dass. 2008. Nanostructured hematite for photoelectrochemical generation of hydrogen. Int. J. Hydrog. Energy 33(1): 312-318.

Satsangi, V. R., S. Dass and R. Shrivastav. 2010. Nanostructured α-Fe_2O_3 in PEC generation of hydrogen. pp. 349. *In*: Lionel Vayssieres (ed.). On Solar Hydrogen & Nanotechnology. John Wiley & Sons, Ltd, Singapore.

Sauvage, F., F. D. Fonzo, A. L. Bassi, C. S. Casari, V. Russo, G. Divitini, et al. 2010. Hierarchical TiO_2 photoanode for dye-sensitized solar cells. Nano Lett. 10(7): 2562-2567.

Schmuki, P. 2002. From Bacon to barriers: a review on the passivity of metals and alloys. J. Solid State Electrochem. 6(3): 145-164.

Schrebler, R., L. A. Ballesteros, H. Gómez, P. Grez, R. Cordova, E. Munoz, et al. 2014. Electrochemically grown self-organized hematite nanotube arrays for photoelectrochemical water splitting. J. Electrochem. Soc. 161(14): H903-H908.

Sivula, K., F. L. Formal and M. Grätzel. 2011. Solar water splitting: progress using hematite (α-Fe_2O_3) photoelectrodes. ChemSusChem 4(4): 432-449.

Sivula, K. 2012. Nanostructured α-Fe_2O_3 photoanodes. pp. 121. *In*: R. V. D. Krol and M. Grätzel (eds.). Photoelectrochemical Hydrogen Production. Springer, New York.

Souza, F. L., K. P. Lopes, P. A. P. Nascente and E. R. Leite. 2009a. Nanostructured hematite thin films produced by spin-coating deposition solution: application in water splitting. Sol. Energy Mater. Sol. Cells 93(3): 362-368.

Souza, F. L., K. P. Lopes, E. Longo and E. R. Leite. 2009b. The influence of the film thickness of nanostructured α-Fe_2O_3 on water photooxidation. Phys. Chem. Chem. Phys. 11(8): 1215-1219.

Sreekantan, S., R. Hazan and Z. Lockman. 2009. Photoactivity of anatase–rutile TiO_2 nanotubes formed by anodization method. Thin Solid Films 518(1): 16-21.

Stojadinović, S., N. Tadić, N. Radić, P. Stefanov, B. Grbić and R. Vasilić. 2015. Anodic luminescence, structural, photoluminescent and photocatalytic properties of anodic oxide films grown on niobium in phosphoric acid. Appl. Surf. Sci. 355: 912-920.

Sulka, G. D. 2008. Highly ordered anodic porous alumina formation by self-organized anodizing. pp. 1. *In*: Ali Eftekhari (ed.). Nanostructured Materials in Electrochemistry. Wiley-VCH Verlag GmbH & Co. KGaA, Weinheim.

Tahir, A. A., K. G. U. Wijayantha, S. Saremi-Yarahmadi, M. Mazhar and V. McKee. 2009. Nanostructured α-Fe_2O_3 thin films for photoelectrochemical hydrogen generation. Chem. Mat. 21(16): 3763-3772.

Tilley, S. D., M. Cornuz, K. Sivula and M. Grätzel. 2010. Light-induced water splitting with hematite: improved nanostructure and iridium oxide catalysis. Angewandte. Chemie. 122(36): 6549-6552.

Torres, G. R., T. Lindgren, J. Lu, C. G. Granqvist and S. E. Lindquist. 2004. Photoelectrochemical study of nitrogen-doped titanium dioxide for water oxidation. J. Phys. Chem. B 108(19): 5995-6003.

Uosaki, K., K. Okazaki, H. Kita and H. Takahashi. 1990. Preparative method for fabricating a microelectrode ensemble: electrochemical response of microporous aluminum anodic oxide film modified gold electrode. Anal. Chem. 62(6): 652-656.

Vincent, T., M. Gross, H. Dotan and A. Rothschild. 2012. Thermally oxidized iron oxide nanoarchitectures for hydrogen production by solar-induced water splitting. Int. J. Hydrog. Energy 37(9): 8102-8109.

Walter, M. G., E. L. Warren, J. R. McKone, S. W. Boettcher, Q. Mi, E. A. Santori, et al. 2010. Solar water splitting cells. Chem. Rev. 110(11): 6446-6473.

Wang, D., Y. Liu, B. Yu, F. Zhou and W. Liu. 2009. TiO_2 nanotubes with tunable morphology, diameter and length: synthesis and photo-electrical/catalytic performance. Chem. Mater. 21(7): 1198-1206.

Wang, J. and Z. Lin. 2009. Anodic formation of ordered TiO_2 nanotube arrays: effects of electrolyte temperature and anodization potential. J. Phys. Chem. C 113(10): 4026-4030.

Wang, L., C. Y. Lee, R. Kirchgeorg, N. Liu, K. Lee, S. Kment, et al. 2015. Anodic self-organized transparent nanotubular/porous hematite films from Fe thin-films sputtered on FTO and photoelectrochemical water splitting. Res. Chem. Intermediat. 41(12): 1-9

Wheeler, D. A., G. Wang, Y. Ling, Y. Li and J. Z. Zhang. 2012. Nanostructured hematite: synthesis, characterization, charge carrier dynamics and photoelectrochemical properties. Energy Environ. Sci. 5(5): 6682-6702.

Wilhelm, S. M., K. S. Yun, L. W. Ballenger and N. Hackerman. 1979. Semiconductor properties of iron oxide electrodes. J. Electrochem. Soc. 126(3): 419-424.

Zandi, O. and T. W. Hamann. 2015. The potential versus current state of water splitting with hematite. Phys. Chem. Chem. Phys. 17(35): 22485-22503.

Zhang, Z., M. F. Hossain and T. Takahashi. 2010. Self-assembled hematite (α-Fe_2O_3) nanotube arrays for photoelectrocatalytic degradation of azo dye under simulated solar light irradiation. Appl. Catal. B: Environ. 95(3-4): 423-429.

Zwilling, V., E. Darque-Ceretti, A. Boutry-Forveille, D. David, M. Y. Perrin and M. Aucouturier. 1999. Structure and physicochemistry of anodic oxide films on titanium and TA6V alloy. Surf. Interface Anal. 27(7): 629-637.

10

Synthesis Processes, Characterization Methods and Energy Related Applications of Nano-Crystalline Titanium Dioxide

Sanjeev K. Gupta,* Abhinav Sharma[a] and A. K. Garg[b]

ABSTRACT

Energy production and energy storage are some of the enormous challenges in the 21st century. Development of new and renewable sources of energy may influence a significantly larger fraction of existing energy supply chain in the view of continuous expansion of the energy needs of our rapidly growing population worldwide. As electricity demand escalates with supply, depending largely on fossil fuels along with some hydro power and then nuclear energy source, concerns about greenhouse gas emission and its impact on climate change energy, Research and Development (R&D) into solar cells that efficiently convert the sun's energy into electricity has generated great interest. In recent years, nanostructured materials have attracted huge attention in the scientific community because of their spectacular physical and chemical properties. Extensive R&D work dedicated to develop noble devices and systems using TiO_2 nanomaterials have been most fascinating after observation of the phenomenon of photocatalytic water splitting on TiO_2 electrodes. A lot of R&D on TiO_2 nanomaterials including nanotubes, nanoparticles, nanorods, nanowires etc. has been widely carried out. Nanostructured TiO_2 can be used to provide an environmentally friendly and much cleaner energy due to its unique properties that hold as a semiconductor as well as nanomaterial. This chapter deals with the most favourable synthesis process of nanocrystalline TiO_2 (chemical routes/physical routes), key characterization techniques (XRD, UV-Vis, Raman, AFM, SEM, TEM,

Department of Electronics and Information Technology (DeitY), Ministry of Communications and Information Technology, New Delhi-11003.

[a] E-mail: abhinav036@gmail.com

[b] E-mail: ajai.dit@gmail.com

* Corresponding author: sanjeevgkp@gmail.com

FTIR) and photocatalytic mechanisms are reviewed, energy related applications (Hydrogen for production and generation of Photovoltaics electricity).

Keywords: Nano-crystalline TiO_2, synthesis, characterization, energy, photocatalysis hydrogen, photovoltaic.

1. INTRODUCTION

Recently, the scientific and research community has shown enormous interest on a nanostructured material in the family of multifunctional metal oxide nanostructures and also trying to find out their applications due to their easy, safe, environmental friendly, cheap synthesis procedure and high-end technological applications in the fabrication of devices particularly for energy harvesting and storage, photonics, sensors as well as medical and biological applications. Titanium dioxide (TiO_2) is a well-known material in the family of wide bandgap (>3eV for all crystalline phase) semiconductor (Gupta et al. 2013). TiO_2 exist in three phases in nature: anatase, rutile and brookite. In the semiconductor electronics industry, the rutile phase of TiO_2 is used as a high dielectric constant (high-k) material. High electrical resistance of this material is a key advantage in the fabrication of power devices, capacitive devices, filter and temperature compensating condensers. There are many conventional deposition methods that have been used to prepare nanocrystalline TiO_2 films, such as Direct Current (DC) or Radio Frequency (RF) magnetron sputtering (Haseeb et al. 2010, Sung and Kim 2007, Suhail et al. 1992), electron-beam evaporation (Lu et al. 2011, Yao et al. 2007), chemical vapour deposition (Sun et al. 2008), plasma enhanced chemical vapour deposition (Yang and Wolden 2006), ion-beam assisted deposition (Lin et al. 2009, Yang et al. 2008), sol-gel methods (Verma et al. 2005) etc.

In recent years, Titanium Dioxide (TiO_2) has become a strategic raw material for the development of TiO_2 based photosensitive semiconductor material for solar energy production in wide range of environmentally friendly applications. When TiO_2 deposited as a thin film, its refractive index and natural colour make it an outstanding reflective optical coating for dielectric mirrors. TiO_2, particularly in the anatase form, shows photocatalyst behaviour under ultraviolet light. The inherent structure of TiO_2 has the strong oxidative potential of the positive holes which oxidizes water to create hydroxyl radicals. It can also oxidize oxygen or organic materials directly. Due to this, TiO_2 is added to paints, windows, tiles, cements or other products for sterilizing, deodorizing and anti-fouling properties. As TiO_2 is exposed to UV light, it shows increasingly hydrophilic behaviour and is being used for anti-fogging coatings or self-cleaning windows.

Over the last decades, copious and renewable energy resources such as geothermal, solar and wind have drawn significant attention as alternative sources for clean and green energy. However, the proper utilization of these resources is significantly lower than fossil fuel. Solar energy, having radiant light and heat from the Sun, seems to be the most copious clean energy source available at present. The solar energy, which strikes the earth surface in 1 hour, is relatively higher than the energy consumed by the human population in an entire year. Hence, extensive research and development (R&D) of a material that can efficiently convert the harvested solar irradiation into clean and renewable energy is highly desirable.

The development of a special class of new photosensitive materials for efficient and clean conversion of solar energy is the most critical issue in the development of solar-hydrogen (Nowotny and Sheppard 2007). The development of such materials, which must associate with functional properties, requires the application of the most recent progress done in the area of surface science and interface. Subsequent to this, because of the ultra low density and high surface area, TiO_2 aerogels have become an attractive choice for applications in solar energy conversion (Yao et al. 2009, Baia et al. 2006).

Photoelectrochemical cell and dye-sensitized solar cell are being used as potential technologies to convert solar energy into hydrogen (H_2) fuel and electricity, respectively. There are several types of photocatalysts materials (Titanium Dioxide (TiO_2), Zinc Oxide (ZnO), Zirconium Dioxide (ZrO_2), Vanadium Oxide (V_2O_5), Tungsten Oxide (WO_3), *Niobium Oxide* (Nb_2O_5), Ferric Oxide (FeO_3)) from the family of multifunctional material that have been used in photoelectrochemical cell and dye-sensitized solar cell. Among them, TiO_2 is one of the most promising candidates because of its superior properties such as, light absorption capabilities, chemical inertness and stability (Bettinelli et al. 2007). However, the major obstacles for achieving high efficiency TiO_2 for photoelectrochemical cell and dye-sensitized solar cell are the poor visible-light absorption and quick recombination of charge carriers. The improved TiO_2 photocatalysts properties have been obtained by doping with non-metal atoms such as nitrogen (Sakthivel and Kisch 2003, Reddy et al. 2005), carbon (Chen et al. 2007), sulphur or using codoped materials (Sun et al. 2006). Moreover, several attempts have also been made to decrease the bandgap energy by doping with suitable transition metals ions (Wilke and Breuer 1999) and lanthanides (Xu et al. 2002). In order to commercialize the feasibility of these aforesaid cells in terms of performance and cost effectiveness, substantial research on the development of high quality TiO_2 has become a major topic of current research.

This chapter elaborates on the most favourable synthesis method, possible characterization techniques of nanocrystalline TiO_2 material and energy related applications of TiO_2.

2. CRYSTAL STRUCTURE AND PHYSICAL PROPERTIES OF TITANIUM DIOXIDE

TiO_2 exists in both crystalline as well as amorphous. Anatase, rutile and brookite are the three most studied crystalline phases of TiO_2. Each crystalline structure of these phases exhibits specific physical properties such as bandgap, surface states, refractive index, density etc. which governs their applications and uses in a particular sector. Out of these three phases of TiO_2, rutile has evolved as the novel material in the development of metal oxides based on structure and systems. Moreover, anatase and rutile are well known for photocatalysts while anatase generally show much higher photocatalytic activity (Zhang et al. 2014). Rutile is commonly used as a white pigment in paints and brookite is not yet of commercial interest but a Dye Sensitized Solar Cell (DSSC) has been reported in literature (Magne et al. 2011). Anatase and rutile are both tetragonal in structure while the brookite structure is orthorhombic. In case of all the crystalline forms of TiO_2, each Ti^{4+} ions are surrounded by an irregular octahedron of oxide ions. In the rutile structure each octahedron is in contact with 10 neighbour octahedrons (two sharing edge oxygen pairs and eight sharing corner oxygen atoms) while the in anatase structure each octahedron is in contact with eight neighbours (four sharing an edge and four sharing a corner) (Linsebigler et al. 1995). The octahedral connection in brookite is such that three edges are shared per octahedron. Figure 10.1 shows the schematic diagram of all crystalline form of TiO_2.

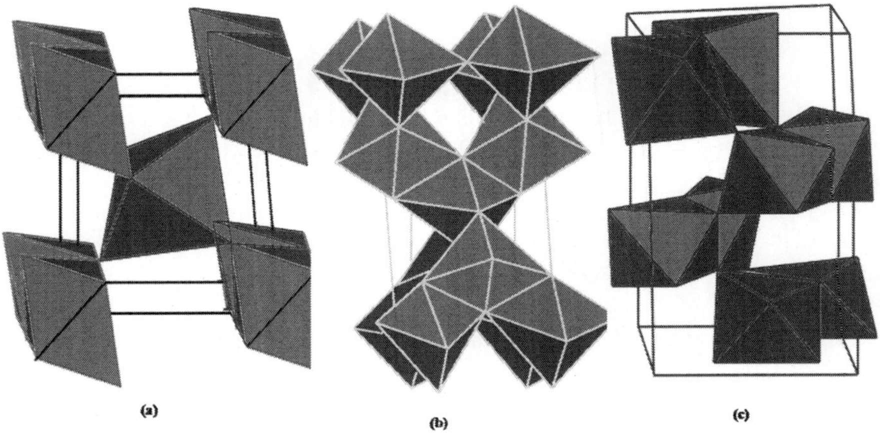

Fig. 10.1 Crystal structure of (a) Rutile TiO_2. (b) Anatase TiO_2. (c) Brookite TiO_2 (Carp et al. 2004).

Furthermore, the structure of rutile and anatase can be described in terms of chains of (TiO_6) octahedra, where each Ti^{4+} ion is surrounded by an octahedron of six O^{2-} ions. The octahedron in rutile is not regular, showing a slight orthorhombic distortion while the octahedron in anatase is significantly distorted so that its symmetry is lower than orthorhombic (Burdett 1985, Burdett et al. 1987, Fahmi et al. 1993, Gupta and Tripathi 2011). A comparative study of different crystalline modifications of TiO_2 is summarized in Table 10.1.

Table 10.1 Crystal Structure data of TiO_2 (Mittal V. 2015).

Properties	Rutile	Anatase	Brookite
Crystal Structure	Tetragonal	Tetragonal	Orthorhombic
Lattice Constant (Å)	a = 4.5936 c = 2.9587	a = 3.784 c = 9.515	a = 9.184 b = 5.447 c = 5.154
Molecule (cell)	2	4	8
Density (g cm^{-3})	4.13	3.79	3.99
Ti–O bond length (Å)	1.949 (4) 1.980 (2)	1.937 (4) 1.965 (2)	1.87-2.04
O–Ti–O bond angle	81.2° 90.0°	77.7° 92.6°	77.0°-105°

2.1 Physical Properties

TiO_2 is insoluble in dilute acid, dilute alkali, but soluble in hot concentrated sulphuric acid, hydrochloric acid and nitric acid. The solubility of titanium dioxide is related to solutes. Because of high dielectric constant of TiO_2, it has excellent electrical properties. The electrical conductivity increased rapidly with rising temperature, but also is very sensitive to hypoxia. Only the rutile phase of TiO_2 has a melting point and boiling point. The melting point of rutile TiO_2 is 1830-1850°C. The melting point of TiO_2 is related to the purity and contents of the TiO_2. For example this temperature rises to some extent when oxygen contents increases. However, melting and boiling points of anatase and brookite TiO_2 actually does not exist, it can be transformed to form rutile. A summary of other physical properties are summarized in Table 10.2.

Table 10.2 Physical properties.

Physical properties	Critical value
Molar mass	79.8658 g·mol^{-1}
Refractive index	2.76~2.55
Hardness (Mohs)	6-7
The coefficient of linear expansion	25°C
Thermal Conductivity (W/m·K)	1.8-10.3
Particle size	0.2~0.3

3. SYNTHESIS OF NANOCRYSTALLINE TiO_2

3.1 Synthesis of Nanocrystalline TiO_2 by Chemical/Solution Routes

A stable nanocrystalline structure of TiO_2 can be prepared in the form of thin films, powder, crystals, ceramic fibres, aerogel etc. For this, chemical/solution route processing is one of the most convenient and explored method of synthesis of TiO_2 for various applications. This solution route method has the advantage over other methods, as this it allows the control of stoichiometry, creation of homogeneous materials, development of complex and stable shapes and preparation of composite materials. However, this method has some drawbacks such as long processing times, expensive and the presence of carbon as an impurity. There are many methods such as precipitation (Namin et al. 2008, Shchipunova and Postnova 2009, You et al. 2005, Wu and Qi 2007), the hydrothermal method (Kim et al. 2006, Seo et al. 2001, Kasuga et al. 1998, Sun and Li 2003) and the electrochemical process (De Tacconi et al. 2006, Natarajan and Nogami 2006, Therese and Kamath 2000) have been elaborated in literature for synthesis of TiO_2 nanocrystalline material. However the sol-gel method (Uekawa et al. 2002, Li et al. 2003a, Li et al. 2003b, Colon et al. 2002, Maira et al. 2001) is one of the most convenient methods for synthesis of nanocrystalline TiO_2, which is as follows:

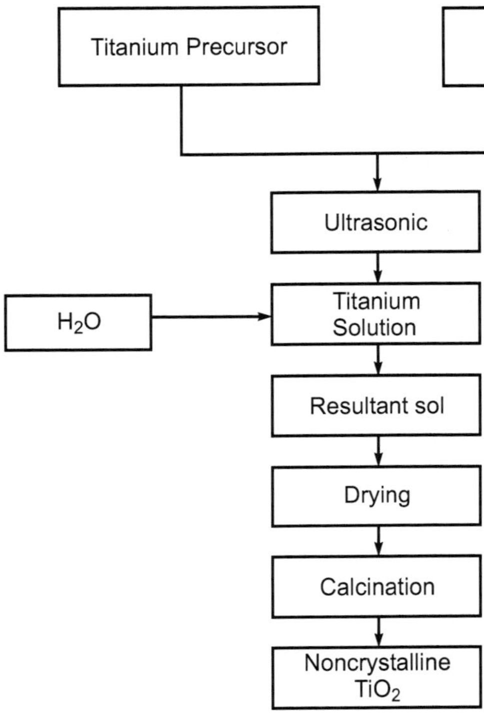

Fig. 10.2 Scheme of the preparation of TiO_2 nanoparticles by sol-gel method.

3.1.1 Sol-gel Method

The sol-gel method is governed by inorganic polymerization reactions. In this method four consultative steps are included: hydrolysis, polycondensation, drying and thermal decomposition. In a typical sol-gel method, a colloidal suspension or a sol pr solution is formed from the hydrolysis and polymerization reactions of the precursors. The 'sol' slowly grows towards the formation of a gel-like diphasic structure containing both a liquid and solid phase whose morphologies range from discrete particles to continuous polymer networks. Complete polymerization and reduction of solvent of the liquid sol results into a solid gel phase. Further treatment of the gel can produce dense thin films, powders, ceramic fibres, aerogel etc. Furthermore, under proper synthesis conditions nanostructured materials can be obtained through this route. For synthesis of nanocrystalline TiO_2, the size of the sol particles can be modified by tuning the solution composition, pH value and process temperature. Figure 10.2 shows the scheme of the preparation of TiO_2 nanoparticles by sol-gel method (Seifried et al. 2000).

3.2 Synthesis of Nanocrystalline TiO_2 by CVD Routes

Chemical Vapour Deposition (CVD) is a generic name widely used in materials-processing technology for a group of processes that involve depositing a solid material from a gaseous phase. The majority of its applications include depositing solid thin-film coatings to semiconductors surfaces, but it is also used to produce high-purity bulk materials and powders, as well as fabricating composite materials by incorporating suitable mixture of gases. CVD is basically used to form coatings to tune the basic properties such as, electrical, thermal, mechanical, optical and wear resistance properties of various substrates. In recent times, it is being used to deposit various nanomaterials on different substrates. This process generally takes place within a horizontal vacuum chamber quartz tube. During the process, thermal energy heats the gases in the quartz chamber and drives the deposition reaction at the surface of substrate. A schematic diagram of conventional CVD process is shown in Fig. 10.3.

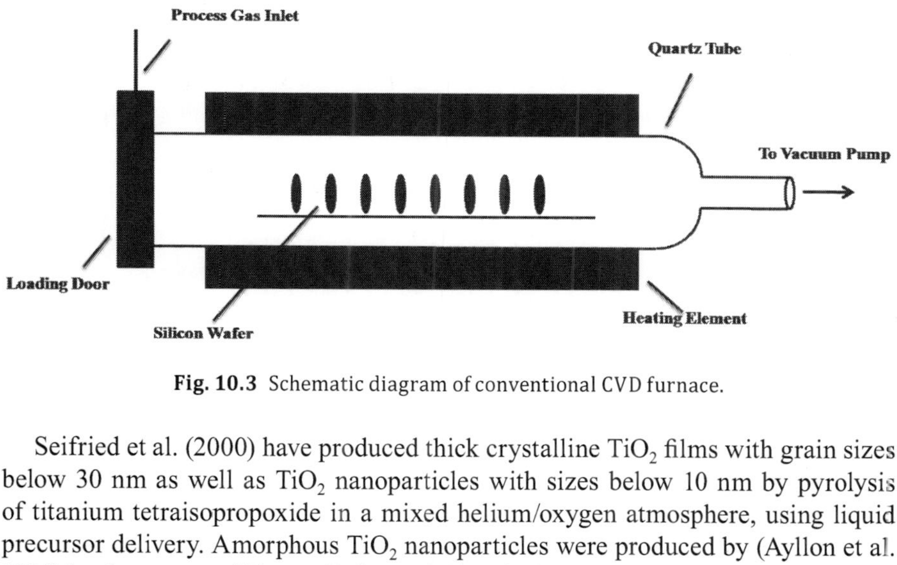

Fig. 10.3 Schematic diagram of conventional CVD furnace.

Seifried et al. (2000) have produced thick crystalline TiO_2 films with grain sizes below 30 nm as well as TiO_2 nanoparticles with sizes below 10 nm by pyrolysis of titanium tetraisopropoxide in a mixed helium/oxygen atmosphere, using liquid precursor delivery. Amorphous TiO_2 nanoparticles were produced by (Ayllon et al. 1999) by the means of Plasma Enhanced Chemical Vapour Deposition (PECVD) by depositing it on the cold areas of the reactor at temperatures below 90°C. Moreover, the crystallize phase was also obtained when the amorphous TiO_2 nanoparticles annealed at high temperatures.

3.3 Synthesis of Nanocrystalline TiO_2 by PVD routes

Physical Vapour Deposition (PVD) is a vacuum deposition method to deposit thin films by the condensation of a vapourized form of the material onto surface of various substrates in the vacuum chamber. The deposition method involves physical processes, which includes vacuum evaporation by e-beam source (gun) or plasma sputter bombardment of the material to be deposited rather than using a chemical reaction at the substrate surface. There are various PVD methods for the deposition or synthesis of TiO_2 thin films or nanostructured materials include e-beam physical vapour deposition, sputtering (magnetron and RF sputtering), cathodic arc deposition and pulsed laser deposition etc. are available in literature (Kaczmareket al. 2012, Heo et al. 2005). The electron beam evaporation method is described below for deposition of nanocrystalline TiO_2 thin film in the chapter.

3.3.1 e-beam Evaporation Method

Gupta et al. (2013) have demonstrated a method to deposit a rutile nanocrystalline TiO_2 by an e-beam evaporation system on oxidized Si wafers. Schematics diagram of the vacuum system used for such deposition is shown in Fig. 10.4. A solid form of TiO_2 target in cylindrical shape of pallet was put in a graphite crucible over the hearth of one e-gun. The oxidized wafers were loaded in a vacuum chamber. The base pressure of chamber achieved to be 6×10^{-8} Torr using vacuum pumps. It was noticed that a beam current of 20 mA was sufficient to evaporate TiO_2 solid source in a vacuum of the order of 8×10^{-7} Torr. The deposition rate in the range of 0.1-0.3 Å/sec was monitored *in situ* employing quartz crystal detector. The rate of deposition was controlled by varying the beam current and the rate was continuously monitored through *in situ* thickness monitor.

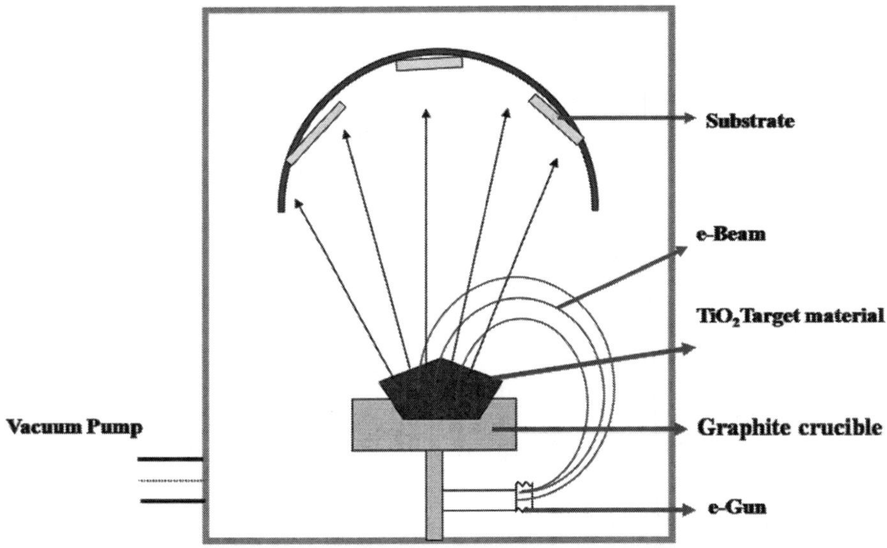

Fig. 10.4 Schematic diagram of e-beam evaporation system used for deposition of nanocrystalline thin film (Gupta et al. 2013).

The most common merits and demerits of various synthesis methods have been summarized in Table 10.3.

Table 10.3 Merits and demerits of various synthesis methods of Nanocrystalline TiO_2.

		Merit	Demerit
Chemical Route	Sol-gel	• Low temperature processing and consolidation is possible. • Smaller particle size and morphological control in powder synthesis. • Sintering at low temperature also possible. • Better homogeneity and phase purity compared to traditional ceramic method.	• The cost of raw materials is expensive. • Products contain high carbons when organic reagents are used in preparative steps. • There is often a large volume shrinkage and cracking during drying. • Since several steps are involved, close monitoring of the process is needed.
Physical Route	e-Beam Evaporation	• Low step coverage. • Low damage to the wafer. • The process is more environmentally friendly. • Excellent purity of the film because of the high vacuum condition.	• Materials with low vapour pressures are very difficult to evaporate. • Poor film adhesion and uniformity. • X-ray damage caused by electron beam evaporation can occur. • High vacuum is required. • More difficult to control the film composition as a result poor stochiometry. • Directional deposition of film.
	Sputtering	• Large-size targets, simplifying the deposition of films with uniform thickness over large wafers. • Composition of film and thickness is easily controlled by fixing the operating parameters. • Control of the alloy composition, step coverage, grain structure is easier obtained. • Low temperature process. • Materials with very high melting points can easily be sputtered and deposited. • Low vacuum is required.	• High capital expenses are required. • Rates of deposition of some dielectric materials are relatively low. • Processes requiring large amounts of heat require appropriate cooling systems. • The rate of coating deposition is usually quite slow. • Deposition operate at high vacuums and temperatures requiring skilled operators.
	CVD	• LPCVD-Excellent cleanliness, conformity, high density, high throughput, low vacuum and uniformity. • PECVD–Low temperature processing, better control of film properties and low vacuum. • APCVD-Simple and fast and no vacuum system.	• LPCVD-High Temperature, low deposition rate, toxic precursor. • PECVD-Risk for participle and chemical contamination Toxic precursor, low throughput. • APCVD-Poor step coverage, particle formation, Toxic gases required.

4. MAJOR CHARACTERIZATION TECHNIQUES TO CHARACTERIZE NANOCRYSTALLINE TiO_2

4.1 X-ray Diffraction Analysis

The X-Ray Diffraction (XRD) technique plays a crucial role in the analysis of structure determination of crystalline solids. In the XRD technique, a beam of X-rays with a wavelength ranging from 0.5 to 2 Å, incident on a specimen is further diffracted by the crystalline phases in a specimen according to Bragg's law.

$$2d\sin\theta = n\lambda \tag{1}$$

where, d is the interplanar spacing, θ is Bragg angle, n is the order of reflection and λ is the wavelength of X-Rays.

In the XRD analysis, diffraction pattern is provided as position, intensity and the shape of diffraction peaks. The comprehensive libraries available in the form of characteristic d-spacings and intensities (JCPDS-files) of previously studied solids are being used for comparative purposes.

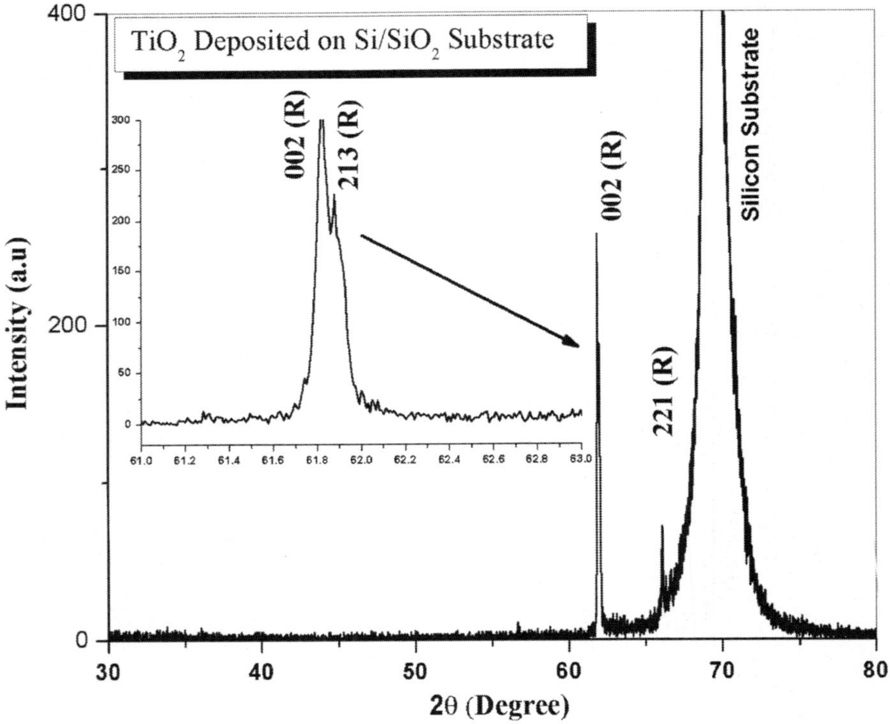

Fig. 10.5 XRD spectra of nanocrystalline TiO_2 prepared by e-beam method (Gupta et al. 2013).

Figure 10.5 shows the wide angle X-ray diffraction patterns of anatase and rutile titania respectively in which some prominent peaks, indicating the crystallinity of deposited TiO_2 thin film prepared by e-beam method (Gupta et al. 2013). However, the average crystalline size of deposited TiO_2 has been determined using a well known Scherrer's equation (Pawar et al. 2011).

$$D = k\lambda/B\cos\theta \tag{2}$$

where B is the FWHM of the XRD spectrum, k is the constant related to the shape of the crystallites and indices of the reflecting plane, λ is the wave length of X-ray and B is the crystallite size. Considering the value of k to be 0.94, as explained in (Akl et al. 2006), Eq. (1) takes the following form:

$$D = 0.94\lambda/B\cos\theta \tag{3}$$

The average crystallite size was determined and is found to be in the range of 80-90 nm (Gupta et al. 2013).

4.2 UV-Vis Spectroscopy

Ultraviolet-visible spectroscopy (UV-Vis) also known as absorption spectroscopy or reflectance spectroscopy in the ultraviolet-visible spectral region. Basically, UV-Vis absorption spectroscopy measures the percentage of radiation that is absorbed at each wavelength. Typically, this is done by scanning the wavelength in ultraviolet-visible range and recording the absorbance. The technique has been successfully expanded to analyze the behaviour of gases and solids and also beyond absorption, to include measured reflected rather than transmitted light. It is also used for the measurement of electronic bandgap of semiconductor thin films.

Figure 10.6 shows the UV-Vis absorption spectra of rutile TiO_2 thin film (Gupta et al. 2013). In the investigated film, more than 94% transmittance of the film has been observed for all acquired spectral range. However, a very sharp decrement of absorption edge, towards the shorter wavelength of the transmittance spectra, revealed a tremendous optical property of the films. The optical absorption coefficient (α) of the deposited TiO_2 film was estimated from transmittance data using the following relation.

$$\alpha = \frac{1}{d}\left(\text{In}\frac{1}{T}\right) \tag{4}$$

where T is the transmittance and d is the thickness of film. The absorption coefficient (α) obeys the following relation for high photon energies (hv):

$$\alpha = \frac{A(hv - Eg)^n}{hv} \tag{5}$$

where E_g is the optical bandgap of the film and A is the constant having the value between 10^5 and 10^6 cm^{-1} (49). The exponent n takes values of 1, 2, 3 and ½ depending on the types of electronic transition taking place in the k-space. Further, the absorption data for the thin film can be analyzed on the basis of well-known relations for direct and indirect transitions (Langford and Wilson 1978).

$$(\alpha hv)^2 = A_1(hv - Eg_1) \tag{6}$$

$$(\alpha hv)^{1/2} = A_2(hv - Eg_2) \tag{7}$$

where A_1 and A_2 are constant and Eg_1, Eg_2 are the direct and indirect bandgap, respectively. The intercept of the extrapolated linear fit to the experimental data of a plot of $(\alpha h\nu)^2$ versus $h\nu$ is determined as the optical direct bandgap, which has been found in the range of 3.64 eV. While the indirect transition bandgap energy (Eg_2) estimated by extrapolation of the linear portion of the plot $(\alpha h\nu)^{1/2}$ as a function of $h\nu$, was found to be 3.04 eV (Gupta et al. 2013).

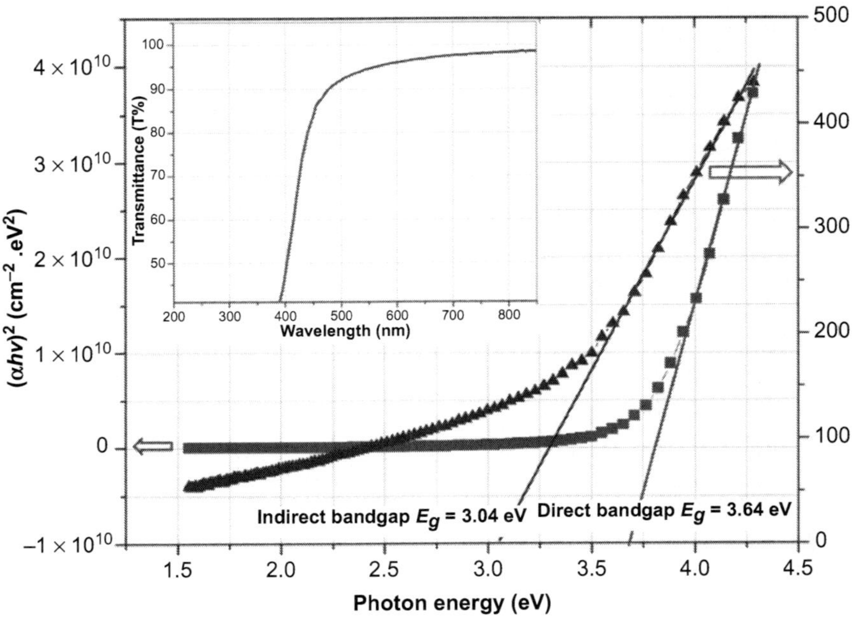

Fig. 10.6 UV-Vis absorption spectra of rutile TiO_2 thin film prepared by e-beam method (Gupta et al. 2013).

4.3 Scanning Electron Microscopy

Scanning Electron Microscopy (SEM) is a well established technique used to study the topography, texture and surface features of thin films, powders etc. In SEM analysis, an electron gun emits a beam of electrons which then interacts with the surface leading to emission of electrons from the surface of the specimen during scanning with the electron beam. The electron from the beam interacts with the sample resulting in deflection of secondary particles to a detector which subsequently converts the signal to voltage and amplifies it. Lui et al. (2005) have demonstrated SEM image of anatase TiO_2 film prepared by Liquid-Phase Deposition (LPD). They observed that TiO_2 films consist of very fine particles with diameters of a few tens of nanometer, typically in the range 30-40 nm. Also, the grains of the TiO_2 film stack closely together.

4.4 Transmission Electron Microscopy

Transmission Electron Microscopy (TEM) is a useful technique in determining crystal morphology and particle size of materials. In TEM a beam of electrons is propagated through a solid sample in a vacuum. Electrons transmitted through the sample are then detected to produce three dimensional images which represent the relative extent of penetration of electrons in a specific sample. TEM can be coupled with Energy Dispersive X-ray (EDX) analysis to obtain information about the chemical nature and composition of the sample. High Resolution TEM (HRTEM) allows the determination of lattice spacings and direction of particles.

4.5 Fourier Transform Infrared (FTIR) spectroscopy

Infrared spectroscopy has been extensively used for identifying the various functional groups on the catalyst itself, as well as for identifying the adsorbed species and reaction intermediates on the catalyst surface. It is one of the few techniques capable of exploring a catalyst both in its bulk and its surface and under actual reaction conditions. It is widely used for characterizing the acid sites of the catalyst, which are responsible for their catalytic properties. A typical FTIR spectra of TiO_2 sample (Gupta et al. 2013) synthesized by synthesize by e-beam method is given in Fig. 10.7.

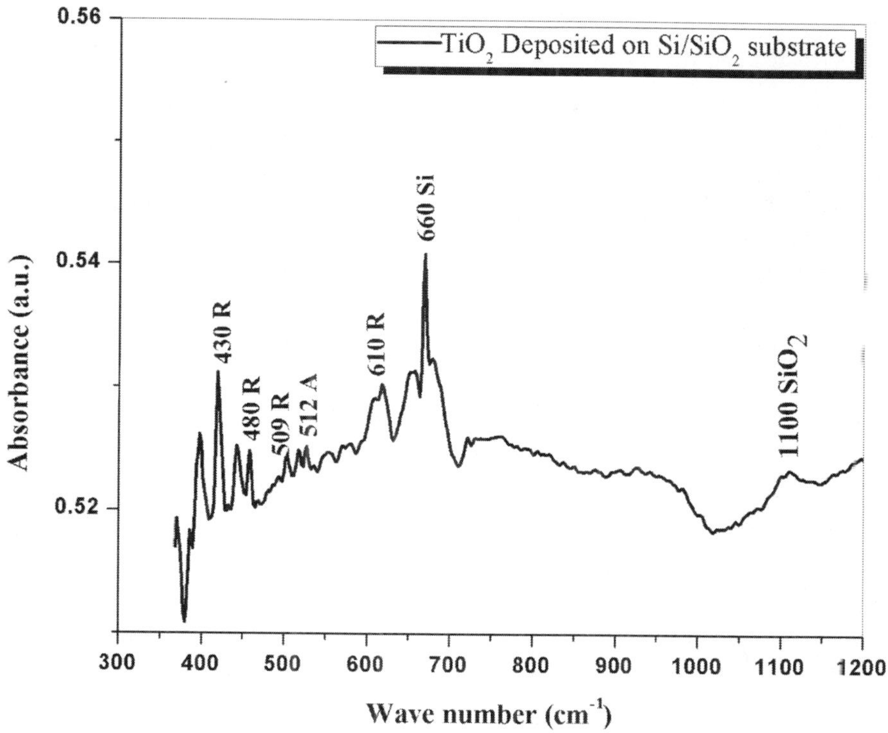

Fig. 10.7 FTIR spectra of as deposited TiO_2 thin film on oxidized Si substrate (Gupta et al. 2013).

4.6 Atomic Force Microscopy

Atomic Force Microscopy (AFM) is a technique used for the surface characterization of thin films. This method has the advantage of producing high quality image of non-conducting surface without any chemical processing. Moreover, this method has another unique feature as compared to the other microscopy technique, such as the mechanical properties of material may also be studied. Moreover, the process does not involve the use of electron beam radiation that may damage the surface to be studied. AFM is based on the principle that when a tip, integrated to the end of a spring cantilever, is brought within inter atomic distance between the tip and sample, inter atomic potential are developed between the atoms of tip and atoms of surface to be studied. As the tip moves across the surface, the inter-atomic potentials, forces the cantilever to the bounce up and down with a change in contours of the surface. Therefore, by measuring the deflection of the cantilever, the topographic features of the surface can be determined.

Figure 10.8(a) and (b) show 3D images of 5 μm × 5 μm scan area and 100 nm × 100 nm scan area, respectively (Gupta et al. 2013). The Root Mean Square (RMS) roughness and peak to peak height were calculated to be ~1.29 nm ~13.30 nm, respectively for TiO_2 film. AFM images obtained at different locations of the wafer showed that the films reveal a homogeneous bulbous structure. The whole film surface is fashioned by small grains of the deposited TiO_2 clusters. The average dimensions of grain size was estimated to be in range of 100-125 nm, while grain boundary was found to be ~20 nm without any annealing treatment of the film.

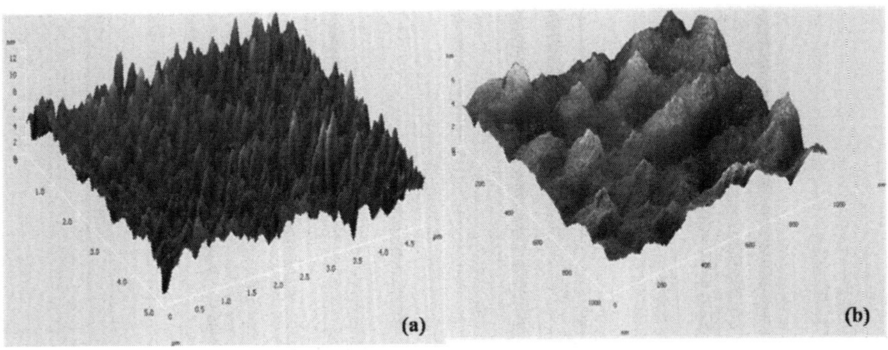

Fig. 10.8 AFM atomic topography of the TiO_2 thin film produced by e-beam evaporation on oxidized Si (100) substrates: (a) 3D image of 5 μm × 5 μm scan area. (b) 3D image of 100 nm × 100 nm scan area (Gupta et al. 2013).

4.7 Raman Spectroscopy

Raman spectroscopy can be thought of as a means to measure the inelastic light scattering which results from the excitation of the vibrations in molecular and crystalline materials all as a function of wave number. The monochromatic light used in Raman scattering is usually from a laser in the visible, near infrared or near ultraviolet region. Excitations in the system are absorbed or emitted by the laser light, resulting in the energy of the laser photons being shifted up or down. The shift in energy gives information about the photon modes in the system.

The phase purity and structural information of deposited TiO_2 thin film as investigated by Gupta et al. (2013) and was characterized using Renishaw in VIA reflex micro-Raman spectrometer. Ar^+ laser of 20 mW was used as the excitation source. The micro-Raman spectra of the as deposited TiO_2 film along with the target material used in e-beam physical vapour deposition are represented in Fig. 10.9. A high intense peak at 300 and 520 cm^{-1} are also observed due to presence of silicon substrate as a base material. TiO_2 anatase phase, which is tetragonal with two units per unit cell (space group D^{19}_{4h}, exhibits six Raman active modes (1 A_{1g} at 519 cm^{-1}, $2B_{1g}$ at 399 and 519 cm^{-1} and 3 E_g at 144, 197 and 639 cm^{-1}, while the rutile structure, which is tetragonal with two TiO_2 units per unit cell and space group D^{14}_{4h}, has four Raman active modes (A_{1g} at 612 cm^{-1}, B_{1g} at 144 cm^{-1}, B_{2g} at 826 cm^{-1} and E_g at 447 cm^{-1} (Rossella et al. 2010). The black line of spectra in Fig. 10.9 was acquired for TiO_2 thin film. The two pure rutile active modes (E_g at 442 cm^{-1} and A_{1g} at 612 cm^{-1}) were obtained in the deposited film on placed substrate. In addition one extra pure rutile phase was also observed (B_{2g} at 824 cm^{-1}). A common peak at 250 cm^{-1} was also identified, which showed the rutile nature but the space group of this peak cannot be classified.

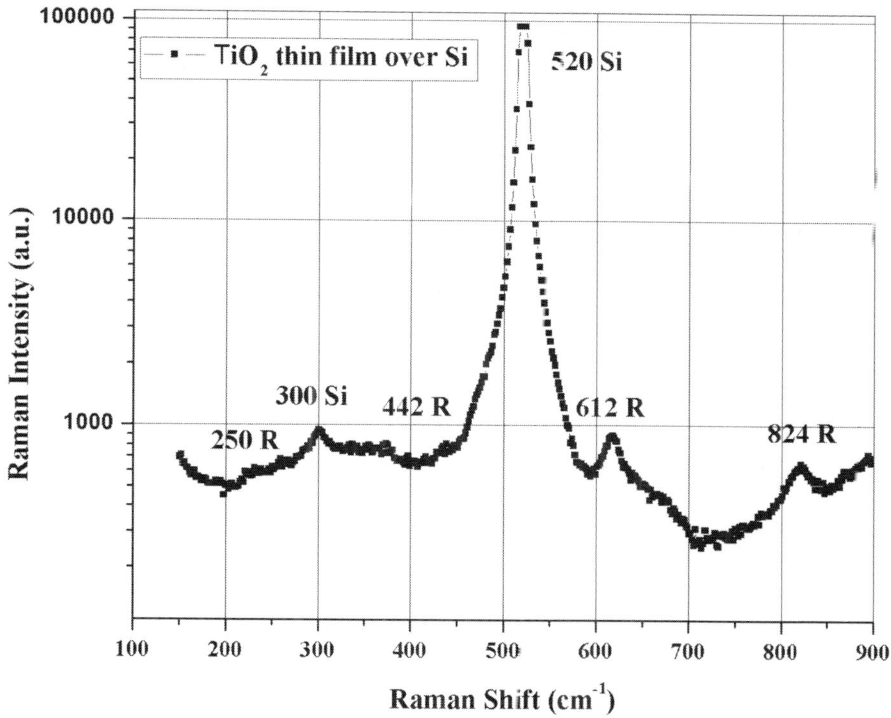

Fig. 10.9 Raman spectra of deposited TiO_2 thin film (Gupta et al. 2013).

5. PHOTOCATALYSTS PROPERTIES OF NANOCRYSTALLINE TIO_2

A semiconductor material is of great technological advantage in the electronics industry due to their ability to generate charge carriers (electrons and holes) under UV radiation from sunlight or illuminated light source (fluorescent lamps). The catalytically favorable combination of electronic structure, ability of light absorption, charge transport characteristics and sufficiently long life lifetimes of some semiconductors allow them to be used as photocatalysts. The photocatalytic mechanism on the surface of nanocrystalline TiO_2 is initiated when the electron of the valence band of TiO_2 becomes excited be means of absorption of the photon (*hv*) with energy equal to or greater than the bandgap of TiO_2 producing an electron-hole pair (Zaleska 2008) as shown in Fig. 10.10. Subsequent to this, an electron is promoted to the conduction band of TiO_2 while a positive hole is formed in the valence band. This stage is known as the semiconductor's 'photo-excitation' state. Further, the excited-state electrons and holes can recombine and dissipate the input energy as heat, get trapped in metastable surface states or react with electron donors and electron acceptors adsorbed on the TiO_2 surface. The positive-hole of TiO_2 breaks apart the water molecule to form hydrogen gas and hydroxyl radical with high redox oxidizing potential. The negative-electron reacts with oxygen molecule to form a super oxide anion. This cycle continues when light is available.

TiO_2 is a well-known and promising candidate for photocatalysis because it is cheap and non-toxic, chemically and biologically inert, easy to produce/use and photo-catalytically active. However, inspite of these outstanding properties which TiO_2 shows, there are many drawbacks of TiO_2 as photocatalysts. The key problems are as follows:

(a) The major drawback of TiO_2 is its wide bandgap energy. This clearly reflects that only a small part of the solar light can be absorbed in the UV region, which inhabits only 5% of the total solar spectrum (Cui et al. 2008). Therefore, proficient absorption of the visible light which comprises the major part of solar spectrum is prevented and subsequently the photocatalytic efficiency of TiO_2 is mired.

(b) The fast recombination rate of photo generated electron-hole carriers decreases photo quantum efficiency of TiO_2.

(c) A power TiO_2 was used in the initial stage of the study of photocatalysis. After photocatalytic reaction, a filtration step was necessary to separate photocatalysts from slurry. This process adds extra cost to the development of commercial applications.

In order to address these drawbacks, many methods have been proposed which confine the wide usage of TiO_2 photocatalysts. Some of these methods are briefly reviewed here.

5.1 Improvement of TiO_2 Photocatalysts Properties by Doping

TiO_2 shows relatively high chemical stability and reactivity under illumination of UV light (λ <387 nm). However, the development of photocatalysts to exhibit a high reactivity under visible light (λ > 400 nm), is one of the major challenges of the time. In this context, doping with suitable dopants has proved to be a great method to enhance the photocatalytic properties of TiO_2. Several approaches for TiO_2 modification have been reported to improve the photocatalysts properties by means of metal doping by copper (Chen et al. 2009), cobalt (Hsieh et al. 2009), nickel (Wang et al. 2008), barium (Atashfaraz et al. 2007), manganese (Zhang et al. 2006), iron (Deng et al. 2009), non-metal doping by (C, S, N, B, P, F) (Dong et al. 2011, Hahn et al. 2007, Choi et al. 2004, Periyat et al. 2008, Asahi et al. 2014, Dozzi et al. 2013 and by co-doping of metal in matrix of TiO_2 (Rahimi et al. 2012, Shang et al. 2014, Chen et al. 2012).

Metal-doping/implantation was the leading method at the initial stage of the study on the doping effect on the photocatalytic properties of TiO_2. It was reported that a new energy level produced in the bandgap of TiO_2 by the dispersion of metal nanocrystalline structure in the TiO_2 matrix significantly influenced photoactivity, charge carrier recombination rates and interfacial electron-transfer rates. These mechanisms can be seen in Fig. 10.1, in which electron excited from the defect state to the TiO_2 conduction band by photon with energy equals hv_1. Moreover, the trapping of electrons to reduce electron-hole recombination during irradiation is the additional benefit of transition metal doping. Thus, a decrease of charge carrier's recombination results in enhanced photo-activity at the surface of nanocrystalline TiO_2.

In literature, there are three different major views regarding modification mechanism of TiO_2 photocatalytic properties doped with non-metals i.e. (i) Bandgap narrowing (ii) Impurity energy levels and (iii) Oxygen vacancies. Asahi et al. (2001) found that when nitrogen is doped in anatase TiO_2, N-2p state hybrids with O-2p states because their energies are very close to each other and thus the bandgap of N-TiO_2 is narrowed and was able to absorb visible light. Irie et al. (2003) reported that oxygen sites of TiO_2 substituted by nitrogen atom form isolated impurity energy levels just above the valence band. It was observed that illumination with visible light only excites electrons in the impurity energy level while irradiation with UV light excites electrons in both the VB and the impurity energy levels. Ihara et al. (2003) reported that oxygen-deficient sites formed in the grain boundaries of nanocrystalline TiO_2 are important to emerge visible activity and nitrogen doped in part of oxygen-deficient sites are important as a blocker for reoxidation. Creation of impurities energy level by suitable do-pants is shown in Fig. 10.10 in contrast to pure TiO_2.

Recently, to improve the low visible light absorption and high charge carrier recombination rate, co-doped TiO_2 were synthesized and studied. Guo et al. (2013) demonstrated that co-doping (Fe + Mo) can further increase the visible absorption and improve the photocatalytic property of TiO_2 compared with Fe mono-doping. They reported that Fe mono-doping improves the photocatalytic property of TiO_2 only at very low doping level (Fe concentration less than 1.0%), while by co-doping a small amount of Mo with Fe, the effective doping concentration of Fe can be pushed to a higher level and the photocatalytic property of TiO_2 can be further improved.

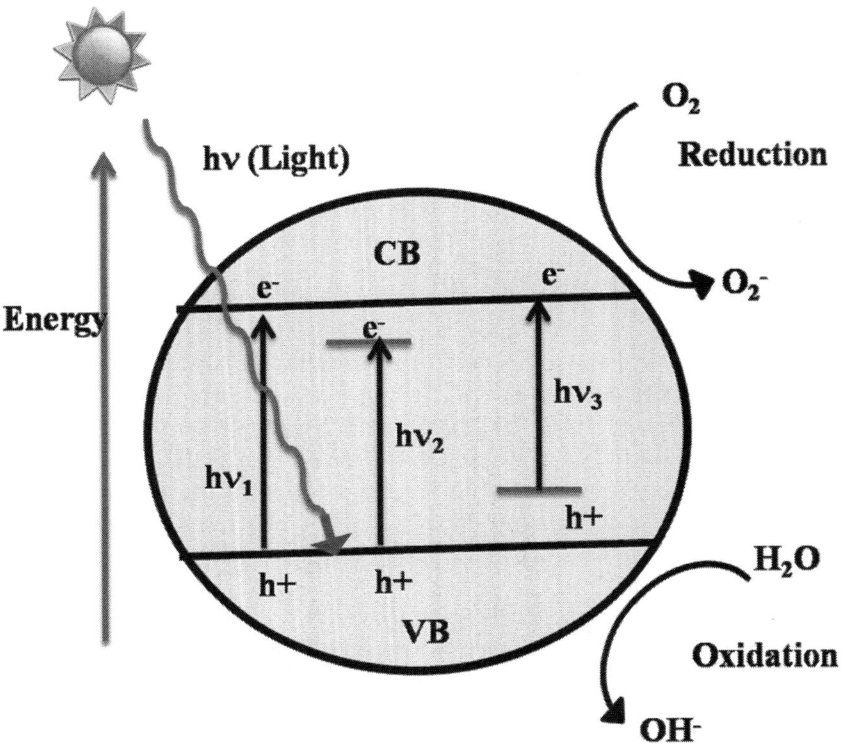

Fig. 10.10 Photocatalysis mechanism of TiO_2: $h\nu_1$-pure TiO_2, $h\nu_2$-metal doped TiO_2, $h\nu_3$-non-metal doped TiO_2.

6. APPLICATIONS OF NANOCRYSTALLINE TiO_2

6.1 Photoelectrochemical Generation of Hydrogen (Solar Hydrogen)

In recent years, there has been a constant search for the development of clean and renewable energy. Hydrogen has been found as the best substitute for clean energy. The production of hydrogen can reduce the dependency on commercial oil and natural gas. Hydrogen can be produced through various routes particularly such as photo-electro-chemical and water splitting by photocatalytic mechanism. In 1972, Fujishima and Honda successfully demonstrated photoelectrolysis of water on n-type TiO_2 single crystal electrode for solar energy conversion and storage in the form of hydrogen. They described an electrochemical cell consisting of a n-type Rutile-TiO_2 as anode and a Pt black cathode. When the electrochemical cell was irradiated with UV light (λ <415 nm) from a 500 W Xe lamp, hydrogen was evolved at the Pt electrode, while oxygen was evolved at the TiO_2 electrode. A basic schematic diagram for the hydrogen production using TiO_2 electrode is illustrated in Fig. 10.11. Several researches have been carried out for production of hydrogen by using the same basic set up but with some variations to the electrode materials. Cowan et al. (2010) demonstrated water splitting by nanocrystalline TiO_2 in a complete photoelectrochemical cell while others have been produced by chemically modifying TiO_2 by controlled combustion of Ti metal in a natural gas flame. In order to increase the water splitting efficiency, modified processes has also been reported by using a chemically modified n-type nanostructured TiO_2 (Khan et al. 2002), dye-sensitized oxide semiconductors (Youngblood et al. 2009) etc. When TiO_2 absorbs photonic light with energy larger than it's bandgap, electrons and holes are generated in the conduction and valence bands, respectively as discussed earlier. Water molecules are reduced by the electrons to form H_2 and oxidized by the holes to form O_2, leading to overall water splitting (Chen 2009).

Production of hydrogen using a highly active TiO_2 photocatalyst was reported by Kitano et al. (2008). They prepared TiO_2 thin films were by Radio Frequency (RF) magnetron sputtering deposition method. As in the sputtering technique substrate temperature may be varied, so synthesis of nanostructured TiO_2 films by this method is able to absorb visible light by manipulating the substrate temperature. They observed a UV-VIS transmission spectra, which showed a lower transmittance when the substrate temperature was at 873 K, showing a higher absorbance in the visible spectrum. The nanocrystalline TiO_2 phase present at this temperature is most likely to be in a rutile phase, which has lower bandgap energy. The TiO_2 electrode prepared by this film was mounted on an H-type container filled with water. These experimental demonstrations were found to have enough potential for the separate evolution of H_2 and O_2 from water under sunlight irradiation.

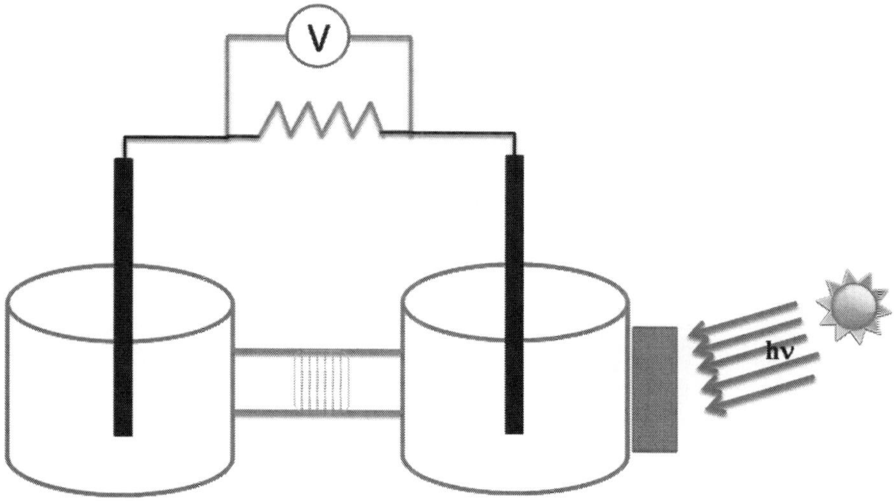

Fig. 10.11 Basic schematic diagram for the hydrogen production using TiO_2 electrode.

There is a lot of literature that is are available on the hydrogen production using nanostructured TiO_2 electrode (Tay et al. 2013, Cha et al. 2015, Badawy et al.

2015). Recently, several groups have used thin films consisting of TiO_2 nanotube as photoanode for the hydrogen production (Cho et al. 2011, Ye et al. 2012, Sunm et al. 2011). As the length of the tube and doping of species (which can lower the bandgap) play a major role in bandgap modification of TiO_2 nanotube, several groups have reported improved rates of hydrogen production using TiO_2 nanotube photoelectrode in comparison to TiO_2 nanoparticulate system (Park et al. 2006, Mohapatra et al. 2007, Allam and Grimes 2007).

6.2 Generation of PV Electricity

TiO_2 nanocrystalline electrodes based photovoltaics have been widely studied for the generation of PV electricity (Ni et al. 2007, Varghese et al. 2009, Chen et al. 2009). The mesoporosity and nanocrystallinity of the semiconductor are important parameters not only because of the large amount of dye that can be absorbed by means of large surface area but they also allow the semiconductor small particles to deplete fully upon immersion in the electrolyte. Also, the proximity of the electrolyte to all nanoparticles enables screening of injected electrons and thus making their transport possible. In 1991, O'Regan and Gratzel reported a low-cost, high-efficiency solar cell based on dye-sensitized colloidal TiO_2 films. They demonstrated a Dye Sensitized Solar Cells, in which the mesoporous nanocrystalline TiO_2 film coated with monolayer of a charge transfer dye was used to enable the sensitized electrochemical photovoltaic device with a conversion efficiency of 7.1-7.9% in simulated solar light and 12% in diffuse daylight. Subsequent to this, several groups have reported the improvement in dye sensitized solar cells using different types of dyes and also by modifying the morphological structures of the nanostructured TiO_2. Using hybrid TiO_2 nanocrystalline electrode such as anatase-rutile TiO_2 nanocrystalline electrode (Kavan et al. 1996), nanocrytalline TiO_2 electrode with a buffer layer (Salim et al. 2011), core-shell structured nanocystalline TiO_2 electrodes (Karuppuchamy and Brundha. 2015) and TiO_2 nanocrystalline electrode coupled with photonic crystals also enhances the efficiency of the cell (Halaoui et al. 2005). A schematic diagram of the Dye Sensitized Solar Cell is illustrated in Fig. 10.12. By absorption of a photon energy (excitation), the dye changes from the electronic ground state (S^0) to the excited state (S^*). Subsequent to this, it injects an excited electron into the Conduction Band (CB) of the TiO_2 electrode and thus, becomes oxidized (S^+) dye. The original state of the dye is then restored by electron donation from the electrolyte. The recombination of the electron in the CB and the hole on the oxidized dye is much slower than the reduction from the S^+ state by the reduced state of the mediator that is present in the solution. As a result, charge separation becomes efficient and oxidized mediator is reduced at the counter electrode, the circuit being completed via electron migration through the external load. The voltage generated under photonic illumination corresponds to the difference between the Fermi level of the electron in the solid and the redox potential of the electrolyte. Finally the device generates electric power from light without suffering any permanent chemical transformation (Peter 2007).

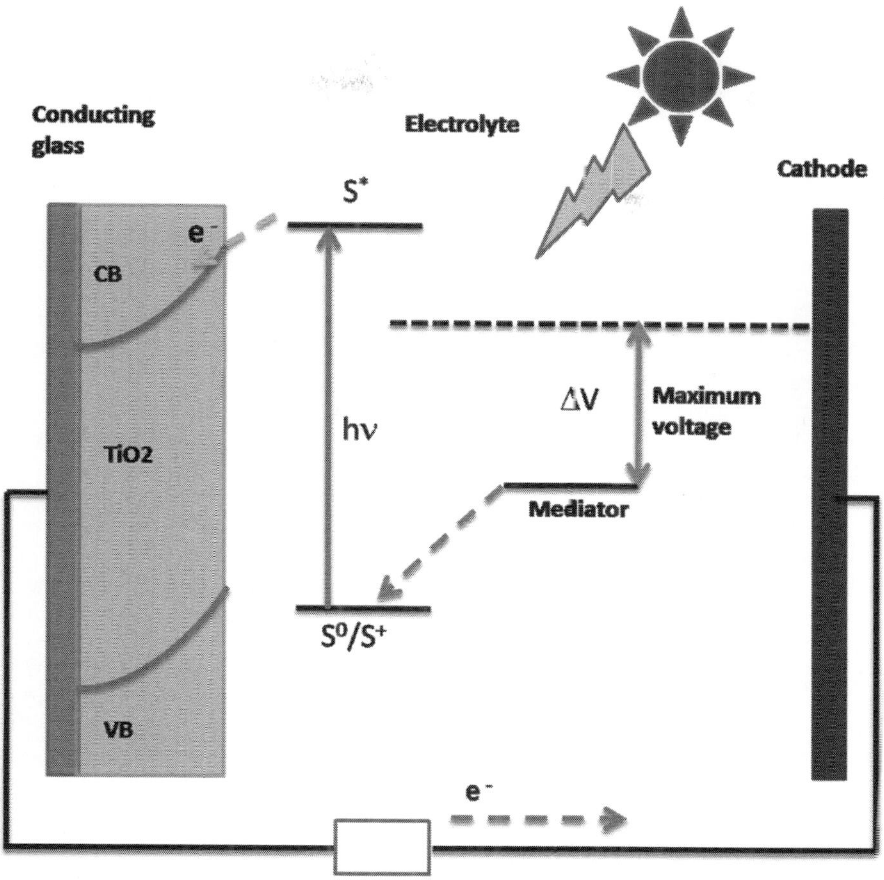

Fig. 10.12 Schematic diagram of TiO_2 based Dye Sensitized Solar Cells.

7. SUMMARY

Nanocrystalline TiO_2 material has been described in detailed with focus on their use as energy related applications particularly for production of hydrogen and PV electricity. Various physical and chemical routes synthesis mechanism have been described to realize reliable and cost effective methods, especially e-beam evaporation techniques by which a thin film of a rutile TiO_2 may be explored to deposit for various applications. The most convenient methods that are being used to characterize the nanocrystallime TiO_2 material have also been reviewed. The importance of nanocrystalline TiO_2 electrode is being widely used in the generation of hydrogen and PV electricity.

8. REFERENCES

Akl, A. A., H. Kamal and K. Abdel-Hady. 2006. Fabrication and characterization of sputtered titanium dioxide films. Appl. Surf. Sci. 252(24): 8651-8656.

Allam, N. K. and C. A. Grimes. 2007. Formation of vertically oriented TiO_2 nanotube arrays using a fluoride free HCl aqueous electrolyte. J. Phys. Chem. C 111(35): 13028-13032.

Asahi, R., T. Morikawa, T. Ohwaki, K. Aoki and Y. Taga. 2001. Visible-light photocatalysis in nitrogen-doped titanium dioxide. Science 293: 269-271.

Asahi, R., T. Morikawa, H. Irie and T. Ohwaki. 2014. Nitrogen-doped titanium dioxide as visible-light-sensitive photocatalyst: designs, developments and prospects. Chem. Rev. 114(19): 9824-9852.

Atashfaraz, M., M. Shariaty-Niassar, S. Ohara, K. Minami, M. Umetsu, T. Naku, et al. 2007. Effect of titanium dioxide solubility on the formation of $BaTiO_3$ nanoparticles in supercritical water. Fluid Phase Equilib. 257(2): 233-237.

Ayllon, J. A., A. Figueras, S. Garelik, L. Spirkova, J. Durand and L. Cot. 1999. Preparation of TiO_2 powder using titanium tetraisopropoxide decomposition in a plasma enhanced chemical vapor deposition (PECVD) reactor. J. Mater. Sci. Lett. 18(16):1319-1321.

Badawy, M. I., M. E. M. Ali, M. Y. Ghaly and M. A. El-Missiry. 2015. Mesoporous simonkolleite–TiO_2 nanostructured composite for simultaneous photocatalytic hydrogen production and dye decontamination. Process Saf. Environ. Prot. 94: 11-17.

Baia, L., A. Peter, V. Cosoveanu, E. Indrea, M. Baia, J. Popp, et al. 2006. Synthesis and nanostructural characterization of TiO_2 aerogels for photovoltaic devices. Thin Solid Films 511-512: 512-516.

Bettinelli, M., V. Dallacasa, D. Falcomer, P. Fornasiero, V. Gombac, T. Montini, et al. 2007. Photocatalytic activity of TiO_2 doped with boron and vanadium. J. Hazard. Mater. 146: 529-534.

Burdett, J. K. 1985. Electronic control of the geometry of rutile and related structures. Inorg. Chem. 24(14): 2244-2253.

Burdett, J. K., T. Hughbands, J. M. Gordon, J. W. Richardson Jr. and J. V. Smith. 1987. Structural-electronic relationships in inorganic solids: powder neutron diffraction studies of the rutile and anatase polymorphs of titanium dioxide at 15 and 295 K. J. Am. Chem. Soc. 109 (12): 3639-3646.

Carp, O., C. L. Huisman and A. Reller. 2004. Photoinduced reactivity of titanium dioxide. Prog. Solid State Chem. 32(1-2): 33-177.

Cha, G., K. Lee, J. E. Yoo, M. S. Killian and P. Schmuki. 2015. Topographical study of TiO_2 nanostructure surface for photocatalytic hydrogen production. Electrochimica Acta 179: 423-430.

Chen, D., Z. Jiang, J. Geng, Q. Wang and D. Yang. 2007. Carbon and nitrogen co-doped TiO_2 with enhanced visible-light photocatalytic activity. Ind. Eng. Chem. Res.: Mater. Interfaces 46(9): 2741-2746.

Chen, H., H. Jin and B. Dong. 2012. Preparation of magnetically supported chromium and sulfur co-doped TiO_2 and use for photocatalysis under visible light. Res. Chem. Intermed. 38(9): 2335-2342.

Chen, R.-F., C.-X. Zhang, J. Deng and G.-Q. Song. 2009. Preparation and photocatalytic activity of Cu^{2+}-doped TiO_2/SiO_2. Int. J. Miner. Metall. Mater. 16(2): 220-225.

Chen, X. 2009. Titanium dioxide nanomaterials and their energy applications. Chinese J. Catal. 30(8): 839-851.

Cho, I. S., Z. Chen, A. J. Forman, D. R. Kim, P. M. Rao, T. F. Jaramillo, et al. 2011. Branched TiO_2 nanorods for photoelectrochemical hydrogen production. Nano Lett. 11(11): 4978-4984.

Choi, Y., T. Umebayashi and M. Yoshikawa. 2004. Fabrication and characterization of C-doped anatase TiO_2 photocatalysts. J. Mater. Sci. 39(5): 1837-1839.

Colon, G., M. C. Hidalgo and J. A. Navio. 2002. A novel preparation of high surface area TiO_2 nanoparticles from alkoxide precursor and using active carbon as additive. Catal. Today 76(2-4): 91-101.

Cowan, A. J., J. Tang, W. Leng, J. R. Durrant and D. R. Klug. 2010. Water splitting by nanocrystalline TiO_2 in a complete photo electrochemical cells exhibits efficiencies limited by charge recombination. J. Phys. Chem. C 114(9): 4208-4214.

Cui, Y., H. Du and L. Wen. 2008. Doped-TiO_2 Photocatalysts and synthesis methods to prepare TiO_2 films. J. Mater. Sci. Technol. 24(5): 675-688.

De Tacconi, N. R., C. R. Chenthmarakshan, G. Yogeeswaran, A. Watcherenwong, R. S. de Zoysa, N. A. Basit, et al. 2006. Nanoporous TiO_2 and WO_3 films by anodization of titanium and tungsten substrates: influence of process variables on morphology and photoelectrochemical response. J. Phys. Chem. B 110(50): 25347-25355.

Deng, L., S. Wang, D. Liu, B. Zhu, W. Huang, S. Wu, et al. 2009. Synthesis, characterization of Fe-doped TiO_2 nanotubes with high photocatalytic activity. Catal. Lett. 129(3) 513-518.

Dong, F., S. Guo, H. Wang, X. Li and Z. Wu. 2011. Enhancement of the visible light photocatalytic activity of C-Doped TiO_2 nanomaterials prepared by a green synthetic approach. J. Phys. Chem. C 115(27): 13285-13292.

Dozzi, M. V., C. D'Andrea, B. Ohtani, G. Valentini and E. Selli. 2013. Fluorine-doped TiO_2 materials: photocatalytic activity vs time-resolved photoluminescence. J. Phys. Chem. C 117(48): 25586-25595.

Fahmi, A., C. Minot, B. Silvi and M. Causa. 1993. Theoretical analysis of the structures of titanium dioxide crystals. Phys. Rev. B 47(18): 11717-11724.

Fujishima, A. and K. Honda. 1972. Electrochemical photolysis of water at a semiconductor electrode. Nature 238: 37-38.

Gupta, S. K., J. Singh, K. Anbalagan, P. Kothari, R. R. Bhatia, P. K. Mishra, et al. 2013. Synthesis, phase to phase deposition and characterization of rutile nanocrystalline titanium dioxide (TiO_2) thin films. Appl. Surf. Sci. 264: 737-742.

Gupta, S. M. and M. Tripathi. 2011. A review of TiO_2 nanoparticles. Chinese Sci. Bull. 56(16): 1639-1657.

Guo, J., Z. Gan, Z. Lu, J. Liu, J. Xi, Y. Wan, et al. 2013. Improvement of the photocatalytic properties of TiO_2 by (Fe+Mo) co-doping—A possible way to retard the recombination process. J. Appl. Phys. 114: 104903(1)-104903(7).

Hahn, R., A. Ghicov, J. Salonen, V.-P. Lehto and P. Schmuki. 2007. Carbon doping of self-organized TiO_2 nanotube layers by thermal acetylene treatment. Nanotechnology 18(10): 105604.

Halaoui, L. I., N. M. Abrams and T. E. Mallouk. 2005. Increasing the conversion efficiency of dye-sensitized TiO_2 photoelectrochemical cells by coupling to photonic crystals. J. Phys. Chem. B 109(13): 6334-6342.

Haseeb, A. S. M. A., M. M. Hasan and H. H. Masjuki. 2010. Structural and mechanical properties of nanostructured TiO_2 thin films deposited by RF sputtering. Surf. Coat. Technol. 205(2): 338-344.

Heo, C. H., S.-B. Lee and J.-H. Boo. 2005. Deposition of TiO_2 thin films using RF magnetron sputtering method and study of their surface characteristics. Thin Solid Films 475(1-2): 183-188.

Hsieh, C.-T., W.-S. Fan, W.-Y. Chen and J.-Y. Lin. 2009. Adsorption and visible-light-derived photocatalytic kinetics of organic dye on co-doped titania nanotubes prepared by hydrothermal synthesis. Sep. Purif. Technol. 67(3): 312-318.

Ihara, T., M. Miyoshi, Y. Iriyama, O. Marsumato and S. Sugihara. 2003. Visible-light-active titanium oxide photocatalyst realized by an oxygen-deficient structure and by nitrogen doping. Appl. Catal. B 42(4): 403-409.

Irie, H., Y. Watanabe and K. Hashimoto. 2003. Nitrogen-concentration dependence on photocatalytic activity of $TiO_{2-x}N_x$ powders. J. Phys. Chem. B 107(23): 5483-5486.

Kaczmarek, D., J. Domaradzki, D. Wojcieszak, E. Prociow, M. Mazur, F. Placido, et al. 2012. Hardness of nanocrystalline TiO_2 thin films. J. Nano Res. 18-19: 195-200.

Karuppuchamy, S. and C. Brundha. 2015. Fabrication of core-shell structured TiO_2/MgO electrodes for dye-sensitized solar cells. Appl. Mech. Mater. 787: 3-7.

Kasuga, T., M. Hiramatsu, A. Hoson, T. Sekino and K. Niihara. 1998. Formation of titanium oxide nanotube. Langmuir. 14(12): 3160-3163.

Kavan, L., M. Gratzel, J. Rathousky and A. Zukalb. 1996. Nanocrystalline TiO_2 (Anatase) electrodes: surface morphology, adsorption and electrochemical properties. J. Electrochem. Soc. 143(2): 394-400.

Khan, S. U. M., M. Al-Shahry and W. B. Ingler. 2002. Efficient photochemical water splitting by a chemically modified n-TiO_2. Science 297: 2243-2245.

Kim, G.-S., Y.-S. Kim, H.-K. Seo and H.-S. Shin. 2006. Hydrothermal synthesis of titanate nanotubes followed by electrodeposition process. Korean J. Chem. Eng. 23(6): 1037-1045.

Kitano, M., K. Tsujimaru and M. Anpo. 2008. Hydrogen production using highly active titanium oxide-based photocatalysts. Top. Catal. 49: 4-17.

Langford, J. I. and A. J. C. Wilson. 1978. Scherrer after sixty years: a survey and some new results in the determination of crystallite size. J. Appl. Cryst. 11: 102-113.

Li, B., X. Wang, M. Yan and L. Li. 2003a. Preparation and characterization of nano-TiO_2 powder. Mater. Chem. Phys. 78(1): 184-188.

Li, Y., T. White and S. H. Lim. 2003b. Structure control and its influence on photoactivity and phase transformation of TiO_2 nano-particles. Rev. Adv. Mater. Sci. 5: 211-215.

Lin, S.-S., Y.-H. Hung and S.-C. Chen. 2009. Optical properties of TiO_2 thin films deposited on polycarbonate by ion beam assisted evaporation. Thin Solid Films 517(16): 4621-4625.

Linsebigler, A. L., G. Lu and J. T. Yates. 1995. Photocatalysis on TiO_2 surfaces: principles, mechanisms and selected results. Chem. Rev. 95(3): 735-758.

Liu, C., Q. Fu, J. B. Wang, W. K. Zhao, Y. L. Fang, T. Mihara, et al. 2005. Structural characterization of nanocrystalline TiO_2 and TiO_2: SiO_2 powders and thin films at 35°C. J. Korean Phys. Soc. 46: S104-S108.

Lu, Z., X. Jiang, B. Zhou, X. Wu and L. Lu. 2011. Study of effect annealing temperature on the structure, morphology and photocatalytic activity of Si doped TiO_2 thin films deposited by electron beam evaporation. Appl. Surf. Sci. 257(24): 10715-10720.

Magne, C., S. Cassaignon, G. Lancel and T. Pauporte. 2011. Brookite TiO_2 nanoparticle films for dye-sensitized solar cells. Chem. Phys. Chem. 12: 2461-2467.

Maira, A. J., K. L. Yeung, J. Soria, J. M. Coronado, C. Belver, C. Y. Lee, et al. 2001. Gas-phase photo-oxidation of toluene using nanometer-size TiO_2 catalysts. Appl. Catal. B 29(4): 327-336.

Mittal, V. 2015. Surface Modification of Nanoparticle and Natural Fiber Fillers. Wiley-VCH Weinheim, Germany.

Mohapatra, S. K., M. Misra, V. K. Mahajan and K. S. Raja. 2007. Design of a highly efficient photoelectrolytic cell for hydrogen generation by water splitting: application of $TiO_{2-x}C_x$ nanotubes as a photoanode and Pt/TiO_2 nanotubes as a cathode. J. Phys. Chem. C 111(24): 8677-8685.

Namin, H. E., H. Hashemipour and M. Ranjbar. 2008. Effect of aging and calcination on morphology and properties of synthesized nanocrystalline TiO_2. Inter. J. Mod. Phys. B 22: 3210-3215.

Natarajan, C. and G. Nogami. 2006. Cathodic electrodeposition of nanocrystalline titanium dioxide thin films. J. Electrochem. Soc. 143(5): 1547-1550.

Ni, M., M. K. H. Leung, D. Y. C. Leung and K. Sumathy. 2007. A review and recent developments in photocatalytic water-splitting using TiO_2 for hydrogen production. Renew. Sustain. Energy Rev. 11(3): 410-425.

Nowotny, J. and L. R. Sheppard. 2007. Solar-hydrogen. Int. J. Hydrogen Energy 32(14): 2607-2608.

O'Regan, B. and M. Gratzel. 1991. A low-cost, high-efficiency solar cell based on dye-sensitized colloidal TiO_2 films. Nature 353: 737-740.

Park, J. H., S. Kim and A. J. Bard. 2006. Novel carbon-doped TiO_2 nanotube arrays with high aspect ratios for efficient solar water splitting. Nano Lett. 6(1): 24-28.

Pawar, S. G., S. L. Patil, M. A. Chougule, D. M. Jundale and V. B. Patil. 2011. Synthesis and characterization of nanocrystalline TiO_2 thin films. J. Mater. Sci.: Mater. Electron. 22: 260-264.

Periyat, P., S. C. Pillai, D. E. McCormack, J. Colreavy and S. J. Hinder. 2008. Improved high-temperature stability and sun-light-driven photocatalytic activity of sulfur-doped anatase TiO_2. J. Phys. Chem. C 112(20): 7644-7652.

Peter, L. M. 2007. Dye synthesized nanocrystalline solar cells. Phys. Chem. Chem. Phys. 9(21): 2630-2642.

Rahimi, R., S. S. Moghaddam and M. Rabbani. 2012. Comparison of photocatalysis degradation of 4-nitrophenol using N, S co-doped TiO_2 nanoparticles synthesized by two different routes. J. Sol-Gel Sci. Technol. 64(1): 17-26.

Reddy, K. M., B. Baruwati, M. Jayalakshmi, M. M. Rao and S. V. Manorama. 2005. S-, N- and C-doped titanium dioxide nanoparticles: synthesis, characterization and redox charge transfer study. J. Solid State Chem. 178(11): 3352-3358.

Rossella, F., P. Galinetto, M. C. Mazzati, L. Malavasi, Y. D. Fernandez, G. Drera, et al. 2010. TiO_2 thin films for spintronics application: a Raman study. J. Raman Spectrosc. 41(5): 558-565.

Sakthivel, S. and H. Kisch. 2003. Photocatalytic and photoelectrochemical properties of nitrogen-doped titanium dioxide. Chem. Phys. Chem. 4(5): 487-490.

Salim, T., Z. Yin, S. Sun, X. Huang, H. Zhang and Y. M. Lam. 2011. Solution-processed nanocrystalline TiO_2 buffer layer used for improving the performance of organic photovoltaics. ACS Appl. Mater. Interfaces 3(4): 1063-1067.

Seifried, S., M. Winterer and H. Hahn. 2000. Nanocrystalline titania films and particles by chemical vapor synthesis. Chem. Vap. Deposition 6(5): 239-244.

Seo, D. S., J. M. Lee and H. Kim. 2001. Preparation of nanotube-shaped TiO_2 powder. J. Cryst. Growth 229(1-4): 428-432.

Shang, X., M. Zhang, X. Wang and Y. Yang. 2014. Sulphur, nitrogen-doped TiO_2/graphene oxide composites as a high performance photocatalyst. J. Exp. Nanoscience 9(7): 749-761.

Shchipunova, Y. and I. Postnova. 2009. One-pot biomimetic synthesis of monolithic titania through mineralization of polysaccharide. Coll. Surf. B: Biointer. 74(1):172-177.

Suhail, M. H., G. M. Rao and S. Mohan. 1992. dc reactive magnetron sputtering of titanium-structural and optical characterization of TiO_2 films. J. Appl. Phys. 71: 1421-1427.

Sun, H., Y. Bai, Y. Cheng, W. Jin and N. Xu. 2006. Preparation and characterization of visible light-driven carbon–sulfur co-doped TiO_2 photocatalysts. Ind. Eng. Chem. Res. 45(14): 4971-4976.

Sun, H., C. Wang, S. Pang, X. Li, Y. Tao, H. Tang, et al. 2008. Photocatalytic TiO_2 films prepared by chemical vapor deposition at atmosphere pressure. J. Non-Cryst. Solids 354(12-13): 1440-1443.

Sun, X. and Y. Li. 2003. Synthesis and characterization of ion-exchangeable titanate nanotubes. Chem. Eur. J. 9: 2229-2238.

Sung, Y. M. and H. J. Kim. 2007. Sputter deposition and surface treatment of TiO_2 films for dye-sensitized solar cells using reactive RF plasma. Thin Solid Films 515(12): 4996-4999.

Sunm, Y., K. Yan, G. Wang, W. Guo and T. Ma. 2011. Effect of annealing temperature on the hydrogen production of TiO_2 nanotube arrays in a two-compartment photoelectrochemical cell. J. Phys. Chem. C 115(26): 12844-12849.

Tay, Q., X. Liu, Y. Tang, Z. Jiang, T. C. Sum and Z. Chen. 2013. Enhanced photocatalytic hydrogen production with synergistic two-phase anatase/brookite TiO_2 nanostructures. J. Phys. Chem. C 117(29): 14973-14982.

Therese, G. H. A. and P. V. Kamath. 2000. Electrochemical synthesis of metal oxides and hydroxides. Chem. Mater. 12(5): 1195-1204.

Uekawa, N., J. Kajiwara, K. Kakegawa and Y. Sasaki. 2002. Low temperature synthesis and characterization of porous anatase TiO_2 nanoparticles. J. Colloid Interface Sci. 250(2): 285-290.

Varghese, O. K., M. Paulose, T. J. LaTempa and C. A. Grimes. 2009. High-rate solar photocatalytic conversion of CO_2 and water vapor to hydrocarbon fuels. Nano Lett. 9(2): 731-737.

Verma, A., A. Basu, A. K. Bakhshi and S. A. Agnihotry. 2005. Structural, optical and electrochemical properties of sol-gel derived TiO_2 films: annealing effects. Solid State Ionics 176(29-30): 2285-2295.

Wang, M., G. Song, J. Li, L. Miao and B. Zhang. 2008. Direct hydrothermal synthesis and magnetic property of titanate nanotubes doped magnetic metal ions. J. Univ. Sci. Technol. B 15(5): 644-648.

Wilke, K. and H. D. Breuer. 1999. The influence of transition metal doping on the physical and photocatalytic properties of titania. J. Photochem. Photobiol. A 121(1): 49-53.

Wu, J. M. and B. Qi. 2007. Low-temperature growth of a nitrogen-doped titania nanoflower film and its ability to assist photodegradation of Rhodamine B in water. J. Phys. Chem. C 111(2): 666-673.

Xu, A. W., Y. Gao and H. Q. Liu. 2002. The preparation, characterization and their photocatalytic activities of rare earth-doped TiO_2 nanoparticles. J. Catal. 207(2): 151-157.

Yang, C., H. Fan, Y. Xi, J. Chen and Z. Li. 2008. Effects of depositing temperatures on structure and optical properties of TiO_2 film deposited by ion beam assisted electron beam evaporation. Appl. Surf. Sci. 254(9): 2685-2689.

Yang, W. and C. A. Wolden. 2006. Plasma-enhanced chemical vapor deposition of TiO_2 thin films for dielectric applications. Thin Solid Films 515(4): 1708-1713.

Yao, J., J. Shao, H. He and Z. Fan. 2007. Optical and electrical properties of TiO_x thin films deposited by electron beam evaporation. Vacuum 81(9): 1023-1028.

Yao, N., S. Cao and K. L. Yeun. 2009. Mesoporous TiO_2-SiO_2 aerogels with hierarchal pore structures. Microporous Mesoporous Mater. 117(3): 570-579.

Ye, M., J. Gong, Y. Lai, C. Lin and Z. Lin. 2012. High-efficiency photoelectrocatalytic hydrogen generation enabled by palladium quantum dots-sensitized TiO_2 nanotube arrays. J. Am. Chem. Soc. 134(38): 15720-15723.

You, X., F. Chen, J. Zhang and M. Anpo. 2005. A novel deposition precipitation method for preparation of Ag-loaded titanium dioxide. Catal. Lett. 102(3): 247-250.

Youngblood, W. J., A. S.-H. Lee, K. Maeda and T. E. Mallouk. 2009. Visible light water splitting using dye-sensitized oxide semiconductors. Acc. Chem. Res. 42(12): 1966-1973.

Zaleska, A. 2008. Doped-TiO_2: a review. Recent Patents on Eng. 2(3):157-164.

Zhang, J., P. Zhou, J. Liu and J. Yu. 2014. New understanding of the difference of photocatalytic activity among anatase, rutile and brookite TiO_2 Phys. Chem. Chem. Phys. 16: 20382-20386.

Zhang, K. J., W. Xu, X. J. Li, S. J. Zheng, G. Xu and J. H. Wang. 2006. Photocatalytic oxidation activity of titanium dioxide film enhanced by Mn non-uniform doping. Trans. Nonferrous Met. Soc. China 16(5): 1069-1075.

11

Design of ZnO Nano-Architectures and Its Applications

Wai Kian Tan,[a] Go Kawamura[b] and Atsunori Matsuda*

ABSTRACT

It is well-known that the exceptional properties exhibited by Zinc Oxide (ZnO) are inter-correlated with its morphological states. Therefore, this has prompted intense focus on the design of ZnO nano-architectures and its formation mechanisms especially by low-temperature and environmentally conservative methods. Many literatures have reported on the formation and application of ID ZnO nanorods and nanowires. Recently, a lot of research interest in the controlled design of 2D ZnO nanostructures has been shown by researchers. In this chapter, after the introduction of ZnO general properties, discussion of the formation mechanisms of 2D ZnO nanostructures, its roles and properties as well as its applications are described. By using low-temperature synthesis techniques, multi-functional architecturally designed ZnO nanostructures could be used for fabrication of flexible device applications such as solar cells, photocatalysts and optoelectronics devices.

Keywords: Zinc oxide, nanostructures, nanosheets, photocatalyst, dye-sensitized solar cells.

1. INTRODUCTION

Nanostructured functional materials have been the main focus of many researchers in the design and fabrication of nano-systems for nanotechnology, which have been widely investigated for various applications (Ariga et al. 2012b). Nanoscopic sciences

Department of Electrical and Electronic Information Engineering, Toyohashi University of Technology, 1-1, Hibarigaoka, Tempaku-cho, Toyohashi, Aichi 441-8580 Japan.

[a] E-mail: tanwaikian@cie.ignite.tut.ac.jp

[b] E-mail: gokawamura@ee.tut.ac.jp

* Corresponding author: matsuda@ee.tut.ac.jp

at atomic level have established a new concept or paradigm for construction of functional materials from nanoscale building units with distinguished and enhanced properties (Ariga et al. 2012a). 'Materials nanoarchitonics' is deemed as the latest technique used in the controlled synthesis of various nanostructures of materials such as nanotubes, nanoparticles, nanosheets and nanoporous into hierarchically organized structures to achieve the desired properties of a nano-system (Ariga et al. 2012a, Ariga et al. 2012b). There are researchers calling these multifunctional materials as smart materials.

Zinc oxide (ZnO) is notable as one of the most used material in the studies of semiconductor and electrochemical phenomena. It has created a lot of interest amongst researchers for its processing methods, characterizations and applications due to its unique inter-correlation between morphologies and properties (Fan and Jia 2011, Chen et al. 2006). ZnO which is a wide bandgap semiconductor with large exciton binding energy of 60 meV at room temperature, can be utilized for many applications such as bio-chemical sensors (Al-Hardan et al. 2010, Fan and Jia 2011, Kim et al. 2011, Lokhande et al. 2009), photocatalyst (Tan et al. 2014a, Li et al. 2012, Umar et al. 2011, Liu et al. 2016), light emitting devices (Djurisic and Leung 2006, Djurišić et al. 2010) and Dye Sensitized Solar Cells (DSSCs) (Djurišić et al. 2010, Heo et al. 2004, Lupan et al. 2010, Zhang et al. 2009). ZnO is also a better light emitting material compared to the widely used Gallium Nitride (GaN) due to its easy fabrication of high quality single crystals as opposed to GaN (Djurišić et al. 2010). The unique electronic and photo-electronic properties of ZnO, shown in Table 11.1, are one of the many reasons this material is largely utilized in many applications (Zhang 1996). Besides this, ZnO is also environmentally friendly.

Table 11.1 General properties of ZnO.

Molecular weight	**Zn: 65.38, O: 16.00, ZnO 81.38**
Lattice	Hexagonal wurtzite
Lattice constants	a = 0.324 nm, c = 0.519 nm, c/a = 1.60
Density	5.78 g/cm^3 or 4.21 × 10^{22} ZnO molecules/cm^3
Dielectric constant	8.54
Refractive index	2.008
Energy bandgap	3.2~3.37 eV (Direct band-gap)
Enthalpy of formation	$Zn(s) + 1/2O_2(g) \rightarrow ZnO(s) = -83.17$ kcal/mol
Solubility of H_2O	1.6×10^{-6} g per gram of H_2O at 25°C

As the unique properties ZnO are often inter-correlated with its morphologies, many researchers have focused on its nano-architecture design such as well oriented ZnO crystals in 1D, 2D nanosheets and hierarchically structured 3D ZnO nanostructures in order to achieve the desired properties required for specific applications. For example, 2-D and 3-D nanostructures have been explored for bio/chemical sensors as well as light sensing materials due to their higher surface area which could enhance detection and sensing ability (Matsuda et al. 2008). A systematic overview of ZnO would be useful for better understanding this multifunctional oxide material. Therefore, in this chapter, a brief introduction of ZnO and its properties will be discussed prior to further mentioning its development of 2D structure and its vast applications. Figure 11.1 shows some of the interesting SEM images of the vast range of ZnO nanostructures.

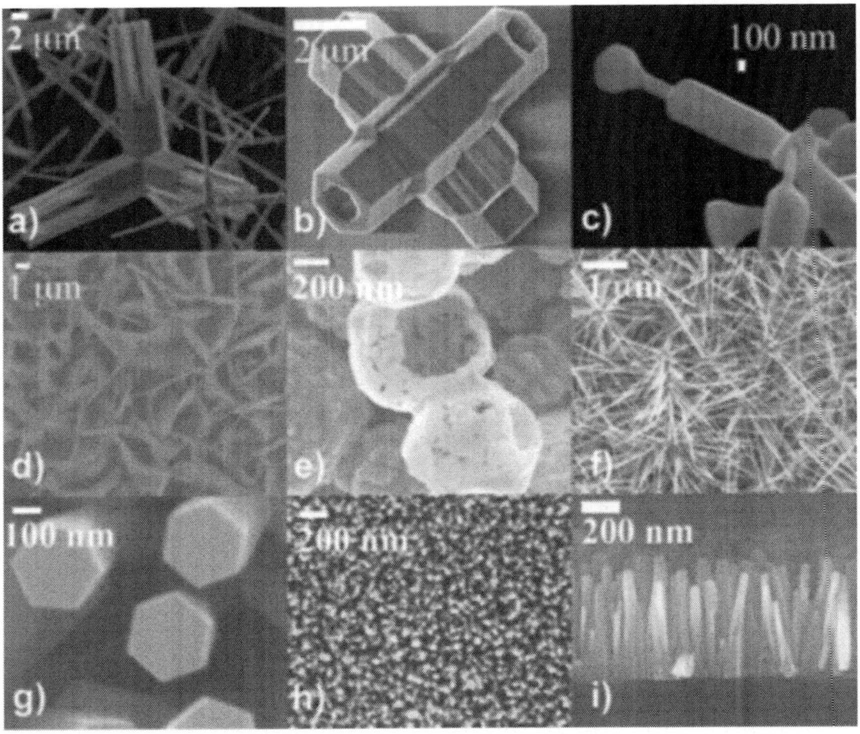

Fig. 11.1 Various nano-sized ZnO morphologies obtained using different methods: (a), (b) tetrapod structures, (c) dumb-dell like structure, (d) 2D nanosheets, (e) nanoshells, (f) multipods, (g), (h), and (i) nanorods. Reproduced with permission (Djurišić et al. 2010).

1.1 ZnO Structures

ZnO has well-defined crystal structures which are commonly found in wurtzite, rocksalt or cubic (zinc blende) structure. The ZnO wurtzite structure has the highest thermodynamic stability among the three structures and therefore the most common structure of ZnO. A rocksalt structure of ZnO can be yielded under high pressure, thus ZnO in this structure is quite rare. The hexagonal wurtzite crystal structure has two lattice parameters, *a* and *c* with values of 0.3296 nm and 0.52065 nm respectively, at ambient pressure and temperature (Pearton 2006). This ZnO hexagonal wurtzite structure belongs to the $P6_3$ mc space group and exhibits a non-centrosymmetric structure which causes ZnO to be piezoelectric and pyroelectric.

The wurtzite ZnO consists of atoms forming hexagonal-close-pack sub-lattices which stack alternatively along the *c*-axis. Each Zn^{2+} sub-lattice contains four Zn^{2+} ions and are surrounded by four O^{2-} ions and vice versa, coordinated at the edges of a tetrahedron (Sugunan et al. 2006). This tetrahedral coordination will form a polar symmetry along the hexagonal axis which induces the effect of piezoelectricity and spontaneous polarization in the ZnO wurtzite crystal. The polarization effect is one of the major factors influencing the crystal growth during the synthesis of ZnO nanostructures especially in the formation of 1D nanostructures, such as ZnO nanowires and nanorods. As Zn^{2+} and O^{2-} sizes differ, these ions only fill approximately 44% of the volume in ZnO crystal and leave a relatively large space (Zhang 1996). The typical ZnO wurtzite structure (inset) and the XRD pattern is shown in Fig. 11.2.

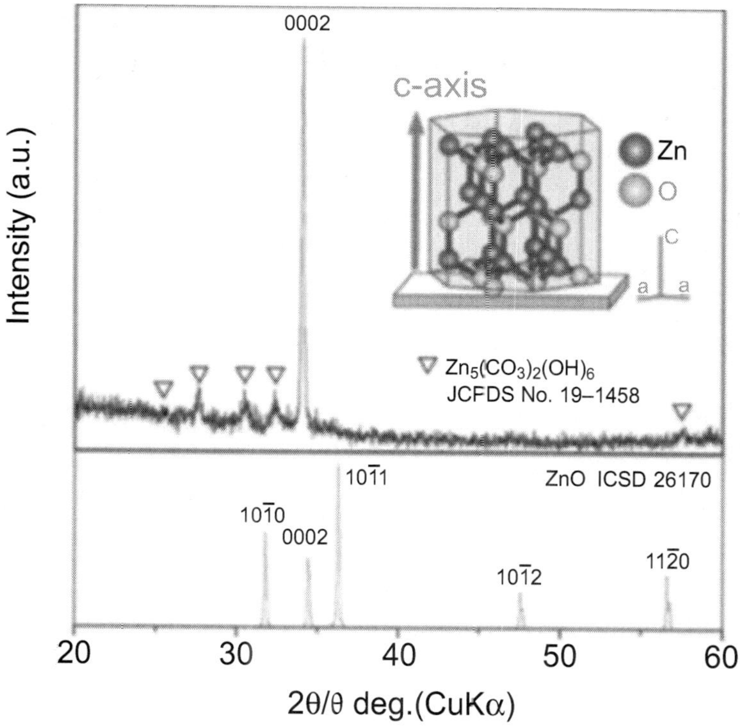

Fig. 11.2 ZnO wurtzite crystal structure (inset) and the typical XRD pattern of *c*-axis oriented ZnO nanostructures. Reproduced with permission (Masuda and Kato 2008) Copyright (2008) American Chemical Society.

1.2 Optical Properties of ZnO

The natural color of ZnO is white and tends to change with heat-treatment and doping or incorporation of impurities into the crystals (Zhang 1996). ZnO changes from reflector to absorber at UV and near UV regions. The alteration of its crystal structure by heat treatment or doping could enable the application of light in various colors. Generally, optical properties of ZnO can be characterized using photoluminescence (PL) and Raman spectrometry. Typical PL spectra exhibited by ZnO nanostructures consist of two regions which are the ultraviolet (UV) and visible region. UV emission observed is also termed as deep-level emission which is attributed to the recombination of excitons (electron-hole pair recombination or band-to-band recombination). Highly crystalline ZnO would exhibit strong UV emission. The origin of the green band in the visible region in ZnO is attributed to various impurities and defects (Pearton 2006). The visible emission could be related to the recombination of electrons with oxygen vacancies or with photo-excited holes in the valence band; hence higher defect concentrations may lead to higher emission intensity in this region (Kurbanov et al. 2011). The ability for ZnO nanostructures to exhibit PL emissions at low or room-temperature PL as well as the feasibility of altering the optical properties by doping especially with rare-earth elements have attracted tremendous interest for researchers to further explore this category with the vision of using it in optoelectronic devices (Li et al. 2008, Tan et al. 2013b). Detail reading is available from the works of Djurisic et al. (2010) and Djurisic and Leung (2006) who have reported on a comprehensive studies and reviews of the ZnO nanostructures' optical properties.

Lattice dynamics of a single crystal wurtzite ZnO can be detected by using a Raman spectroscopy. In a perfect wurtzite ZnO crystal, four atoms per unit cell will correspond to 12 phonon modes. The modes are one Longitudinal-Acoustic (LA), two Transverse-Acoustic (TA), three Longitudinal-Optical (LO) and six Transverse-Optical (TO) branches. Table 11.2 shows the value of the general Raman modes obtained from single crystal wurtzite ZnO at room temperature (Ashrafi and Jagadish 2007, Pearton 2006). The A_1 and E_1 branches are Raman and infra-red active, while the two E_2 branches (non-polar) are only Raman active. The E_2^{low} mode is associated with the vibrations of the Zn sub-lattice whilst the E_2^{high} mode is associated with the oxygen atoms only. The B_1 branches are always inactive. Both A_1 and E_1 modes are polar and split into TO and LO phonons as shown in Table 11.2. Non-polar modes with symmetry E_2 have two frequencies: High (H) and Low (L) which is associated with oxygen atoms and with Zn sub-lattices respectively.

For ZnO single crystal materials, among the eight sets of optical modes, A_1, E_1 and E_2 are Raman active with the corresponding band shift as listed in Table 11.2. It is typical to have a band at around 570 cm^{-1} i.e. the LO phonon mode which is often associated with the presence of surface defects.

Table 11.2 Phonon modes of wurtize ZnO at room temperature (Zhang et al. 2009, Pearton 2006).

Phonon mode	Value (cm^{-1}): single crystal
E_2^{low}	101
E_2^{high}	437
TO(A_1)	380
LO(A_1)	574
TO(E_1)	591

1.3 Electronic Properties of ZnO

The electrical properties of ZnO nanostructures exhibit different properties than in bulk form. ZnO is an intrinsic n-type semiconductor as electrons are excited from ionized zinc interstitials existing in the crystal lattice. As ZnO has a wide direct bandgap of 3.3 eV, this enables the design and control of its semiconducting properties by inducing the fermi level within the wide bandgap intrinsically or extrinsically. At room temperature, intrinsic generation of fermi level generally lies midway between the conduction band and valence band of the bandgap. As ZnO has low formation energies, large amounts of point defects and native donors contribute to its intrinsic fermi level (Djurišić et al. 2010). In extrinsic doping of n-type or p-type materials in ZnO, the fermi level could be altered to be nearer to the conduction or valence band respectively. Table 11.3 below shows the general electrical properties of ZnO.

Table 11.3 Electrical properties of single crystal wurtzite crystal (Pearton 2005, Pearton 2006).

Properties	Values
Effective electron mass (M*)	0.24-0.30 m_e
Effective hole mass (m_h*)	0.45-0.60 m_e
Electron Hall mobility at 300K for n-type (μ_e)	200 cm^2 V^{-1} s^{-1}
Electron Hall mobility at 300K for p-type	5-50 cm^2 V^{-1} s^{-1}
Intrinsic carrier concentration (n)	<10^6 cm^{-3}
Background carrier doping	n-type: ~10^{20} electron cm^{-3} p-type: 10^{19} holes cm^{-3}
Optical transmission, T(1/α)	80%-95%

ZnO has conductivity of approximately 10^{-17} to 10^3 Ω^{-1} cm^{-1} which is also dependent on its sample preparations. ZnO crystals have low conductivity and its semiconducting properties are largely dependent on the defects present in the oxide lattice. Two distinguished defects are interstitial and substitutional Zn atoms. In interstitial, the partial reduction by reactive agents causes the Zn atoms to enter the void space by becoming interstitial atoms in the form of Zn, Zn^+ or Zn^{2+}. Meanwhile, in substitutional, with the presence of other metals or salts, the Zn atoms in ZnO crystal lattice are replaced and these Zn atoms will diffuse to the surface and vaporize. The conductivity of ZnO can be either increased or decreased depending on the substitution metallic atoms induced.

2. SYNTHESIS AND DESIGN OF 2D ZnO NANOSTRUCTURES, ITS PROPERTIES AND APPLICATIONS

One of the advantages of ZnO material is the vast synthesis techniques ranging from both low and high temperature methods such as sol-gel (Jongnavakit et al. 2012, Singh et al. 2012), hydrothermal (Chen and Wu 2011, Cheng et al. 2009, Dutta et al. 2012, Guo et al. 2005a, Guo et al. 2005b, Ismail et al. 2005, Jiang et al. 2012, Kiomarsipour and Shoja Razavi 2013, Kiomarsipour and Shoja Razavi 2012, Ko et al. 2011, Li et al. 2012, Pei et al. 2010, Shi et al. 2013, Tan et al. 2014a, Tan et al.

2013c, Tan et al. 2011a, Tao et al. 2010, Wang et al. 2009, Tong et al. 2006, Yu et al. 2011), electrodeposition (Guo et al. 2008, Hou et al. 2012, Mahalingam et al. 2005, Qiu et al. 2011, Wang et al. 2010, Guerin et al. 2010, Guérin et al. 2012), thermal oxidation (Hong et al. 2012, Kim et al. 2006, Li et al. 2005, Martínez et al. 2011, Rusu et al. 2007, Tan et al. 2014b, Xu et al. 2011, Zou et al. 2007), sputtering (Zhang et al. 2010), molecular beam epitaxy and etc. Physical methods which generally require higher processing temperature are more precise with good reproducibility but are less likely to be able to be integrated in the ever increasing demand of flexible devices possessing low thermal stability. As for solution-based methods, despite the persistent reproducibility issue, they are more attractive due to the lower cost, higher yield, compatible with flexible organic substrates and needless of metal catalysts which enable integration with well-developed silicon technologies (Xu and Wang 2011).

In this chapter, the methods that focus on formation of 2D ZnO nanostructures by thermal oxidation, hydrothermal and hot-water treatment will be selectively discussed. The properties of these 2D nanostructures as well as potential applications are also described.

2.1 Thermal Oxidation of Metallic Zn

Thermal oxidation of metallic Zn is deemed as one of the facile methods to form ZnO nanostructures because this process is simple, inexpensive and no catalyst is required. During thermal oxidation, the oxide formation involves transportation, diffusion and absorption of oxygen to the surface of metallic zinc. From the study on the rapid oxidation of Zn foils conducted by Tan et al. 2011b, the evolution of ZnO nanostructures as a function temperature from 100 to 500°C for 30 minutes to 1 hour in air was reported as shown in Fig. 11.3 (Tan et al. 2011b). It is important to note that the melting point of metallic Zn is 419.5°C and the melting point of ZnO is at 1975°C, therefore thermal oxidation of metallic Zn for a long period could lead to the deformation of Zn foils or the coated metallic Zn layer.

Prior to the oxidation process, the Zn foils were ground and etched in order to remove the native layer of oxide and also to reveal the grain boundaries on the metallic Zn as shown in the schematic diagram in Fig. 11.3(a). During the initial stages of growth, preferential growth at the grain boundaries occurred and its continuous growth generated ZnO nanoparticles through the surface of the Zn foils. At a later stage, ZnO nanostructures formed evolved from nanoparticles on the metallic surface to form 2D ZnO porous and solid nanosheets at 350°C before forming random-orientated ZnO nanorods at 400°C and above (Tan et al. 2011b). In the work reported by Hong et al. (2012), rapid oxidation of sputtered Zn layer under ambient condition occurred via a different route compared to conventional oxidation of metallic Zn films which enable the fabrication of amorphous transparent oxide thin films from low-melting-point metals. This phenomenon could promote the aerosphere near the surface of the Zn foils and the re-deposition may lead to the generation of the nanostructures. The schematic diagram in Fig. 11.4 shows different mechanisms between rapid oxidation and conventional thermal oxidation. Rapid oxidation could alter the diffusion route of Zn onto the surface to form ZnO nanostructures.

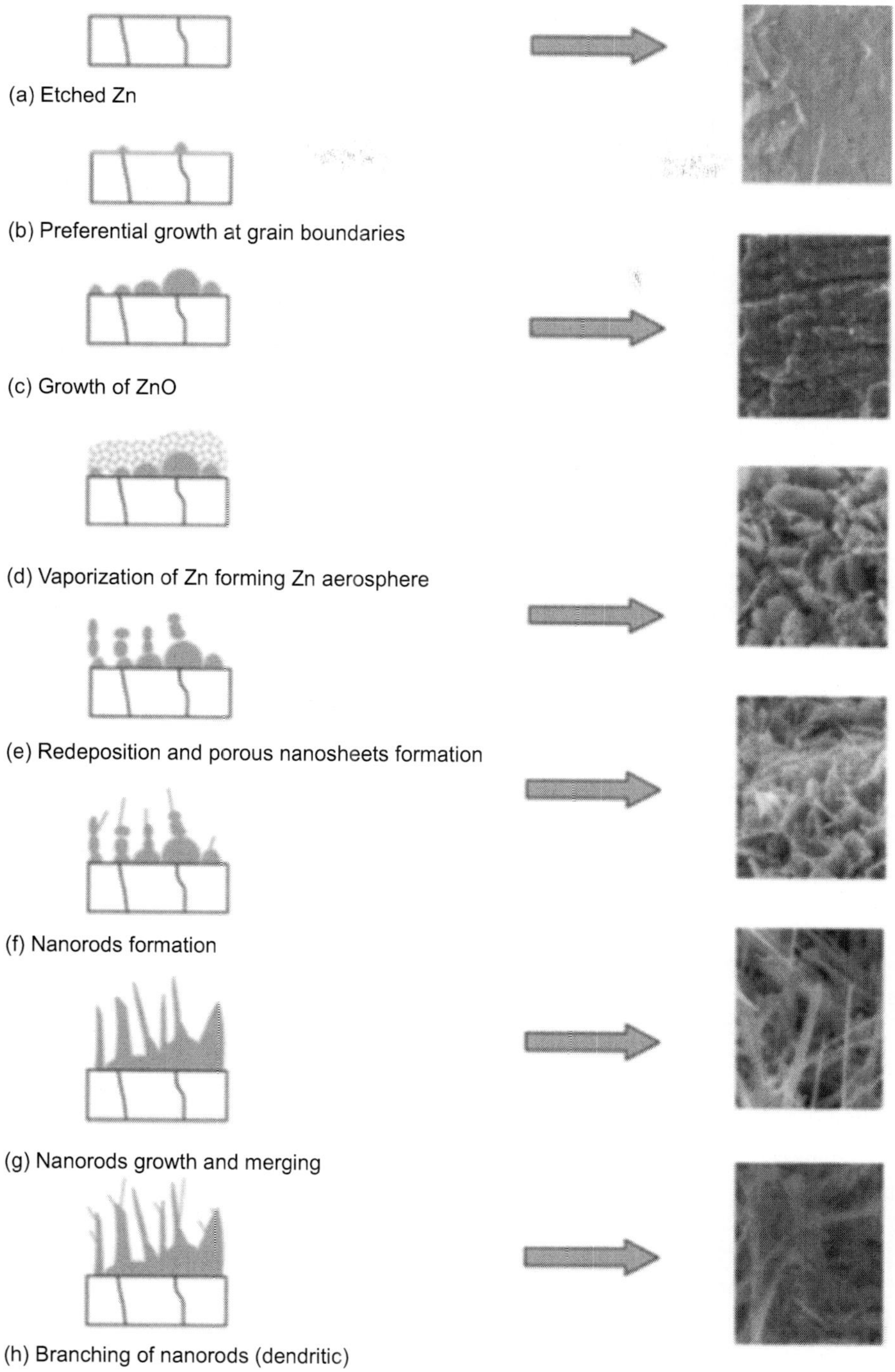

Fig. 11.3 Schematic illustrations and the corresponding SEM images of the ZnO nanorods and nanosheets obtained after thermal oxidation of metallic Zn foils in air. Reproduced with permission (Tan et al. 2011b).

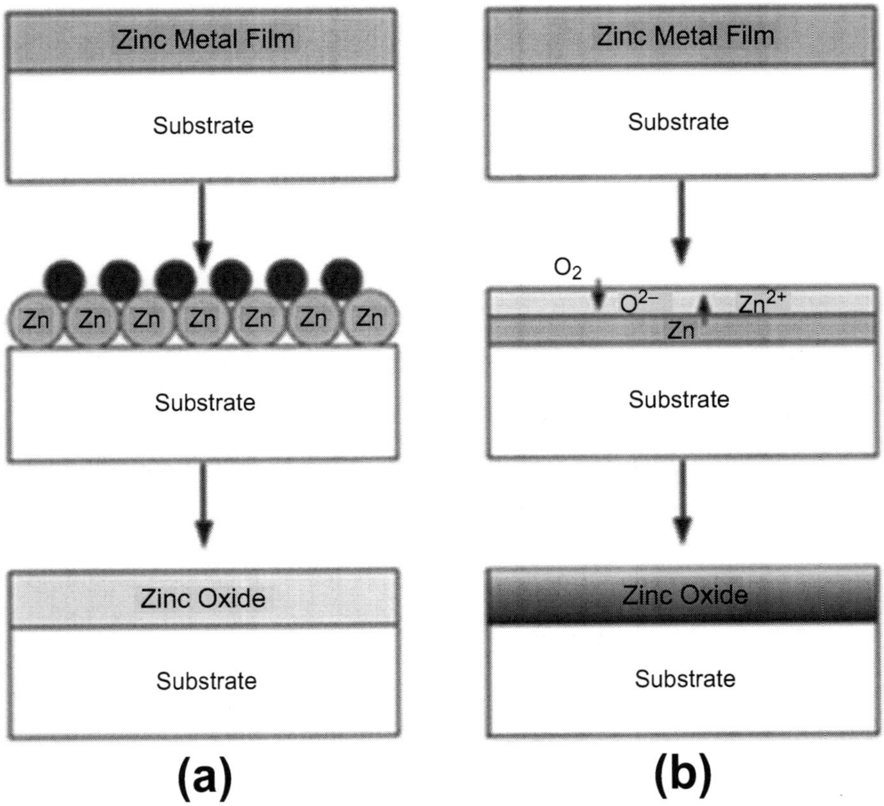

Fig. 11.4 Schematic diagram of mechanisms of metallic Zn films oxidation by (a) Rapid thermal oxidation process. (b) Conventional thermal oxidation process. Reproduced with permission from (Hong et al. 2012).

In the study by Tan et al. (2014b), oxidation of etched metallic Zn foils with the influx of excess oxygen during rapid thermal oxidation led to the formation of crystalline 2D ZnO nanosheets at 300°C and 400°C at a short oxidation time of 10 to 30 minutes. In controlled conditions such as controlled partial pressure of oxygen, nanocrystal ZnO thin films can be formed on metallic zinc substrate (Tan et al. 2014b, Noothongkaew et al. 2012). Reports on the variation in oxygen partial pressure during evaporation of Zn resulted in the formation of Zn microcrystal and nanowires (Fan and Jia 2011, Yuvaraj et al. 2009). By adjusting the partial pressure of oxygen, ZnO nanostructures changed from 2D microdisks to 1D nanowires as reported by Yuvraj et al. (2009). Further thermal annealing of the ZnO nanowires resulted in the formation of nano-canal structures as the Zn core sublimed due to the low melting point of Zn (419.5°C). The summary of their findings are shown in Fig. 11.5.

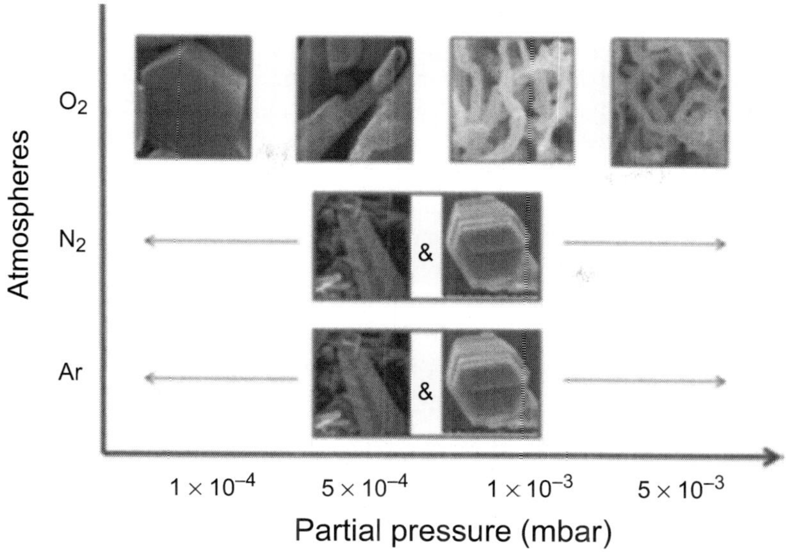

Fig. 11.5 The ZnO morphologies obtained at difference partial pressure of O_2, N_2 and Ar during thermal oxidation of metallic Zn evaporated on Si. Reproduced with permission (Yuvaraj et al. 2009).

Changes in morphologies and optical properties of the ZnO nanosheets are observed against oxidation temperature and time. The thickness and surface roughness of the nanosheets increased with temperature and oxidation time as observed in the higher magnification SEM images reported in the work of Tan et al. (2014b). Despite the short oxidation time of 10 to 30 minutes, these 2D ZnO nanosheets exhibited typical ZnO properties as the PL results obtained showed UV emission at approximately 370 nm corresponding from the free exciton emission while visible emission peaked from deep level emissions at the green and red regions that were attributed to defects in the crystal structure such as O-vacancy, Zn-vacancy, O-interstitial, Zn-interstitial and substitution were also observed (Fan and Jia 2011, Tian et al. 2003). The sharp and strong UV emissions with suppressed green emission indicated better crystalline quality with less structural defects. Further Raman analysis showed band at 439 cm^{-1} corresponds to the E_2 non-polar phonon of the hexagonal wurtzite phase was observed for the samples oxidized at 400°C. On confirmation of the photocatalytic activity, all the 2D ZnO exhibited degradation of methyl orange under UV irradiation.

The mechanism of formation is described as follows, prior to oxidation, similar to the work reported by Lockman et al. (2011), the Zn foils were etched to remove the native oxides on the surface and to reveal the grain boundaries. Images of the Zn foils before and after etching are available in their reported work (Tan et al. 2014b).

The grain boundaries on the surface of the Zn foils would provide an easy diffusion path for Zn ions to diffuse from the substrate to the surface during oxidation (Tan et al. 2011b). Higher magnification image available in their reported work clearly show that the surface morphologies of the etched Zn foils appear to be rougher with ridge-like appearance compared to unetched Zn foils. Thick and dense ZnO nanosheets with rough surface are observed after oxidized for only 10 minutes at 400°C. Compared to oxidation in air, the presence of excess oxygen inhibited the growth of dendritic ZnO nanorods and promoted the formation of ZnO nanosheets with rough ridge-like surfaces. With large areas of exposed polar basal planes of the nanosheets and the high roughness factor, these ZnO nanosheets are believed to be the main factor for the enhancement of photocatalytic properties during the decomposition of volatile organic compounds (Tian et al. 2003).

Thermal oxidation mechanism involves transportation, diffusion and absorption of oxygen to the surface of the zinc substrate (Li et al. 2005, Rusu et al. 2008). During rapid oxidation, the temperature was rapidly elevated from room temperature to the 300°C and 400°C in a short time. The rapid oxidation would result in the formation of nanosize metallic Zn vapor aero-sphere (Fan and Jia 2011). At the initial stage, the activation energy barrier is low as the interaction between the metal and the oxide layer is weak during the sublimation of Zn (Hong et al. 2012). Subsequently, Zn ions would diffuse along the grain boundaries to grow preferentially at sites of the revealed grains. Due to the relatively low melting point of Zn (419.5°C), quick temperature elevation during rapid oxidation at 300°C and 400°C is believed to be adequate for the sublimation of Zn to occur and create a Zn vapor aero-sphere at the vicinity of the surface (Tan et al. 2011b, Djurisic and Leung 2006). The reaction between the sublimed metallic Zn with the oxygen led to the formation of ZnO nuclei which then grow on reaching super saturation and agglomerate into larger particles before being re-deposited onto the surface in the preferred hexagonal sheet-like structure of ZnO. This is clearly observed as nanosheets were already formed after oxidized at 300°C for 10 minutes. However, they appeared to be consisting of loosely bound ZnO particles forming a sheet-like structure. At higher oxidation temperature of 400°C, with higher degree of sublimation nearing the melting point of Zn, a similar phenomenon occurred but at higher rate generating denser ridge-like ZnO nanosheets. The overall mechanism of ZnO nanosheets formation is illustrated in their reported work (Tan et al. 2014b).

2.2 Hot-Water Treatment

2.2.1 Hot-Water Treatment of Metallic Zn

Generally, many metal oxides hydrolyze in water forming hydroxide layers at the surface. Water molecules are adsorbed on the surface generation either physically or chemically. Due to the amphoteric behavior of H_2O, the surface of the oxides or the hydroxide would be partially charged with the presence of H^+ or OH^- ions (Degen and Kosec 2000).

Similarly, when ZnO is immersed in aqueous solution, H^+, OH^- ions adsorb onto the surface. In pure water, OH^- ions are pulled to the Zn sites and H: ions are attracted to the oxygen sites of the ZnO surface depending ions that are on the terminated surface. Depending on the reaction equilibrium the excess of either species would render the surface to be positively or negatively charged (Zhang 1996). As the surface of the ZnO begins to hydrolyze, zinc hydroxide ions which are water soluble are formed. The solubility of ZnO varies depending on the pH and the state of the hydrolyzed species differ according to Fig. 11.6 (Degen and Kosec 2000).

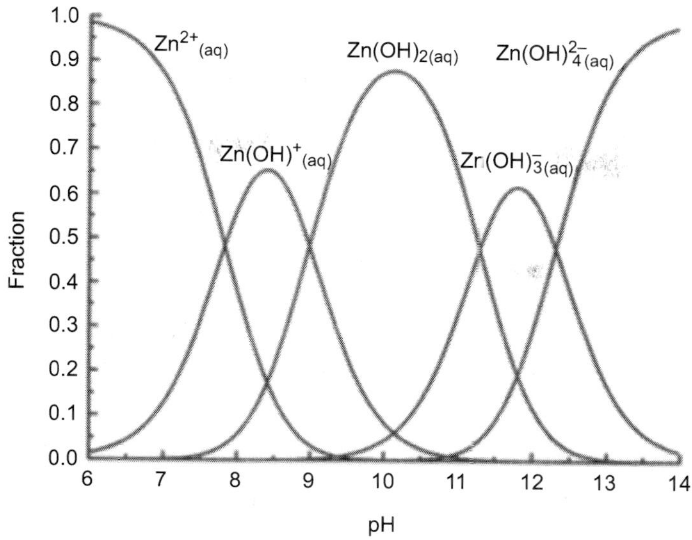

Fig. 11.6 Zn ions species existing as Zn^{2+}, $Zn(OH)^+$, $Zn(OH)_2$, $Zn(OH)^-_3$ and $Zn(OH)^{2-}_4$ over the range of pH at 25°C. Reproduced with permission (Degen and Kosec 2000).

Several researchers had reported the effect of water or water vapor effect on ZnO during its growth. For example, Chen et al. (2011) reported the formation of ZnO nanostructures using a wet-oxidation process and demonstrated that water vapor plays an important role on the initial and final dimensions of the ZnO nanostructures formation (Chen et al. 2011). Meanwhile, Matsuda et al. (2008) have been studying low temperature crystallization of metal oxide from alkoxide-derived gel films including ZnO by Hot-Water Treatment (HWT) (Matsuda et al. 2013, Prastomo et al. 2009, Prastomo et al. 2010, Prastomo et al. 2011). The feasibility of ZnO nanostructures formation by hot-water treatment at low temperature of 90°C not only enables the generation on flexible substrates by also avoids unnecessary contamination.

By using facile hot-water treatment, feasible formation of ZnO nanostructures is demonstrated by Tan et al. (2013a) In their study, 1D ZnO nanorods and 2D ZnO nanosheets were observed on etched and unetched metallic Zn foils. The main challenge of this method is the reproducibility and precise controlled formation of the nanostructures. In the hot-water treatment of etched Zn foils, 1D ZnO nanowires and ZnO nanorods formation were formed. By adjusting the hot-water treatment time from 4 to 24 hours, ZnO nanostructures obtained transformed from nanowires and nanorods. Throughout the transformation, the ZnO nanowires evolved to flat-topped hexagonal ZnO nanorods after HWT of 8 hours and on HWT at prolonged times of 12 and 24 hours, coarse surfaces ZnO nanorods were observed. This resulted from the massive dissolution of ZnO to form Zn^{2+} ions, followed by subsequent re-deposition of Zn^{2+} ions as ZnO. At this stage, coalescence of neighboring nanorods forming rods with a larger diameter occurred due to Ostwald ripening.

However, HWT treatment in the similar condition unetched Zn foils, as there is no easy diffusion channel for the Zn ions compared to that of etched Zn foils, 2D

ZnO nanosheets were obtained. The formation mechanism of the ZnO nanosheets is believed to deviate into a different route which encourages the lateral 2D growth of the ZnO nanosheets as shown in Fig. 11.7. Hu et al. (2010) also demonstrated that by using this simple dissolution-recrystallization technique of neural hot-water treatment, $Zn(OH)_2$/ZnO nanosheet and ZnO nanoarrays hybrid films were obtained. In their findings, HWT of ZnO nanowhiskers formed on FTO glass substrates by chemical bath deposition support the hypotheses that HWT time and temperature would influence the dimension of ZnO nanostructures as this facile formation without any addition of catalyst and additives is feasible. One of the interesting observations reported by Hu et al. (2010) is that HWT at 65°C for a prolonged period of 5 days to 2 months could generate a bilayer of morphologies as shown in Fig. 11.7. A clear gradual transformation of dissolved nanowhiskers into a new layer of ZnO nanosheets on top of the ZnO nanowhiskers could be observed.

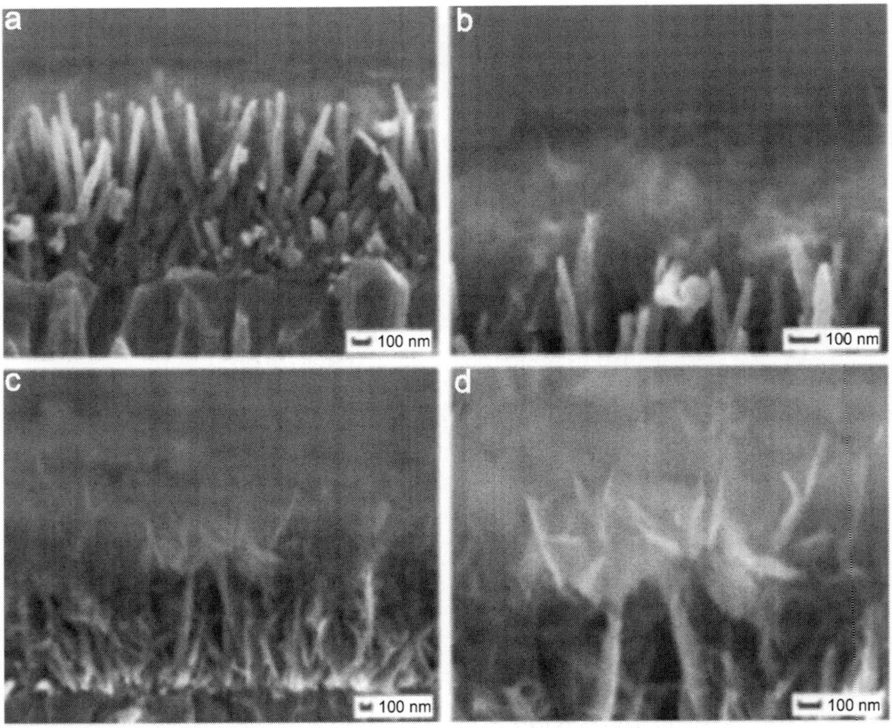

Fig. 11.7 Cross sectional images of the as-formed bilayer nanostructures that consisted of ZnO nanosheets on top of ZnO nanowhiskers by hot-water treatment at 65°C after prolonged period of (a), (b) 25 days and (c) (d) 2 months. Reproduced with permission (Hu et al. 2010).

Post-treatment of ZnO nanowires also reported by Lupan and Pauporté (2010), Lupan et al. 2010 using hot-water treatment and hot-water vapor treatment. It was found that these treatments improved the ZnO optical properties by decreasing the density of the non-radiative recombination centers near the surface of the nanowires. Hot-water vapor treatment was found to be more efficient in the reduction of structural defects compared to reactive treatment in oxygen whereas the dissolution of the nanostructures during HWT was too severe and tends to destroy the as-grown ZnO nanowires.

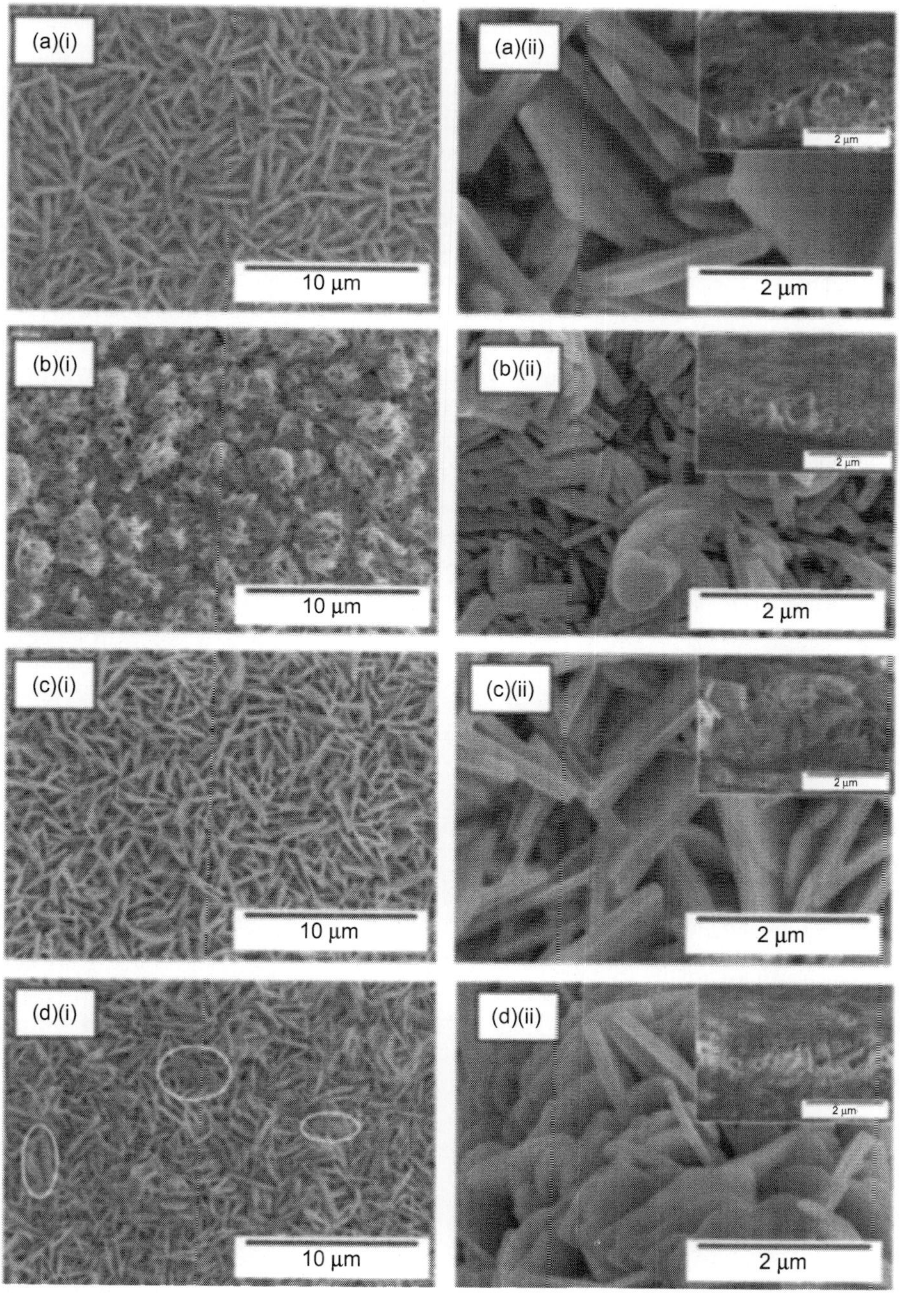

Fig. 11.8 Surface morphologies and cross sectional images of the ZnO nanosheets formed after hot-water treatment at 90°C for (a) 4 hours. (b) 8 hours. (c) 12 hours. (d) 24 hours of the unetched Zn foils. Reproduced with permission (Tan et al. 2013d).

The ZnO nanosheets obtained after HWT of unetched Zn foils portrayed good optical properties as determined by PL and Raman and exhibited typical ZnO properties. However, no emission peak within the green region was observed after HWT for a prolonged time of 24 hours and this could be due to crystal structure rearrangement during dissolution and recrystallization of the ZnO surface that reduced or eliminated some amount of oxygen vacancies that contribute to the deep-level green emission which is further supported by its Raman results (Djurisic and Leung 2006, Tan et al. 2013d). This unique interconnected 2D structure is believed to be an advantage for optoelectronic applications (Deng et al. 2005).

Although the mechanism involved in the formation of ZnO nanostructures by HWT is almost similar, changes in the initial nucleation stage have promoted different routes for the subsequent oriented growth. As for the etched Zn foils, surface oxidation using hot water is a surface diffusion dominant process which promotes non-planar growth perpendicular to the metallic Zn surface. In wet-oxidation processes, nucleation sites for the growth of ZnO nanostructures are required where the edges of the Zn grains of the etched foils acted as the nucleation sites. As water molecules exhibit polarity with a negative partial charge on the O side, the water molecules are likely to be attracted by the Zn atoms and adsorb onto the surface of the foil (Noothongkaew et al. 2012). The fast adsorption and breakup of the water molecule on the Zn surface will accelerate the kinetics of oxidation to form ZnO nanocrystals that serve as a seed layer. More Zn ions then diffuse to the surface to react with water molecules and form ZnO nanostructures. Generally, ZnO nanocrystals grow into a preferential hexagonal structure. Polar faces with surface dipoles (Zn^{2+} or O^{2-}) are thermodynamically less stable than non-polar faces and tend to minimize their surface energy through rearrangement during the dissolution-redeposition process (Zhang 1996). Interestingly, the continuous rearrangement results in the growth of ZnO nanowires and nanorods with an ideal hexagonal ZnO crystal structure. At the initial stage, the growth of the ZnO nanowire occurred through the diffusion of Zn ions to the tips of the ZnO nanostructures where they reacted with the water molecules to form ZnO crystals. As the ZnO nanowires became too long for ions diffusion to the tips, sufficient supply of Zn ceased and this changed the formation mechanism. Subsequently, the dissolution of ZnO formed resulted in the formation of hydroxide at the surface of the specimens. When the pH of water was 7 (neutral), $Zn(OH)^+$ ionic species were formed which can be deduced from the graph in Fig. 11.6. This polar hydroxide layer then subsequently dissolved into the hot-water bath as the surface of the oxide continued to be hydrolyzed (Degen and Kosec 2000). Hereafter, the area near the vicinity of the surface would eventually become supersaturated and the dissolved hydroxide would then redeposit onto the surface forming ZnO as shown in Eq. (1-3). This phenomenon could be clearly observed from the surface of the nanostructures after a prolonged hot-water treatment time of 24 hours where rough and coarse surface of nanorods were formed.

$$Zn^{2+} + OH^- \leftrightarrow Zn(OH)^+ \tag{1}$$

$$Zn(OH)^+ + OH^- \leftrightarrow Zn(OH)_2 \tag{2}$$

$$Zn(OH)_2 \leftrightarrow ZnO + H_2O \tag{3}$$

As there is no easy diffusion channel for the Zn ions in unetched Zn foils compared to surface of etched Zn foils, the formation mechanism of the ZnO nanosheets is expected to follow an epitaxial growth route. The epitaxial growth of the ZnO nanosheets can be divided into two stages, (1) the initial stage of ZnO buffer formation and (2) growth of the nanosheets by epitaxial diffusion which is also reported by Jia et al. (2008). During the formation of the buffer layer, the fast adsorption and breakup of the water molecule on the Zn surface will accelerate the kinetics of oxidation to form the hexagonal ZnO nanosheets as ZnO nanocrystals prefer a hexagonal structure. The subsequent continuous growth of the nanosheets edges then lead to the formation of interconnected network of the nanosheets as the areas between two adjacent nanosheets or the nanosheets' edges possess lower surface energy compared to the energy required to initiate a new nanostructure. Due to the lower epitaxial diffusion rate of the Zn ions from the substrate to the surface compared to the higher diffusion of Zn ions along the grain boundaries of etched Zn foils, no significant morphological change was observed for all the hot-water treated samples from 4 to 24 hours as shown in Fig. 11.8. However, Ostwald ripening phenomenon is also observed at prolonged HWT time of 24 hours as highlighted in Fig. 11.8(d). The compactness of the oxide formed is dependent on the hot-water temperature and at 90°C; with higher oxidation rate during the second stage oxidation compared to those oxidized at lower temperatures (Noothongkaew et al. 2012), compact and dense $Zn(OH)_2$ is formed during the initial stage and finally converted to form ZnO similar to Eq. (1) to (3).

In another work by Kuan et al. (2012) hierarchical ZnO nanostructures evolution from nanosheets to hierarchical nanorods and nanowires were demonstrated by chemical solution growth of the micro-tips prepared by anodic etching of Zn foils. The subsequent chemical growth was achieved by varying the precursors composition that consist of KOH and $Zn(NO_3)_2$. The 2D hierarchical nanostructures were obtained at room temperature at a short time of only 10 minutes chemical growth. Figure 11.9 shows the morphologies obtained in their work against the precursor concentrations.

This emphasizes the potential of mass production of ZnO nanostructures short duration under ambient conditions. As the main factor for the various morphologies still remains to attribution of pH values that is influenced by the concentration of Zn^{2+} and OH^- in the solutions, similar Eqs. of (1) to (3) are applicable upon reaching supersaturation for crystallization of ZnO.

2.2.2 Hot-Water Treatment of Sol-gel Derived Layer

In works related to HWT, various authors have demonstrated that this technique is applicable to fabricate films from alkoxide-derived sol-gel coatings at low-temperature (Katagiri et al. 2008, Matsuda et al. 2008, Matsuda et al. 2013, Prastomo et al. 2009, Prastomo et al. 2010, Prastomo et al. 2011, Tan et al. 2013b). They also reported on control formation and design of ZnO crystals into flower-like structures at low-temperature of 50°C by Electric Field Hot-Water Treatment (EF-HWT) of sol-gel derived coatings (Matsuda et al. 2013). In their report, alteration of ZnO nanostructures with applied electric field is also influenced by the substrates used. On FTO coated glass, hexagonal column of ZnO nanostructures were obtained compared to granular crystallites while in Si water, EF-HWT of ZnO gel films resulted in the formation of flower-like hexagonal ZnO crystals where the branching degree increased with the applied voltage during HWT (Matsuda et al. 2013). A similar mechanism of HWT occurred where the dissolution of the ZnO gel film will form Zn^{2+} and Zn $(OH)_2$ species, after that, the subsequent redeposition of the particulate ZnO crystals takes place by dehydration-condensation process on the surface of Zn $(OH)_2$.

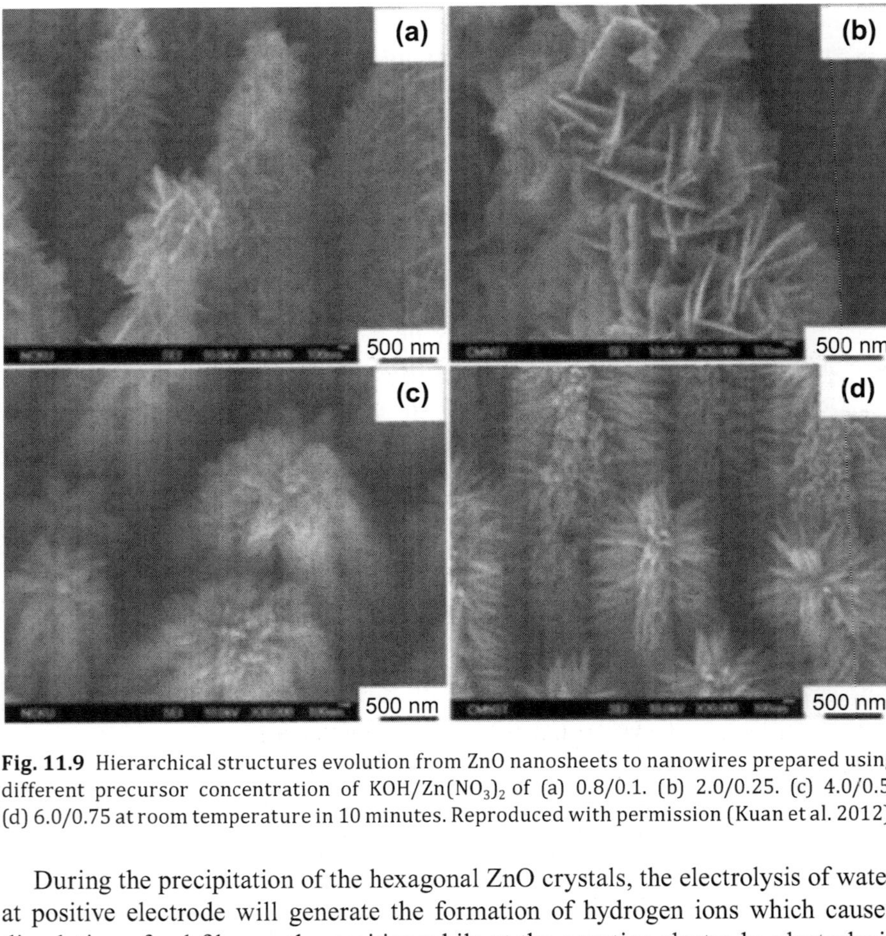

Fig. 11.9 Hierarchical structures evolution from ZnO nanosheets to nanowires prepared using different precursor concentration of KOH/Zn$(NO_3)_2$ of (a) 0.8/0.1. (b) 2.0/0.25. (c) 4.0/0.5. (d) 6.0/0.75 at room temperature in 10 minutes. Reproduced with permission (Kuan et al. 2012).

During the precipitation of the hexagonal ZnO crystals, the electrolysis of water at positive electrode will generate the formation of hydrogen ions which causes dissolution of gel film on the positive while at the negative electrode, electrolysis of water near the vicinity of the substrate would form a basic condition due to the formation of hydroxide ions. These hydroxide ions play an important role during the heterogeneous nucleation during the reprecipitation of the ZnO crystals on the surface of the substrate. Therefore, the heterogeneous nucleation will promote the preferential growth *c*-axis (0 0 2) plane depending on the polarity of the terminating ions on the hexagonal ZnO columns (Tan et al. 2013a).

The optical transmission of the films on FTO were investigated and was found to be reduced after longer HWT time as a result of light scattering on the surface of the film due to the higher amount of ZnO crystal precipitates. However, when Si wafers were used as substrates, HWT with applied voltage of 10 V/cm for 3 hours led to the formation of significant flower-like ZnO crystals compared to hexagonal columnar ZnO crystals formed on FTO. It is noteworthy that the type of substrate used during EF-HWT affects ZnO crystals formation of the ZnO gel films. By changing the electric field voltages from 0-20 V/cm, the ZnO morphologies evolved from hexagonal columnar to flower-like nanostructures as shown in Fig. 11.10. With higher degrees of branching as stronger voltage was applied during HWT, indicating that external field applied during HWT of the ZnO films on the Si substrate would affect the concentration of the dissolved species forming different nucleation planes that leads to the formation of 2D flower-like structures. Yu et al. (2011) also reported in their investigations of ZnO nanostructures formation by hydrothermal that different substrates used would affect the morphologies obtained. ZnO nanorods and nanosheets were obtained on glass and aluminium substrates, respectively after hydrothermal growth with different precursors.

Fig. 11.10 Degree of branching increased when higher electric field was applied as shown in the surface morphologies of ZnO films obtained on Si substrates after HWT at 50°C for 3 hours: (a) 0 V/cm. (b) 5 V/cm. (c) 10 V/cm. (d) 20 V/cm. Reproduced with permission (Matsuda et al. 2013).

Almost similar flower-like ZnO microstructures formation is also reported by Venkatesha et al. (2012) by using electrochemical method. The colloidal precipitates formed were gathered by centrifugation after electrolysis of sodium nitrate at difference pH and current densities. It is important to note that the control growth of the flower-like ZnO nanostructures was feasible at low current densities and electrolyte concentration where higher current densities generated irregular growth. According to their report, the formation of twin nuclei caused by the generated OH during electrolysis promoted polyhedral core that later developed into multifaceted structure further supporting the phenomenon report by Matsuda et al. (2013).

2.3 Hydrothermal Growth of ZnO Nanostructures

Hydrothermal growth is one of the chemical bath deposition methods that are commonly used in the formation of ZnO nanostructures especially 1D ZnO nanorods and nanowires. Generally, seed layer is required for the oriented growth of these 1D ZnO nanostructures which can be formed using electrodeposition or sol-gel derived coatings prior to subsequent hydrothermal growth (Chen and Wu 2011, Cheng et al. 2009, Guo et al. 2005a, Guo et al. 2005b, Guo et al. 2005c, Kiomarsipour and Shoja Razavi 2012, Ko et al. 2011, Qin et al. 2010, Shi et al. 2013, Tan et al. 2013c, Tan et al. 2011a, Yu et al. 2011). It is also reported that the Zn precursors that consist of different anions used during the hydrothermal growth will have an effect on the final nanostructures obtained (Lin and Jiang 2011).

Recently, simultaneous growth of 1D ZnO nanorods and 2D ZnO nanosheets in a single hydrothermal growth was demonstrated by Tan et al. (2014a). The growth mechanism involved the redeposition of the ZnO nanosheets generated in the hydetalrothermal bath on reaching supersaturation onto the surface of ZnO nanorods that also grew simultaneously. With longer hydrothermal exposure time, longer ZnO nanorods were formed and after 24 hours, the width of the nanosheets increased to approximately 1~2 μm. The density and structure of the 2D ZnO nanosheets also consolidated on longer hydrothermal time. Predominant (0 0 2) peak was observed from its XRD patterns indicating that the ZnO nanorod array-nanosheet composite films exhibited *c*-axis orientation (Qin et al. 2012). A schematic illustration of the formation mechanism of the aforementioned ZnO composite nanostructures is shown in Fig. 11.11 which involved self-assembly. In this mechanism, the presence of hexamethyltetramine (HMTA) which is a non-ionic ligand, encourages the oriented growth of the ZnO nanorod arrays. HMTA also hinders the growth of the six prismatic side planes of wurtzite ZnO crystals while allowing growth along the *c*-axis and thus a rod-like structure is formed (Lockman et al. 2010, Sugunan et al. 2006). The chelation of HMTA on the non-polar surfaces of the nanorods prevents radial growth by inhibiting the absorption of Zn^{2+} ions. It is also important to note that the temperature used for the hydrothermal reaction must be high enough to allow the decomposition of HMTA to supply hydroxyl ions for the formation of ZnO, as described in Eq. (4) and (5) below. When a sufficient quantity of OH^- is present, the hydrothermal solution reaches supersaturation and the Zn ions will react with OH^- to form ZnO, as shown in Eq. (6) and (7).

$$C_6H_{12}N_4 + 6H_2O \leftrightarrow 6HCHO + 4NH_3 \quad (4)$$

$$NH_3 + H_2O \leftrightarrow NH_4^+ + OH^- \quad (5)$$

$$2OH^- + Zn^{2+} \leftrightarrow Zn(OH)_2 \quad (6)$$

$$Zn(OH)_2 \leftrightarrow ZnO + H_2O \quad (7)$$

The presence of the ZnO nanocrystal seed layer during the initial stage of the process is crucial because it provides heterogeneous nucleation sites for ZnO formation. As ZnO is a polar crystal, each Zn atom is tetrahedrally coordinated to four O atoms and vice versa. The alternating arrangement of Zn^{2+} and O^{2-} ions along the *c*-axis gives either positive or negative charges, depending on the terminating ions. These electrostatic charges attract ions of the opposite charge and lead to the accumulation of Zn $(OH)_2$ or Zn $(OH)_2(NH_3)_4$ on the existing grains. In the report by Yu et al. (2011), the controlled formation of either ZnO nanorods, nanowires or nanosheets could be done by changing the precursors used such as zinc chloride, zinc nitrate and zinc acetate during the hydrothermal process.

In the simultaneous growth, the use of aqueous solution promoted the formation of the ZnO nanosheets due to the homogenization of the reactants in the medium which could affect the numbers of individual nuclei formed (Fig. 11.11b) as well as the nucleus preferential growth direction (Giri et al. 2010, Jiang et al. 2012). Dissolution of the sol-gel seed layers during hydrothermal reaction could also contribute to the generation of ZnO clusters (Jiang et al. 2012). Agglomerations of these smaller clusters would then form larger clusters (Fig. 11.11c) and the self-assembly of the active sites could eventually promote the formation of sheet-like structures. After that, the as-formed ZnO nanosheets were redeposited back onto the substrate surface and were further consolidated with the hydrothermal reaction time forming the composite structure as shown in the SEM images (Fig. 11.11d-f) (Xu and Wang 2011). Jiang et al. (2012) also observed the formation of nanosheets by micro-based hydrothermal synthesis that was based on the micro-restriction effect of the emulsions. They proposed that small clusters of ZnO were first formed and then these clusters bonded together forming sheet-like aggregations where the preferential growth would lead to larger sheet formation. They also mentioned that water content and concentration of reactants, $Zn(Ac)_2$ and NaOH had an effect on the diameter and the thickness of the nanosheets generated. The schematic formation mechanism proposed by Jiang et al. (2012) is shown in Fig. 11.12. Formation of ZnO flakes is also reported by Venkatesha et al. (2012) where higher pH of solution was used during their electrochemical synthesis of flower-like ZnO nanostructures where more presence of OH^- increased the rate of precipitation leading to the formation of flakes. Therefore, autogenous pressure generated during the hydrothermal process could promote more OH presence in the solution that prompted a higher ZnO nanosheet formation rate. The negative nature of $[Zn(OH)_4]^{2-}$ could lead to different growth rates for different planes during hydrothermal processing and HMTA is expected to act as an organic template during the heating process up to 80°C and dynamically modify the nucleation process. With increasing Zn^{2+} and OH^- concentrations, the $Zn(OH)_2$ or ZnO nuclei developed under low precursor concentrations and the action of HMTA. Also according to Guo et al. (2010) the self-assembly of a number of active sites that trigger nucleation could promote the formation of petal crystals that extend from the interface and result in the sheet-like structure (Ehrentraut et al. 2012).

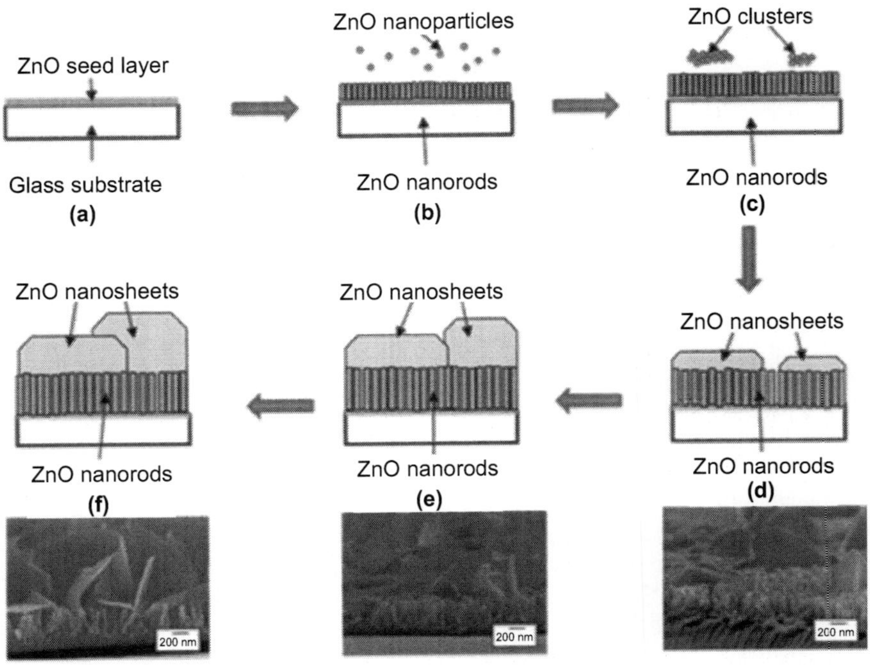

Fig. 11.11 Schematic of the formation mechanism of ZnO composite nanostructures that composed of 1D ZnO nanorods and 2D ZnO nanosheets. Reproduced with permission from (Tan et al. 2014a).

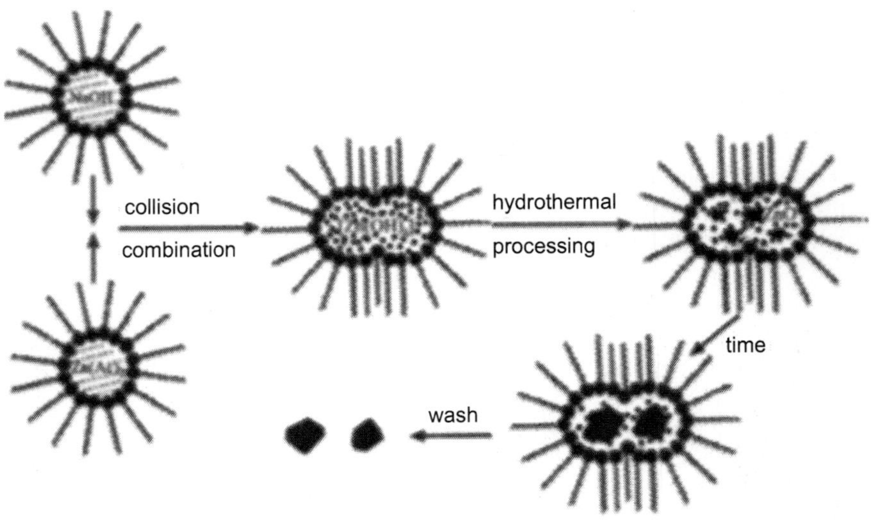

Fig. 11.12 Schematic formation of the ZnO nanosheets proposed by Jiang et al. that involves the combination of clusters and preferential growth to form a sheet-like structure. Reproduced with permission (Jiang et al. 2012).

In an interesting discovery by Pan et al. (2005), comparison on the formation of ZnO nanostructures obtained by hydrothermal and sclvothermal with and without the influence ultrasonic pre-treatment on the final outcome of the ZnO nano-architectures were reported. The schematic adopted from their findings is shown in Fig. 11.13. As ZnO nanosheets were generated due to the anisotropic growth, it is also expected that the optical properties also exhibit asymmetrical emission band and Stokes shift attributed to strong quantum confinement and anisotropic energy band structure of the nanosheets (Pan et al. 2005). By using ultrasonic mediated rinsing of the ZnO seed layer prepared using aqueous deposition method, ZnO hierarchical nanostructures growth was promoted compared to those without and this supports that ultrasonic pre-treatment influence on the subsequent morphologies formation of ZnO nanostructures (Kumar et al. 2010).

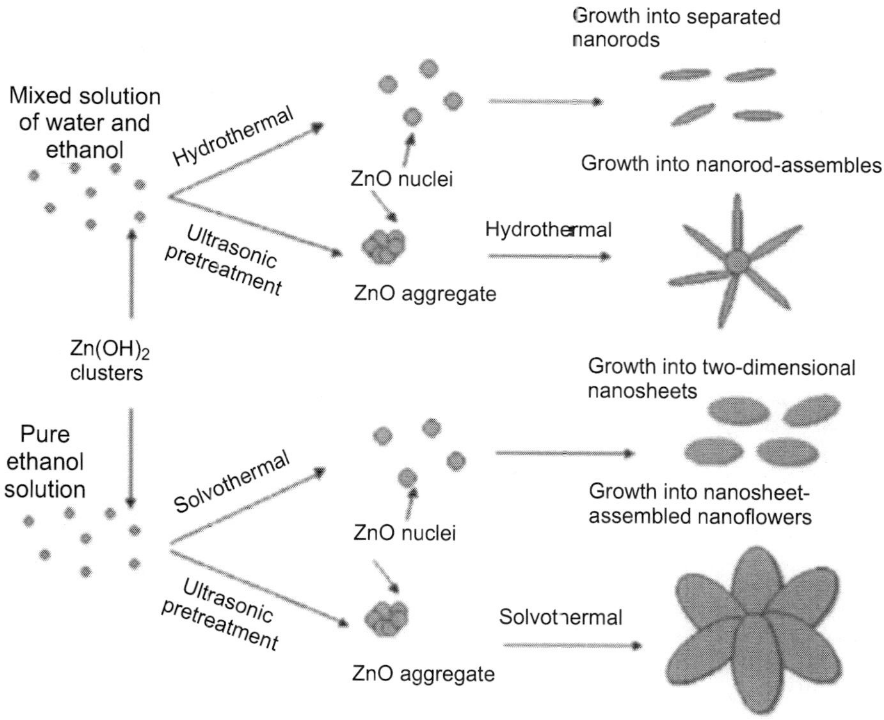

Fig. 11.13 The controlled growth of ZnO nanostructures by solvothermal and hydrothermal with ultrasonic pre-treatment as proposed by Pan et al. Reproduced with permission (Pan et al. 2005).

The ZnO nanorod-nanosheet composite exhibits PL emissions at approximately 380 nm (UV), 445 nm (blue) and 575 nm (green). The observed blue emission was attributed to the nanosheets which is also reported by other researchers that this is considered as having been caused by electron transitions from the shallow donor levels of oxygen vacancies and zinc interstitials to the valence band (Li et al. 2010). Wang et al. (2010a) also reported on multiple emissions in the blue-green region for ZnO nanosheets formed by a mediated hydrothermal method. The increasing visible emission with increasing hydrothermal process time could be caused by an increase in the surface defects in the nanosheets (Jiang et al. 2012). The photocatalytic activity of the ZnO composite on the photo-degradation of MB under visible light irradiation against ZnO nanorods is shown in Fig. 11.14.

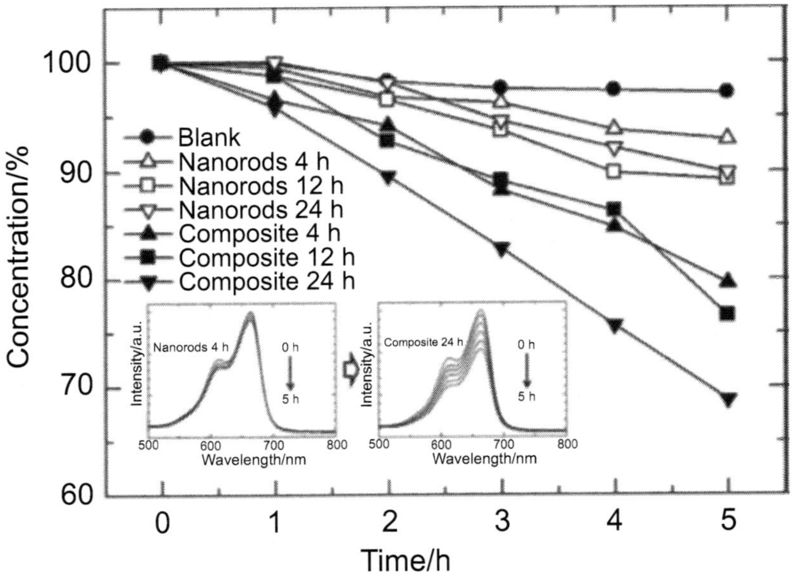

Fig. 11.14 Comparison on the photo-degradation properties of MB under visible light irradiation of ZnO nanorod-nanosheet composite structure against ZnO nanorods with the inset showing absorption spectra of both obtained after 24 hours of hydrothermal growth. Reproduced with permission (Tan et al. 2014a).

After 5 hours of visible light exposure, the photo-degradation exhibits approximately 8 to 11% for the ZnO nanorods grown from 4 to 24 hours, respectively. Meanwhile, ZnO composite nanostructures demonstrated higher photo-degradation activities of 20, 24 and 32% under visible-light irradiation. When the ZnO nanocrystals are irradiated by light with energy higher or equal to the bandgap, an electron (e_{cb}^-) in the valence band can be excited to the conduction band with the simultaneous generation of a hole (h_{vb}^+) in the valence band (Han et al. 2012). These excited e_{cb}^- and h_{vb}^+ can recombine and get trapped in metastable surface states or react with electron donors and acceptors adsorbed on the semiconductor surface. The photoelectron is easily trapped by electron acceptors like adsorbed O_2, whereas the photo-induced holes can be easily trapped by electronic donors such as OH^- or organic pollutants which lead to the oxidation of the organic elements. The photocatalytic ability is also dependent on the surface adsorption ability and concentration of oxygen defects on the surface of the nanostructures. Therefore, the oxygen vacancies generated on the surface of the nanosheets can serve as electron capturing centers to restrain the recombination of e_{cb}^- and h_{vb}^+ and could generate active species on the surface of the semiconductor which is beneficial for photo-degradation. Tong et al. (2006) reported that nanosheets structure with more surface defects generated could increase the photo-reactivity.

However, solution based methods without the assistance of any seeds, catalysts and surfactants have sparked great interest from many researchers. Reported works using self-assembly method in the design of its hierarchical nano-architecture have been attempted and successful lateral growth and secondary nucleation have enabled the fabrication of 2D ZnO nanostructures for optoelectronic applications (Kim et al. 2010). Yue et al. (2009) reported on the assembly of ZnO nanosheets on nanorods forming a hierarchical structure by solution based method. Figure 11.15 shows the SEM images and the illustration of the hierarchical nanostructures obtained by Yue et al. (2009). They managed to assemble the hierarchical ZnO nanosheet-nanorod structure on Zn substrate which also exhibited UV, blue and green emissions aiming to be used for optoeletronics applications.

(a)

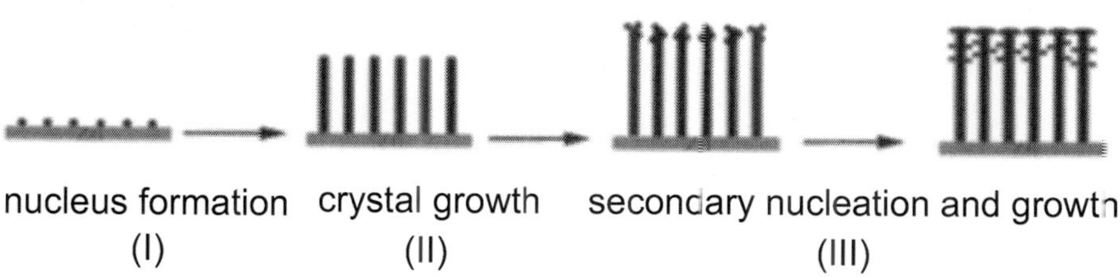

(b)

Fig. 11.15 Hierarchical ZnO nanosheets on nanorods and its proposed formation mechanism by wet chemical method. Reproduced with permission (Yue et al. 2009).

3. APPLICATION OF 2D ZnO NANOSTRUCTURES

Due to the many unique properties of ZnO structures and its easy control in design of the nanostructures, it has been widely used in energy-related application such as in the DSSCs. Many works reported on the increment of the conversion efficiency by increasing the active surface area through nano-architecture design of the nanostructures. Despite this, the highest DSSCs efficiency obtained using ZnO to date is approximately 5% compared to TiO_2 which is at approximately 11%. For comparison, the silicon solar cells which are commercially available recorded the highest efficiency of approximately 20%. Although ZnO exhibits many superior characteristics compared to TiO_2, there are factors that limit its performance which will be explained later. Research works using ZnO nanosheets were also reported to improve the conversion efficiency in comparison to ZnO nanoparticles due to lower recombination rate and higher scattering effect (Zhang et al. 2009, Elkhidir Suliman et al. 2007) while not as good as 1D ZnO nanostructures like nanorods and nanowires which are largely used. By using 2D ZnO sheets, Ameen et al. (2012) demonstrated the usefulness of these ZnO nanosheets generated using hydrothermal method as shown in Fig. 11.16 as the backbone for the subsequent generation of nanospikes on the sheets for solar cell application (Ameen et al. 2012). This nano-architecture design improved the overall surface area which is crucial for the dye adsorption that led to enhanced photo-conversion efficiency to 2.51% (Ameen et al. 2012).

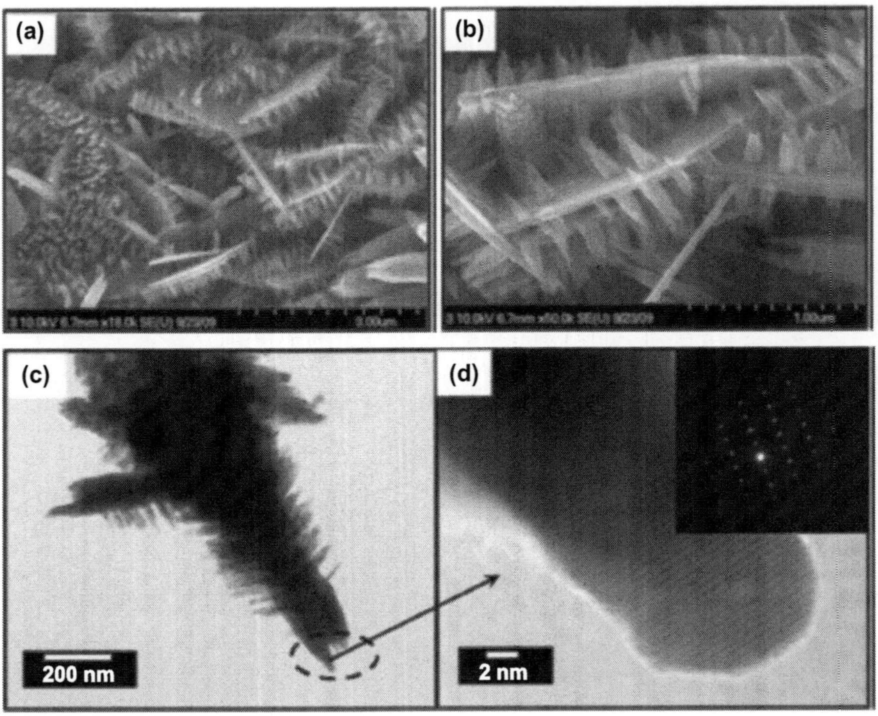

Fig. 11.16 SEM (a, b) and TEM (c, d) images of the nanospikes decorated ZnO sheets that show the feasibility of hierarchical design of ZnO for solar cell application. Reproduced with permission (Ameen et al. 2012).

By using the two-step electrodeposition, Qiu at al. (2011) developed the fabrication of DSSCs based on hierarchical ZnO nanorod-nanosheet structure that achieved a higher conversion efficiency of 3.12%. Similarly, ZnO nanosheets were first deposited as the primary backbone structure and followed by the secondary deposition of the ZnO nanorods. In their detailed systematic studies, factors that influence the DSSCs efficiencies such as the composition of the nanosheets on the formation of its morphologies as well as the first and secondary step deposition time were highlighted (Qiu et al. 2011). From the work reported by Yu et al. (2011), 2D ZnO nanosheets demonstrated the highest PL intensities compared to only 1D ZnO nanorods and nanowires. However, the photocurrent density for the ZnO nanorods was the highest (almost six times higher). Therefore, the decoration of 1D nano-spikes formed on 2D ZnO nanosheets would be beneficial as the preferential orientation of ZnO nanospikes could enhance the electron transfer to the maximized surface area 2D ZnO nanosheets.

Fig. 11.17 Low (a), (b) and high magnification (c), (d) SEM images of the ball-like ZnO that consist of fluffy nanosheets obtained by Umar et al. Reproduced with permission (Umar et al. 2011).

Despite possessing almost a similar bandgap with TiO_2 and showing higher electron mobility, lower recombination rate and excellent crystallinity of various nano-architectures obtainable at low-temperature, the maximum photo-conversion efficiencies of ZnO is still relatively low compared to TiO_2. The main factor is the instability of ZnO in acidic dye that is commonly used in DSSCs where the H^+ ions presence resulted in the dissolution of Zn atoms at the surface of ZnO. This could lead to the formation of excessive complex consisting of Zn^{2+}/dye agglomerates and reduce the electron injection kinetics from the dye to ZnO (Zhang et al. 2009,

Ke et al. 2010). Therefore, it is more favorable to use a shorter sensitization time and the formation of the complex layer could be avoided when the immersing time is less than 3 hours. There are other methods such as buffer layer coating of ZnO to form a shell to improve its stability or application of new types of photosensitizers (Guerin et al. 2010).

The 2D surface of ZnO nanosheets have been widely used as photocatalysts for decomposition of organic pollutant under irradiation of UV and visible light. Intrinsic and extrinsic doping of these ZnO nanosheets are also performed in order to fully exploit the use of visible light in the photo-degradation as ZnO is mostly responsive in the UV region (Tan et al. 2014a, Yousefi et al. 2011, Abe 2010). For example, the large surface area of ZnO nanosheets was utilized by Umar et al. (2011) as a photocatalyst in the degradation of methylene blue. In their work, large scale synthesis of ZnO balls that consist of fluffy thin ZnO nanosheets by a simple solution method was obtained as shown in Fig. 11.17. The photocatalytic performance of these ZnO fluffy balls exhibit superior efficiency in comparison of the commercially available TiO_2-UV-100 in 70 minutes as shown in Fig. 11.18 (Umar et al. 2011).

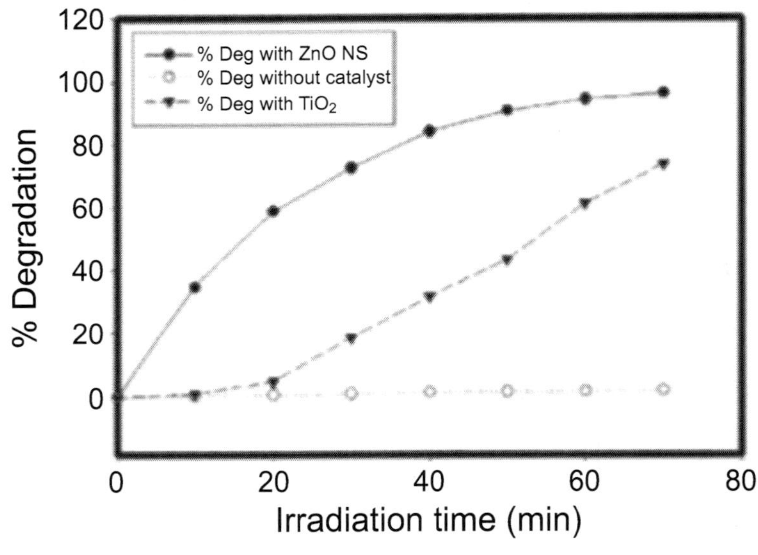

Fig. 11.18 The comparison between the more superior photocatalytic property of ZnO fluffy nanosheets with the commercially available TiO_2 during the degradation of methylene blue under UV light irradiation (Umar et al. 2011).

Lei et al. (2012) fabricated ZnO microspheres that consist of porous nanosheets which are almost similar to those obtained by Umar et al. (2011) and this further emphasizes the ease of nano-architecture engineering of ZnO. The porous microspheres were formed by a two step process involving hydrothermal preparation followed by thermal decomposition of zinc hydroxide carbonate precursor and significant enhancement of photocatalytic activity was observed compared to the commercially available ZnO and TiO_2 (Lei et al. 2012). Methods that involve sudden thermal heat application during its formation process could lead to the formation of a porous-like structure such as porous microspheres and nanosheets. Lockman et al. (2012) demonstrated this through rapid thermal oxidation of metallic Zn foils in air and in oxygen and the porous ZnO nanosheets also exhibited good photodegradation properties (Lockman et al. 2012, Sreekantan et al. 2009, Tan et al. 2014b, Tan et al. 2011b). In a recent reported work, Liu et al. (2016) successfully fabricated 2D ZnO porous nanosheets using a colloidal template approach by controlling the infiltration speed for the ethylene glycol capped ZnO nanoparticles and polymer colloids. These unique 2D structures were reported to be able to accelerate and electron transportation from the single crystal nature and at the same time enhanced the selective adsorption of organic molecules which increased its photocatalytic performance (Liu et al. 2016).

To further demonstrate the versatility of applications of ZnO nanosheets, Geng et al. (2010) reported on the novel ZnO multi-layer architectures by using poly (sodium 4-styrenesylfonate) (PSS) as structure directing agent in controlling the growth mechanism for application as gas sensors. The multi-layer architecture formation mechanism is shown in Fig. 11.19 and exhibited good responses towards acetone and ethanol.

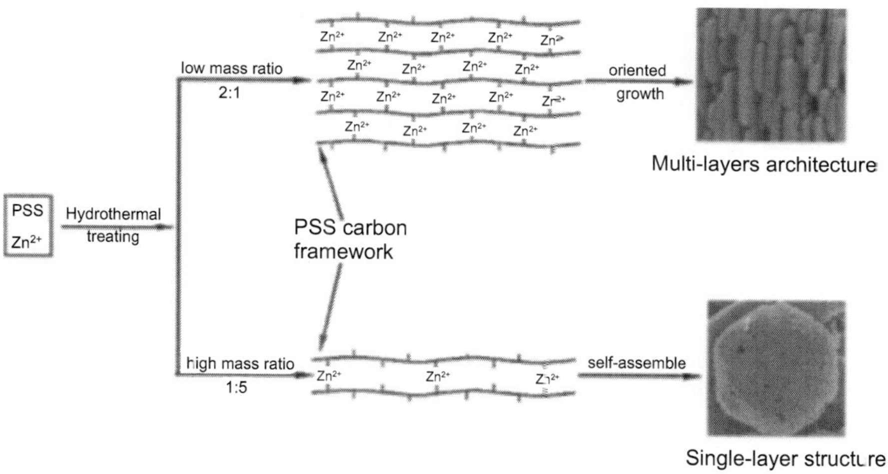

Fig. 11.19 The mechanism of multilayer architecture formation of 2D ZnO nanostructures reported by Geng et al. for application as gas sensors. Reproduced with permission (Geng et al. 2010).

As ZnO possesses the unique optical properties in the UV and blue spectral range as well as exhibiting excitonic emission at room temperature, 2D ZnO nanostructures have been widely used in optoelectronics applications such as UV light emitter and interestingly ZnO lasers (Djurisic and Leung 2006). Numerous works and studies on the 2D nanostructures related to its spontaneous PL emission have been reported (Tan et al. 2013d, Kiomarsipour and Shoja Razavi 2012, Kiomarsipour and Shoja Razavi 2013, Al-Gaashani et al. 2013, Cheng et al. 2009, Jia et al. 2008) and investigations of doping of ZnO on its optical properties still remain scarce compared to undoped and bulk ZnO nanostructured films. The reason could be that a clear understanding of different defects luminescence of undoped ZnO remains unresolved. Kim et al. (2010) demonstrated the nanoscale design of 2D design of ZnO on Si (0 0 1) substrates by laser interference photography followed by low-temperature hydrothermal process at 90°C as illustrated in Fig. 11.20. With the feasibility of controlled 2D 'bottom-up' approach on formation of ZnO on Si in various 2D Bravais lattices such as square, rectangular, centered rectangular, hexagonal and oblique; practical advanced functional electronic or photonic are envisaged.

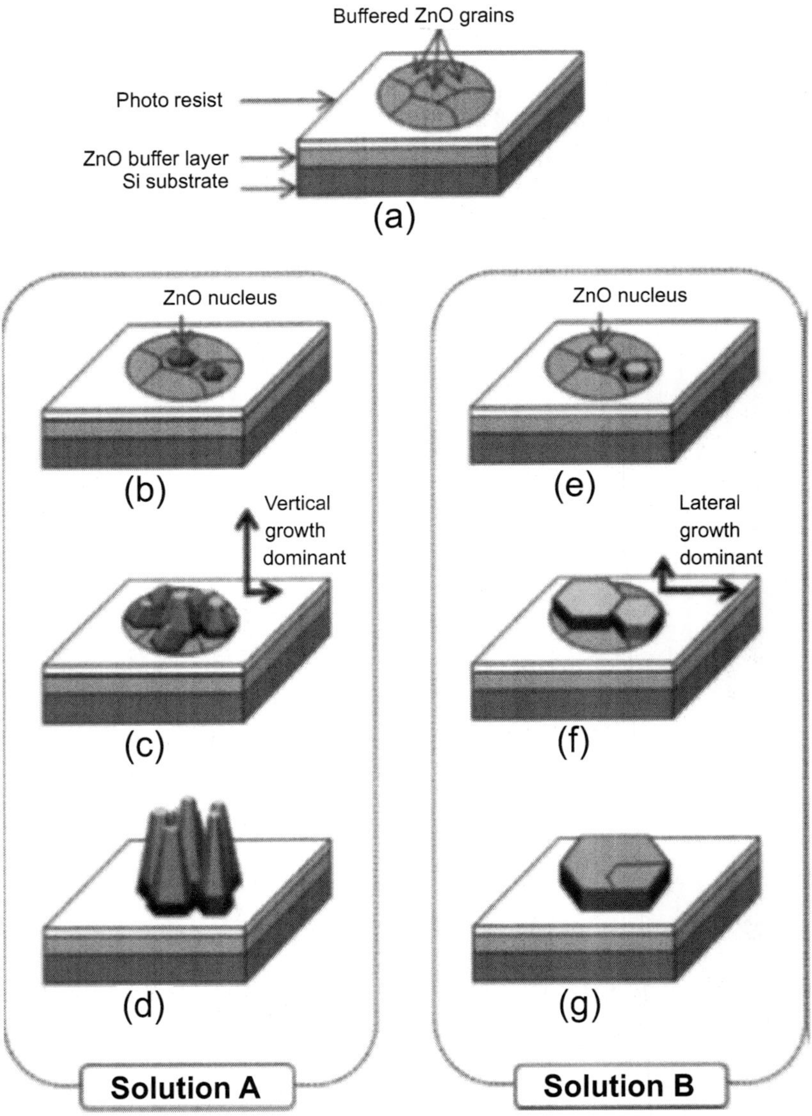

Fig. 11.20 Schematic of the plausible growth mechanism of ZnO in nanopatterned ZnO/Si (0 0 1) in achieving 2D ZnO nanostructures. Reproduced with permission (Kim et al. 2010). Copyright (2010) American Chemical Society.

4. CONCLUSIONS

ZnO nanostructures in multiple dimensions have been demonstrated to exhibit unique and exceptional properties in a wide range of applications. ZnO nano-architectures designs by various methods have been widely used to increase its specific surface area for electron transport and light scattering effect as well as optimizing its optical properties. The design and formation of 2D ZnO nanostructures, its properties and application by thermal oxidation, hot-water treatment and hydrothermal as well as other related methods were discussed. The feasibility of design in ZnO nano-architectures in multiple dimensions using facile routes especially in low-temperature would be an advantage for the development of flexible energy conversion and optoelectronic devices.

5. ACKNOWLEDGEMENTS

The author is profoundly grateful to the support of the members in Matsuda-Muto-Kawamura laboratory as well as Prof. Akihiro Wakahara and Prof. Masanobu Izaki of Toyohashi University of Technology for the research carried out in part of the studies reported in this chapter.

6. REFERENCES

Abe, R. 2010. Recent progress on photocatalytic and photoelectrochemical water splitting under visible light irradiation. J. Photochem. Photobiol., C 11: 179-209.

Al-Gaashani, R., S. Radiman, A. R. Daud, N. Tabet and Y. Al-Douri. 2013. XPS and optical studies of different morphologies of ZnO nanostructures prepared by microwave methods. Ceram. Int. 39: 2283-2292.

Al-Hardan, N. H., M. J. Abdullah and A. Abdul Aziz. 2010. Electron transport mechanism of thermally oxidized ZnO gas sensors. Physica. B: Condensed Matter. 405: 4509-4512.

Ameen, S., M. Shaheer Akhtar and H. S. Shin. 2012. Growth and characterization of nanospikes decorated ZnO sheets and their solar cell application. Chem. Eng. J. 195-196: 307-313.

Degen, Andrej and Marija Kosec. 2000. Effect of pH and impurities on the surface charge of zinc oxide. J. Eur. Ceram. Soc. 20: 667-673.

Ariga, K., Q. Ji, J. P. Hill, Y. Bando and M. Aono. 2012a. Forming nanomaterials as layered functional structures toward materials nanoarchitectonics. NPG Asia Mater. 4: e17.

Ariga, K., T. Mori and J. P. Hill. 2012b. Mechanical control of nanomaterials and nanosystems. Adv. Mater. 24: 158-176.

Ashrafi, A. and C. Jagadish. 2007. Review of zincblende ZnO: stability of metastable ZnO phases. J. Appl. Phys. 102: 071101.

Chen, R., C. Zou, X. Yan, A. Alyamani and W. Gao. 2011. Growth mechanism of ZnO nanostructures in wet-oxidation process. Thin Solid Films 519: 1837-1844.

Chen, S. J., Y. C. Liu, Y. M. Lu, J. Y. Zhang, D. Z. Shen and X. W. Fan. 2006. Photoluminescence and Raman behaviors of ZnO nanostructures with different morphologies. J. Cryst. Growth 289: 55-58.

Chen, S.-W. and J.-M. Wu. 2011. Nucleation mechanisms and their influences on characteristics of ZnO nanorod arrays prepared by a hydrothermal method. Acta Materialia 59: 841-847.

Cheng, C., B. Liu, H. Yang, W. Zhou, L. Sun, R. Chen, et al. 2009. Hierarchical assembly of ZnO nanostructures on SnO_2 backbone nanowires: low-temperature hydrothermal preparation and optical properties. ACS Nano 3: 3069-3076.

Deng, G., A. Ding, W. Cheng, X. Zheng and P. Qiu. 2005. Two-dimensional zinc oxide nanostructure. Solid State Commun. 134: 283-286.

Djurisic, A. B. and Y. H. Leung. 2006. Optical properties of ZnO nanostructures. Small 2: 944-961.

Djurišić, A. B., A. M. C. Ng and X. Y. Chen. 2010. ZnO nanostructures for optoelectronics: material properties and device applications. Prog. Quantum Electron. 34: 191-259.

Dutta, K., S. Das and A. Pramanik. 2012. Concomitant synthesis of highly crystalline Zn-Al layered double hydroxide and ZnO: phase interconversion and enhanced photocatalytic activity. J. Colloid Interface Sci. 366: 28-36.

Ehrentraut, D., K. Fujii, J. Riegler, K. Byrappa, M. Nikl and T. Fukuda. 2012. Functional one, two and three-dimensional ZnO structures by solvothermal processing. Prog. Cryst. Growth Charact. Mater. 58: 51-59.

Elkhidir Suliman, A., Y. Tang and L. Xu. 2007. Preparation of ZnO nanoparticles and nanosheets and their application to dye-sensitized solar cells. Sol. Energy Mater. Sol. Cells 91: 1658-1662.

Fan, H. and X. Jia. 2011. Selective detection of acetone and gasoline by temperature modulation in zinc oxide nanosheets sensors. Solid State Ionics 192: 688-692.

Geng, B., J. Liu and C. Wang. 2010. Multi-layer ZnO architectures: polymer induced synthesis and their application as gas sensors. Sens. Actuators, B 150: 742-748.

Giri, P. K., S. Dhara and R. Chakraborty. 2010. Effect of ZnO seed layer on the catalytic growth of vertically aligned ZnO nanorod arrays. Mater. Chem. Phys. 122: 18-22.

Guerin, V. M., C. Magne, T. Pauporte, T. LeBahers and J. Rathousky. 2010. Electrodeposited nanoporous versus nanoparticulate ZnO films of similar roughness for dye-sensitized solar cell applications. ACS Appl. Mater. Interfaces 2: 3677-3685.

Guerin, V. M., J. Rathousky and T. Pauporte. 2012. Electrochemical design of ZnO hierarchical structures for dye-sensitized solar cells. Sol. Energy Mater. Sol. Cells 102: 8-14.

Guo, M., P. Diao and S. Cai. 2005a. Hydrothermal growth of perpendicularly oriented ZnO nanorod array film and its photoelectrochemical properties. Appl. Surf. Sci. 249: 71-75.

Guo, M., P. Diao and S. Cai. 2005b. Hydrothermal growth of well-aligned ZnO nanorod arrays: dependence of morphology and alignment ordering upon preparing conditions. J. Solid State Chem. 178: 1864-1873.

Guo, M., P. Diao, X. Wang and S. Cai. 2005c. The effect of hydrothermal growth temperature on preparation and photoelectrochemical performance of ZnO nanorod array films. J. Solid State Chem. 178: 3210-3215.

Guo, M., C. Yang, M. Zhang, Y. Zhang, T. Ma, X. Wang, et al. 2008. Effects of preparing conditions on the electrodeposition of well-aligned ZnO nanorod arrays. Electrochimica Acta 53: 4633-4641.

Guo, X. D., H. Y. Pi, Q. Z. Zhao and R. X. Li. 2012. Controllable growth of flowerlike ZnO nanostructures by combining laser direct writing and hydrothermal synthesis. Mater. Lett. 66: 377-381.

Han, Z., L. Liao, Y. Wu, H. Pan, S. Shen and J. Chen. 2012. Synthesis and photocatalytic application of oriented hierarchical ZnO flower-rod architectures. J. Hazard. Mater. 217-218: 100-106.

Heo, Y. W., D. P. Norton, L. C. Tien, Y. Kwon, B. S. Kang, F. Ren, et al. 2004. ZnO nanowire growth and devices. Mater. Sci. Eng., R 47: 1-47.

Hong, R., L. Xu, H. Wen, J. Chen, J. Liao and W. You. 2012. Control and characterization of structural and optical properties of ZnO thin films fabricated by thermal oxidation Zn metallic films. Opt. Mater. 34: 786-789.

Hou, Q., L. Zhu, H. Chen, H. Liu and W. Li. 2012. Growth of porous ZnO nanosheets by electrodeposition with $Zn_4SO_4(OH)_6{\cdot}4H2O$ as precursor. Electrochimica Acta 85: 438-443.

Hu, X., Y. Masuda, T. Ohji and K. Kato. 2010. Fabrication of $ZN(OH)_2$/ZnO nanosheet-ZnO nanoarray hybrid structured films by a dissolution-recrystallization route. J. Am. Ceram. Soc. 93: 881-886.

Ismail, A. A., A. El-Midany, E. A. Abdel-Aal and H. El-Shall. 2005. Application of statistical design to optimize the preparation of ZnO nanoparticles via hydrothermal technique. Mater. Lett. 59: 1924-1928.

Jia, C., X. Zhang, Y. Chen, Y. Su, Q. Zhou, M. Xin, et al. 2008. Liquid phase epitaxial growth and optical property of flower-like ZnO nanosheets on Zinc foil. Appl. Surf. Sci. 254: 2331-2335.

Jiang, J., Z. Huang, S. Tan, Y. Li, G. Wang and X. Tan. 2012. Preparation of ZnO nanosheets by a novel microemulsion-based hydrothermal method. Mater. Chem. Phys. 132: 735-739.

Jongnavakit, P., P. Amornpitoksuk, S. Suwanboon and T. Ratana. 2012. Surface and photocatalytic properties of ZnO thin film prepared by sol-gel method. Thin Solid Films 520: 5561-5567.

Katagiri, K., T. Suzuki, H. Muto, M. Sakai and A. Matsuda. 2008. Low temperature crystallization of TiO_2 in layer-by-layer assembled thin films formed from water-soluble Ti-complex and polycations. Colloids Surf., A 321: 233-237.

Ke, L., S. B. Dolmanan, L. Shen, P. K. Pallathadk, Z. Zhang, D. M. Ying Lai and H. Liu. 2010. Degradation mechanism of ZnO-based dye-sensitized solar cells. Sol. Energy Mater. Sol. Cells 94: 323-326.

Kim, K.-M., H.-R. Kim, K.-I. Choi, H.-J. Kim and J.-H Lee. 2011. ZnO hierarchical nanostructures grown at room temperature and their C_2H_5OH sensor applications. Sens. Actuators, B 155: 745-751.

Kim, S., M.-C. Jeong, B.-Y. Oh, W. Lee and J.-M. Myoung. 2006. Fabrication of Zn/ZnO nanocables through thermal oxidation of Zn nanowires grown by RF magnetron sputtering. J. Cryst. Growth 290: 485-489.

Kim, T.-U., J.-A. Kim, S. M. Pawar, J.-H. Moon and J. H. Kim. 2010. Creation of nanoscale two-dimensional patterns of ZnO nanorods using laser interference lithography followed by hydrothermal synthesis at 90°C. Cryst. Growth Des. 10: 4256-4261.

Kiomarsipour, N. and R. Shoja Razavi. 2012. Characterization and optical property of ZnO nano-, submicro- and microrods synthesized by hydrothermal method on a large-scale. Superlattices Microstruct. 52: 704-710.

Kiomarsipour, N. and R. Shoja Razavi. 2013. Hydrothermal synthesis and optical property of scale- and spindle-like ZnO. Ceram. Int. 39: 813-818.

Ko, S. H., D. Lee, H. W. Kang, K. H. Nam, J. Y. Yeo, S. J. Hong, et al. 2011. Nanoforest of hydrothermally grown hierarchical ZnO nanowires for a high efficiency dye-sensitized solar cell. Nano Lett. 11: 666-671.

Kuan, C. Y., M. H. Hon, J. M. Chou and I. C. Leu. 2012. Growth characteristics of hierarchical ZnO structures prepared by one-step aqueous chemical growth. Ceram. Int. 38: 1255-1260.

Kumar, R. S., P. Sudhagar, P. Matheswaran, R. Sathyamoorthy and Y. S. Kang. 2010. Influence of seed layer treatment on ZnO growth morphology and their device performance in dye-sensitized solar cells. Mater. Sci. Eng., B 172: 283-288.

Kurbanov, S. S., H. D. Cho and T. W. Kang. 2011. Effect of excitation and detection angles on photoluminescence spectrum from ZnO nanorod array. Opt. Commun. 284: 240-244.

Lei, A., B. Qu, W. Zhou, Y. Wang, Q. Zhang and B. Zou. 2012. Facile synthesis and enhanced photocatalytic activity of hierarchical porous ZnO microspheres. Mater. Lett. 66: 72-75.
Li, G. R., X. H. Lu, W. X. Zhao, C. Y. Su and Y. X. Tong. 2008. Controllable electrochemical synthesis of Ce^{4+}-doped ZnO nanostructures from nanotubes to nanorods and nanocages. Cryst. Growth Des. 8: 1276-1281.
Li, J., G. Lu, Y. Wang, Y. Guo and Y. Guo. 2012. A high activity photocatalyst of hierarchical 3D flowerlike ZnO microspheres: synthesis, characterization and catalytic activity. J. Colloid Interface Sci. 377: 191-196.
Li, Q., J. Bian, J. Sun, J. Wang, Y. Luo, K. Sun, et al. 2010. Controllable growth of well-aligned ZnO nanorod arrays by low-temperature wet chemical bath deposition method. Appl. Surf. Sci. 256: 1698-1702.
Li, Z. W., W. Gao and R. J. Reeves. 2005. Zinc oxide films by thermal oxidation of zinc thin films. Surf. Coat. Technol. 198: 319-323.
Lin, Y. and Q. Jiang. 2011. Effect of substrates and anions of zinc salts on the morphology of ZnO nanostructures. Appl. Surf. Sci. 257: 8728-8731.
Liu, J., Z.-Y. Hu, Y. Peng, H.-W. Huang, Y. Li, M. Wu, et al. 2016. 2D ZnO mesoporous single-crystal nanosheets with exposed (Tan et al.) polar facets for the depollution of cationic dye molecules by highly selective adsorption and photocatalytic decomposition. Appl. Catal., B 181: 138-145.
Lockman, Z., Y. Pet Fong, T. Wai Kian, K. Ibrahim and K. A. Razak. 2010. Formation of self-aligned ZnO nanorods in aqueous solution. J. Alloys Compd. 493: 699-706.
Lockman, Z., K. A. Razak, T. K. Huat, T. W. Kian, L. C. Li, G. Kawamura, et al. 2012. Formation of 1-dimensional (1D) and 3-dimensional (3D) ZnO nanostructures by oxidation and chemical methods. Materialwiss. Werkstofftech. 43: 457-460.
Lokhande, C. D., P. M. Gondkar, R. S. Mane, V. R. Shinde and S.-H. Han. 2009. CBD grown ZnO-based gas sensors and dye-sensitized solar cells. J. Alloys Compd. 475: 304-311.
Lupan, O. and T. Pauporte. 2010. Hydrothermal treatment for the marked structural and optical quality improvement of ZnO nanowire arrays deposited on lightweight flexible substrates. J. Cryst. Growth 312: 2454-2458.
Lupan, O., V. M. Guerin, I. M. Tiginyanu, V. V. Ursaki, L. Chow, H. Heinrich, et al. 2010. Well-aligned arrays of vertically oriented ZnO nanowires electrodeposited on ITO-coated glass and their integration in dye sensitized solar cells. J. Photochem. Photobiol., A 211: 65-73.
Mahalingam, T., V. S. John, M. Raja, Y. K. Su and P. J. Sebastian. 2005. Electrodeposition and characterization of transparent ZnO thin films. Sol. Energy Mater. Sol. Cells 88: 227-235.
Martínez, O., V. Hortelano, J. Jiménez, J. L. Plaza, S. de Dios, J. Olvera, E. Diéguez, R. Fath, J. G. Lozano, T. Ben, D. González and J. Mass. 2011. Growth of ZnO nanowires through thermal oxidation of metallic zinc films on CdTe substrates. J. Alloys Compd. 509: 5400-5407.
Masuda, Y. and K. Kato. 2008. High *c*-axis oriented stand-alone ZnO self-assembled film. Cryst. Growth Des. 8: 275-279.
Matsuda, A., K. Kobayashi, T. Kogure, M. Sakai, K. Tadanaga, T. Minami, et al. 2008. Characterization of ramiform precipitates formed on SiO_2–TiO_2 gel coatings by electric field hot water treatment. J. Non-Cryst. Solids 354: 1263-1266.
Matsuda, A., W. K. Tan, S. Furukawa and H. Muto. 2013. Morphology-control of crystallites precipitated from ZnO gel films by applying electric field during hot-water treatment. Mater. Sci. Semicond. Process. 16: 1232-1239.

Noothongkaew, S., H. Nakajima, A. Tong-On, W. Meevasana and P. Songsiritthigul. 2012. Oxidation of Zn in UHV environment at low temperature. Appl. Surf. Sci. 258: 1955-1957.

Pan, A., R. Yu, S. Xie, Z. Zhang, C. Jin and B. Zou. 2005. ZnO flowers made up of thin nanosheets and their optical properties. J. Cryst. Growth 282: 165-172.

Pearton, C. J. A. S. 2006. Zinc Oxide Bulk, Thin Films and Nanostructures: Processing, Properties and Applications. Elsevier, Amsterdam.

Pearton, S. 2005. Recent progress in processing and properties of ZnO. Prog. Mater. Sci. 50: 293-340.

Pei, L. Z., H. S. Zhao, W. Tan, H. Y. Yu, Y. W. Chen, C. G. Fan, et al. 2010. Hydrothermal oxidization preparation of ZnO nanorods on zinc substrate. Physica E 42: 1333-1337.

Prastomo, N., Y. Daiko, T. Kogure, H. Muto, M. Sakai and A. Matsuda. 2009. Formation mechanism of titania nanosheet cryatallites on silica-titania gel films by vibration hot-water treatment. Mater. Sci. Eng., B 161: 170-174.

Prastomo, N., H. Muto, M. Sakai and A. Matsuda. 2010. Formation and stabilization of tetragonal phase in sol-gel derived ZrO_2 treated with base-hot-water. Mater. Sci. Eng., B 173: 99-104.

Prastomo, N., N. H. B. Zakaria, G. Kawamura, H. Muto, M. Sakai and A. Matsuda. 2011. High surface area $BaZrO_3$ photocatalyst prepared by base-hot-water treatment. J. Eur. Ceram. Soc. 31: 2699-2705.

Qin, Z., Q. Liao, Y. Huang, L. Tang, X. Zhang and Y. Zhang. 2010. Effect of hydrothermal reaction temperature on growth, photoluminescence and photoelectrochemical properties of ZnO nanorod arrays. Mater. Chem. Phys. 123: 811-815.

Qin, Z., Y. Huang, J. Qi, H. Li, J. Su and Y. Zhang. 2012. Facile synthesis and photoelectrochemical performance of the bush-like ZnO nanosheets film. Solid State Sci. 14: 155-158.

Qiu, J., M. Guo and X. Wang. 2011. Electrodeposition of hierarchical ZnO nanorod-nanosheet structures and their applications in dye-sensitized solar cells. ACS Appl. Mater. Interfaces 3: 2358-2367.

Rusu, G. G., M. Girtan and M. Rusu. 2007. Preparation and characterization of ZnO thin films prepared by thermal oxidation of evaporated Zn thin films. Superlattices Microstruct. 42: 116-122.

Rusu, M., G. G. Rusu, M. Girtan and S. Dabos Seignon. 2008. Structural and optical properties of ZnO thin films deposited onto ITO/glass substrates. J. Non-Cryst. Solids 354: 4461-4464.

Shi, R., P. Yang, X. Dong, Q. Ma and A. Zhang. 2013 Growth of flower-like ZnO on ZnO nanorod arrays created on zinc substrate through low-temperature hydrothermal synthesis. Appl. Surf. Sci. 264: 162-170.

Singh, N., P. Pandey and F. Z. Haque. 2012. Effect of heat and time-period on the growth of ZnO nanorods by sol-gel technique. Optik 123: 1340-1342.

Sreekantan, S., L. R. Gee and Z. Lockman. 2009. Room temperature anodic deposition and shape control of one-dimensional nanostructured zinc oxide. J. Alloys Compd. 476: 513-518.

Sugunan, A., H. C. Warad, M. Boman and J. Dutta. 2006. Zinc oxide nanowires in chemical bath on seeded substrates: role of hexamine. J. Sol-Gel Sci. Technol. 39: 49-56.

Tan, W. K., K. A. Razak, K. Ibrahim and Z. Lockman. 2011a. Formation of ZnO nanorod arrays on polytetraflouroethylene (PTFE) via a seeded growth low temperature hydrothermal reaction. J. Alloys Compd. 509: 820-825.

Tan, W. K., K. A. Razak, K. Ibrahim and Z. Lockman. 2011b. Oxidation of etched Zn foil for the formation of ZnO nanostructure. J. Alloys Compd. 509: 6806-6811.

Tan, W. K., K. Abdul Razak, Z. Lockman, G. Kawamura, H. Muto and A. Matsuda. 2013a. Formation of highly crystallized ZnO nanostructures by hot-water treatment of etched Zn foils. Mater. Lett. 91: 111-114.

Tan, W. K., K. Abdul Razak, Z. Lockman, G. Kawamura, H. Muto and A. Matsuda. 2013b. Photoluminescence properties of rod-like co-doped ZnO nanostructured films formed by hot-water treatment of sol-gel derived coating. Opt. Mater. 35: 1902-1907.

Tan, W. K., Z. Lockman, K. Abdul Razak, G. Kawamura, H. Muto and A. Matsuda. 2013c. Enhanced dye-sensitized solar cells performance of ZnO nanorod arrays grown by low-temperature hydrothermal reaction. Int. J. Energy Res. 37(15): 1992-2000.

Tan, W. K., K. A. Razak, Z. Lockman, G. Kawamura, H. Muto and A. Matsuda. 2013d. Optical properties of two-dimensional ZnO nanosheets formed by hot-water treatment of Zn foils. Solid State Commun. 162: 43-47.

Tan, W. K., K. Abdul Razak, Z. Lockman, G. Kawamura, H. Muto and A. Matsuda. 2014a. Synthesis of ZnO nanorod-nanosheet composite via facile hydrothermal method and their photocatalytic activities under visible-light irradiation. J. Solid State Chem. 211: 146-153.

Tan, W. K., L. C. Li, K. A. Razak, G. Kawamura, H. Muto, A. Matsuda, et al. 2014b. Formation of two-dimensional ZnO nanosheets by rapid thermal oxidation in oxygenated environment. J. Nanosci. Nanotechnol. 14: 2960-2967.

Tao, Y., M. Fu, A. Zhao, D. He and Y. Wang. 2010. The effect of seed layer on morphology of ZnO nanorod arrays grown by hydrothermal method. J. Alloys Compd. 489: 99-102.

Tian, Z. R., J. A. Voigt, J. Liu, B. Mckenzie, M. J. Mcdermott, M. A. Rodriguez, et al. 2003. Complex and oriented ZnO nanostructures. Nat. Mater. 2: 821-826.

Tong, Y. H., Y. C. Liu, L. Dong, D. X. Zhao, J. Y. Zhang, Y. M. Lu, et al. 2006. Growth of ZnO nanostructures with different morphologies by using hydrothermal technique. J. Phys. Chem. B 110: 20263-20267.

Umar, A., M. S. Chauhan, S. Chauhan, R. Kumar, G. Kumar, S. A. Al-Sayari, et al. 2011. Large-scale synthesis of ZnO balls made of fluffy thin nanosheets by simple solution process: structural, optical and photocatalytic properties. J. Colloid Interface Sci. 363: 521-528.

Venkatesha, T. G., Y. Arthoba Nayaka, R. Viswanatha, C. C. Vidyasagar and B. K. Chethana. 2012. Electrochemical synthesis and photocatalytic behavior of flower shaped ZnO microstructures. Powder Technol. 225: 232-238.

Wang, D., Y. Zhao and C. Song. 2010a. Synthesis and properties of cuboid-shaped ZnO hierarchical structures. Solid State Sci. 12: 776-782.

Wang, H., Y. Liu, M. Li, H. Huang, H. Xu and H. Shen. 2010b. Fabrication of three-dimensional ZnO with hierarchical structure via an electrodeposition process. Appl. Phys. A 103: 463-466.

Wang, Y., X. Fan and J. Sun. 2009. Hydrothermal synthesis of phosphate-mediated ZnO nanosheets. Mater. Lett. 63: 350-352.

Xu, C. H., H. F. Lui and C. Surya. 2011. Synthetics of ZnO nanostructures by thermal oxidation in water vapor containing environments. Mater. Lett. 65: 27-30.

Xu, S. and Z. L. Wang. 2011. One-dimensional ZnO nanostructures: solution growth and functional properties. Nano Res. 4: 1013-1098.

Yousefi, M., M. Amiri, R. Azimirad and A. Z. Moshfegh. 2011. Enhanced photoelectrochemical activity of Ce doped ZnO nanocomposite thin films under visible light. J. Electroanal. Chem. 661: 106-112.

Yu, J., B. Huang, X. Qin, X. Zhang, Z. Wang and H. Liu. 2011. Hydrothermal synthesis and characterization of ZnO films with different nanostructures. Appl. Surf. Sci. 257: 5563-5565.

Yue, S., J. Lu and J. Zhang. 2009. Controlled growth of well-aligned hierarchical ZnO arrays by a wet chemical method. Mater. Lett. 63: 2149-2152.

Yuvaraj, D., K. Narasimha Rao and K. K. Nanda. 2009. Effect of oxygen partial pressure on the growth of zinc micro and nanostructures. J. Cryst. Growth 311: 4329-4333.

Zhang, Q., C. S. Dandeneau, X. Zhou and G. Cao. 2009. ZnO Nanostructures for dye-sensitized solar cells. Adv. Mater. 21: 4087-4108.

Zhang, X. G. 1996. Corrosion and Electrochemistry of Zinc. Plenum Publishing Corporation, New York.

Zhang, Z., M. F. Hossain, T. Arakawa and T. Takahashi. 2010. Facing-target sputtering deposition of ZnO films with Pt ultra-thin layers for gas-phase photocatalytic application. J. Hazard. Mater. 176: 973-978.

Zou, G., W. Chen, R. Liu and Z. Xu. 2007. Orientation enhancement of polycrystalline ZnO thin films through thermal oxidation of electrodeposited zinc metal. Mater. Lett. 61: 4305-4308.

12

Thermally Grown Native Oxide Thin Films on SiC

Banu Poobalan[1,*] and Kuan Yew Cheong[2,a]

ABSTRACT

Integration of high quality functional thin layer of oxides on semiconductor, in particular wide-bandgap silicon carbide (SiC), is of extreme importance in order to generate metal-oxide-semiconductor based devices for high power, high temperature and/or high radiation applications in the future. This chapter reviews the current understanding of surface modification of SiC by thermally grown native oxide thin films. In the first section, importance of SiC as a semiconductor substrate and the need of gate oxide are presented. This is followed by a discussion on growth mechanisms, types and origin of oxide trap charges and interfacial and near interfacial traps in SiC/SiO_2 system. Besides, the current technological process for the improvements of SiO_2/SiC system and the promising techniques for achieving device-quality interfaces that is required for commercial applications are also reviewed. Finally, factors influencing a high quality oxide on SiC, namely process ambient, oxidation/nitridation parameters and semiconductor substrate are also systematically compared, discussed and reviewed. Based on the knowledge and understanding of these topics, further improvements on the fabrication technology of gate oxide can be obtained.

Keywords: Silicon carbide, thermal oxidation, gate oxide, surface modification, oxide trap charges, interfacial traps, process technology.

1. INTRODUCTION

In the early years of semiconductor electronics industry, Germanium (Ge) was the original material used to fabricate semiconductor devices such as diodes and transistors. However, the narrow bandgap (0.66 eV) characteristic of Ge causes its pn junctions to be in reverse-biased which eventually contribute to large leakage current.

[1] Green Technology and Energy Efficiency Research Group, School of Electrical Systems Engineering, Universiti Malaysia Perlis, Pauh Putra Campus, 02600 Arau, Perlis Malaysia.

[2] Electronic Materials Research Group, School of Materials & Mineral Resources Engineering, Engineering Campus, Universiti Sains Malaysia, 14300 Nibong Tebal, Penang, Malaysia.

[a] E-mail: srcheong@usm.my

* Corresponding author: banu@unimap.edu.my

This limits the operation temperature of Ge-based device to be lower than 100°C. In addition, planar technology in integrated circuit requires a high quality of passivation layer on a semiconductor surface. Native oxide of Ge (GeO_2) is able to provide a natural passivation layer on Ge but such a layer is water soluble and can easily be dissociated at the processing temperature of 800°C. As a result, Ge is being replaced by silicon (Si) for semiconductor devices fabrication (Stanley and Richard 2000).

Si has a larger bandgap (1.12 eV) in comparison with Ge, which results in smaller leakage current and thereby allows Si-based devices to be built with maximum operating temperature of about 150°C. Si has dominated as the main stream semiconductor material in the electronics industry due to its feasibility to form chemically stable silicon dioxide (SiO_2), which is the most critical requirement for the formation of gate oxide (Nicollian and Brews 1982, Stanley and Richard 2000). However, Si is not suitable for high temperature, high power and high switching frequencies applications as the bulk of its properties are unable to withstand high breakdown field (Si critical avalanche electric field is 0.3 MV/cm) (Zhao 2005). To overcome these limitations, wide bandgap semiconductors are presently switching from research and development into real world applications. Wide bandgap semiconductors such as silicon carbide (SiC), gallium nitride (GaN) and indium nitride (InN) can be categorized into one group, while diamond, boron nitride (BN) and aluminum nitride (AlN) into another because the former has a bandgap of 2-3.5 eV and the latter 5.5-6.5 eV (Chow and Agarwal 2006). As compared to Si, wide bandgap semiconductors have superior physical properties, which offer a lower intrinsic carrier concentration (10 to 35 orders of magnitude), higher electric breakdown field (4-20 times), a higher thermal conductivity (3-13 times) and a larger saturated electron drift velocity (2-2.5 times) (Siergiej et al. 1999, Wang and Zhong 2002, Dimitrijev and Jamet 2003, Chow and Agarwal 2006).

Of all wide bandgap semiconductors, SiC has become the material of choice for semiconductor devices because of its superior properties. This is because the other wide bandgap nitride materials (GaN, InN, BN and AlN) need to be grown on substrates such as sapphire to get their thermal advantage properties. On the other hand, diamond is a hard material (10 Mohs) (Ted Pella, Inc., 2015) which needs higher processing temperature (sublimation point at 3642°C) (Greenville Whittaker 1978; Committee on Materials for High-Temperature Semiconductor Devices 1995). SiC has the ability to grow SiO_2 using conventional thermal oxidation (like Si) and is able to withstand a harsh environment such as at elevated temperature (~1700°C) (Strife and Sheehan 1988, Gupta et al. 2013, Roy et al. 2014, Cheong and Wong 2014). This makes SiC as the choice of material for the development of power semiconductor devices applications (Fujihira et al. 2004, Fraga et al. 2012).

2. SiC AS A SEMICONDUCTOR SUBSTRATE

SiC was found in the early 1900s, but its use was limited only for niche applications (Zetterling 2002). There are several reasons for this, the most critical problem is making SiC material of sufficient quality for power semiconductor devices applications (Zolper and Skowronski 2005). The possibility to exploit the materials for electronic devices became a reality in the late 1980s after SiC wafers became available from a commercial vendor (Zetterling 2002). The significant progress achieved in developing SiC material for semiconductor devices in the recent years has boosted the interest of many researchers (Gupta et al. 2013, Fraga et al. 2012).

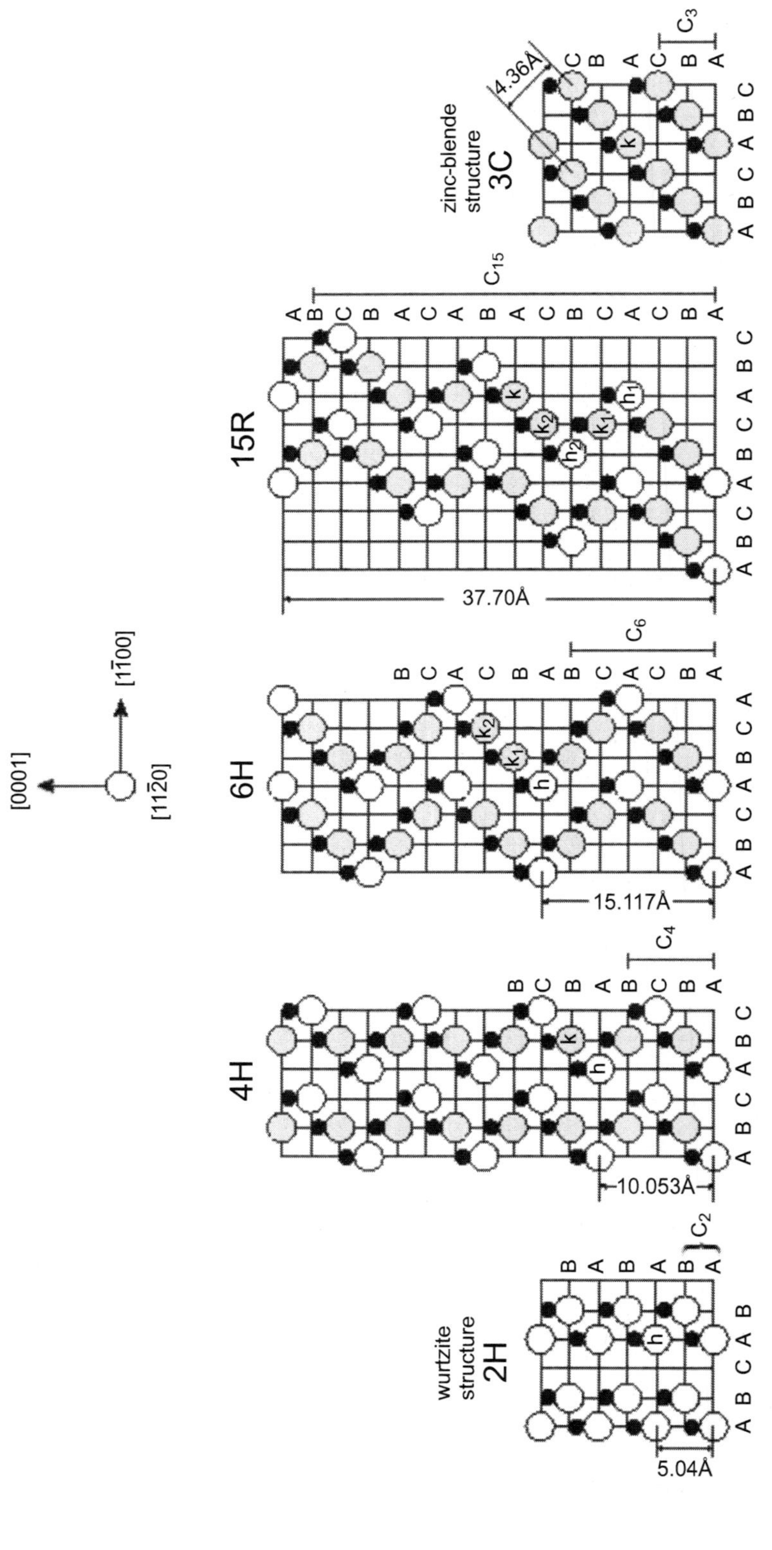

Fig. 12.1 Stacking sequences for different SiC polytypes in the [1120] direction (Ayalew 2004).

SiC crystal are comprised of two atoms, which are silicon (Si) and carbon (C) and normally each Si atom has exactly four C atoms as its neighbor and vice versa. SiC generally has several hundred stacking orders being identified in nature (Fig. 12.1). Figure 12.2 shows the layer structure of SiC in the [0001] direction. In this structure, carbon atoms are tetrahedrally bonded and linked to three Si atoms within the bilayer with a single bond linked to a Si atom in the layer beneath. The locations of the carbon atoms within a bilayer forms a hexagonal structure, labelled 'A' in the Fig. 12.2. The next bilayer then has the possibility of positioning its carbon atom in the 'B' or the 'C' lattice sites. Thus, the stacking sequence defines the material polytype (Zetterling 2002, Ayalew 2004).

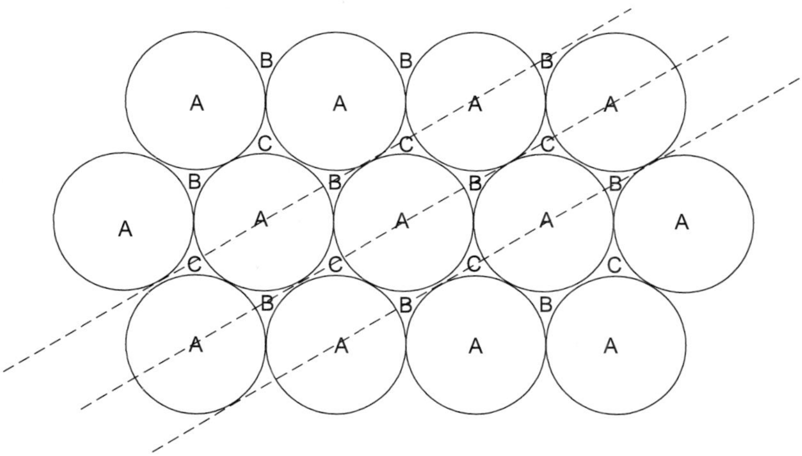

Fig. 12.2 Site locations of C atoms in the [1100] direction (Ayalew 2004).

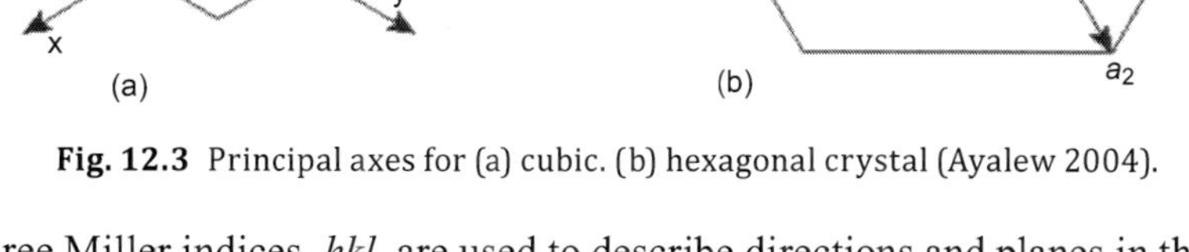

Fig. 12.3 Principal axes for (a) cubic. (b) hexagonal crystal (Ayalew 2004).

Three Miller indices, *hkl*, are used to describe directions and planes in the cubic crystal. These are integers with the equal ratio as the reciprocals of the intercepts with *x*-, *y*- and *z*-axes, respectively, as shown in Fig. 12.3(a). Four principal axes are commonly used: a_1, a_2, a_3 and c in hexagonal structures. Only three axes are used to clearly identify a plane or a direction, since the sum of the reciprocal intercepts with a_1, a_2 and a_3 is zero. The three a-vectors (with 120° angles between each other) are all in the close-packed plane called a-plane, whereas the *c*-axis is perpendicular to this plane as shown in Fig. 12.3(b) (Ayalew 2004).

SiC has a polar crystal structure whereby looking perpendicularly at the *a*-plane (which is between a_1 and *c* in Fig. 12.3(b)), *C* atoms will be either directly arranged on top of Si atoms or vice versa. The former is called silicon face orientation, the latter one is termed carbon face orientation. Between these two faces, silicon face is the most commonly produced and used to for device manufacture. In order to dope the intrinsic SiC semiconductor, aluminum and boron are used as dopants to form p-type SiC and nitrogen and phosphorous are used as dopants for n-type SiC (Zetterling 2002).

SiC has emerged as a semiconductor device substrate, due to several reasons related to the electrical property. The key reason to almost all advantages of using SiC in devices is its wide bandgap characteristics. The ability to withstand a harsh environment such as at elevated temperatures adds an advantage for the development of SiC for semiconductor devices (Choyke et al. 2004). SiC with its wide bandgap, depending on polytype and doping concentrations has an intrinsic temperature around 1000°C (Zetterling 2002).

SiC as the wide bandgap material has high impact ionization energy, which means that the electric field can become very high without avalanche multiplication of ionized carriers. SiC offers approximately 10 times higher electric breakdowns field than Si for the same depletion width which offers advantages for high-voltage devices (Zetterling 2002).

Basically, SiC-based device can be made smaller for the same breakdown voltage or in other words the signal has a shorter distance to travel that makes the device operate faster. The relative dielectric constant is also lower for SiC if compared with other semiconductors and since the capacitance is directly proportional to the dielectric constant, parasitic capacitance value will also become smaller.

Besides, it is often stated that higher power densities are achieved with a high thermal conductivity (λ) of SiC. According to Eq. (1), three times higher power flow θ can be accommodated with the same temperature increases ΔT at the junction:

$$\Delta T = \theta\ (1/\lambda) \tag{1}$$

Ten times higher power flow (θ) is expected if the higher critical field is exploited and the devices are made 10 times smaller where l is the device thickness. This means a higher junction temperature is achieved for SiC devices, even though the thermal conductivity is higher (Zetterling 2002).

Recent developments in SiC device technology have opened up its applications to fields such as the aerospace and aircraft sectors, where by SiC based power electronics devices could be utilized for substantial weight savings and enhanced jet engine performance (Dixit 2008). SiC material can be used to replace Si as the substrate in power circuits of electric motors and power control for electric vehicles, robotics and power supplies (Dhar et al. 2005). It offers much higher efficiency than Si in these applications. Figure 12.4 shows that by replacing Si-based devices to SiC-based devices, the power conversion loss can be reduced to one-third and by 2030, approximately 5.8 million kW of energy can be saved. The facts clearly show that SiC-based devices will have a major impact on the size, efficiency and application of power electronics (Arai 2011).

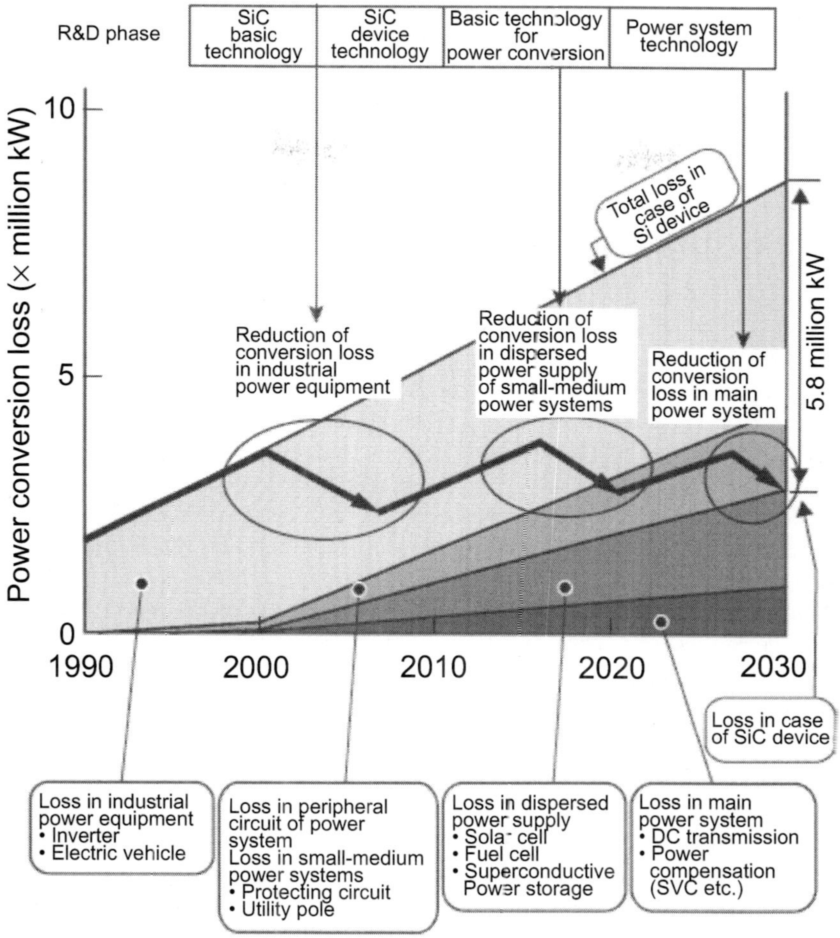

Fig. 12.4 Effect on energy saving when the SiC device is implemented in Japan (Arai 2011).

Table 12.1 Electrical Properties of common polytypes of SiC (Park 1998, Levinshteir and Shur 2001).

Properties	3C-SiC	4H-SiC	6H-SiC
Breakdown Field (V cm^{-1})	~10^6	6×10^5	6×10^5
Mobility (cm^2 V^{-1} s^{-1}): holes	≤ 320	≤ 120	≤ 90
Mobility (cm^2 V^{-1} s^{-1}): electrons	≤ 800	≤ 900	≤ 400
Diffusion coefficient (cm^2 s^{-1}): holes	≤ 8	≤ 3	≤ 2
Diffusion coefficient (cm^2 s^{-1}): electrons	≤ 20	≤ 22	≤ 10
Electron Thermal Velocity (m s^{-1})	2×10^5	1.9×10^5	1.5×10^5
Hole Thermal Velocity (m s^{-1})	1.5×10^5	1.2×10^5	1.2×10^5
Saturated Electron Drift Velocity (cm s^{-1})	2×10^7	2×10^7	2×10^7

Table 12.1 shows some of the important electrical properties of common polytypes of SiC. In terms of application purposes, the wafer has to be single crystal and only a few polytypes are stable enough for large wafer production. There has been much research conducted on 4H-SiC or 6H-SiC which the polytypes are commercially available at large (Zetterling 2002, Gupta et al. 2013, Fraga et al. 2012, Liu et al. 2015).

3. GATE OXIDE ON SiC SUBSTRATE

Gate oxide is a dielectric layer that isolates gate terminal of a Metal-Oxide-Semiconductor (MOS)-based transistor from the underlying source and drain terminals as well as the conductive channel that links the source and drain when the transistor is turned on (Fig. 12.5). When no bias is applied to the gate with respect to the substrate, source and drain are isolated. In contrast, if the applied gate voltage is high enough, a thin conductive layer of electrons is induced in the substrate and the source and drain is connected via a channel. In this condition, current can flow from the source to drain with a drain voltage being also applied (Zeghbroeck 2011). From the above description, it appears clearly that the operation of a MOS transistor is based on the perfect insulating properties of the oxide layer. Therefore, it is crucial to understand the properties of thermally grown gate oxide on semiconductor substrate, particularly SiC substrate. Generally, a MOS structure (Fig. 12.6) is used to investigate the quality and reliability of the thermally grown oxide as it is a simple structure and easy to fabricate. Besides, it is also considered as the first step towards integration of Metal–Oxide–Semiconductor Field-Effect Transistor (MOSFET). By using MOS as the test structure, it may provide considerable information regarding the properties of its dielectric sandwiched between the metal electrode and semiconductor (Bentarzi 2011, Liu et al. 2015).

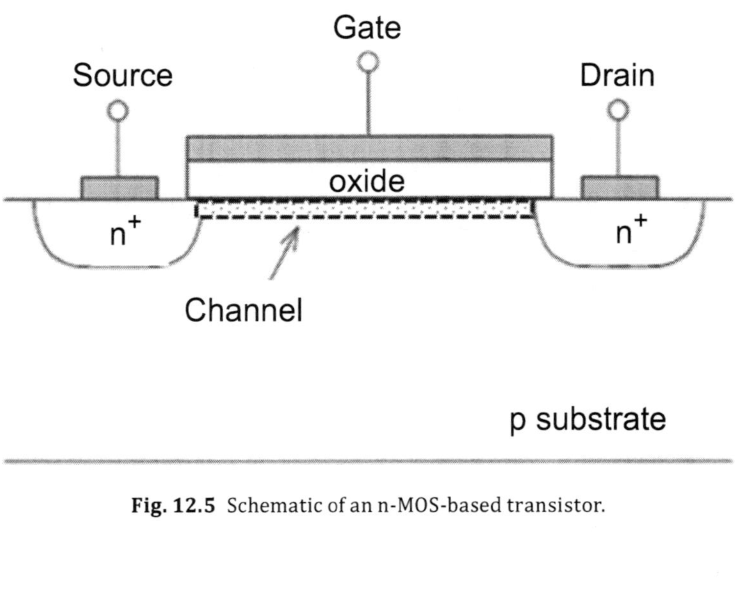

Fig. 12.5 Schematic of an n-MOS-based transistor.

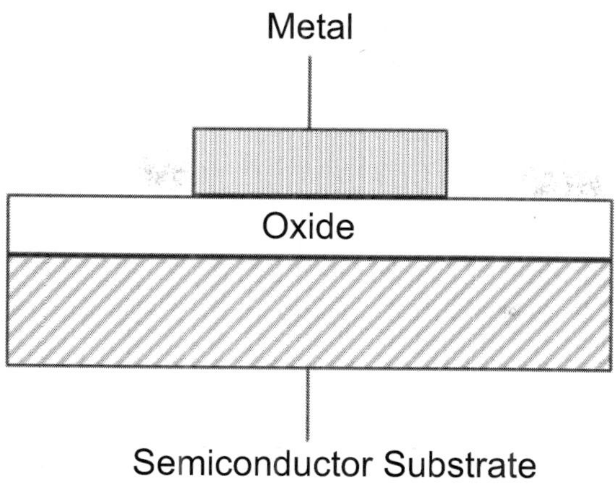

Fig. 12.6 Schematic of a MOS test structure.

Table 12.2 shows major applications of SiC-based MOSFET and their respective voltage ratings. In terms of power MOSFETs for 600 V and higher voltage applications such as switch-mode power supplies, AC motor drives, solar inverters and automotive electronics (Hamada et al. 2010, Casady et al. 1998), (Fairchild Semiconductor 2007), (CREE Inc., 2012), a high-quality with a relatively thick gate oxide that may be subjected to a high electric field is much needed (Takaya et al. 2013), (Agarwal et al. 2004). The equation below shows the dielectric constant ratio between the 4H-SiC and oxide:

$$\varepsilon_{\text{ratio}} = \frac{\varepsilon_{r,\text{SiC}}}{\varepsilon_{r,\text{ox}}} \tag{2}$$

where dielectric constant of SiC, $\varepsilon_{r,\text{SiC}} = 10$ and dielectric constant, $\varepsilon_{r,\text{ox}} = 3.9$. From Eq. (2), it can be calculated that the electric field in the gate oxide would be roughly 2.5*x* higher than in SiC when the device is in the blocking mode during operation (Agarwal et al. 2004, Krishnaswami et al. 2004).

The electric field in SiC and gate oxide is given by the following equation:

$$E_{\text{SiC}} = E_{\text{ox}}\left(\frac{\varepsilon_{r,\text{ox}}}{\varepsilon_{r,\text{SiC}}}\right) \tag{3}$$

where E_{SiC} is the electric field in SiC and E_{ox} is the electric field in gate oxide. In order to insure long-term device reliability in SiO_2/4H-SiC system, the maximum surface electric field for SiC, E_{SiC}, has to be kept approximately 1.2 MV/cm which is equivalent to 3 MV/cm of electric field in the gate oxide, E_{ox} (Agarwal et al. 2004).

The gate oxide thickness is related to the gate voltage and electric field in the gate oxide by the following expression:

$$t_{\text{ox}} = \frac{(V_G - \phi\text{ms})}{E_{\text{ox}}} \tag{4}$$

where t_{ox} is the gate oxide thickness, V_G is the gate voltage and ϕms is the metal –semiconductor work-function difference. Considering E_{ox} = 3 MV/cm as the maximum electric field limitation value for 4H-SiC based gate oxide and ϕms = 0.45 V, a minimum gate oxide thickness of 50 nm is required in order to support gate voltage, V_G, of 15 V (Agarwal et al. 2004, Gurfinkel et al. 2008). On the other hand, as the gate voltage increases, the electric field in the oxide also rises, resulting in a large current flow into the oxide through various charge conduction mechanisms (Ranua′rez et al. 2006, Cheong et al. 2008, Alialy et al. 2014), which eventually leads to oxide breakdown (Agarwal et al. 1997, Agarwal et al. 2004, Gupta et al. 2010). Hence, the need to thermally grow a thick SiO_2 gate (>50 nm) with good quality is crucial for high power devices applications.

Table 12.2 Major Applications of SiC-MOSFET in Motor Controls and Power Supplies (Majumdar 2013).

Applications	Voltage ratings
Home Appliances (Refrigerator, Air-conditioner and Washing Machines)	600 V
Automotive	600-1200 V
Elevators, UPS and Factory Automation, Power supplies, Alternative Energy Sources	600-1700 V
Electric Railway Systems, Metal and Heavy Industries	1200-6500 V
Power network and Utilities	>10 kV

4. GROWTH MECHANISMS OF THERMALLY GROWN SiO_2 THIN FILMS ON SiC

The knowledge of thermal oxidation mechanisms on SiC is important in order to produce reliable and high-quality gate oxides for high power devices applications. Different quality of SiO_2/SiC interfaces are created by various oxide growth processing techniques. The mechanisms of thermally grown SiO_2 films on SiC is explained well using two common oxidation techniques which are dry and wet oxidation models. It is important to understand the effects associated to each models and use this knowledge for development of optimized process conditions in particular for gate oxide growth in the industry-preferred environment.

4.1 Dry Oxidation Model

In general, oxidation rate of SiC is more than one order of magnitude lower than that of Si (Harris and Afanas'ev 1997). As a compound semiconductor, reaction of SiC to form SiO_2 results in a by-product containing C. It is well known that C plays a detrimental role in the formation of a high quality oxide (Harris and Afanas'ev 1997, Liu et al. 2015). It has been shown that the oxidation kinetics of SiC is described by the same kinetics rules as oxidation of Si, as defined by the Deal and Grove model. In the beginning, oxidation is limited by reactions at the SiO_2/SiC interface. Then, once an oxide layer is formed, the oxidizing species have to pass through this oxide layer (Harris and Afanas'ev 1997). For increasing oxide thickness there will be a point at which the reaction is oxygen deficient and a parabolic dependence sets in leading to some speculation that the reaction should be limited by the out-diffusion of CO. This growth law could be written as:

$$x^2 + Ax = B(t + \tau) \tag{5}$$

where x is the oxide thickness and t is the oxidation time. The time constant, τ, determines the thickness of the initial oxide layer. Ratio B/A is called the linear rate and is proportional to the reaction rate of the slowest reaction. Constant B is proportional to the diffusion coefficient and is called the parabolic rate. These two constants are thermally activated as exp $(-E_{B/A}/kT)$ and exp $(-E_B/kT)$ for the linear and parabolic rate, respectively (Raynaud 2001). Contrary to the relatively simple oxidation of Si, there are five major steps in the thermal oxidation of SiC (Gupta and Akhtar 2011):

1. Transport of molecular oxygen gas to the oxide surface.
2. In-diffusion of oxygen through the oxide film.
3. Reactions with SiC at the oxide/SiC interface
4. Out-diffusion of product gases (e.g., CO_2) through the oxide film.
5. Removal of product gases away from the oxide surface.

The last two steps are not involved in the oxidation of Si. The first and last steps are rapid and the fact that the reaction rate is not constant over time shows that this is not the rate limiting step but diffusion limited model. The principal chemical reactions that can occur at the SiC interface are as following expressions (Harris and Afanas'ev 1997):

$$\mathrm{SiC} + \frac{3}{2}\mathrm{O_2} \leftrightarrow \mathrm{SiO_2} + \mathrm{CO} \tag{6}$$

$$\mathrm{SiC} + \mathrm{O_2} \leftrightarrow \mathrm{SiO_2} + \mathrm{C} \tag{7}$$

In addition there are a number of secondary reactions that will determine the equilibrium at the reaction interface:

$$\mathrm{SiC} + 2\mathrm{CO} \leftrightarrow 3\mathrm{C} + \mathrm{SiO_2} \tag{8}$$

$$2\mathrm{CO} + \mathrm{O_2} \leftrightarrow 2\mathrm{CO} \tag{9}$$

The oxidation rate is a strong dependence on crystal orientation, face, polytype, doping density type and concentration (Harris and Afanas'ev 1997). However, the Deal-Grove model cannot explained the growth rate in thin oxide regime (< 20 nm) (Hijikata et al. 2009) as reported in Si. A good fitting could not be achieved with any values B/A and B, though the fittings are well in the oxide thickness larger than 20 nm (Hijikata et al. 2009).

On the other hand, 'Si-C emission' model describes the SiC oxidation process better than the Deal-Grove model (Hijikata et al. 2009). In 'Si-C emission' model, the interfacial Si and C emission accompanied by the oxidation of SiC, showed that the model well reproduced the oxide growth rates of SiC at the entire thickness range both for the (0001) Si-face and (000-1) C-face.

Taking into account that Si and C atoms emitted from the interface during the oxidation, the reaction equation for SiC oxidation can be written as (Hijikata et al. 2009):

$$SiC + \left(2 - v_{Si} - v_c - \frac{\alpha}{2}\right) O_2 \rightarrow (1 - v_{Si})\, SiO_2 + v_{Si}Si + v_cC + \tag{10}$$

$$\alpha CO + (1 - v_c - \alpha)\, CO_2$$

where v and α denote the interfacial emission rate and the production rate of CO, respectively and the subscripts, Si and C, denote the values for the corresponding atoms. Interfacial reaction rate for SiC oxidation (k) is thought to be suppressed by the accumulation of Si atoms and C atoms emitted near the interface, k is given by multiplying decreasing functions for Si and C (Hijikata et al. 2009):

$$k = k_0 \left(1 - \frac{C^1_{Si}}{C^0_{Si}}\right)\left(1 - \frac{C^0_{C}}{C^0_{C}}\right) \tag{11}$$

where C^1 and C^0 are the interfacial concentration of corresponding interstitials and the solubility limit the corresponding interstitials in the oxide, respectively and k_0 is the interfacial reaction rate when oxide thickness nearly equals zero, i.e. intrinsic interfacial interaction rate without influence of the accumulation of emitted Si and C atoms.

Higher oxidation temperatures result in the higher areal densities of Si interstitials regardless of Si- and C-face. The decline of C_{Si} with respect to the distance from the interface for Si-face is dependent on oxidation temperature, in contrast, that for C-face decreases remarkably with decreasing oxidation temperature. In comparison of Si- and C-face, C_{Si} for Si-face is higher than that for C-face regardless of temperature. The concentration of C interstitials C_C is nearly constant against the distance from the interface at any temperature for both of Si- and C-faces and this indicates that C interstitials rapidly diffuse through the oxide. Likewise in the case of Si interstitials, a higher oxidation temperature may bring about a higher concentration of C interstitials regardless of Si- and C-face. In comparison to Si- and C-face, the C_C for Si-face is lower than that of C-face at any temperature, which is the opposite in the case of Si interstitials. The SiC–oxide interface structure can also be discussed on the basis of Si-C emission model, leading to the estimation of interface trap density (D_{it}) (Hijikata et al. 2009).

Besides Deal-Grove and Si-C emission models, first principles simulation suggests that oxygen molecules are dissociated in the SiO_2 layers or by Si atoms at the SiO_2 interface. The O atoms from the O_2 molecule oxidize the C atoms at the SiC interface and initiate the formation of Si-C-O or CO_2-C complexes. CO_x ($x = 1$ or 2) molecules are desorbed from these complexes by thermal motion (Ohnuma et al. 2007). CO_x molecules diffuse through the SiO_2 layers when they do not react with dangling bonds. The CO_x molecule that formed during C-face oxidation is diffused with less effort than those formed during Si-face oxidation in the interface region (Ohnuma et al. 2007). A single carbon residing within the interfacial region of the SiO_2 results in a Si-C-Si bridge and an oxygen protrusion. This carbon atom has a three-fold bandgap and an empty level close to the conduction band edge associated with the carbon dangling bond (Rozen et al. 2009).

4.2 Wet Oxidation Model

In the case of wet oxidation, only a little literature has been reported on the mechanisms involved during oxide growth on SiC. Based on the reported literature (Raynaud 2001), the growth rate can be fitted by linear-parabolic functions independently of the SiC terminal face. During linear growth, activation energy, $E_{B/A}$, is 3 eV, which is within the range of the binding energy of SiC. Therefore, the disruption of Si-C bonds explains rather well the activation energies. During parabolic growth, E_B is much larger than the activation energy of 0.79 eV found for the diffusion of H_2O. This proves that the oxidation process is not limited by the diffusion of H_2O, whereas, the limiting factor is the out-diffusion of CO or the diffusion of O_2 (Raynaud 2001).

Other literature (Tortorelli and More 2003) have reported that a principal effect is the influence of H_2O on the parabolic rate constant, k_p, that often governs the growth of SiO_2 on SiC:

$$x^2 = k_p t \tag{12}$$

where x is the thickness of the oxide product and t is the oxidation time. As shown by Deal and Grove in Si and as extensively characterized by Opila (1999) for SiC, water vapor significantly increases k_p, even at relatively low concentrations in the environment, through its direct influence as an oxidant.

Fundamentally, this is because k_p is dependent on the product of oxidant diffusivity (D) and solubility (C) in the growing oxide and while D for H_2O in SiO_2 is somewhat less than that of oxygen, C is substantially higher such that [Eq. (13)]:

$$k_p(H_2O, SiO_2)/k_p(O_2, SiO_2) = \sim 50 \tag{13}$$

Reaction between the incoming water molecules and the SiO_2 network is observed during diffusion, leading to the formation of silanol groups (Si–OH) [Eq. (14)]. This can take place even at low temperatures (<250°C) due to the low activation energy of the following reversible reaction (Soares et al. 2009):

$$H_2O + Si - O - Si \leftrightarrow Si - OH \tag{14}$$

With conditions whereby significant volatility of the SiO_2 can occur, the SiO_2 can grow by solid-state oxidation process governed by k_p, but after a certain duration, its thickness is restricted by the simultaneous loss of SiO_2 by formation of gaseous products. These joined reactions can be described by para linear oxidation kinetics in terms of both k_p and a linear rate constant, k_l, related to the volatilization rate of the SiO_2. The equilibrium SiO_2 layer is created at an equal rate where it is removed by formation of the volatile products and the material beneath is consumed at a linear (recession) rate set by k_l. Since k_l is related to the volatilization rate, it should be dependent on the pressure and velocity of the oxidizing gas in well-prescribed ways (Tortorelli and More 2003).

4.3 Comparison Between Thermally Grown Oxides in Dry and Wet Ambient

Generally, thicker oxide could be obtained by growing oxides in wet ambient rather than in dry ambient due to a much higher solid solubility of H_2O in SiO_2 than O_2 in

SiO_2, which in addition provides hydrogen passivation of electrically active defects near the SiO_2/SiC interface (Harris and Afanas'ev 1997, Xu et al. 2003, Benfdila and Zekentes 2010). Figure 12.7 shows the experimentally measured thickness of thermal oxide at different temperatures by the method of wet and dry oxidation on Si-face 4H-SiC. It shows that thermally grown oxides in wet ambient exhibit thicker oxides than oxides grown in dry ambient at respective processing duration and temperature (Gupta and Akhtar 2011).

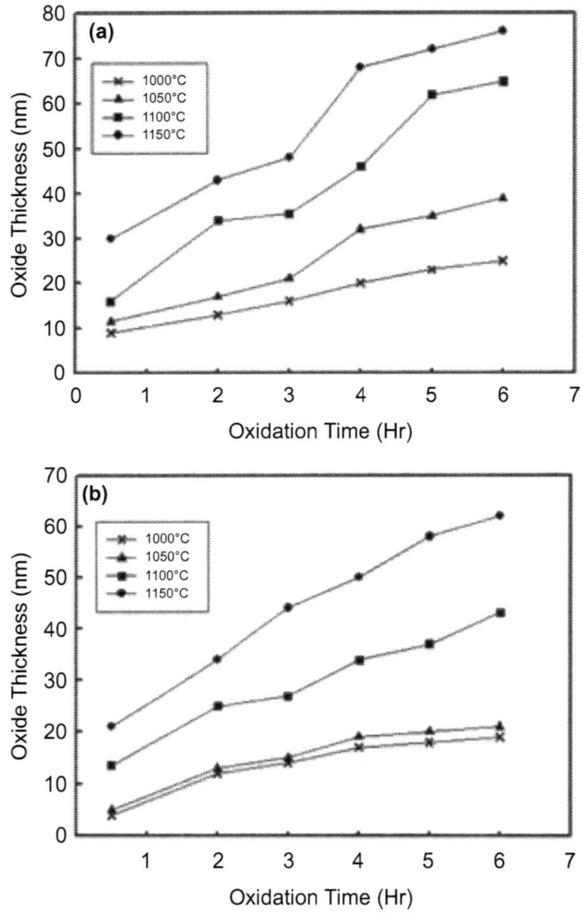

Fig. 12.7 Plots of oxide growth profile on Si-face by (a) Wet oxidation. (b) Dry oxidation (Gupta and Akhtar 2011).

A plot of oxide thickness (t) versus t/x from Eq. (1) should yield a straight line with intercept $-A$ and slope B. It has been observed that parabolic rate constant value of B rises with increasing oxidation temperature. At the same condition, the linear rate constant value of (B/A) also heightens with increasing temperature. Measured values of these constants are listed in Tables 12.3 and 12.4 for wet and dry oxidation on Si-face 4H-SiC, respectively.

Table 12.3 Experimentally measured Parabolic Rate Constant (*B*) (Gupta and Akhtar 2011).

Temperature (°C)	1000	1050	1110	1150
Si-face (Dry oxidation)	0.0000748	0.0001035	0.0003021	0.0006130
Si-face (Wet oxidation)	0.000158	0.00033088	0.0008830	0.00120

Table 12.4 Experimentally measured Linear Rate Constant (*B/A*) (Gupta and Akhtar 2011).

Temperature (°C)	1000	1050	1110	1150
Si-face (Dry oxidation)	0.01533	0.02728	0.14081	0.58161
Si-face (Wet oxidation)	0.01022	0.02916	0.10441	0.88499

5. TYPES AND ORIGIN OF TRAP CHARGES IN SiO_2/SiC SYSTEM

Oxide grown on SiC substrate exhibits a much less perfect interface than the oxidized Si (Afanas'ev 1999). Defects located in the near-interfacial oxide layer may appear in the energy gap of SiC, because SiC width exceeds that of Si by a factor between 2 and 3 depending on the SiC polytype. The presence of carbon implies additional chemical complexity of the oxidation of SiC and may potentially promote the generation of defects in the SiO_2 layer (Afanas'ev and Stesmans 1997a, b).

There are five reported types of charges or traps associated with the SiO_2/SiC system. They are interface trap charges, near interface trap charge, fixed oxide charge, oxide trapped charge and mobile oxide charge (Li et al. 2000, Dimitrijev et al. 2004a, b). The presence of negative or positive effective oxide charge results from the balance of all kinds of charges existing at the SiO_2/SiC interface and in the bulk of SiO_2. Mobile charge and oxide trapped charge are distributed throughout the oxide (Li et al. 2000). Mobile charge is positive and is caused primarily by ionic impurities such as Na^+, Li^+, K^+ and possibly H^+, whereas the oxide trapped charges may be positive or negative due to holes or electrons trapped in the bulk of the oxide. The fixed oxide charge is generally positive and located near SiC/SiO_2 interface. Fixed oxide charge does not exchange charge with the substrate when the gate is biased (Li et al. 2000). Interface traps are located at SiC/SiO_2 interface (Li et al. 2000), whereas near interface traps or known as 'border traps' or 'slow traps' are located within the oxide of distance about 1.5-2.0 nm from the interface (Zippelius et al. 2010) and both traps communicate with the underlying SiC over a wide range of time scales. Their polarities can be positive or negative depending on their relative positions to the Fermi level. Thus, the effective oxide charge density obtained from the flat band voltage can be expressed as follows:

$$Q_{eff} = N_f + N_m + D_{it} + N_{ot} + N_{it} \tag{15}$$

where Q_{eff} is the effective oxide charge density, N_f is the fixed oxide charge density, N_m is the mobile oxide charge density, D_{it} is the interface trap density, N_{it} is the near interface trap density and N_{ot} is the oxide trapped charge density. If the sum of the positive charge is lower than the total of the negative charge, the effective oxide charge is negative and vice versa (Li et al. 2000).

N_{ot} and N_f charges can be minimized by annealing treatment, although neutral traps may remain. N_m which may originated form intrinsic nature or processing conditions are minimized by improving the processing ambient in a less contaminated condition. The major barriers for researchers to produce high quality and reliability gate oxide on SiC substrate are the D_{it} and N_{it} charges (Palmieri et al. 2009).

D_{it} exists at the lower of the energy gap which are thought to be related to carbon compounds (Dimitrijev et al. 2004a, b, Pensl et al. 2010) (Fig. 12.8). N_{it} exists near the conduction band of 4H-SiC and the energy position is round 2.77 eV below the SiO_2 conduction band edge (Afanas'ev and Stesmans 1997a). In terms of applications, D_{it} reduces the channel mobility of the MOS-based transistor (Stanley and Richard 2000, Pantelidas et al. 2006) whereas N_{it} limits MOS radiation response and long-term reliability (Fleetwood et al. 1995).

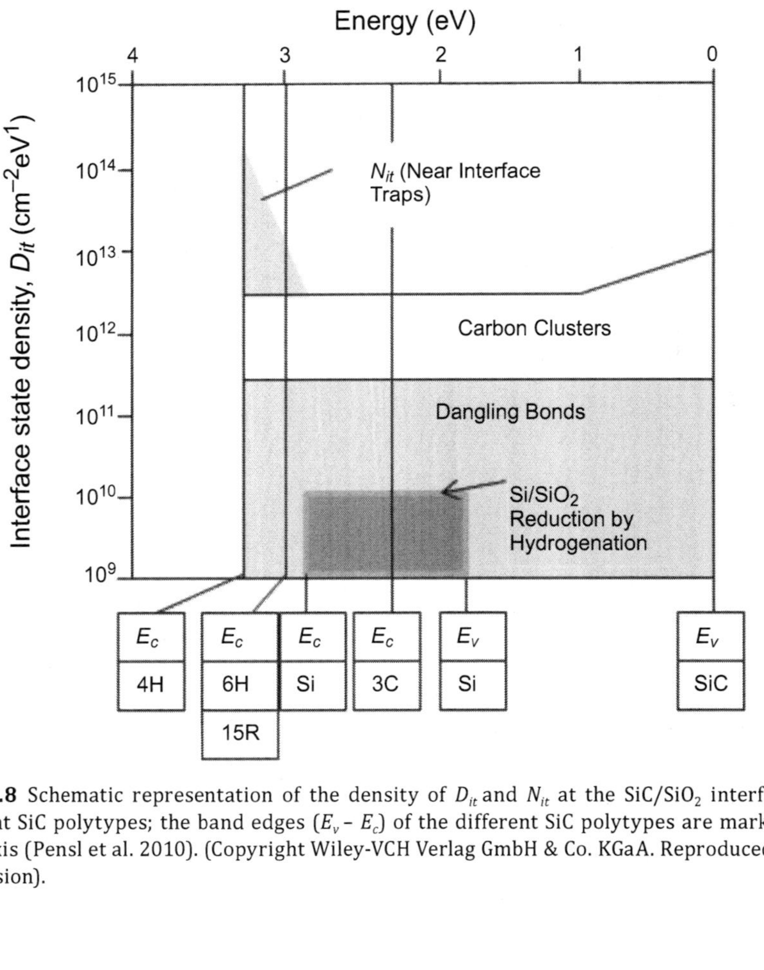

Fig. 12.8 Schematic representation of the density of D_{it} and N_{it} at the SiC/SiO_2 interface of different SiC polytypes; the band edges ($E_v - E_c$) of the different SiC polytypes are marked on the *x*-axis (Pensl et al. 2010). (Copyright Wiley-VCH Verlag GmbH & Co. KGaA. Reproduced with permission).

The major contributions of D_{it} charges are due to the structural mismatch between Si and the oxide (Dimitrijev et al. 2004a, b, Dhar et al. 2005, Zolper and Skowronski 2005), Si dangling bonds, Si-C-O interlayer (Pantelidas et al. 2006, Pitthan et al. 2013, Liu 2014, Liu et al. 2015) isolated C atoms (Dimitrijev et al. 2004a, b, Dhar et al. 2005, Liu et al. 2015) or carbon clusters with sp^2-bonded fragments (Dimitrijev et al. 2004a,b, Dhar et al. 2005), residual carbon complexes (Dimitrijev et al. 2004a, b) and O-vacancy defects (Dimitrijev et al. 2004a, b, Dhar et al. 2005). On the other hand, the N_{it} charge is expected due to the H-related defects (Fleetwood 1996, Dimitrijev et al. 2004a), isolated C atoms (Dimitrijev et al. 2004a, b) or carbon clusters (Dimitrijev et al. 2004a, b), structural mismatch between Si and the oxide (Afanas'ev and Stesmans 1997a), O-vacancy defects (Fleetwood 1996, Afanas'ev and Stesmans 1997b, Pantelidas et al. 2006, Fukushima et al. 2015), Si interstitials and strained Si bonds (Liu et al. 2015). The following section reviews the above-mentioned origin of D_{it} and N_{it} charges in SiO_2/SiC system.

5.1 Structural Mismatch between Si and the Oxide

The P_b-centers caused by structural misfit between semiconductor and oxide, are expected at SiC/SiO_2 interface as well as at Si/SiO_2-interfaces while their density should be comparable in both systems. They play only a marginal role compared to the total density of traps at SiC/SiO_2-interfaces (Pensl et al. 2010). N_{it} correlates with the presence of excess Si or an oxygen deficiency in the oxide (Dimitrijev et al. 2004a, b, Fukushima et al. 2015). It is reasonable to assume that the mismatch of lattice constants between SiC and SiO_2 causes internal stress at the interface, which provides the driving force for Si atoms to diffuse into the SiO_2 and to form (Si-Si)-pairs acting as acceptors (Pensl et al. 2010). The image of structure between SiO_2/SiC interface and SiO_2 bulk can be observed via HRTEM characterization.

5.2 Si-Si Bonding

Calculations show that the Si-Si bonding level which is located near the top of the valence bands tends to move into the bandgap location for longer Si-Si bonds (Pantelidas et al. 2006). On the other hand, the anti-bonding level remains in the conduction bands near the edge. The uncertainty in the calculations is too large to be determined. The difficultly is that normal-length (2.35 Å) or slightly-long Si-Si bonds are not passivated by hydrogen, however only the very long ones are able to be passivated by hydrogen. Si-Si bonds that are farther from the nominal interface may behave more like O vacancies and contribute to so-called slow interface traps (Pantelidas et al. 2006).

5.3 Si-C-O Interlayer

First principle calculations demonstrate that Si-C-O bonded interlayer with threefold-coordinated C atoms may account for defective states at the interface, which can be passivated by N (Pensl et al. 2010). Mainly Si-C-O compounds are reportedly formed in the first stage of oxidation, in agreement with previous work (Radtke et al. 2001). As the O_2 molecule approaches the interface through the growing oxide, O-O bonds readily break apart to form the stronger Si-O-C bonds (Liu et al. 2015). As oxidation progresses in time, stoichiometric SiO_2 becomes the main compound (Liu 2014).

5.4 Isolated C atoms/Carbon Clusters with sp^2-Bonded Fragments

There is substantial evidence that some of the excess carbon released during thermal oxidation remains at or near the SiO_2/SiC interface, either as isolated atoms or in the form of clusters (Afanas'ev et al. 1998). The presence of excess carbon at SiO_2/SiC interface was revealed by electron spectroscopy (Hornetz et al. 1994, Vathulya et al. 1998) and by electron spin resonance (Afanas'ev et al. 1996a), Besides, unidentified centers, believed to be related to C, were detected at the SiO_2/SiC interface by electron paramagnetic resonance (Macfarlane and Zvanut 1999, Son et al. 1999). Carbon clusters are neutral when the valence states are filled, but become positively charged when one of the valence electrons is emitted. In the case of n-type SiC, the Fermi level is always energetically located well above the p-bonded states of carbon clusters, consequently, no significant interface charge is expected. In the case of p-type SiC MOS structures, positively charged carbon clusters are available, which act as interface traps and are able to trap electrons. The empty states of the carbon clusters are energetically well above the lower conduction band edge of SiC and it is unlikely to affect the electronic properties of the SiC/SiO_2 interface (Afanas'ev et al. 1997c).

5.5 Residual Carbon Complexes

Analysis of an as-oxidized SiO_2/4H-SiC interface using Atomic-Force Microscopy (AFM) images reveals the presence of nanometer-sized protrusions (Dimitrijev et al. 2004a, b, Liu et al. 2015). These platelet-shaped in homogeneities were suggested to be carbon protrusions formed during thermal oxidation of SiC. These residual carbon complexes have been identified as the main reason for the problems observed with the electronic properties of the SiO_2/SiC interface and the gate oxide itself. D_{it} originated from residual carbon from the surface of substrate prior to oxidation and carbon generated at the interface during oxidation (Dimitrijev et al. 2004a, b).

The carbon cluster model associates the high density of interface defects with the partial oxidation of C at the interface between SiC and SiO_2. When SiC react with O_2, the Si-C bond is broken and the Si atom forms a bond with oxygen, whereas the remaining C atom is able to form a stable bond to a neighboring C atom. Depending on conditions, a number of C atoms that will bond to form a stable complex or cluster that have an increased oxidation resistance. The total reaction to form SiO_2 and a C complex is energetically complimentary with the C bond strength.

The work of Owman et al. (1995) has shown that when annealed above 950°C, Si evaporation occurs and the SiC surface reconstructs to form a graphitic like layer. In most environments, the C will simply oxidize, given the normal lattice separation between equivalent C atoms the formation of a cluster is unfavorable. On the other hand, the C defect formation is strongly dependent on the initial surface, in particular the presence of excess C that is acting as a nucleation site for cluster formation during the oxidation process. The carbon may be either present at the surface prior to the oxidation and/or be generated at the SiC/SiO_2, interface during the oxidation process. Angular resolved photoelectron spectroscopy investigations (ARXPS) conducted by Hornetz et al. (1994) reveal evidence of the accumulation of carbon at the interface during the oxidation process. These authors describe the formation of a thin interface oxide layer that is composed of $Si_4C_{4-x}O_2$ ($x<2$) and the composition is increased with the growing oxide thickness (Bassler et al. 1997).

If the C complex being formed has indeed a higher oxidation resistance then, it is possible to observe the extension of C islands above the plane of SiC/SiC_2 interface. Further evidence for such roughness comes from the work of Wahab et al. (1995) using Auger Electron Spectroscopy (AES) that shows the extension of C beyond the interface and penetrating into the oxide by approximately 10 Å. Indeed, this observation is close to the experimental resolution of such a system reported by Afanas'ev et al. (1996a) which relates the SiC/SiO_2 interface to that between amorphous C(αC:H) and SiO_2. It is found that the energy of a trap is related to the cluster size. A distribution of cluster sizes, which leads to a corresponding distribution in the energy of defect states lies around the mid-gap of SiC. The Fermi-level position in the SiC determines the charge state of the interface defects. For p-type material, the defects are positively charged since they are able to trap a hole. On the other hand, for n-type material, the states are donor-like in character and the defects are largely neutral. However, the defect state energy is fixed with respect to the valence band energy as the exact Fermi-level position is dependent on polytype of SiC. For 3C, it is found that the flat band voltage is shifted to indicate a net positive charge even for n-type material. This behavior expresses that the Fermi-level in n-type 3C material is still below the neutrality level for the interface defects. In 6H and 4H-SiC, the defects are neutral (or negatively charged) as the Fermi-level is higher than in 3C (Harris and Afanas'ev 1997).

5.6 H-Related Defects

Donor-type states may also originate from oxide defects. Afanas'ev and Stesmans (1997d) reported on the existence of an H-complexed oxygen vacancy ($O_3\equiv SiH$ $HSi\equiv O_3$) acting as an electron trap with an energy level at 3.1 eV below the SiO_2 conduction band. It has been reported that the creation of more acceptor-like interface states in the wet oxidation ambient which is speculated to result from hydrogen-related species such as -OH bonds in the steam may become hydrogen-bonding sites (Lai et al. 2002).

5.7 O-Vacancy and O Atoms Clustering Defects

The E'_y-center related to O-vacancy defects observed in SiO_2-SiC structures may also contribute to the interface trap distribution (Afanas'ev et al. 1996b). The energy of positively charged oxygen vacancy centers is predicted to be within the range of 4.3-5 eV below the SiO_2 conduction band (Chu and Fowler 1990).

Contrary to Si, O in cubic SiC has a very low solubility. The clustering of O atoms proceeds through the formation of electrically active O complexes (thermal donors), which are, however, metastable. Instead, the O atoms can bind to the central C of the thermal donor-like complex and then diffuse away as a CO molecule, leaving behind a perfectly bonded SiO_2 precipitate (Massimiliano and Pantelides 2000).

In the (1 1 0) plane, it has been reported by Massimiliano and Pantelides (2000) that two oxygen atoms bind with one of them surrounded by two Si atoms and one surrounded by two carbons. Each oxygen atom consists of binding energy of about 0.3 eV. On the other hand, both oxygen atoms which are surrounded by two Si atoms respectively have a lower binding of about 0.1 eV per oxygen atom. A formation of a metastable complex reminiscent of the thermal donor in Si is formed by a third O atom addition. The three oxygen atoms bind by about 0.5 eV and the complex is electrically active. Nevertheless, it is metastable and can be transformed into electrically inactive complex. Of the two complexes, one can transform to the stable defect through emission of a CO molecule.

5.8 Other Defects

Defects in the SiO_2/SiC system also may be originated from various dopants incorporation into SiC. The severity of the defects may vary according to polytype, face and crystallographic orientation of the SiC substrate. Despite substrate effects, equipment contaminations and processing conditions such as process ambient, temperature, oxidation duration also contribute to the chemical complexity and generate defects in the SiO_2/SiC system.

6. IMPROVEMENT OF SiO_2/SiC SYSTEM

The optimization of the SiO_2/SiC interface is critical for the development of SiC MOS-based devices. Improvement in quality of the grown oxide and the resulting interface would have an impact on SiC MOS devices. Hence, much effort has been focused on reducing oxide traps, particularly interface and near-interface traps. For this purpose, nitridation, hydrogenation and other species incorporation into SiO_2/SiC system have been established and the following sections are devoted to provide a better understanding on these mechanisms in the SiO_2/SiC system.

6.1 Nitridation

Nitridation is involved in two sets of mechanisms at the SiO_2/SiC interface (Jamet et al. 2001a): (1) creation of strong Si≡N bonds that passivate interface traps due to dangling and strained bonds and (2) removal of carbon. The mechanism which leads to the creation of strong Si≡N bonds is completely analogous to the case of the SiO_2/Si interface. In the case of both SiO_2/Si and SiO_2/SiC interfaces, there is dangling Si bonds passivated by N and strained Si-O bonds that are substituted by strong Si≡N bonds during the nitridation. In both cases, the formation of these bonds improves the gate-oxide reliability. An important observation in the case of SiC is that not only is the reliability improved by the strong Si≡N bonds but also the initial interface-trap density is significantly reduced. This is because the energy levels of the much stronger Si≡N bonds are outside the energy gap of SiC, meaning that they are electrically passive defects (Jamet et al. 2001a, Jamet and Dimitrijev 2001b, Dimitrijev and Jamet 2003, Fujihara et al. 2004, Moon et al. 2007).

The second role of nitridation is related to carbon removal from the interface. Most of the carbon atoms released by oxidation of the SiC substrate react with the existing oxygen and nitrogen to create CO/CO_2 or CN molecules that diffuse out of the oxide. However, some of the released carbon atoms accumulate into carbon clusters acting as interface traps and initiating the appearance of complex silicon oxycarbon compounds at the interface. Nitridation not only passivates the carbon-related interface traps but more importantly it removes interstitial carbon as well as carbon from already formed clusters. As a result, nitridated SiO_2/SiC interfaces appear free of defects due to silicon oxycarbon bonds (Moon et al. 2007). The abstraction of C atom by N atoms occurs but the formed CN readily reacts with a C or Si atom in the interface to form a C-C-N or Si-C-N complex, respectively (Ohnuma et al. 2007). The existence of C-N bonds near the interface is observed using SIMS characterization not only for NO annealed oxides but also in the case of N_2O annealed oxides. Quantitatively, however, the effects of the two nitridation gases are not the same: much lower C-C graphitization appears for NO annealed relative to N_2O postoxidation treatment (Dimitrijev et al. 2004a, b, Pantelidas et al. 2006, Moon et al. 2007).

C-C graphitization signal represents 14% in the XPS C 1 s spectrum at the SiO_2/SiC interface of NO-annealed oxide, whereas 27% in the case of N_2O. This implies that more efficient removal of interfacial carbon clusters is taking place in the case of NO as compared to that of N_2O (Moon et al. 2007). Significant reductions of the interface trap density near the conduction band edge by annealing standard thermal oxides in NO atmosphere (Chakraborty et al. 2001). According to the calculations of Gupta et al. (1998), N_2O molecule dissociates into NO (15%), O_2 (25.7%) and N_2 (59.3%) at 1130°C. Obviously, the NO component may cause the same chemical reactions and effects that occur in the cases of pure NO nitridation. It has been reported that at high temperatures, N_2O decompose via the following two reactions (Chakraborty et al. 2001):

$$N_2O \rightarrow N_2 + O \tag{16}$$

$$N_2O + O \rightarrow 2NO \tag{17}$$

Most of N_2O decomposes following the reaction (16) while only very small amount of N_2O obeys the reaction (17). De Meo et al. (1994) found that in each case there was only 0.5-0.75% pile up of nitrogen at the SiO_2/SiC interface. Nitrogen piled up at the interface during N_2O nitridation neutralizes the Si dangling bonds and thus decreases interface states. On the other hand, for n-type SiC normally doped by nitrogen, a large number of Si≡N bonds may be already formed at the SiO_2/n-type SiC interface. During N_2O nitridation, incorporated nitrogen is doubly bonded to Si atoms in the bulk of the oxide but triply bonded to Si at or very close to the interface. The triply bonded state provides a means of accommodating a Si bond that would otherwise be unsatisfied, thus constituting an active interface state, if the nitrogen atom is not replaced by an oxygen atom at the interface (Chakraborty et al. 2001). However, in the bulk of the oxide, the doubly bonded N may be replaced by atomic oxygen during its exposure to atomic oxygen. Ellis and Buhrman (1996) argued that incorporated nitrogen during the initial stage of nitridation is partially removed at the end of the nitridation step. Therefore, it is assumed that the atomic oxygen generated during dissociation of N_2O following the reaction (16) could reach the wafer surface and remove nitrogen already existing at the interface, thus creating more interface states and fixed charge and deteriorating the oxide quality (Chakraborty et al. 2001).

In terms of growth rate, nitridated oxide has lower thickness than non-nitridated counterparts. The effects of nitrogen on the oxidation rate are complex. Two possibilities are that the nitrogen-containing region may (i) form a diffusion barrier which makes oxygen delivery to the interface more difficult and/or (ii) modify the reaction mechanism for oxide growth near the interface (Lu et al. 1997). The first hypothesis is supported by the very low oxygen diffusivity in silicon nitride and oxynitride. The second could result for example from nitrogen bonding with reaction sites near the interface or suppression of Si suboxides (Si interstitial injection) in the near-interfacial region of the dielectric film. The preliminary experiments on oxidation of an oxynitride film containing nitrogen near the outer surface also show significant retardation of the interfacial growth reaction, favoring the 'diffusion hypothesis' (Lu et al. 1997). According to an N_2O-growth model developed for silicon, the creation of Si≡N bonds at the interface causes reduced concentration of growth sites, which is the reason for reduced oxidation rate in NO and N_2O. However, the diluted-NO oxidation, which gives a lower oxidation rate is beneficial for obtaining a good equilibrium between carbon accumulation and carbon removal and thus a much smoother, less disordered and strained surface (Jamet et al. 2001a, Jamet and Dimitrijev 2001b). When carbon-accumulation and carbon-removal rates are equal, the residual carbon atoms or clusters would be the least near/at the interface, leading to an excellent SiO_2/SiC interface with low interface-state density (Lai et al. 2005).

For the oxide grown between 2 to 20 nm in 10%-N_2O ambient at oxidation temperature of 1175°C, the ultrathin region can be well fitted in the following model (Cheong et al. 2007):

$$\frac{dt_{ox}}{dt} = C + a_0 e^{-t/\tau \frac{-E_A}{kT}} \tag{18}$$

where the first and second terms are representing the linear-(C) and initial-rate constants, respectively. On the other hand, α_0, τ, E_A, k and T represents constant, time constant, activation energy, Boltzmann's constant and temperature, respectively. It has been reported that a cyclic trend of effective oxide charge, Q_{eff}, is observed for the above grown nitridated oxide. During the initial oxidation (t_{ox} = 2-5 nm), carbon is assumed to accumulate at SiC/SiO_2 interface (increasing trend of N_{it}), thus a non equilibrium and disordered oxide is produced. When t_{ox} increases, the accumulated carbon at the bulk and interface is reduced as the oxidation and nitridation processes are in equilibrium. However, the reduction has a limitation. As the t_{ox} increases further, the well passivated interface has to be depassivated in order for the subsequent oxidation to take place. The forming and breaking of the bonds may affect arrangement of the bulk SiO_2 network. As a result, Q_{eff} again demonstrated a negative value. This cycle continues as the oxide gets thicker and thicker (Cheong et al. 2007).

6.2 Hydrogenation

It has been reported that the H_2 passivation in SiC/SiO_2 interface defects is very slow in comparison to that on the Si/SiO_2 interface for the same oxide thickness.

It can be concluded that the kinetics of the underlying reactions is not limited by the H_2 diffusion through the oxide layer but rather by the activation energies of the said reactions (Knaup et al. 2009). Therefore, the H_2 concentration at the interface remains constant, while the concentration of the passivated defects may change over time. Such reactions are described by the Arrhenius law (Knaup et al. 2009):

$$\frac{-d[A]}{dt} = k_1 . [A]; \; k_1 \alpha e^{\frac{E_A}{kT}} \tag{19}$$

where $[A]$ denotes the concentration of defect being passivated and E_A is the activation energy of the passivation reaction. Obviously, the reaction efficiency depends strongly upon the activation energy.

Dangling bond defects at the Si surface are well known to be easily passivated by hydrogen at 230°C to 260°C and to become electrically inactive (Afanasev et al. 1997c). A similar effect may be expected for dangling bonds, which contributes to the SiC/SiO_2 interface traps density. Annealing at 400°C, usually sufficient to electrically passivate Si dangling bonds at Si/SiO_2 interfaces, hardly affects charged defects at the SiC/SiO_2 interface as revealed by marginal shifts of the Capacitance-Voltage (C-V) and conduction-voltage (G-V) characteristics. An increase of the annealing temperature to 700°C and further, to 1000°C results in two effects:

(i) Density of interface states in upper part of the SiC bandgap decreases, as indicated by the decreased peak height of the G-V curves of the n-type sample.
(ii) Build-up of positive charge at the SiC/SiO_2 interface, which causes nearly equal voltage shifts of C-V and G-V curves in both the n- and p-type samples.

By comparing the annealing behavior of defects at the SiC/SiO_2 interface with the annealing behavior of Pb-type defects at the Si/SiO_2 interface (Brower 1990), two conclusions can be reached. Firstly, the nonappearance of any passivation effect at 400°C proposes an insignificant contribution of dangling bonds to the total SiC/SiO_2 interface defect density. Secondly, the dangling bonds at Si/SiO_2 interfaces are electrically amphoteric which means, they trap holes when the Fermi level is near the SiC valence band and they trap electrons when the Fermi level is near the Si conduction band (Uren et al. 1996). Both phenomena contribute to the vanishing of D_{it} spectrum subsequent to hydrogen passivation. With respect to a H_2 treatment, the interface states at the SiC/SiO_2 interface in the upper and lower parts of the energy gap show distinctly different reactivity which suggests different structural defect elements. This observation can be related to the framework of the carbon clusters. Thus, the hydrogenation experiments report that the contribution of the dangling bond defects to the total interface state density at SiC/SiO_2 interfaces is likely to be insignificant (Afanasev et al. 1997c).

H_2 annealing is not effective in D_{it} reduction in the case of SiC, only a shift of their energy away from the midgap and they stay within the energy gap (Afanas'ev and Stesmans 1997d, Von Kamienski et al. 1995). The reason for that is the wider energy gap of SiC, particularly the 4H polytypes, which encompasses the energy levels of Si–H bonds. In principle, atomic H should passivate $(C_i)_2$ by attacking the C=C double bond and forming C–H bonds. However, it has long been known that H has either no effect or minimal effect on either the measured interfacial defect density or carrier mobility. This result is consistent with $(C_i)_2$ in the substrate being a major cause for mobility degradation: in SiC, the stable form of hydrogen is H^+ in p-type samples and H_2 in intrinsic and n-type samples. H^+ has a small diffusion barrier of 0.6 eV, while H_2 has a diffusion barrier of >2.5 eV, which exceeds its dissociation energy. On the other hand, in n-type MOS structures, H_2 cannot be incorporated into the substrate to passivate the $(C_i)_2$ defects; in p-type MOS structures, H^+ can be incorporated, but the $(C_i)_2$ defects are also positively charged, whereby passivation is ineffective (Shen and Pantelides 2011).

However, when hydrogen annealing is done after the deposition of a Pt metal layer, which breaks H_2 molecules into monatomic H, the resulting reduction of D_{it} is substantial (Pantelidas et al. 2006). Hydrogen atoms passivate carbon related dangling bonds at the interface (Cho et al. 2000, Endo et al. 2009, Allerstam and Sveinbjornsson 2009). When two H atoms react with the defect, one H atom is bonded to the O atom (bond length: 0.98 Å), while the other is bonded to the C atom (bond length: 1.1 Å). No states are found in the bandgap. The defect state is thus removed by the two H atoms. The defect is also passivated by two H atoms if they are bonded to the dangling C atoms. The binding energy of the two H atoms relative to the gas-phase molecule is calculated to be 2.2 eV, suggesting a stable bonding configuration.

6.3 Others

Besides nitrogen and hydrogen, chlorine can also be incorporated into the SiO_2/SiC system by Trichloroethylene (TCE) solution (Xu et al. 2005) and $POCl_3$ (Kr′ol et al. 2015). TCE decomposition at high temperature (~1100°C) in O_2 could result in the following species:

$$Cl_2C = CHCl + O_2 \rightarrow Cl_2 + HCl + CO_2 + H_2O \quad (20)$$

The resulting Cl_2 and HCl molecules react with SiC in the O_2 ambient at high temperature (~1100°C), presumably producing the following species:

$$Cl_2 + SiC + O_2 \rightarrow SiO_2 + SiCl_4 + CO_2 + Cl_2O + CO + CCl_4 \quad (21)$$

and

$$HCl + SiC + O_2 \rightarrow SiO_2 + SiCl_4 + CO_2 + Cl_2O + CO + CCl_4 + H_2O \quad (22)$$

Therefore, the following explanations can be proposed: (1) a passivation effect probably occurs when chlorine reacts with residual carbon atoms or clusters that remain at the interface during oxidation and also there are possibly other passivation effects on the structural defects at/near the interface (e.g., dangling Si and O bonds) by Cl_2 or HCl, (2) negligible carbon is generated from TCE decomposition during the TCE oxidation because CO_2 is easily formed in the O_2-rich ambient, thus creating few interface states and small oxide charges and (3) Cl_2 and HCl can getter the ion contamination from the out diffusion of the SiC dopant and oxidation system, leading to a reduction of oxide charges. It is proposed that these improvements are likely associated with the passivation effects of Cl_2 and HCl on the structural defects at/near the SiC/SiO_2 interface (formation of stronger Si–Cl bonds) and also their gettering effects on ion contamination (Xu et al. 2005).

On the other hand, fluorine atoms can passivate the isolated point defects containing a single carbon atom or correlated carbon atoms at the SiC/SiO_2 interface as deduced from first-principles calculations (Liu et al. 2010). The quality of the interface may be improved by incorporating F in the form of molecules, which could react with the interface defects and dissociate into atoms. The defect state is removed by two fluorine atoms with two Si–C bonds which are broken and the two F atoms are bonded to the Si and C atoms with bond lengths of 1.59 and 1.40 Å, respectively.

The F–C bond lengths are about 1.45 Å while the C–C bond between the two central carbon atoms has a bond length of 1.50 Å. The geometry itself suggests that the two C dangling bonds are saturated by the F atoms and confirms that the defect states in the gap are indeed removed. The binding energy of two F atoms with the defect, relative to F_2 molecule, is 4.6 eV, which indicates a strong bonding. Such strong F–C bonds replace the C dangling bonds on the fluorine passivation results in the shifts of the defect states out of the bandgap. A similar effect is also found for passivation of the defect in the SiC side near the interface. Thus, two F atoms can passivate the defect caused by two correlated C atoms that are right at or near the interface (Liu et al. 2010).

Besides chlorine and fluorine, phosphorous (P) atoms are reported to relax the strained bonds at the interface in the case of SiO_2/(001) Si structures (Morino et al. 1997). Therefore, it is expected that the incorporation of P atoms in SiO_2/SiC system also may reduce the strain at the interface and reduces the N_{it} density (Okamoto et al. 2010b, Sharma et al. 2012). In fact, a strong reduction in density of N_{it} at the SiO_2/4H-SiC interface after dry oxidation in the presence of potassium has also been reported (Hermannsson and Sveinbjornsson 2011). This is accompanied by a significant enhancement of the oxidation rate. The results are in line with recent investigations of the effect of sodium on oxidation of 4H-SiC. It is evident that both alkali metals enhance the oxidation rate of SiC and strongly influence energy distribution of interface states (Hermannsson and Sveinbjornsson 2011).

6.4 Summary of Various Types of Species Incorporated into SiO_2/SiC System

Figure 12.9 shows the summary of roles for various types of species incorporated into SiO_2/SiC system during/after oxidation and their disadvantages. Generally, Si and C passivation/removal/oxidation mechanisms occur during oxidation/annealing process by species incorporation into SiO_2/SiC system. Si oxidation and removal mechanisms occur by the reaction of O^- species with SiC substrate. Si passivation occurs by H^+, N^- and Cl^- incorporation into SiO_2/SiC. On the other hand, C is being passivated by incorporating H^+, F^-, N^- or Cl^- species during oxidation/annealing process. The C removal occurs by CO/CO_2 or CN out-diffusion via the reaction of O^- and N^- species with SiC during oxidation/annealing process. The summary of the roles of each species incorporated into SiO_2/SiC and their respective passivation/removal/oxidation mechanisms is tabulated in Table 12.5.

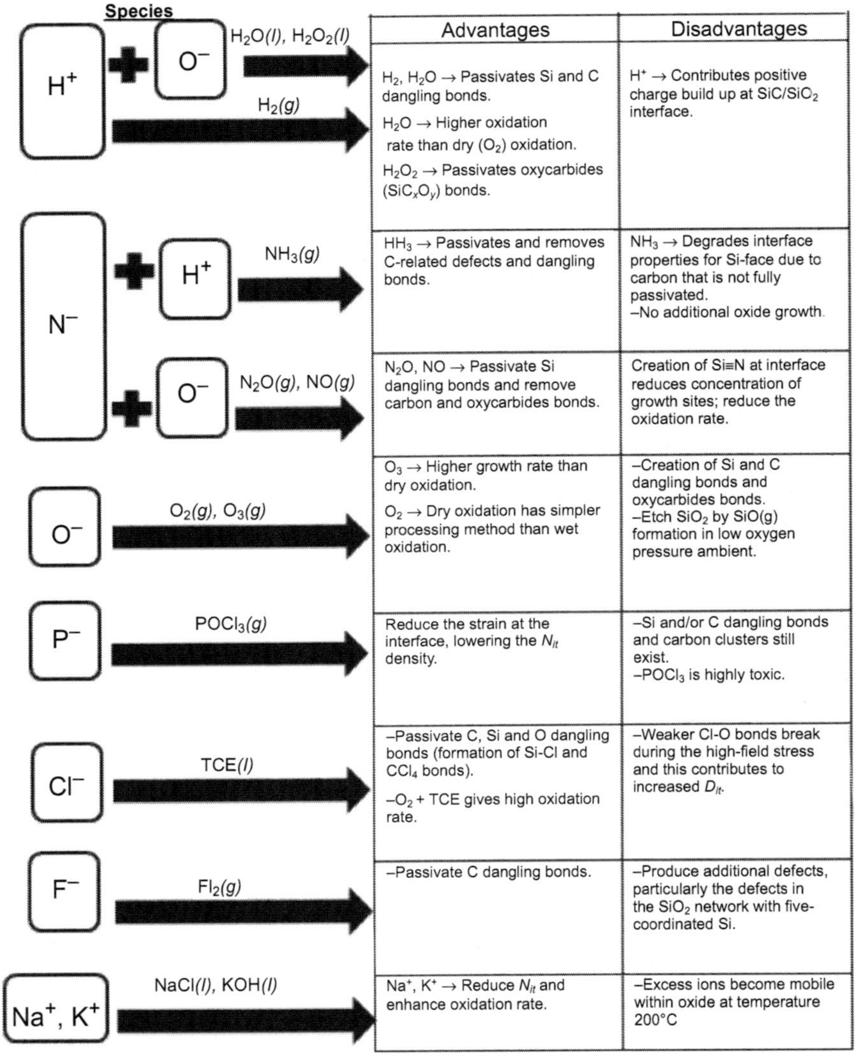

Fig. 12.9 The roles of various types of species incorporated into SiO_2/SiC system during/after oxidation and their disadvantages.

Table 12.5 Si and C passivation/removal/oxidation mechanisms by species incorporation into SiO_2/SiC system.

Species	Si oxidation	Si removal	Si passivation	C passivation	C removal
H^+	×	×	✓	✓	×
N^-	×	×	✓	✓	✓
O^-	✓	✓	×	×	✓
Cl^-	×	×	✓	×	×
F^-	×	×	×	✓	×

7. FACTORS INFLUENCING THE FABRICATION OF THERMALLY GROWN SiO_2 FILMS ON SiC SUBSTRATE

In order to thermally grow SiO_2 films on SiC substrate, a furnace with the necessary equipment set up is required. Figure 12.10 shows the equipment set up for thermal oxidation/annealing in dry and wet ambient. A controller in the furnace allows the creation of recipes and selection of temperature for thermal oxide growth process. A quartz tube is attached to the furnace in which samples are placed in a boat. Dilution gases, carrier gases, direct/re-oxidation species and post oxidation anneal species such as oxygen, nitrogen, argon, ammonia and forming gas are connected to the quartz tube. Gas flow and gas ratio are controlled by gas gauge and mass flow meters. In general, the oxidation procedure should be carried out following certain steps and schemes (Benfdila and Zekentes 2010). After a slow wafer load at around 600°C, the temperature is ramped to the required oxidation value (generally > 1000°C). It is desirable that SiC wafer surface faces the gas inlet in order to obtain a thicker oxide growth. The thermal oxidation is followed by post-oxidation annealing under the same temperature or varied temperature. During dry oxidation, various species are flown into the quartz tube. Inlet gas flow rate and gas concentration play an important role in the fabrication of SiO_2 films. On the other hand, wet oxidation is performed with carrier gas by turning on the valve to carry the vapor to the quartz tube for oxidation.

Table 12.6 Dry and wet ambient species and their role during direct and post oxidation/nitridation/hydrogenation anneal.

Dry ambient species	
Direct/Re-oxidation/Post Oxidation Anneal Species	Post Oxidation Anneal Non-Oxidizing Species
$-O_2(g)$, $-O_3(g)$ } Direct/ Re-oxidation $-N_2O(g)$, $-NO(g)$ } Direct/ Post Oxidation	$-He(g)$ $-Ar(g)$, $-N_2(g)$ } Gases used for dilution during SiO_2 direct growth: $-H_2(g)$ $-NH_3(g)$ $-POCl_3(g)$ $Ar(g)/H_2(g)$ $N_2(g)/H_2(g)$ $N_2(g)/N_2O(g)$ $N_2(g)/NO(g)$ $N_2(g)/POCl_3(g)$
Wet ambient species	
Direct/Re-oxidation Species	Direct Anneal Species
$-H_2O(l)$ } Carrier gas: $-O_2(g)–N_2(g)$ $-N_2O(g)$ $-H_2O_2(l)$ } Carrier gas: $-O_2(g)$	$-TCE(l)$ } Carrier gas: $-O_2(g)$
SiO_2 grown in dry/wet ambient followed by dipping sample into species based solution and re-oxidation	
Species based Solution Enhanced Oxidation	
$-KOH(l)$ } Potassium $-NaOH(l)$ } Sodium	

Table 12.6 shows the common dry and wet ambient species and their role during direct oxidation and post oxidation/anneal. In addition, Table 12.6 also includes species based solutions in which SiO_2 grown in dry/wet ambient are dipped into the solutions before re-oxidation process. SiO_2 films fabrication is greatly influenced by process parameters such as carrier gas flow rate, carrier gas concentration, solution concentration and solution boiling temperature during wet oxidation/nitridation. After post oxidation annealing process, generally the temperature is decreased with the same rate as it was increased prior to the main oxidation step. This step is necessary to lower the interface trap density. Lastly, the temperature should be decreased at the same rate before the sample is removed (Benfdila and Zekentes 2010). The SiO_2 films quality and thickness is influenced by the oxidation/annealing temperature and duration. Despite process parameters, SiC substrate parameters such as polytype, face and dopant also play crucial roles on the SiO_2 films fabrication.

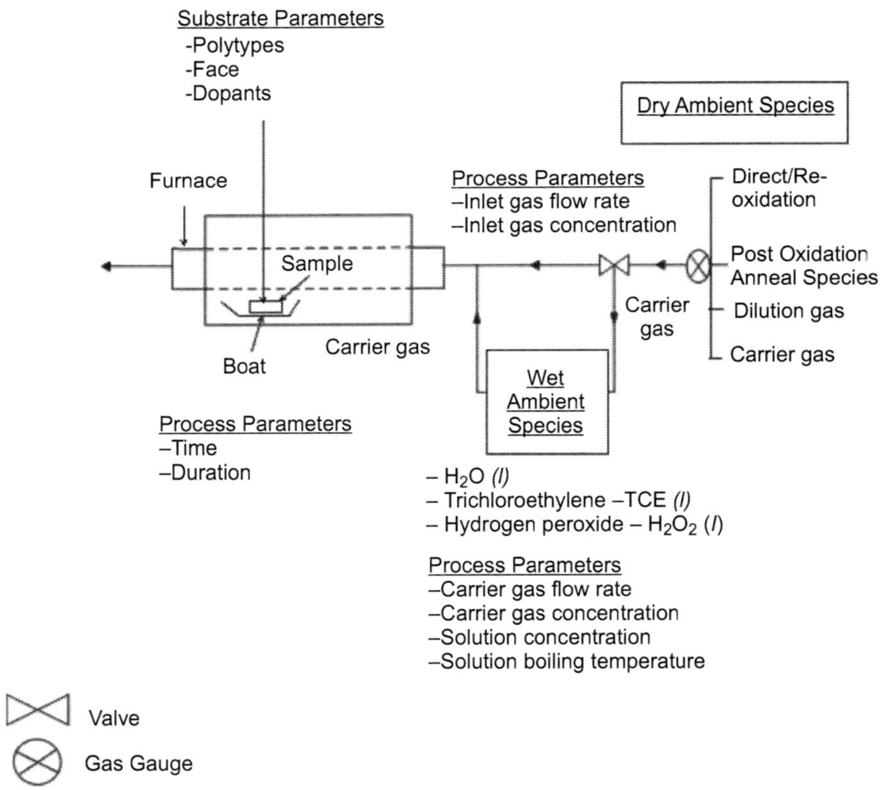

Fig. 12.10 Furnace design for dry/wet direct and post oxidation thermal oxidation/anneal species with relevant process/substrate parameters.

7.1 Process Recipes and Ambient

In the early fabrication of SiO_2 thin films on SiC (1990-1996), conventional dry/wet oxidation methods were used followed by post oxidation annealing in non-oxidizing ambient. Figure 12.11 presents the development of SiO_2 films grown on SiC by thermal oxidation and post oxidation annealing techniques. Dry oxidation (sometimes called pyrogenic oxidation for temperatures greater than 1200°C) has a simpler processing method than wet oxidation. It consists of supplying an oxygen gas to the high temperature dry environment, typically 1000°C-1200°C (Benfdila and Zekentes 2010). Wet oxidation consists of supplying a mixture of saturated H_2O vapor and oxygen to the oxidation atmosphere (furnace). Wet ambience has a much faster growth rate than in a dry ambiance (the parabolic rate constant for wet oxidation is about 25 times larger than that of dry oxidation at 1050°C) (Harris and Afanas'ev 1997, Benfdila and Zekentes 2010). This is primarily due to a much higher solid solubility of H_2O in SiO_2 than O_2 in SiO_2, hence providing a much higher amount of oxidizing species to the SiC-surface.

It has been reported that a lower interface trap density and reduced fixed charge were observed for wet oxidation conditions (Harris and Afanas'ev 1997). Wet oxidation consists of additional hydrogen passivation (due to hydrogen species in steam) and this helps to reduce residual carbon atoms or clusters, dangling bonds and structural defects. Moreover, it effectively eliminates carbon-related species due to steam-enhanced out-diffusion of CO and removal of interstitial carbon as well as carbon clusters (Xu et al. 2003). For dry oxidation, interface trap density in SiO_2 is higher compared to wet oxidation process (Fig. 12.11).

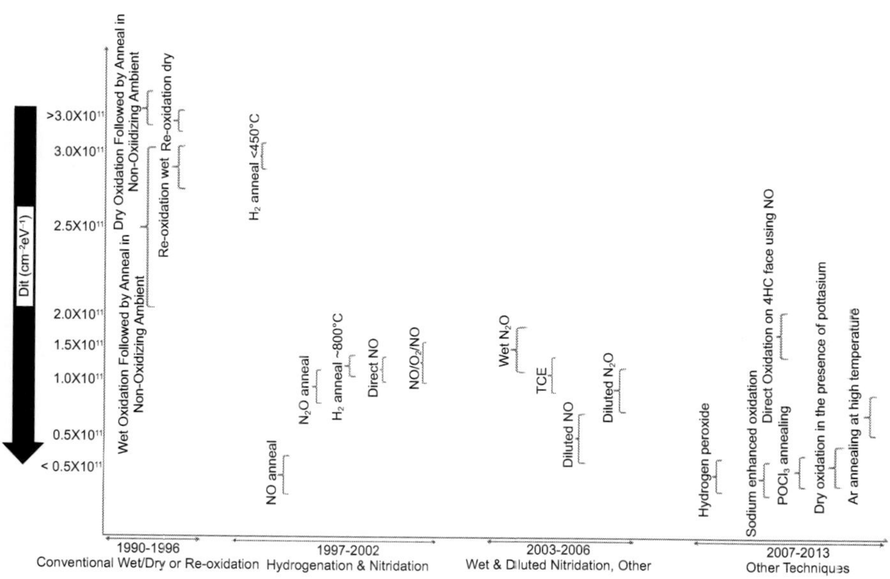

Fig. 12.11 Development of SiO_2 films grown on SiC by thermal oxidation/nitridation and post-oxidation annealing techniques.

A post-oxidation annealing process as a means of 'healing' interface traps and fixed oxide charge is a common processing techniques for SiC (Harris and Afanas'ev 1997). Post-oxidation annealing in non-oxidation ambient such as helium increases fixed charge and interface state densities in both 4H- and 6H-MOS capacitors (Shenoy et al. 1996). *In situ* post-oxidation annealing in argon results in lower interface trap densities than annealing in helium (Shenoy et al. 1996). Fluorine annealing has no effect on the interface quality (Das et al. 1998), however fluorine atoms can passivate the isolated point defects containing a single carbon atom or correlated carbon atoms at the SiC–SiO_2 interface (Liu et al. 2010). Improvements to the post annealing scheme have been made by (Lipkin and Palmour 1996) who used a wet re-oxidation (950°C) method. Overall passivation of defects using post-oxidation annealing has been shown to significantly reduce both interface defects and fixed oxide charge. Re-oxidation improves channel mobility and reliability (Suzuki et al. 2007). However, under charge injection conditions such as high electric field across the oxide, there is also significant charge build up due to the presence of hydrogen related trapping centers (Harris and Afanas'ev 1997). The dry re-oxidation anneal doubles the interface trap densities, as compared with the oxide that received no post oxidation treatment (Lipkin and Palmour 1996).

The conventional oxidation is further developed to hydrogenation and nitridation techniques (1997-2002). Hydrogen annealing is performed at lower temperatures, which is approximately 450°C. However, the lower hydrogen annealing has very low impact on the D_{it} improvement (Afanasev et al. 1997c). Hydrogen annealing at temperatures above 800°C reduces D_{it} drastically compared with argon annealing due to the termination of the dangling bonds of Si and C with hydrogen (Fukuda et al. 2000). It is considered that the reduction in D_{it} occurs above 700°C because of the band energy of C–H which is higher than that of Si–H and the D_{it} reduction saturates above 800°C because this temperature is high enough for termination of carbon atoms by hydrogen atoms (Fukuda et al. 2000). Hydrogen annealing helps the reduction of D_{it} by termination of silicon and carbon dangling bonds (Cho et al. 2000). Fukuda et al. (2000) reduced the interface-trap density near E_c[for n-type 4H-SiC MOS structures and increased the channel mobility on the by post-oxidation hydrogen annealing (Xu et al. 2005).

Nitridation annealing is performed by nitrous oxide (N_2O) or nitric oxide (NO) gases which give drastic improvement on the D_{it} (Sweatman et al. 1997). While both techniques lead to incorporation of nitrogen at the interface, the annealing in N_2O leads to new oxide growth, as opposed to the annealing in NO which nitrides the interface with virtually no new oxide growth. The interface characteristics can be improved by NO annealing than by N_2O annealing probably due to the influence of oxygen (Sweatman et al. 1997). It has been reported that SiO_2 thin films grown directly in N_2O and NO show a considerably improved device reliability under high-field stress compared to annealed counterparts and wet thermal oxides on 6H-SiC (Jamet et al. 2001a). The improvements of the electrical characteristics of grown nitrided oxide may be associated with the nitrogen accumulation at the interface.

Nitridation is believed to progressively remove interfacial carbon clusters and complex silicon–oxycarbon compounds from the SiO_2/SiC interface. In the case of direct growth in either NO or N_2O, the process releases the carbon at the interface of SiO_2/SiC system as the carbon-removal action of nitrogen begins at the initial surface and continues simultaneously with SiC consumption by the oxidation (Jamet et al. 2001a). At the initial SiC surface, any carbon clustering seeds exist on it are removed and the creation of new ones by the release of carbon is maintained at a low level. Post oxidation nitridation does remove the carbon clusters and the associated silicon–oxycarbon compounds, however it creates conditions of additional carbon release by the simultaneous oxidation and the state of the interface may never be improved to the levels achieved by direct growth in NO or N_2O (Jamet et al. 2001a). In addition NO/O_2/NO sandwich is also performed to fabricate thicker oxide (Cheong et al. 2004).

In order to solve the gate oxide thickness and reliability issues, from the years 2003 to 2006, wet and diluted nitridation was introduced. The growth rate of wet N_2O oxidation is higher than that of dry N_2O oxidation, probably due to the higher oxidation rate in steam atmosphere than in dry ambient and enhanced out-diffusion of CO (one of the factors limiting the growth rate in wet oxidation) (Xu et al. 2003). Samples prepared in wet N_2O nitridation (bubbling N_2O gas through de-ionized water at 95°C) (Lai et al. 2002) greatly reduced interface-state density and enhanced reliability of both n- and p-types on 6H-SiC MOS devices. However, there is no claim about wet nitridation on 4H-SiC as yet. On the other hand, it has been reported that diluted NO gives lower D_{it} in comparison to diluted N_2O.

As compared to conventional 100%-NO oxidation, the diluted-NO (50 and 23%) oxidations lead to lower interface-state, near-interface and oxide-charge densities. This is due to the fact that carbon-accumulation and carbon removal rates are closer when oxidation is performed in diluted NO, giving a smoother, less disordered and strained interface. Furthermore, less degradation of the diluted-NO samples than 100%-NO samples is observed during high-field stressing (7 MV/cm), representing that stronger Si ≡ N bonds are created near/at the SiC/SiO_2 interface for oxide grown in diluted NO ambient (Lai et al. 2002). Jamet et al. (2001a) and (Cheong et al. 2003) proposed methods of annealing and growing oxides in diluted N_2O which reported to an improvement for gate-oxide reliability and significant reduction of interface-state density.

Besides NO and N_2O as nitridation agents, it was found that ammonia (NH_3) treatment before oxidation is beneficial for the interface-quality improvement. Moreover, combination of NH_3 treatment followed by N_2O annealing helps to improve the SiC/SiO_2 interface and its resistance against high field and high-temperature stressing (Lai et al. 1999). This is attributed to the formation of a nitrogen/hydrogen-passivated layer on SiC surface during NH_3 treatment and nitrogen incorporation near the interface during N_2O annealing. NH_3 treatment produces a nitrogen- and hydrogen-terminated and atomically flat surface (Iwasaki et al. 2010). This gives superior device reliability under high-field stress as compared to thermal-oxide devices. Significant improvement in reliability of thermal oxide has also been observed using NH_3 as post oxidation annealing species (Senzaki et al. 2010). However, NH_3 does not contribute to any oxide growth. Recently, it has been revealed that HNO_3 and H_2O vapor can be utilized as direct thermal oxidation or post oxidation annealing agent, as they play a major role in oxidation/nitridation/hydrogenation mechanisms at the SiO_2/SiC interface and bulk oxide at temperature 1050°C. It has been shown that a passivation effect of nitrogen occurs, which reacts with Si-dangling bonds and removes carbon atoms at the SiO_2/SiC interface during direct thermal oxidation/annealing process.

Gate oxide also can be fabricated by using oxynitridation in 100% N_2O by Rapid Thermal Processing (RTP). This process has the advantage to reduce the thermal budget as compared to classical oxidation and produces a significant improvement in MOSFET performance (Constant et al. 2011). However, (Ellis and Buhrman 1996) showed that exposing an oxynitride film to N_2O in a RTP furnace or in a standard oxidation furnace can lead to the removal or depletion of nitrogen from the interface. Under these conditions, atomic oxygen reacts with the oxynitride film. In one of their experiments, (Ellis and Buhrman 1996) an oxynitride film was exposed to ozone at 875°C, where the decomposition of ozone is rapid and atomic oxygen is generated, which causes nitrogen to be completely removed after this exposure. They concluded that the atomic oxygen was responsible for the removal of nitrogen (Ellis and Buhrman 1996). In contrast, a sufficient level of nitrogen incorporation can be achieved by RTP when using high energetic photons as the source of thermal and optical energies. It was found from the C-V measurements that nitrogen incorporation during oxide formation provides a better SiO_2/SiC structure by reducing the effective oxide charge. Rapid thermal oxidation in 100% N_2O followed by RTP Ar reveals the best improvements of the SiC/SiO_2 interface (Constant et al. 2010). In recent study (Kagei et al. 2010), an improvement in interface quality and MOSFET channel mobility was reported when n-type 4H-SiC surface is being nitridated by a plasma source prior to dry oxidation.

Other techniques, which have been reported to reduce the D_{it} are atomic oxygen oxidation (Benfdila and Zekentes 2010) and trichloroethylene (TCE) annealing (Xu et al. 2005). The atomic oxygen technique uses a mixture of ozone (O_3) and oxygen (O_2) in a heated quartz tube. Atomic oxygen is formed by the decomposition of ozone at high temperature to form O and O_2. It is reported that the growth rate is increased and the interface quality is also improved (Benfdila and Zekentes 2010). As compared to the conventional dry O_2 oxidation, the dry O_2 + TCE oxidation results in lower interface-state density, reduced oxide-charge density and enhanced reliability. The passivation effects of Cl_2 and HCl on the structural defects at/near the SiC/SiO_2 interface and also their gettering effects on ion contamination could be the reason of the mentioned improvement. An increased oxidation rate induced by can be useful for reducing the normally high thermal budget of oxide growth. A passivation effect probably occurs when chlorine reacts with residual carbon atoms or clusters that remain at the interface during oxidation. There are also possibly other passivation effects on the structural defects at/near the interface (e.g., dangling Si and O bonds) by Cl_2 or HCl (Xu et al. 2005).

From the year 2007 till today, researchers are diverting nitridation and hydrogenation techniques to other elements based techniques. Research has been performed on hydrogen peroxide (Palmieri et al. 2009), sodium enhanced oxidation (Hermannsson and Sveinbjornsson 2011), $POCl_3$ annealing (Kotake et al. 2011, Kr′ol et al. 2015), dry oxidation in the presence of potassium (Hermannsson and Sveinbjornsson 2011) and nitic acid vapor oxidation (Banu et al. 2013, Banu et al. 2014). These techniques reported to produce D_{it} lower than the nitridation techniques which are very promising for the SiO_2 based SiC fabrication (Figs. 12.11 and 12.12). In addition, Ar annealing in high temperature (1300°C) (Kato et al. 2011) has also been reported to result in reduced D_{it} values (Fig. 12.12). On the other hand, channel mobility of MOSFETs is improved by extending the Ar annealing time (Kato et al. 2011). The incorporation of phosphorus atoms into SiO_2/SiC interface are also able to improve MOSFET performance for the Si-face and C-face SiC (Okamoto et al. 2010b). The presence of sodium during oxidation led to the improvements in mobility and the reduction in interface density and hence this technique is named Sodium-Enhanced Oxidation (SEO). The presence of sodium in the samples leads to severe threshold voltage instabilities and makes this technique of gate-oxide growth for MOSFETs commercially unfeasible (Hermannsson and Sveinbjornsson 2011). It has been reported that there is a strong reduction in the density of N_{it} at the SiO_2/4H-SiC interface after dry oxidation in the presence of potassium. This is followed by a significant enhancement of the oxidation rate (Hermannsson and Sveinbjornsson 2011).

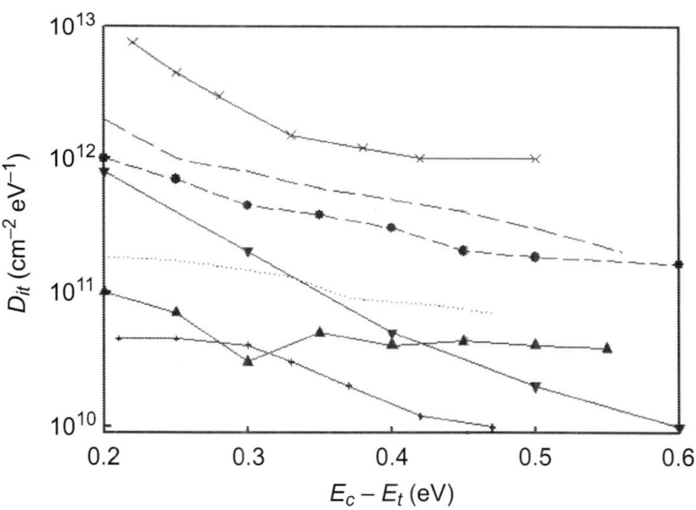

Fig. 12.12 D_{it} distributions for 4H n-type SiC samples with varied process ambient: (—✕—) Dry O_2 & Diluted N_2O Anneal, (—●—) Dry O_2 & NO (C-face), (— — -) Dry O_2 & H_2 anneal, (—▼—) Wet H_2O_2 & Ar Anneal, (..........) Dry O_2/K/Dry O_2, (—▲—) Dry + $POCl_3$, (—+—) Dry O_2/Na/Dry O_2.

7.2 Oxidation/Annealing Temperatures and Durations

Earlier studies have revealed that high temperature annealing of SiC in ultrahigh vacuum before oxidation process tends to leave a carbon-rich surface resulting from the evaporation of Si (Pehrsson and Kaplan 1989). This Si evaporation is significantly reduced at temperatures below 900°C, hence this effect could be minimized by loading in O_2 at temperature below 900°C and then raising the furnace to the final oxidation temperature of above 1000°C. (Lipkin and Palmour 1996) have reported that wet oxidation at 1050°C results in improved interfaces compared to wet oxidation at 1150°C. Longer anneal times are slightly better than the shorter anneal times for reducing interface state densities of oxides grown at 1100°C (Lipkin and Palmour 1996).

The D_{it} of SiC-oxidized (SiO_2) in dry O_2 samples at thermal oxidation temperature of 1300°C, is lower than SiC-oxidized (SiO_2) samples at and below 1200°C. This suggests that a decrease of the C component in SiO_2 film and SiO_2/SiC interface by higher oxidation temperature improves the MOS characteristics (Kurimoto et al. 2006, Moon et al. 2007). Based from the data of the XPS analyses, the XPS spectra of the samples at 1150°C and 1200°C shows a remarkable presence of SiC_xO_y and formation of SiO_2 of which the binding energy of the Si 2p core level is lower than that of the SiO_2 formed by oxidation of Si. On the other hand, a decrease of SiC_xO_y and existence of the pure SiO_2 formed due to oxidation of Si at 1300°C is observed in the XPS spectrum. It is suggested that the D_{it} is strongly associated to the presence of C component in the interfacial layer of the SiO_2/SiC structure. It is proposed that the thermal oxidation at a temperature higher than 1200°C leads to an improvement of MOS characteristics (Kurimoto et al. 2006). These may be attributed to two factors (i) oxide decomposition [$Si + SiO_2 \rightarrow 2SiO$ (volatile)] and (ii) reactions and diffusion of atomic oxygen in the oxide network. The decomposition of SiO_2 structure by breaking of Si-O bonds with the assistance of accumulated carbon clusters at the interface (Moon et al. 2007). Figure 12.13 shows the D_{it} distributions for 4H n-type samples with varied oxidation/nitridation temperatures.

The nitridation process using N_2O so far has been mostly limited at above 1200°C (Lipkin et al. 2002), because 4H-SiC(0001) is directly oxidized in N_2O at such elevated temperatures (Fujihira et al. 2005). Despite the interface quality is improved by nitridation, interface traps are generated simultaneously by oxidation during the N_2O treatment. At annealing temperature below 1200°C, the oxidation rate is expected to be higher if compared to the nitridation rate (Lipkin et al. 2002). It has been reported that there is an initial decrease in D_{it} from 900°C to 1000°C, which may be attributed to the nitridation process, while the following decrease from 1000°C to 1100°C may be due to the further oxidation which enhanced N_2O decomposition to O_2. Therefore, there is a trade-off between nitrogen incorporation at the interface and further oxidation of the SiC substrate. On the other hand, the high temperature dynamic decomposition process of N_2O (Zhao et al. 2006) explains that the higher temperature nitridation increases the trap emission rate. It is found that higher temperature nitridation reduces the trap time constant and increases the trap capture cross-section.

It has been reported that the effect of N_2O anneal is not confirmed in 900°C annealed sample, suggesting that the nitridation hardly occurs below 900°C (Fujihira et al. 2005). There is no significant difference in the results between the samples annealed at 900°C and without N_2O annealing; however the MOSFETs annealed at above 1000°C shows a lower threshold voltage (V_{th}) and a higher drain current (I_d) (Fujihira et al. 2005). The origins of interface traps such as C clusters, dangling bonds of Si and/or C and $Si_xC_yO_z$ are expected to make a chemical reaction with NO decomposed from N_2O and form Si–N or C–N bonds, resulting in the reduction of D_{it} near the conduction band edge of 4H-SiC (Fujihira et al. 2005, Jamet et al. 2001a). Longer NO annealing time decreases the change in effective oxide charges (Li et al. 2000, Jamet et al. 2001a).

In the case of re-oxidation process, the interface state density of these oxides is greatly affected by the re-oxidation temperature grown at 1100°C in wet ambient. Longer time but fixed at 950°C may produce better oxide, which results in lowering net oxide charge density by a factor of 2.5 and interface trap densities by a factor of 4, as compared to samples which received no re-oxidation (Lipkin and Palmour 1996). However, negative fixed charges appear in the oxide near the interface. The origin of negative fixed charges might be OH-, which cannot diffuse out from the interface at 950°C, however the mechanisms is not fully understood yet (Yano et al. 1999). On the other hand, removal of the nitrogen increases with the increment of re-oxidation temperature with complete removal at 1100°C (Chatty et al. 1999). At lower re-oxidation temperatures, there is only partial depletion of nitrogen from the interface (Chatty et al. 1999).

In terms of hydrogen passivation, (Afanas'ev 1999) reported that hydrogen annealing at 400°C has no effect on the C-V characteristics of 6H-SiC-based MOS structures. They suggested that D_{it} was not caused by the dangling bonds of Si atoms at the SiO_2/Si interface, but attributed to the carbon clusters. In another investigation, (Tsuchida et al. 1997) reported that clear absorption bands of Si–H and C–H stretching vibrations were noticed from the 6H-SiC surface after hydrogen annealing around 1000°C using Fourier-transformed infrared-attenuated total reflection. This leads to the conclusion that hydrogen terminates the dangling bonds of Si and C atoms on the 6H-SiC surface at the SiO_2/SiC interface. As the hydrogen annealing temperatures increases, D_{it} also decreases and saturates around 800°C when compared with Ar annealing. This confirms that the temperature of 800°C is high enough for termination of carbon atoms by hydrogen atoms.

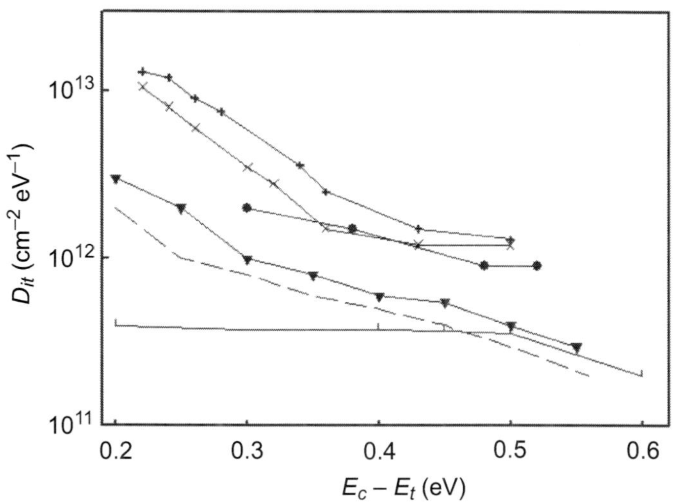

Fig. 12.13 D_{it} distributions of 4H n-type SiC samples with varied oxidation/nitridation temperatures: (—▾—) O_2(1200°C) + H_2(400°C), (— —) O_2(1200°C) + H_2(800°C), (—+—) O_2(1100°C) + Diluted N_2O(900°C), (—×—) O_2(1100°C) + Diluted N_2O(1100°C), (—●—), O_2(1400°C) + NO (1175°C), (—⊥—) O_2(1300°C) + NO (1175°C).

7.3 Dopants, Polytypes and Face Types of SiC Substrate

Laukhe et al. (1981) reported that the substrate quality plays an important role in oxidation kinetics. It is observed that oxides grown on n-type SiC wafers have lower D_{it} and Q_{eff} than those grown on p-type (Lipkin and Palmour 1996). The trapping of majority carriers by defects in the near interfacial layer could be the reason for this difference between n- and p-type charge densities. In the p-type semiconductor-oxide structures, the positive trapped carriers are added to the fixed charges in oxide; while in the n-type semiconductor-oxide structure, positive fixed charges in the oxide are neutralized by trapped negative charge from majority carriers (Palmieri et al. 2009). As for dry nitridation ambient, stable nitrided interface is created during the oxidation of nitrogen-doped n-type SiC substrates. It is also demonstrated that the oxides grown on p-type can be improved by NO annealing, but not by N_2O annealing (Dimitrijev et al. 1997) probably due to the influence of oxygen in the N_2O ambient (Sweatman et al. 1997).

Type and concentration of dopant in SiC has also been shown to affect the oxidation rate. In general, high dopant concentration has a tendency to increase the oxidation rate. Besides, a certain increase in oxide charge and trap density is obvious as the doping concentration is increased (Das et al. 1998). Palmour et al. (1989) have shown that during thermal oxidation of Al doped SiC, Al atoms are incorporated into the oxide at approximately the same density as the dopant density in the semiconductor, which leads to relatively large fixed charge and interface state densities at the SiO_2 p-type SiC interface (Shenoy et al. 1996). The p-type dopant density adversely affects the interface at doping levels on the order of 10^{17} cm^{-2} and above (Shenoy et al. 1996). The dopant species is seen to have a slight effect on the interface quality between B and Al for doping levels in the low range of 10^{16} cm^{-3} (Shenoy et al. 1996). The oxide on a more heavily doped substrate has a lower electric field strength than lightly doped counterpart (Lipkin and Palmour 1996).

To date, vast majority of SiC devices and applications have been implemented using two stable hexagonal polytypes, namely 4H and 6H (Roccaforte et al. 2010). Further developments observed that oxidation rate has a strong dependence on polytype. Cubic (3C)-SiC oxidizes faster than the rhombohedral (15R) while the hexagonal 4H and 6H oxidize at a still lower rate (Harris and Afanas'ev 1997). However, 3C has a narrower energy gap and the lowest breakdown field when compared to 4H and 6H (Dimitrijev and Jamet 2003). Amongst these three polytypes, 3C-SiC needs to be grown heteroepitaxially on Si substrates and does not have a technology for the growth of bulk crystals. On the other hand, the electron mobility in 4H-SiC is twice as high in the direction perpendicular to the *c*-axis and almost 10 times as high in the direction parallel to the *c*-axis than 6H. This fact alone explains the increased importance of 4H-SiC (Dimitrijev and Jamet 2003). It is reported that oxidation rates, interface state and effective fixed charge densities for 6H- and 4H-SiC are similar in wet O_2 at 1150°C. For equivalent nitrogen content, 3C-based devices exhibit more charge trapping than the 4H-based counterparts (Harris and Afanas'ev 1997).

In terms of SiC faces, (Muench and Pfaffender 1975, Roy et al. 2014) noticed that oxidation kinetics for C-face depends on the substrate conductivity. The Si-face (0001) of SiC wafers oxidizes slower but has lower oxide charge densities than the C-face (Benfdila and Zekentes 2010, Lipkin and Palmour 1996). The cause of the low oxide growth rate for Si-face if compared to C-face can be due to the low intrinsic interfacial reaction rate and to the high Si emission rate for Si-face (Hijikata et al. 2009). It is reported that the activation energy of Si-face is larger than that of C-face. This confirms that the Si-face intrinsically needs higher energy to oxidize than of C-face. D_{it} of Si-face has little dependence on oxidation temperature, while D_{it} depends on the temperature of C-face. The lower temperatures bring about a lower D_{it} value (Hijikata et al. 2010). It is verified that the oxidation rate on Si-face increases when the degree of hexagonality decreases. It is independent on hexagonality in the case of C-face (Benfdila and Zekentes 2010).

Hornetz et al. (1995) reported that the oxidation has some problems and limitations. The first layer on Si-faced SiC is an intermediate layer of the form $Si_4C_{x-4}O_2$ with average thickness of 1 nm. This layer is followed by a SiO_2 layer, which gives good interface quality compared to C-face. On the other hand, for C-faced SiC, the transitional layer is much thicker (several monolayers) and its exact composition is unknown (Benfdila and Zekentes 2010). It was reported that nitridation of the C-face 4H-SiC MOS interface generates N_{it} in the oxide. These traps capture channel mobile electrons and worsen the performance of MOSFETs. Unloading the samples at room temperature after oxidation is one of the methods to reduce this type of traps (Okamoto et al. 2010a).

The surface roughness and surface energy of the SiC after removal of an as-oxidized sample can be determined by using AFM (Guy et al. 2008) and Goniometer (Yasuhiro and Masahiko 1994, Nelson Martinez 2009) characterizations respectively. The presence of high density of carbon protrusion may results in high surface roughness and surface energy on the SiC surface after oxide removal. The interface roughness can also arise from compositional gradients between SiC and SiO_2 which can be observed via TEM characterization (Pongracz et al. 2007, Chuang and Hwu 2009) which provides the image of interfacial layer of SiO_2/SiC and SiO_2 bulk.

7.4 Other Factors

Besides temperatures and oxidation/nitridation parameters, other process parameters such as gas flow rate, gas concentration, gas pressure, solution boiling temperature and concentration also play important roles on the properties of SiO_2. However, only limited research has been carried out on these parameters (Yamamoto et al. 2009, Starodub et al. 1999, Sasse and Konig 1990, Hermannsson and Sveinbjornsson 2011, Kotake et al. 2011, Kr′ol et al. 2015). In terms of gas flow rate, it was found that oxygen partial-pressure range from 0.1 to 1 atm have an enhancement of growth rate for SiC (0001) Si-face as well as for (000-1) C-face. Moreover, the parabolic and linear rate constants are dependent on the oxygen partial-pressure (Yamamoto et al. 2009). It has been reported that the etching of the SiO_2 surface via volatile SiO decomposition occurs due to the gas deficient of oxygen ambient at low vapor pressure (Starodub et al. 1999, Sasse and Konig 1990).

As for nitridation gas concentrations, a lower interface-trap, near-interface trap and oxide-charge densities have been reported for nitrided oxide grown in diluted

NO gas with a mixture of nitrogen as compared to conventional 100%-NO oxidation. This is attributed to the fact that carbon-accumulation and carbon removal rates are closer when oxidation is performed in diluted NO, giving a smoother, less disordered and strained interface (Lai et al. 2005).

In the case of solution boiling temperature and concentration, wet environment is created by bubbling $O_2/N_2/N_2O$ through de-ionized water at 95°C (Lipkin and Palmour 1996, Xu et al. 2003). As for TCE solution, dry O_2 gas (250 ml/min) is bubbled through TCE solution at room temperature. In this set of experiment, a 0.017 TCE/O_2 mole ratio or equivalent 0.025 Cl_2/O_2 mole ratios was employed (Xu et al. 2005). On the other hand, 68% HNO_3 solution heated at 110°C was found to be an optimum heating temperature during oxide growth via direct and post oxidation annealing techniques (Banu et al. 2013, Banu et al. 2014).

8. SUMMARY

The improvement of oxide quality is essential for power devices applications. This can be achieved if appropriate process technology is implemented. The biggest challenge is to reduce traps in the SiO_2 thin films, particularly D_{it} and N_{it} because they have significant influence on the carrier mobility in the inversion channel. Hence, selection of a suitable oxidation process is very important for obtaining a high-quality gate oxide with low density of traps. In this chapter, recent passivation elements and process techniques have been compiled and compared. Besides, it is equally important to understand oxide growth mechanisms in order to eliminate and/or passivate the traps by feasible technological processes. By selecting a suitable process ambient, oxidation/nitridation parameters and semiconductor substrate, a high quality and reliable gate oxides could be realized in SiC MOS-based devices.

9. ACKNOWLEDGEMENTS

The authors would like to express their appreciation for the contributions of several people, who helped them to complete this edition successfully. Heartfelt thanks to Universiti Sains Malaysia (USM) for providing funds to one of the authors through the Post Graduate Fellowship Program (RU(1001/441/CIPS/AUPE001)), which enabled her to complete this edition without obstacles. This work was also supported by the Ministry of Science, Technology and Innovation (MOSTI), Malaysia and Universiti Malaysia Perlis (UniMAP). Besides, the authors also would like to take this opportunity to acknowledge the School of Materials and Mineral Resources Engineering, USM and School of Electrical Systems Engineering, UniMAP departments' staff for their direct and indirect support.

10. REFERENCES

Afanas'ev, V. V., A. Stesmans and M. O. Anderson. 1996a. Electron states and microstructure of thin a-C:H layers. Phys. Rev. B 54: 10820.

Afanas'ev, V. V., M. Bassler, G. Pensl, M. J. Schulz and E. Stein von Kamienski. 1996b. Band offsets and electronic structure of SiC/SiO_2 interfaces. J. Appl. Phys. 79(6): 3108-3113.

Afanas'ev, V. V. and A. Stesmans. 1997a. Interfacial defects in SiO_2 revealed by photon stimulated tunnelling of electrons. Phys. Rev. Lett. 78(12): 2437-2440.
Afanas'ev V. V. and A. Stesmans. 1997b. Analysis of near-interfacial SiO_2 traps using photon stimulated electron tunnelling, Microelectron. Eng. 36: 149.
Afanas'ev, V. V., M. Bassler, G. Pensl and M. Schulz. 1997c. Intrinsic SiC/SiO_2 interface states. Phys. Stat. Sol. (a) 162: 321-336.
Afanas'ev, V. V. and A. Stesmans. 1997d. H-complexed oxygen vacancy in SiO_2: energy level of a negatively charged state. Appl. Phys. Lett. 71: 3844-3846.
Afanas'ev, V. V., A. Stesmans and C. I. Harris. 1998. Observation of carbon clusters at the 4 H-SiC/SiO_2 interfaces. Mater. Sci. Forum 264-268: 857-860.
Afanas'ev, V. V. 1999. Electronic properties of SiO_2/SiC interfaces. Microelectron. Eng. 48: 241-248.
Agarwal, A., S. H. Ryu and J. Palmour. 2004. Power MOSFETs in 4H-SiC: device design and technology. pp. 785-810. *In*: W. J. Choyke, H. Matsunami and G. Pensl (eds.). Silicon Carbide – Recent Major Advances. Springer-Verlag. Berlin.
Agarwal, A. K., S. Seshadri and L. B. Roland. 1997. Temperature dependence of fowler-nordheim current in 6H-and 4H-SiC MOS capacitors. IEEE Elec. Dev. Lett. 18(12): 592-594.
Alialy, S., S. Altindal, E. E. Tanrkulu and D. E. Yldz. 2014. Analysis of temperature dependent current-conduction mechanisms in Au/TiO_2/n-4HSiC (metal/insulator/semiconductor) type schottky barrier diodes. J. Appl. Phys. 116: 083709.
Allerstam, F. and E. O. Sveinbjornsson. 2009. A study of deep energy-level traps at the 4H-SiC/SiO_2 interface and their passivation by hydrogen. Mater. Sci. Forum 600-603: 755-758.
Arai, K. 2011. R&D of SiC semiconductor power devices and strategy towards their practical utilization. Synthesiology 3(4): 245-258.
Ayalew, T. 2004. SiC Semiconductor Devices Technology, Modeling and Simulation [Online]. [Accessed on 2 December 2013].
Available from World Wide Web: http://www.iue.tuwien.ac.at/phd/ayalew/node20.html.
Banu, P., J. H. Moon, S. C. Kim, S. J. Joo, W. Bahng, I. H. Kang, et al. 2013. Investigation of thermally grown oxide on 4H-SiC by a combination of H_2O and HNO_3 vapor with varied HNO_3 solution heating temperatures. Appl. Surf. Sci. 285: 795-804.
Banu, P., J. H. Moon, S. C. Kim, S. J. Joo, W. Bahng, I. H. Kang, et al. 2014. Effects of wet-oxidized 4H-SiC annealed in HNO_3/H_2O vapour. Microelectron. Int. 31(1): 42-53.
Bassler, M., G. Pensl and V. V. Afanas'ev. 1997. Carbon cluster model for electronic states at SiC/SiO_2 interfaces. Diam. Relat. Mater. 6: 1472-1475.
Benfdila, A. and K. Zekentes. 2010. On silicon carbide thermal oxidation. African Phys. Rev. 4(5): 26-29.
Bentarzi, H. 2011 Transport in Metal-Oxide-Semiconductor Structures. Eng. Mater., Springer-Verlag Berlin Heidelberg, p. 1-104.
Brower, K. L. 1990. Dissociation kinetics of hydrogen-passivated (111) Si-SiO_2 interface defects. Phys. Rev. B 42: 3444.
Casady, J. B., A. K. Agarwal, S. Seshadri, R. R. Siergiej, L. B. Rowland, M. F. MacMillan, et al. 1998. 4H SiC power devices for use in power electronic motor control. Solid-State Electron. 42(12): 2165-2176.
Chakraborty, S., P. T. Lai, J. P. Xu, C. L. Chan and Y. C. Cheng. 2001. Interface properties of N_2O-annealed SiC metal oxide semiconductor devices. Solid-State Electron. 45: 471-474.
Chatty, K., V. Khemka, T. P. Chow and R. J. Gutmann. 1999. Re-oxidation characteristics of oxynitrides on 3C-and 4H-SiC. J. Electronic Mater. 28(3): 161-166.

Cheong, K. Y., S. Dimitrijev, J. Han and H. B. Harrison. 2003. Electrical and physical characterization of gate oxides on 4H-SiC grown in diluted N_2O. J. Appl. Phys. 93: 5682.

Cheong, K. Y., S. Dimitrijev and J. Han. 2004. Comparison of charge-retention times in n- and p-type 4H-SiC MOS capacitors as non-volatile memory elements. J. Cryst. Growth 268: 547-553.

Cheong, K. Y., W. Bahng and N. K. Kim. 2007. Effects of thermal nitride gate-oxide thickness on 4H silicon-carbide-based metal-oxide-semiconductor characteristics. Appl. Phys. Lett. 90(1): 1-3.

Cheong, K. Y., W. Bahng and N. K. Kim. 2008. Analysis of charge conduction mechanisms in nitrided SiO_2 film on 4H SiC. Phys. Lett. A 372: 529-532.

Cheong, K. Y. and Y. H. Wong. 2014. Surface modification of semiconductor by simultaneous thermal oxidation and nitridation. pp. 1-28. *In*: A.Y. Ching (ed.). Handbook of Manufacturing Engineering and Technology. Springer-Verlag, London.

Cho, W., R. Kosugi, K. Fukuda, K. Arai and S. Suzuki. 2000. Improvement of charge trapping by hydrogen post-oxidation annealing in gate oxide of 4H-SiC metal–oxide–semiconductor capacitors. Appl. Phys. Lett. 77(8): 1215-1217.

Chow, T. P. and A. K. Agarwal. 2006. SiC BJTs. International J. High Speed Electron and Syst. 16(3): 855-881.

Choyke, W. J., H. Matsunami and G. Pensl. 2004. Silicon Carbide Recent Major Advances. Springer, Germany.

Chu, A. X. and W. B. Fowler. 1990. Theory of oxide defects near the Si–SiO_2 interface. Phys. Rev. B 41: 5061-5075.

Chuang, K. C. and J. G. Hwu. 2009. Thin silicon oxide films on n-type 4H-SiC prepared by scanning frequency anodization method. Microelectron Eng. 86: 2207-2210.

Committee on Materials for High-Temperature Semiconductor Devices, Commission on Engineering and Technical Systems, National Research Council. 1995. Materials for High-Temperature Semiconductor Devices, National Academies Press, p. 4.

Constant, A., N. Camara, P. Godignon, M. Berthou, J. Camassel and J. M. Decams. 2010. Rapid and efficient oxidation process of SiC by *in situ* multiple RTP steps. Mater. Sci. Forum 645-648: 817-820.

Constant, A., N. Camara, J. Montserrat, E. Pausas, J. Camassel and P. Godignon. 2011. Oxidation process by RTP for 4H-SiC MOSFET Gate Fabrication. Mater. Sci. Forum 679-680: 500-503.

CREE, Inc. 2012. CMF20120D- Silicon Carbide Power MOSFET 1200 V, 80 mΩ Z-FET™ MOSFET data sheet [Online]. [Accessed on 2 December 2013].
Available from World Wide Web: http://www.cree.com/~/media/Files/Cree/Power/Data%20Sheets/CMF20120D.pdf.

Das, M. K., J. A. Cooper and M. R. Melloch. 1998. Effect of epilayer characteristics and processing conditions on the thermally oxidized SiO_2/SiC interface. J. Electron. Mater. 24(4): 353-357.

De Meo, R. C., T. K. Wang, T. P. Chow, D. M. Brown and L. G. Matus. 1994. Thermal oxidation of SiC in N_2O. J. Electrochem. Soc. 141: L50-52.

Dhar, S., S. Wang, J. R. Williams, S. T. Pantelides and L. C. Feldman. 2005. Interface passivation for silicon dioxide layers on silicon carbide. MRS Bulletin 30(4): 288-290.

Dimitrijev, S., H. F. Li, H. B. Harrison and D. Sweatman. 1997. Nitridation of silicon-dioxide films grown on 6H silicon carbide. IEEE Elec. Device Lett. 18(5): 175-177.

Dimitrijev, S. and P. Jamet. 2003. Advances in SiC power MOSFET technology. Microelectronics Reliab 43(2): 225-233.

Dimitrijev, S., H. B. Harrison, P. Tanner, K. Y. Cheong and J. Han. 2004a. Oxidation, MOS capacitors and MOSFETs. pp. 348-367. *In*: C. F. Zhe (ed.). SiC Power Materials Devices and Applications. Springer-Verlag, Berlin.

Dimitrijev, S., H. B. Harisson, P. Tanner, K. Y. Cheong and J. Han. 2004b. Properties of nitrided oxides on SiC. pp. 373-386. *In*: W. J. Choyke, H. Matsunami and G. Pensl (eds.). Silicon Carbide – Recent Major Advances. Springer-Verlag, Berlin.

Dixit, S. K. 2008. Radiation-Induced Charge Trapping Studies of Advanced Si and SiC Based MOS Devices [Online]. [Accessed on 2 December 2013].
Available from World Wide Web: http://etd.library.vanderbilt.edu/available/etd-03312008-170923/unrestricted/Sriram_Dixit_Dissertation_final.pdf.

Ellis, K. A. and R. A. Buhrman. 1996. Furnace gas-phase chemistry of silicon oxynitridation in N_2O. Appl. Phys. Lett. 68: 1696-1698.

Endo, T., E. Okuno, T. Sakakibara and S. Onda. 2009. High channel mobility of MOSFET fabricated on 4H-SiC (11-20) using wet annealing. Mater. Sci. Forum 600-603: 691-694.

Fairchild Semiconductor. 2007. FQP12N60C/FQPF12N60C 600 V, 650 80 mΩ, N-Channel MOSFET data sheet [Online]. [Accessed on 2 December 2013].
Available from World Wide Web: http://www.fairchildsemi.com/ds/FQ/FQP12N60C.pdf.

Fleetwood, D. M., M. R. Shaneyfelt, W. L. Warren, J. R. Schwank, T. L. Meisenheimer and P. S. Winokur. 1995. Border traps: issues for MOS radiation response and long-term reliability. Microelectronics Reliab 35(3): 403-405.

Fleetwood, D. M. 1996. Fast and slow border traps in MOS devices. IEEE Trans. Nucl. Sci. 43(3): 779.

Fraga, M. A., R. S. Pessoa, M. Massi and H. S. Maciel. 2012. Applications of SiC-based thin films in electronic and MEMS devices. pp. 313-322. *In*: H. Yasuto (ed.). Physics and Technology of Silicon Carbide Devices, INTECH, London, UK.

Fujihira, K., N. Miura, K. Shiozawa, M. Imaizumi, K. Ohtsuka and T. Takami. 2004. Successful enhancement of lifetime for SiO_2 on 4H-SiC by N_2O anneal. IEEE Electron. Dev. Lett. 25(11): 1.

Fujihira, K., Y. Tarui, M. Imaizumi, K. Ohtsuka, T. Takami, T. Shiramizu, et al. 2005. Characteristics of 4H-SiC MOS interface annealed in N_2O. Solid-State Electron. 49: 896-901.

Fukuda, K., S. Suzuki, T. Tanaka and K. Arai. 2000. Reduction of interface-state density in 4H-SiC n-type metal–oxide–semiconductor structures using high-temperature hydrogen annealing. Appl. Phys. Lett. 76(12): 1585.

Fukushima, Y., A. Chanthaphan, T. Hosoi, T. Shimura and H. Watanabea. 2015. Cathodoluminescence study of radiative interface defects in thermally grown SiO_2/4H-SiC(0001) structures. Appl. Phys. Lett. 106(26): 261604-1-5.

Greenville Whittaker, A. 1978. The controversial carbon solid−liquid−vapour triple point. Nature 276(5689): 695-696.

Gupta, A., S. Toby, E. P. Gusev, H. C. Lu, Y. Li, M. L. Green, et al. 1998. Nitrous oxide gas phase chemistry during silicon oxynitride film growth. Prog. Surf. Sci. 59: 103.

Gupta, S. K., A. Azam and J. Akhtar. 2010. Experimental analysis of I-V and C-V characteristics of Ni/SiO_2/4H-SiC system with varying oxide thickness. Microelectron. Int. 27(2): 106-112.

Gupta, S. K. and J. Akhtar. 2011. Thermal oxidation of silicon carbide (SiC) – experimentally observed facts. pp. 208-228. *In*: M. Mukherjee (ed.). Silicon Carbide – Materials, Processing and Applications in Electronic Devices. INTECH, Brazil.

Gupta, S. K., J. Singh and J. Akhtar. 2013. Materials and processing for gate dielectrics on silicon carbide (SiC) surface. pp. 207-232. *In*: H. Yasuto (ed.). Physics and Technology of Silicon Carbide Devices. INTECH, London, UK.

Gurfinkel, M., J. C. Horst, J. S. Suehle, J. B. Bernstein, Y. Shapira, K. S. Matocha, et al. 2008. Time-dependent dielectric breakdown of 4H-SiC/SiO_2 MOS capacitors. IEEE Trans. Device Mat. Rel. 8(4): 635-640.

Guy, O. J., M. Lodzinski, K. S. Teng, T. G. G. Maffeis, M. Tan, I. Blackwood, et al. 2008. Investigation of the 4H-SiC surface. Appl. Surf. Sci. 254: 8098-8105.

Hamada, K. 2010. Present status and future prospects for electronics in electric vehicles/hybrid electric vehicles and expectations for wide-bandgap semiconductor devices. pp. 1-19. *In*: P. Friedrichs, T. Kimoto, L. Ley and G. Pensl (eds.). Silicon Carbide: Power Devices and Sensors. Wiley-VCH, Germany.

Harris, C. I. and V. V. Afanas'ev. 1997. SiO_2 as an insulator for SiC devices. Microelectron. Eng. 36(1-4): 167-174.

Hermannsson, P. G. and E. O. Sveinbjornsson. 2011. Reduction in the density of interface states at the SiO_2/4H-SiC interface after dry oxidation in the presence of potassium. Mater. Sci. Forum 679-680: 334-337.

Hijikata, Y., T. Yamamoto and H. Yaguchi. 2009. Model calculation of SiC oxidation rates in the thin oxide regime. Mater. Sci. Forum 600-603: 663-666.

Hijikata, Y., H. Yaguchi and S. Yoshida. 2010. Model calculations of SiC oxide growth rate at various oxidation, temperatures based on the silicon and carbon emission model. Mater. Sci. Forum 645-648: 809-812.

Hornetz, B., H. J. Michel and J. Halbritter. 1994. ARXPS studies of SiO_2-SiC interfaces and oxidation of 6H SiC single crystal Si-(001) and C-(00-1) surfaces. J. Mater. Res. 6(12): 3088.

Hornetz, B., H. J. Michel and J. Halbritter. 1995. Oxidation and 6H-SiC-SiO_2 interfaces. J. Vac. Sci. Technol. A 13(3): 767.

Iwasaki, Y., H. Yano, T. Hatayama, Y. Uraoka and T. Fuyuki. 2010. Significant decrease of the interface state density by NH_3 plasma pretreatment at 4H-SiC (0001) surface and its bond configuration. Mater. Sci. Forum 645-648: 503-506.

Jamet, P., S. Dimitrijev and P. Tanner. 2001a. Effects of nitridation in gate oxides grown on 4H-SiC. J. Appl. Phys. 90(10): 5058-5063.

Jamet, P. and S. Dimitrijev. 2001b. Physical properties of N_2O and NO-nitrided gate oxides grown on 4H SiC. Appl. Phys. Lett. 79: 323.

Kagei, Y., T. Kirino, Y. Watanabe, S. Mitani, Y. Nakano, T. Nakamura, et al. 2010. Improved electrical properties of SiC-MOS interfaces by thermal oxidation of plasma nitrided 4H-SiC(0001) surfaces. Mater. Sci. Forum 645-648: 507-510.

Kato, M., Y. Nanen, J. Suda and T. Kimoto. 2011. Improved characteristics of SiC MOSFETs by post-oxidation annealing in Ar at high temperature. Mater. Sci. Forum 679-680: 445-448.

Knaup, J. M., P. Deak and T. Frauenheim. 2009. The Inefficiency of H_2-passivation as a criterion for the origin of SiC/SiO_2 deep interface states – A theoretical study. Mater. Sci. Forum 600-603: 723-726.

Kotake, S., H. Yano, D. Okamoto, T. Hatayama and T. Fuyuki. 2011. Improved MOS interface properties of C-face 4H-SiC by $POCl_3$ annealing. Mater. Sci. Forum 679-680: 425-428.

Krishnaswami, S., M. K. Das, A. K. Agarwal and J. W. Palmour. 2004. Reliability of nitrided oxides in N- and p-type 4H-SiC MOS structures. Mat. Res. Soc. Symp. Proc. 815: J8.4.1-4.6.

Kr′ol, K., M. Sochacki, W. Strupinski, K. Racka, M. Guziewicz, P. Konarski, et al. 2015. Chlorine-Enhanced Thermal Oxides Growth and Significant Trap Density Reduction at SiO_2/SiC Interface by Incorporation of Phosphorus. Thin Solid. Films, In Press, Accepted Manuscript.

Kurimoto, H., K. Shibata, C. Kimura, H. Aoki and T. Sugino. 2006. Thermal oxidation temperature dependence of 4H-SiC MOS interface. Appl. Surf. Sci. 253: 2416-2420.

Lai, P. T., J. P. Xu, C. L. Chan and Y. C. Cheng. 1999. Interface properties of N_2O annealed NH_3-treated 6H-SiC MOS capacitor. IEEE, pp. 46-49.

Lai, P. T., J. P. Xu and C. L. Chan. 2002. Effects of wet N_2O oxidation on interface properties of 6H-SiC MOS capacitors. IEEE Electron Device Lett. 23(7): 410-412.

Lai, P. T., J. P. Xu, C. X. Li and C. L. Chan. 2005. Improved interfacial properties of SiO_2 grown on 6H-SiC in diluted NO. Appl. Phys. A 81: 159-161.

Laukhe, Y., Y. M. Tairov, V. F. Tsvetkov and F. Schepanski. 1981 Oxidation-kinetics of SiC single-crystals. Inorg. Mater. 17(2): 177.

Levinshtein, S. L. R. M. and M. S. Shur. 2001. Properties of Advanced Semiconductors: GaN, AlN, InN, SiC, SiGe. Wiley, New York, p. 1-216.

Li, H. F., S. Dimitrijev, D. Sweatman and H. B. Harrison. 2000. Effect of NO annealing conditions on electrical characteristics of n-type 4H-SiC MOS capacitors. J. Electron. Mater. 29(8): 1027.

Lipkin, L. A. and J. W. Palmour. 1996. Improved oxidation procedures for reduced SiO_2/SiC defects. J. Electron. Mater. 25(5): 909-915.

Lipkin, L. A., M. K. Das and J. W. Palmour. 2002. N_2O processing improves the 4H-SiC:SiO_2 interface. Mater Sci Forum 389-393: 985-988.

Liu, G., B. R. Tuttle and S. Dhar. 2015. Silicon carbide: a unique platform for metal-oxide-semiconductor physics. Appl. Phys. Rev. 2: 1-21.

Liu, P. 2014. Atomic Structure of the Vicinal Interface between Silicon Carbide and Silicon Dioxide, Doctoral Dissertations, University of Tennessee – Knoxville, pp. 87-97.

Liu, Y., M. R. Halfmoon, C. A. Rittenhouse and S. Wang. 2010. Passivation effects of fluorine and hydrogen at the SiC–SiO_2 interface. Appl. Phys. Lett. 97: 1-3.

Lu, H. C., E. P. Gusev, T. Gustafsson and E. Garfunkel. 1997. Effect of near-interfacial nitrogen on the oxidation behavior of ultrathin silicon oxynitrides. J. Appl. Phys. 81(10): 6994.

Macfarlane, P. J. and M. E. Zvanut. 1999. Generation and annealing characteristics of paramagnetic centers in oxidized 3C-SiC and 6H-SiC. J. Electron. Mater. 28: 144-147.

Majumdar, G., J. Donlon, E. Motto, T. Ozeki, H. Yamamoto and M. Seto. 2013. Present Status and Future Prospects of SiC Power Devices, Mitsubishi Electric, [Online]. [Accessed on 2 December 2013].
Available from World Wide Web: http://www.pwrx.com/pwrx/app/ias_04_24_sic pdf.

Massimiliano, D. V. and S. T. Pantelides. 2000. Oxygen stability, diffusion and precipitation in SiC: Implications for thin-film oxidation. J. Electron. Mater. 29(3): 353-358.

Moon, J. H., K. Y. Cheong, H. K. Song, J. H. Yim, M. S. Oh, J. H. Lee, et al. 2007. Effect of Post Oxidation Annealing on High-Temperature Grown SiO_2/4H-SiC Interface. ICSCRM, Otsu, Japan, p. 732.

Morino, K., S. Miyazaki and M. Hirose. 1997. Phosphorous Incorporation in Ultrathin Gate Oxides and Its Impact to the Network Structure, 1997 International Conference on Solid State Devices and Materials (SSDM1997), Hamamatsu, Sept. 16-19 1997, A-2-3, pp. 18-19.

Muench, W. V. and I. Pfaffeneder. 1975. This Journal. J. Electrochem. Soc. Vol. 122, pp. 642.

Nelson Martinez, B. S. 2009. Wettability of Silicon, Silicon Dioxide and Organosilicate Glass, Masters Dissertation, University of North Texas, pp. 1-71.

Nicollian, E. H. and J. R. Brews. 1982. MOS (Metal Oxide Semiconductor) Physics and Technology. John Wiley & Sons. Inc., USA. pp. 1-4.

Ohnuma, T., A. Miyashita, M. Yoshikawa and H. Tsuchida. 2007. Dynamical simulation of SiO_2/4H-SiC interface on C-face oxidation. Mater. Sci. Forum 600-603: 591-596.

Okamoto, D., H. Yano, Y. Oshiro, T. Hatayama, Y. Uraoka and T. Fuyuki. 2010a. Investigation of oxide films prepared by direct oxidation of C-face 4H-SiC in nitric oxide. Mater. Sci. Forum 645-648: 515-518.

Okamoto, D., H. Yano, Y. Oshiro, T. Hatayama and T. Fuyuki. 2010b. Removal of near-interface traps at SiO_2/4H-SiC (0001) interfaces by phosphorus incorporation. Appl. Phys. Lett. 96: 203508.

Opila, E. J. 1999. Variation of the oxidation rate of silicon carbide with water-vapor pressure. J. Am. Ceram. Soc. 82: 625-636.

Owman, F., L. I. Johansson and P. Mirtensson. 1995. Inst. Phys. Conf. Ser. 142, 477 (Proc. Int. Conf. SiC and Rel. Mater. 1995, Kyoto, Japan).

Palmieri, R., C. Radtke, M. R. Silva, H. Boudinov and E. F. Silva. 2009. Trapping of majority carriers in SiO_2/4H-SiC structures. Journal of Physics D: Appl. Phys. 42: 1-6.

Palmour, J. W., R. F. Davis, H. S. Kong, S. F. Corcoran and D. P. Griffis. 1989. Dopant redistribution during thermal oxidation monocrystalline beta-SiC thin films. J. Electrochem. Soc. 136: 502-507.

Pantelidas, S. T., S. Wang, A. Franceschetti, R. Buczko, M. Di Ventra, S. N. Rashkeev, et al. 2006. Si/SiO_2 and SiC/SiO_2 interfaces for MOSFETs-challenges and advances. Mater. Sci. Forum 527-529: 935-948.

Park, Y. S. 1998. SiC Materials and Devices in Semiconductors and Semimetals, Vol. 52. Academic Press, New York.

Pehrsson, P. E. and R. Kaplan. 1989. Excimer laser cleaning, annealing, and ablation of β–SiC. J. Mater. Res. (USA) 4(6): 1480-1490.

Pensl, G., S. Beljakowa, T. Frank, K. Gao, F. Speck, T. Seyller, et al. 2010. Alternative techniques to reduce interface traps in n-type 4H-SiC MOS capacitors. pp. 193-195. *In*: P. Friedrichs, T. Kimoto, L. Ley and G. Pensl (eds.). Silicon Carbide – Volume 2: Power Devices and Sensors. Wiley-VCH Verlag GmbH & Co. KGaA, Weinheim, Germany.

Pitthan, E., R. Palmieri, S. A. Corr^ea, G. V. Soares, H. I. Boudinov and F. C. Stedilea. 2013. The role played in the improvement of the SiO_2/SiC interface by a thin SiO_2 film thermally grown prior to oxide film deposition. ECS Solid State Lett. 2(1): 8-10.

Pongracz, A., G. Battistig, C. Dücso, K. V. Josepovits and P. Deák. 2007. Structural and electronic properties of Si/SiO_2 MOS structures with aligned 3C–SiC nanocrystals in the oxide. Mater. Sci. Eng. C 27: 1444-1447.

Radtke, C., I. J. R. Baumvol, J. Morais and F. C. Stedile. 2001. Initial stages of SiC oxidation investigated by ion scattering and angle-resolved X-ray photoelectron spectroscopies. Appl. Phys. Lett. 78: 3601.

Ranua´rez, J. C., M. J. Deen and C. H. Chen. 2006. A review of gate tunneling current in MOS devices. Microelectronics Reliability 46: 1939-1956.

Raynaud, C. 2001. Silica films on silicon carbide: a review of electrical properties and devices applications. J. Non-Crystalline Solids 280: 1-31.

Roccaforte, F., F. Giannazzo and V. Raineri. 2010. Nanoscale transport properties at silicon carbide interfaces. J. Phys. D: Appl. Phys. 43(22): 1-25.

Roy, J., S. Chandra, S. Das and S. Maitra. 2014. Oxidation behavior of silicon carbide – A review. Rev. Adv. Mater. Sci. 38: 29-39.

Rozen, J., S. Dhar, S. Wang, V. V. Afanasev, S. T. Pantelides, J. R. Williams, et al. 2009. Impact of nitridation on negative and positive charge buildup in SiC gate oxides. Mater. Sci. Forum 600-603: 803-806.

Sasse, H. E. and U. Konig. 1990. SiO diffusion during thermal decomposition of SiO_2. J. Appl. Phys. 67(10): 6194-6196.

Senzaki, J., T. Suzuki, A. Shimozato, K. Fukuda, K. Arai and H. Okumura. 2010. Significant improvement in reliability of thermal oxide on 4H-SiC (0001) face using ammonia post-oxidation annealing. Mater. Sci. Forum 645-648: 685-688.

Sharma, Y. K., A. C. Ahyi, T. Issacs-Smith, X. Shen, S. T. Pantelides, X. Zhu, et al. 2012. Phosphorous passivation of the SiO_2/4H-SiC interface. Solid-State Electron. 68: 103-107.

Shen, X. and S. T. Pantelides. 2011. Identification of a major cause of endemically poor mobilities in SiC/SiO_2 structures. Appl. Phys. Lett. 98: 053507, pp. 1-3.

Shenoy, J. N., J. A. Cooper and M. R. Mellocha. 1996. Comparison of thermally oxidized metal–oxide–semiconductor interfaces on 4H and 6H polytypes of silicon carbide. Appl. Phys. Lett. 68(6): 1.

Siergiej, R. R., R. C. Clarke, S. Sriram, A. K. Agarwal, R. J. Bojko, A. W. Morse, et al. 1999. Advances in SiC materials and devices: an industrial point of view. Mater. Sci. Eng. 61-62: 9-17.

Soares, G. V., I. J. R. Baumvol, S. A. Correa, C. Radtke and F. C. Stedile. 2009. Water vapor interaction with silicon oxide films thermally grown on 6H-SiC and on Si. Appl. Phys. Lett. 95: 191912, pp. 1-3.

Son, N. T., W. M. Chen, J. L. Lindstrom, B. Monemar and E. Janzen. 1999. Carbon-vacancy related defects in 4H- and 6H-SiC. Mater. Sci. Eng. B 61-62: 202-207.

Stanley, W. and N. T. Richard. 2000. Silicon Processing for the VLSI Era, Vol. 1: Process Technology Second Edition. Lattice Press, California, p. 1-660.

Starodub, D., E. P. Gusev, E. Garfunkel and T. Gustafsson. 1999. Silicon oxide decomposition and desorption during the thermal oxidation of silicon. Surf. Rev. and Lett. 6(1): 45-52.

Strife, J. S. and J. E. Sheehan. 1998. Ceramic coating for carbon/carbon composites. Ceramic Bulletin 67: 369.

Suzuki, T., J. Senzaki, T. Hatakeyama, K. Fukuda, T. Shinohe and K. Arai. 2007. Effect of Gate Wet Reoxidation on Reliability and Channel Mobility of Metal-Oxide-Semiconductor Field-effect Transistors Fabricated on 4H-SiC (0001), ICSCRM, Otsu, Japan, p. 792, 2007.

Sweatman, D., S. Dimitrijev, H. F. Li, P. Tanner and H. B. Harrison. 1997. Growth and nitridation of silicon-dioxide films on silicon-carbide. Mat. Res. Soc. Symp. Proc. 470: 413-418.

Takaya, H., J. Morimoto, T. Yamamoto, J. Sakakibara, Y. Watanabe, N. Soejima, et al. 2013. 4H-SiC Trench MOSFET with thick bottom oxide. Mater. Sci. Forum 740-742: 683-686.

Ted Pella, Inc., Microscopy Products for Science and Industry, Hardness Table Online]. [Accessed on 24 September 2015].
Available from World Wide Web: https://www.tedpella.com/company_html/hardness.htm.

Tortorelli, P. F. and K. L. More. 2003. Effects of high water-vapor pressure on oxidation of SiC at 1200°C. J. Am. Ceram. Soc. 86(8): 1249-1255.

Tsuchida, H., I. Kamata and K. Izumi. 1997. Infrared spectroscopy of hydrides on the 6H-SiC surface. Appl. Phys. Lett. 70: 3072.

Uren, M. J., J. H. Stathis and E. Cartier. 1996. Conduction mechanisms on Pb centres at Si/SiO_2 interfaces. J. Appl. Phys. 80: 3915.

Vathulya, V. R., D. H. Wang and M. H. White. 1998. On the correlation between the carbon content and the electrical quality of thermally grown oxides on p-type 6H-silicon carbide. Appl. Phys. Lett. 73: 2161-2163.

Von Kamienski, E. S., E. A. Golz and H. Kurz. 1995. Effects of Ar and H_2 annealing on the electrical properties of oxides of 6H-SiC. Mater. Sci. Eng. B 29: 131-133.

Wahab, Q., L. Hultman, M. Willander and J.-E. Sundgren. 1995. Structural characterization of oxide layers thermally grown on 3C-SiC films. J. Electronic Mater. 24(10): 1345-1348.

Wang, S.R. and L. L. Zhong. 2002. Studies of 6H-SiC devices. Current Appl. Phys. 2(5): 393-399.

Xu, J. P., P. T. Lai and C. L. Chan. 2003. Steam-induced interface improvement of N_2O-nitrided SiO_2 grown on 6H-SiC. Solid-State Electron. 47(8): 1397-1400.

Xu, J. P., P. T. Lai and C. L. Chan. 2005. Effects of chlorine on interfacial properties and reliability of SiO_2 grown on 6H-SiC. Appl. Phys. A 81: 173-176.

Yamamoto, T., Y. Hijikata, H. Yaguchi and S. Yoshida. 2009. Oxygen-partial-pressure dependence of SiC oxidation rate studied by *in situ* spectroscopic ellipsometry. Mater. Sci. Forum 600-603: 667-670.

Yano, H., F. Katafuchi, T. Kimoto and H. Matsunami. 1999. Effects of wet oxidation/anneal on interface properties of thermally oxidized SiO_2/SiC MOS system and MOSFET's. IEEE Trans. Electron Devices 46(3): 504-510.

Yasuhiro, S. and M. Masahiko. 1994. Study of HF-treated heavily-doped Si surface using contact angle measurements. Jpn. J. Appl. Phys. 33: 6508-6513.

Zeghbroeck, B. V. 2011. MOS Field-effect-transistors, principles of semiconductor devices [Online]. [Accessed on 2 December 2013].
Available from World Wide Web: http://ecee.colorado.edu/~bart/book/book/chapter7/ch7_1.htm.

Zetterling, C. M. 2002. Process Technology for Silicon Carbide Devices. KTH, Royal Institute of Technology, Sweden. INSPEC, The Institution of Electrical Engineers, London, United Kingdom, p. 1-163.

Zhao, J. H. 2005. Silicon carbide power field-effect transistors. MRS Bulletin, USA 30(4): 293.

Zhao, P., Rusli, Y. Liu, C. C. Tin, W. G. Zhu and J. Ahn. 2006. Investigation of 4H-SiC MOS capacitors annealed in diluted N_2O at different temperatures. Microelectronic Eng. 83: 61-64.

Zippelius, B., S. Beljakowa, M. Krieger, G. Pensl, S. A. Reshanov, M. Noborio, et al. 2010. High electron mobility achieved in n-channel 4H-SiC MOSFETs oxidized in the presence of nitrogen. *In*: P. Friedrichs, T. Kimoto, L. Ley and G. Pensl (eds.). Silicon Carbide – Volume 2: Power Devices and Sensors. Wiley-VCH Verlag GmbH & Co. KGaA, Weinheim, Germany, p. 2363-2373.

Zolper, J. C. and M. Skowronski. 2005. Advances in silicon carbide electronics. MRS Bulletin, USA 30: 273.

13

Formation, Growth Mechanism and Electronic Structures of Ge Films on Si Substrates

Yoshitaka Fujimoto

ABSTRACT

The high-quality germanium (Ge) films on silicon (Si) substrates have received much interest because of its compatibility with the well-established Si-based technology and their relevant applicability in electronics and opto-electronics devices. The growth of high-quality Ge films on the Si substrate is difficult due to the existence of the lattice mismatch between Ge and Si crystals, which gives rise to a rough surface morphology and a high density of defects. This chapter reviews the first-principles density-functional study that reveals atomic structures, stabilities and electronic structures of various Ge films deposited on Si (001) substrates. The energetics associated with the layer-by-layer growth of the Ge films on the Si substrate is examined, which is used for clarifying the strain-relaxation mechanism and the introduction of the 90° misfit dislocation. The scanning tunneling microscopy images of various Ge films are demonstrated and the possibility to observe the core structure of the 90° dislocation is exhibited.

Keywords: Ge film on Si, epitaxial growth, strain engineering, first-principles total-energy calculation, density-functional theory.

1. INTRODUCTION

The epitaxial growth of a high-quality semiconductor film with a smooth surface, low density of defects and thin buffer layers on substrates is a key fabrication technique and also provides fundamental knowledge about missing links between atomic reactions and morphology of resultant films in the growth process (Voigtländer 2001, Oshiyama 2002). Growth of Ge films on Si (001) substrates is an important example: Ge on Si is a simple example model system for understanding

Department of Physics, Tokyo Institute of Technology, Oh-okayama, Meguro-ku, Tokyo 152-8551, Japan.
E-mail: fujimoto@stat.phys.titech.ac.jp

hetero-epitaxial growth of strained films (Liu et al. 1997, Shiraki and Sakai 2005, Berbezier and Ronda 2009).

Because of its compatibility with the conventional Complementary Metal-Oxide-Semiconductor (CMOS) technology, the high-quality Ge films on Si substrates are also indispensable systems for providing full potential applications in electronic devices. The carrier mobility of electrons as well as holes in Ge crystal is higher than that in Si and the tensile strains in Si crystals also enhance their carrier mobility, compared with unstrained Si, which leads to the improvement of Metal-Oxide-Semiconductor Field-Effect-Transistors (MOSFET) performance (Lee et al. 2005). Thereby, the Ge as well as SiGe layer is often used as conduction channels of carriers in high-performance Field-Effect Transistors (FETs) and they are also used as buffer layers to obtain tensilely strained Si layers (Känel et al. 2002, Goo et al. 2003, Shang et al. 2006).

Owing to its strong interband absorption at near-infrared wavelengths and high bandwidths as well as the Si compatibility, Ge-on-Si materials have also provided optoelectronics applications including photodetectors, light emitters, lasers and solar cells (Masini et al. 2003, Boucaud et al. 2013, Liu et al. 2012, Posthuma et al. 2007). As is well-known, unstrained Ge possesses an indirect-bandgap. On the other hand, the tensile strained Ge becomes a direct-bandgap material and it dramatically increases interband radiative efficiency, which produces possible high efficient optoelectronics devices (Michel et al. 2010, Boztug et al. 2014).

The main problem in the growth of Ge films on Si substrates is attributed to ~4% lattice-mismatch between Ge and Si and this lattice mismatch leads to a generation of a rough surface and an introduction of a high density of defects (Liu et al. 1997, Kahng et al. 2000, Olsen et al. 2002). As is well-known, the initial growth of Ge films on Si substrates usually proceeds under Stranski-Krastanov (SK) mechanism: the Ge film (wetting layer) grows up to a thickness of approximately three monolayers (ML) (Eaglesham and Cerullo 1991). Beyond this thickness, three-dimensional Ge islands with a density of 10^{10}-10^{11} cm^{-2} are formed on the wetting layer (Sheldon et al. 1985, Mo and Lagally 1991, Mo and Lagally 1992), where the islanding is usually accompanied by stress-releasing 60° dislocation which are caused by the <100>/{111} slip in the diamond structure (Arias and Joannopoulos 1994, Goldman et al. 1998, Bolkhovityanov et al. 2008). The coalescence of the islands produces a large number of defects which cause treads in the Ge film and the generation of the 60° dislocation degrades the quality of the Ge film with a rough surface morphology (Speck et al. 1996). Therefore, it is of great importance to suppress the nucleation of threading defects at the situation of island coalescence by delaying a transition from the 2D growth to 3D islanding.

One of the possible ways to form the strain-released 2D Ge films on Si is to grow Ge buffer layers between relaxed Ge films and Si substrates under the condition of low temperatures (Currie et al. 1998, Capellini et al. 2010). Actually, the growth of the buffer layers without accomplishing islanding formation at the low temperature of ~300°C is performed (Eaglesham and Cerullo 1990, Colace et al. 1998, Colace et al. 2000, Luan et al. 1999, Hartmann et al. 2004) and then, the high temperatures with more than ~600°C are used for the growth of the additional Ge layers deposited on the buffer layers (Colace et al. 1998, Colace et al. 2000, Luan 1999,

Bolkhovityanov 2004a, b, Cannon et al. 2007). It is reported that the cyclic thermal annealing technique can decrease the density of threading dislocations considerably at high temperatures of less than ~900°C (Luan et al. 1999, Colace et al. 2000, Liu et al. 2004).

Another effective way to prevent the Ge islanding growth is to deal with the surfactants in the growth process using Molecular Beam Epitaxy (MBE) (Copel et al. 1989, Schmidt et al. 2005, Liu et al. 1999, Liu et al. 2006). It is reported that the formation of 3D islands is strongly suppressed and the 2D growth of Ge films continues beyond the film thickness of 3 Ge ML with the use of surfactants such as arsenic, antimony and atomic hydrogen for growing Ge layers on Si (001) substrates (Sakai and Tatsumi 1994, Hoegen et al. 1994, Dentel et al. 1998, Kahng et al. 1998, Kahng et al. 2000, Poravoce et al. 2004). Under atomic hydrogen and antimony-surfactants mediated epitaxy, the 90° edge dislocation with the Burgers vector along <110> direction is observed near the Ge/Si interface in the Ge/Si heterostructures (Sakai et al. 2005, Sakai et al. 1997, Wietler et al. 2005, Wietler et al. 2006, Mochizuki et al. 2007, Bolkhovityanov et al. 2008). For realization of the high-quality Ge film on Si with smooth surfaces and the low density of defects, it is vital to understand the strain relief mechanism in the film growth process in microscopic viewpoint, especially atomistic level.

This chapter provides a review of the first-principles density-functional study that clarifies atomic geometries, energetics and electronic structures of various strained Ge films on Si (001) surfaces. The film-energy diagrams corresponding to the layer-by-layer epitaxial growth of Ge-on-Si films are shown together with and without the hydrogen surfactant effects and the critical thickness at which the 90° dislocation occurs is estimated. The Scanning Tunneling Microscopy images (STM) of various Ge-on-Si films are also exhibited for identifying the core structure of the 90° dislocation (Fig. 13.1). The 90° Dislocation Core (DC) structure appears to be dark and bright lines in the STM images, which is dependent on the bias voltages and the surfactant effects.

The chapter is organized as follows. In the next section, the theoretical methodology and computational details are presented to perform the first-principles total-energy calculations. In the third section, the film-energy diagram of the $p(2 \times 2)$, the 2×8 Dimer-Vacancy-Line (DVL) and the 90° DC structures without surfactant effects are described. From the result of energetics, the 90° DC structure is found to be favored in energy when sufficiently thick Ge overlayers are deposited on the Si (001). The STM images of the $p(2 \times 2)$, the 2×8 DVL and the 90° DC structures are demonstrated and they are shown to be distinguishable from one another. The fourth section is devoted to discussion on the hetero-epitaxial growth of hydrogen-covered Ge film on Si. The morphology of the H-covered wetting layer is examined and the 2×1 structure is shown to be favorable energetically, compared with the $2 \times N$ reconstructed structures with DVL. The critical thickness of the transition from the 2×1 structure into the 90° DC structure is also estimated. The STM image of the 90° DC structure in which the Ge top-surface is covered with H atoms is exhibited and the 90° DC structure is shown to be observable by STM experiments. The present chapter is concluded with a brief summary in the final section.

2. METHODOLOGY

In this section, theoretical methods and computational details are presented for clarifying atomic structures, growth processes and electronic properties of various Ge films deposited on Si substrates on the basis of the first-principles total-energy calculations within the Density-Functional Theory (DFT) (Hohenberg and Kohn 1964). Firstly, the central equation of the density-functional theory − the Kohn-Sham equation − is briefly introduced to determine the atomic configurations and the electronic structures. Next, the real-space finite-difference approach to solve numerically the Kohn-Sham equation is briefly explained (Chelikowsky et al. 1994a). Then, the film energy associated with energetics of the layer-by-layer growth is introduced and the Tersoff-Hamann approximation for generating the scanning tunneling microscopy (STM) images of the Ge-film morphology is also explained (Tersoff and Hamann 1983). The computational details used for obtaining the electronic structures and the total energies of the various Ge-on-Si systems are finally given.

2.1 First-Principles Total-Energy Calculation

A quantum theoretical approach is an effective way to clarify the properties of the real materials at microscopic levels since behaviors of atoms as well as electrons consisting of the materials are governed by quantum mechanics. The most difficult problem in dealing with realistic materials is to obtain the solutions of the many-electron Schrödinger equation for huge systems composed of a large number of atoms. Hohenberg and Kohn proposed the underlying theorem for obtaining the ground states of many electron systems, i.e. density-functional theory (Hohenberg and Kohn 1964). In this theory, the ground-state energy of the system corresponds to the minimum value of the total-energy functional. Therefore, by moving atoms under calculated forces acting on each atom, the (meta)-stable atomic structures can be determined.

Supposing that $n(r)$ is a single electron density, the total energy of the ground state for the interacting electron system is given by:

$$E[n] = \int v_{ext}(r)\, n(r) + T_s[n] + \frac{1}{2}\iint \frac{n(r)\, n(r')}{|r - r'|}\, dr\, dr' + E_{xc}[n] \qquad (1)$$

Instead of using the many-electron wave function, the ground-state total energy is written in a form of a functional of $n(r)$. Here, v_{ext} is the potential from nuclei and $T_s[n]$ and $E_{xc}[n]$ are the kinetic energy and the exchange-correlation energy, respectively. The Kohn–Sham equation is obtained by minimizing the above total energy for the single-electron density (Kohn and Sham 1965):

$$\left[-\frac{1}{2}\nabla^2 + v_{eff}(r)\right]\psi_i(r) = \varepsilon_i \psi_i(r) \qquad (2)$$

where

$$v_{eff}(r) = v_{ext}(r) + \int \frac{n(r')}{|r - r'|}\, dr' + \frac{\delta E_{xc}[n]}{\delta n[r]} \text{ with } n(r) = \Sigma_i |\psi(r)|^2 \qquad (3)$$

Here, the atomic units $|e| = m = h/2\pi = 1$ are used, where e, m and h are the electron charge, electron mass and Plank's constant, respectively. For the exchange-correlation energy E_{xc}, the Local Density Approximation (LDA) (Perdew and Zunger 1981) and the Generalized Gradient Approximation (GGA) (Perdew et al. 1996) are typically employed.

2.2 Real-Space Finite-Difference Approach

Here the real-space finite-difference method is briefly explained to numerically solve the Kohn-Shame equation. The real-space finite-difference method has the following advantages in computations:

(i) The method has no restrictions regarding the boundary conditions, i.e. arbitrary boundary conditions are available.
(ii) The computational costs can be reduced significantly since most of the Hamiltonian matrix elements are zero.
(iii) The computational algorithm is suitable for massively parallel computers since the real-space method does not need the fast Fourier Transformation used in the conventional plane-wave expansion methods.

In the real-space finite-difference approach, wave functions, electronic charge density and potentials are directly represented on uniform three-dimensional real-space grids. Assuming three-dimensional periodicity for the $r = (x, y, z)$ coordinates, the x, y and z coordinates in the unit cell of periodicity are divided into grids $\{x_i; i = 1, \ldots, N_x\}$, $\{y_j; j = 1, \ldots, N_y\}$ and $\{z_k; k = 1, \ldots, N_z\}$ with equi-spacings h_x, h_y and h_z, respectively. Then, the Kohn-Sham equation is expressed within the real-space finite-difference approach (Chelikowsky et al. 1994a, b, Chelikowsky et al. 1996, Hirose et al. 2005) as:

$$-\frac{1}{2}\sum_{n=-N_f}^{N_f}[c_n^x\psi(x_i + nh_x, y_j, z_k) + c_n^y\psi(x_i, y_j + nh_y, z_k) +$$

$$c_n^z\psi(x_i, y_j, z_k + nh_z) + v_{eff}(x_i, y_j, z_k)\psi(x_i, y_j, z_k)] = E\psi(x_i, y_j, z_k) \tag{4}$$

where $c_\mu^n(\mu = x, y, z)$ are constant parameters due to the finite differentiation of kinetic energy operator $-\frac{1}{2}\Delta^2$. For constant values $c_\mu^n(\mu = x, y, z)$ in the finite-difference formulas, see Chelikowsky et al. 1994b. Owing to the periodicity of the x, y and z directions, the following Bloch conditions are imposed on the wave function: in the x direction

$$\psi(x_0, y_j, z_k) = e^{+ik_xL_x}\psi(x_{N_x}, y_j, z_k) \tag{5}$$

$$\psi(x_{N_{x+1}}, y_j, z_k) = e^{-ik_xL_x}\psi(x_i, y_j, z_k) \tag{6}$$

and in the y direction

$$\psi(x_i, y_0, z_k) = e^{+ik_yL_y}\psi(x_i, y_{N_y}, z_k) \tag{7}$$

$$\psi(x_i, y_{N_{y+1}}, z_k) = e^{-ik_yL_y}\psi(x_i, y_j, z_k) \tag{8}$$

and in the z direction

$$\psi(x_i, y_j, z_0) = e^{+ik_zL_z}\psi(x_i, y_j, z_{N_z}) \tag{9}$$

$$\psi(x_i, y_j, z_{N_{z+1}}) = e^{+ik_zL_z}\psi(x_i, y_j, z_i) \tag{10}$$

where $e^{\pm i k_t L_t}$ $(t = x, y, z)$ represents the Bloch phase factor due to the periodicity of the x, y and z directions. $k = (k_x, k_y, k_z)$ and $L = (L_x, L_y, L_z)$ denote the three-dimensional vectors of Bloch wave numbers and the unit-cell lengths, respectively. For details of the real-space finite-difference approach, see Hirose et al. 2005.

2.3 Formation Energy

To determine the structural stability of Ge layers deposited on Si substrates, the film energy γ_F is introduced and it is defined as:

$$\gamma_F = \frac{E_{\text{tot}} - m_{\text{Si}}\mu_{\text{Si}} - m_{\text{Ge}}\mu_{\text{Ge}} - m_{\text{H}}\mu_{\text{H}}}{\text{A}} - \Gamma_{\text{b}} \tag{11}$$

Here, E_{tot} is the total energy and the chemical potentials μ_{Si}, μ_{Ge} and μ_{H} are the energies per atom in the equilibrium Si bulk, the biaxially compressed Ge bulk (Lu et al. 2005) and the hydrogen molecule, respectively and m_{Si}, m_{Ge} and m_{H} are the number of Si, Ge and H atoms in the slab, respectively. Γ_{b} is the surface energy arising from the bottom surface of the slab and is obtained by independent LDA calculations using different slab models in which both top and bottom surfaces consist of Si atoms with H termination.

2.4 Scanning Tunneling Microscopy Image

The STM images of various Ge/Si (001) structures are calculated based on the Tersoff-Hamann approximation (Tersoff and Hamann 1983, Tersoff and Hamann 1985). Due to its simplicity, the method is widely used and is well known to be valid for many systems (Fujimoto et al. 2001, Fujimoto et al. 2003, Okada et al. 2001). In this method, the tunneling current is assumed to be proportional to the Local Density Of States (LDOS) of the surface at the tip position integrated over an energy range restricted by the applied bias voltage. Consequently, the STM images can be generated from the isosurface of the spatial distribution integrated by the LDOS $\rho(r)$ at spatial points r and energy by several sampling k points of the Brillouin zone over the energy range from $E_F -$ eV to E_F with applied voltage V and the Fermi energy E_F:

$$I(r) \sim \int_{E_F - eV}^{E_F} \rho(r, E)\, dE \tag{12}$$

The negative and positive voltages reflect the occupied and unoccupied electronic states, respectively.

2.5 Computational Detail

First-principles total-energy calculations have been performed using the real-space finite-difference approach in the framework of the Density-Functional Theory (DFT), as implemented in the Real-Space Density-Functional Theory (RSDFT) code (Iwata et al. 2010). The interactions between the ions and the valence electrons are described by the norm-conserving Troullier-Martins pseudopotentials (Troullier and Martins 1991) and exchange-correlation effects are treated using the local density approximation (LDA) parameterized by Perdew and Zunger (Perdew and Zunger 1981).

The 2 × 1 reconstructed, the $p(2 \times 2)$ reconstructed and the (2 × 8)-dimer-vacancy-line (DVL) (Fig. 13.2) reconstructed surfaces are used to examine the Ge layers deposited on Si (001). For the 90° dislocation core (DC) structures, a single core, which consists of a pair of five- and seven-membered rings (solid lines) (Fig. 13.1), is introduced into 24 lateral periodicity along the [110] direction. By introducing the 5-7 membered rings, the number of <110> atomic planes in the Ge films can be reduced and therefore the misfit strains along the [110] direction are released. It is noted that in this model the dislocation core is aligned only along [110] direction. Thereby, the strain-relaxed Ge layers in this model are uniaxially compressed along [1$\bar{1}$0]. For the Ge/Si (001) systems uncovered by H atoms, the top surfaces of Ge films in all calculations are composed of the buckled dimers, whereas for the H-covered Ge/Si systems, the dimers of the top surfaces of the Ge films are unbuckled. In the DVL structure, a dimer in a Ge top surface is missing. The illustration of the DVL structure corresponds to Ge films consisting of two Ge layers deposited on Si substrate (Fig. 13.2).

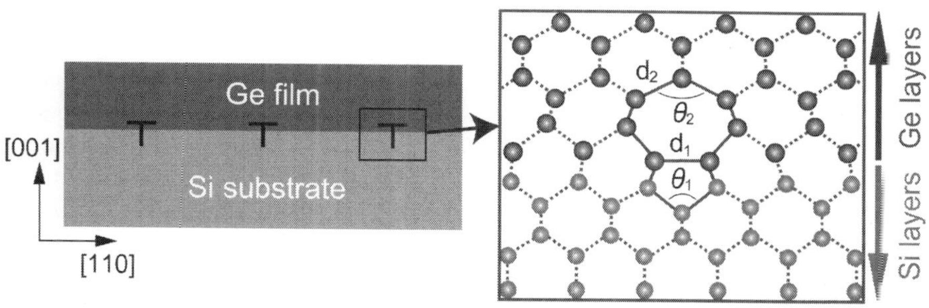

Fig. 13.1 Schematic representation of 90° dislocation in Ge/Si interface (Left panel) and atomic structure of 90° dislocation core (DC) (Right panel). Reproduced with permission from Fujimoto and Oshiyama 2010, copyright 2010 the American Physical Society.

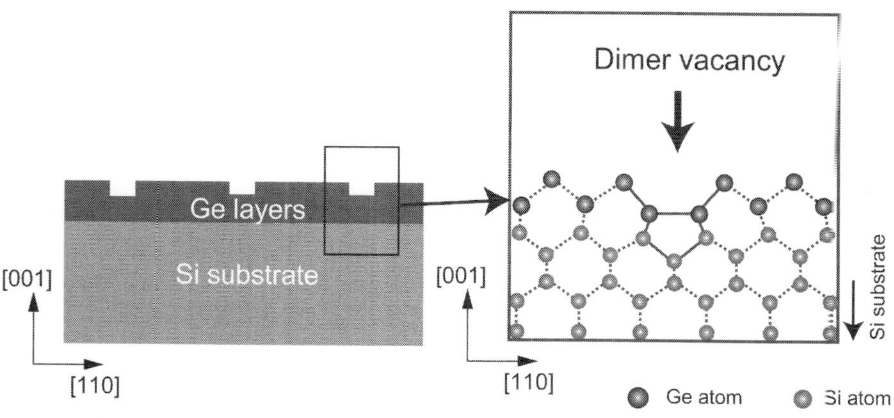

Fig. 13.2 Schematic representation of (2 × N)-reconstructed Ge/Si (001) with a dimer-vacancy line (DVL) (Left panel) and the atomic configuration of the DVL structure (Right panel). Data taken from Fujimoto and Oshiyama 2013.

The lattice parameter is fixed at our calculated Si bulk lattice constant of 5.38 Å and thus the Ge layers on Si (001) are laterally compressed by the calculated lattice mismatch (4.3%) between Si and Ge. The repeating slab models are used, in which the thickness of the vacuum layer is kept more than 16 Å for all cases. The atomic slab treated in this calculation is composed of seven Si atomic layers and the bottom Si layer of the slab is terminated by the hydrogen atoms. The grid spacing in the real-space calculations is taken to be 0.32 Å corresponding to a cutoff energy of 27.5 Ry and the sixth-order finite difference is adopted for the kinetic operator. Brillouin zone integration is performed with (6×12), (6×6), (6×3) and (6×1) k-point grids for the 2×1, the $p(2 \times 2)$, the 2×8 DVL and the 90° DC structures, respectively. The Si atoms in the bottom layer and H atoms attached to the bottom Si atoms are fixed to mimic the Si substrate. Other atoms are fully relaxed until forces acting on the atoms are smaller than 0.05 eV/Å.

3. GROWTH OF Ge FILM ON Si (001)

This section discusses the atomic structures and the energetics of Ge films during expitaxial growth of Ge layers grown on the Si (001) substrates (Fujimoto and Oshiyama 2013). It has been reported experimentally that three dimensional islanding is suppressed under the condition of the low temperatures, typically less than about 300°C and two-dimensional layer-by-layer growth proceeds (Eaglesham and Cerullo 1991). In the first subsection, the atomic structures of plausible Ge surfaces suggested by experimental observations and their film energy diagram corresponding to the two-dimensional layer-by-layer growth are revealed and it is estimated that the critical thickness where the 90° dislocation occurs. In the next subsection, the STM images of various Ge films on Si (001) substrates are exhibited for identifying the dimer vacancy and the 90° dislocation core.

3.1 Structure and Energetics

Let us consider the film energy defined by Eq. (11), corresponding to the growth processes of the Ge film on the Si substrates. In Fig. 13.3, the calculated film energies of the $p(2 \times 2)$, the 2×8 DVL and the 90° DC structures are plotted as a function of the number of the deposited Ge layers on Si (001). In Fig. 13.3, the film energy of Si (001)-$p(2 \times 2)$ surface structure without the deposition of Ge overlayers is also shown.

We begin by examining how the film energy of the reconstructed $p(2 \times 2)$-Ge/Si (001) structure behaves with increasing the number of Ge layers. When the Ge layer is deposited on the Si (001)-$p(2 \times 2)$ surface, the film energy of the Ge-covered Si (001)-$p(2 \times 2)$ reconstructed structure is considerably lower than that of Si (001)-$p(2 \times 2)$ surface without Ge layers. This is because the dangling-bond energy of Ge is lower than that of Si. The film energy of $p(2 \times 2)$ reconstructed Ge-covered Si (001) structure decreases with increasing the number of Ge layers and it approaches an asymptotic value when the number of the Ge layers is more than approximately four layers. In general, any interactions between the top surface of Ge film and the interface between the Ge film and the Si substrate interface are expected to decrease as the number of the Ge layers increases. The film energy defined by Eq. (11) is therefore expected to become close to a sum of the top-surface energy of Ge film and the interface energy between the Ge film and Si substrate with increasing the number of Ge layers. To validate this expectation, the surface energy $\Gamma^{\infty}_{p(2\times2)}$ and the interface energy I_{GeSi} are calculated, where $\Gamma^{\infty}_{p(2\times2)}$ and I_{GeSi} are the surface energy of the biaxially compressed Ge (001)-$p(2 \times 2)$ reconstructed surface and the interface energy of the heterostructure consisting of the equilibrium Si bulk and the biaxially compressed Ge bulk, respectively. Those calculations have been performed using suitable slab models. Here, the value $\gamma^{\infty}_{p(2\times2)}$ defined as $\gamma^{\infty}_{p(2\times2)} = \Gamma^{\infty}_{p(2\times2)} + I_{GeSi}$ is plotted in Fig. 13.3 as the horizontal dotted line. It is thus confirmed that the asymptotic value of the film energy is quantitatively in good agreement with the asymptotic value of when sufficient Ge layers are laid on the Si (001) substrate.

At the initial stage of the epitaxial growth of the Ge film on the Si (001) substrate without the use of the surfactant, it is commonly observed that there are $2 \times N$ DVL reconstructed structures on top surface of the Ge film, where N is typically ~8, because the strain energy accumulates with increasing the Ge film thickness (Köhler et al. 1992). Here, the optimal N value is considered when enough Ge layers are deposited on the Si substrate and the asymptotic values $\gamma^{\infty}_{2\times N\mathrm{DVL}}$ of the $2 \times N$ DVL reconstructed structures for $N = 6$, 8 and 10 have been calculated. The asymptotic value $\gamma^{\infty}_{2\times N\mathrm{DVL}}$ is found to be the lowest when $N = 8$ and this value of $N = 8$ agrees with the result observed experimentally (Köhler et al. 1992). It is interesting that the surface energies of the $2 \times N$ reconstruction of Ge/Si (001) with DVL structures with $N = 4$, 6, 8 and 10 are usually lower than those with $N = 5$, 7, 9 and 11 (Varga et al. 2004). Thereby, the 2×8 DVL structure is chosen as a representative among $2 \times N$ DVL structures and the film energies of the 2×8 DVL structure is plotted in Fig. 13.3 as a function of the number of Ge overlayers. The film energy $\gamma^{\infty}_{2\times8\mathrm{DVL}}$ of the 2×8 DVL structure also decreases with increasing the number of Ge layers as in the case of $p(2 \times 2)$ structure and becomes close to the asymptotic value $\gamma^{\infty}_{2\times8\mathrm{DVL}}$ ($= \Gamma^{\mathrm{Ge}}_{2\times8\mathrm{DVL}} + I_{GeSi}$), where $\Gamma^{\mathrm{Ge}}_{2\times8\mathrm{DVL}}$ is the surface energy of the biaxially compressed Ge(001)-(2×8) DVL surface. The asymptotic value $\gamma^{\infty}_{2\times8\mathrm{DVL}}$ of the 2×8 DVL structure is lower than $\gamma^{\infty}_{p(2\times2)}$ of the $p(2 \times 2)$ structure by ~10 meV/Å^2. The film energy of the 2×8 DVL structure is lower than that of $p(2 \times 2)$ structure when more than two Ge overlayers are laid on the Si substrate.

Let us consider the film energy of the 90° dislocation core (DC) structure which contains five- and seven-membered rings (Fig. 13.1) (Fujimoto and Oshiyama 2013). It is noticed that the relaxation of the 90° DC structure varies depending on the position of the DC in the Ge film (Fig. 13.7(a)) and here the model A2 shown in Fig. 13.7(a) is chosen as the position of the DC. The detailed discussion on how the position of the DC is determined is given in the following section. It can be seen that the film energy of the 90° DC structure decreases as the number of the Ge layer increases in the range from four to 14 Ge layers. The Ge layers deposited above the dislocation core are released from the laterally compressed strains arising from the 4% lattice mismatch between Si and Ge and thereby the atomic structures of the Ge overlayers become close to those of the strain-released Ge films as the number of the deposited Ge layers increases. Thus, the film energy of the 90° DC structure would decrease with increasing the Ge film thickness. The film energy behavior of the 90° DC structure can be further estimated when the Ge layers are sufficiently deposited on Si (001) substrate. The film energy of the 90° DC structure is expected to decrease by $n_l\Delta\gamma$ for enough Ge film thickness. Here, n_l is the number of deposited Ge overlayers and $\Delta\gamma$ is the energy difference per layer between the compressed and the strain-relaxed Ge bulks and the energy difference $\Delta\gamma$ can be written as $\Delta\gamma = \gamma^{\mathrm{rl}}_{\mathrm{Ge}} - \gamma^{\mathrm{St}}_{\mathrm{Ge}}$, where $\gamma^{\mathrm{rl}}_{\mathrm{Ge}} = \mu'_{\mathrm{Ge}}/S$ and $\gamma^{\mathrm{St}}_{\mathrm{Ge}} = \mu_{\mathrm{Ge}}/S$. The chemical potentials μ_{Ge} and μ'_{Ge} are the energies per atom in biaxially compressed and strain-relaxed Ge bulks, respectively and S is the area in Å^2 of the lateral unit cell of the biaxially compressed ideal (001) face. In the above-mentioned way, the energy difference $\Delta\gamma$ is calculated to be –1.18 meV/Å^2 and plot $n_l\Delta\gamma$ in Fig. 13.3 as an oblique dashed line for more than 14 Ge layers.

We are now in a position to discuss the critical thickness of Ge layers at which the 90° dislocation takes place. The film energy $\gamma_{2\times 8\mathrm{DVL}}$ of the reconstructed Ge/Si (001)-(2×8) DVL structure decreases with increasing the number of Ge layers and it takes almost the same value as the asymptotic value of $\gamma^{\infty}_{2\times 8\mathrm{DVL}}$ when more than five Ge layers are deposited on the Si substrate. On the other hand, the film energy of the 90° DC structure decreases continuously with increasing the number of the deposited Ge layers. Thus, it is found that the 90° DC structure becomes energetically favorable for the deposition of more than 14 Ge layers on the Si substrate because the film energy of the 90° DC structure is lower than that of 2 × 8 DVL structure. It is concluded that the critical thickness is 14 Ge layers at which 90° dislocation takes place. It is of interest to compare the value from the present LDA calculations with the values obtained from classical elastic theory which diverge in a range from eight to 71 Ge layers (People and Bean 1985).

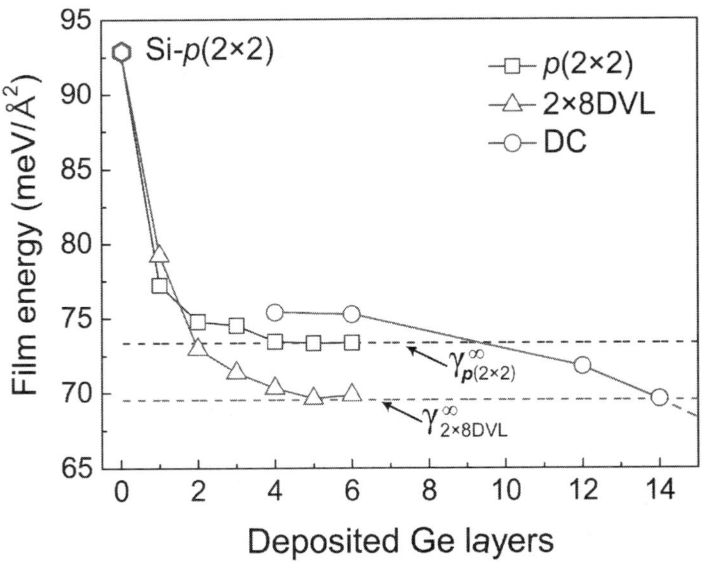

Fig. 13.3 Film energies of various Ge films on Si (001) as a function of Ge layer thickness. The squares, triangles and circles denote the $p(2\times 2)$, 2×8 DVL and 90° DC structures, respectively. The hexagon denotes Si (001)-$p(2\times 2)$ surfaces. Reproduced with permission from Fujimoto and Oshiyama 2013, copyright 2013 the American Physical Society.

3.2 Scanning Tunneling Microscopy Image

Scanning Tunneling Microscopy (STM) images of Ge/Si (001) surfaces with $p(2 \times 2)$ reconstructed, 2 × 8 DVL and 90° DC structures are exhibited. Figure 13.4 shows the simulated STM images of the $p(2 \times 2)$ reconstructed Ge/Si (001) structure at applied bias voltages of (a) $V = -0.5$ eV (filled state) and (b) $V = +0.5$ eV (empty state), respectively. Here, the white arrow denotes the position of an upper atom of a buckled Ge dimer and gray ball denotes a Ge atom. In the filled-state image ($V = -0.5$ eV), the dimer row appears to be a zigzag pattern along the [110] direction, i.e. one atom in a buckled dimer looks bright, while the other looks dark as if the Ge atom is absent since the Ge atoms in the Ge top-surface of Ge/Si (001) structure possess the dangling bonds (Fig. 13.4a) (Fujimoto et al. 2001, Fujimoto et al. 2005, Takagi et al. 2007). On the other hand, in the empty-state image ($V = +0.5$ eV), the zigzag corrugation is considerably different from that in the case of the filled-state image (Fig. 13.4b); the two bright protrusions run along the [110] direction. It is reported that similar features in the STM images are also observed in the cases of Si (001)-$p(2 \times 2)$ as well as unstrained Ge (001)-$p(2 \times 2)$ surfaces (Fujimoto et al. 2001, Takagi et al. 2007).

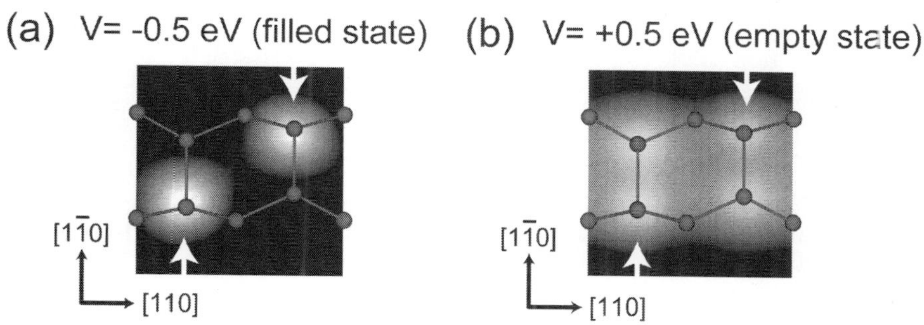

Fig. 13.4 Simulated STM images of the reconstructed $p(2 \times 2)$-Ge/Si (001) where Ge film is composed of six Ge layers. The STM images are shown at bias voltages of (a) −0.5 eV (filled state). (b) +0.5 eV (empty state). The black arrows represent the position of an upper atom in the buckled Ge dimer, the gray ball denotes a Ge atom. Data taken from Fujimoto and Oshiyama 2013.

Figure 13.5 shows the simulated STM images of the 2 × 8 DVL structure at applied bias voltages of (a) $V = -0.5$ eV and (b) $V = +0.5$ eV. The filled-state STM image ($V = -0.5$ eV) also shows the zigzag corrugation pattern along the [110] direction as in the case of the $p(2 \times 2)$ reconstructed structure (Fig. 13.4a); the upper Ge atom in the buckled dimer has a bright spot. However, the STM image of the 2 × 8 DVL structure is different from that of the $p(2 \times 2)$ reconstructed structure; the dark line in the STM image runs along the [1$\bar{1}$0] direction because one dimer in the dimer row is missing and the width of the dark line is ~2.6 Å. It is reported that the dark line in STM images is also observed in the experiments of the epitaxial growth of the Ge layers deposited on the Si (001) substrate (Köhler et al. 1992). On the other hand, the empty-state STM image ($V = +0.5$ eV) has two bright spots along the [110] direction, but the lower atom in the buckled dimer looks brighter than the upper atom. It can be also seen that the dark line runs along the [1$\bar{1}$0] direction as in the case of the filled-state image.

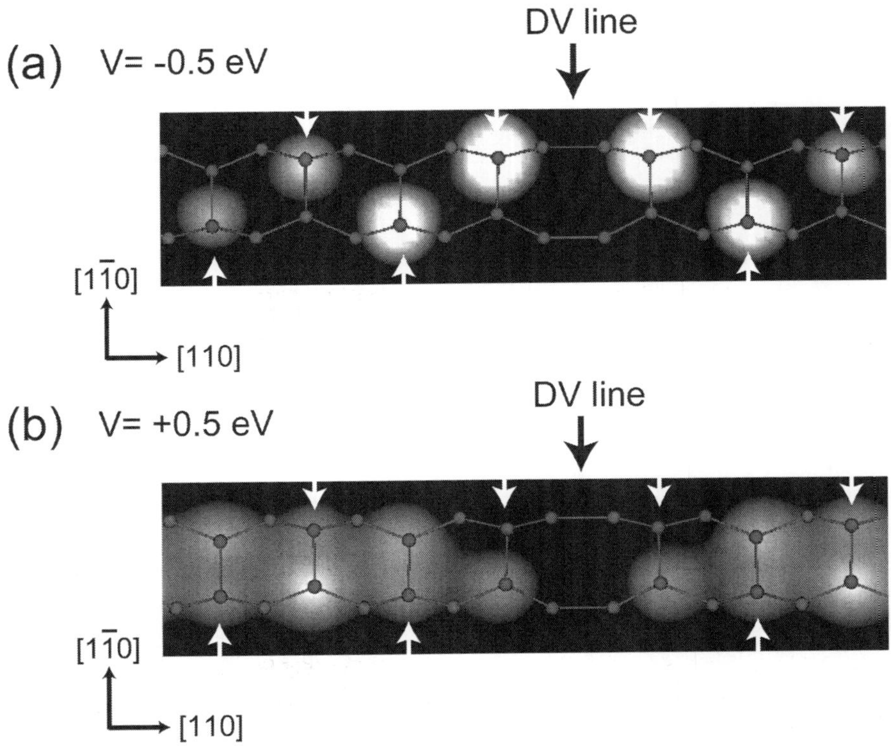

Fig. 13.5 Simulated STM images of the 2 × 8 DVL structure consisting of six Ge layers. The STM images are shown at bias voltages of (a) −0.5 eV (filled state). (b) +0.5 eV (empty state). The black arrows represent the position of an upper atom in the buckled Ge dimer, the gray ball denotes a Ge atom. Reproduced with permission from Fujimoto and Oshiyama 2013, copyright 2013 the American Physical Society.

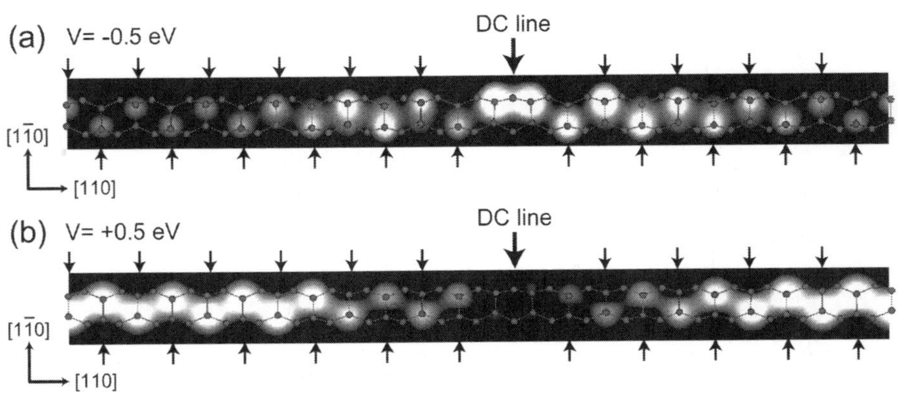

Fig. 13.6 Simulated STM images of the 90° DC structures consisting of 14 Ge layers. The STM images are shown at bias voltages of (a) −0.5 eV (filled state). (b) +0.5 eV (empty state). The black arrows represent the position of an upper atom in the buckled Ge dimer, the gray ball denotes a Ge atom. Reproduced with permission from Fujimoto and Oshiyama 2013, copyright 2013 the American Physical Society.

Figures 13.6a and 13.6b show the STM images of the 90° DC structure at the bias voltages of $V = -0.5$ eV (filled state) and +0.5 eV (empty state), respectively. In the filled-state STM image, the zigzag pattern along the [110] direction appears, where one upper atom appears to be bright, while one lower atom to be dark as in the cases of the $p(2 \times 2)$ reconstructed and 2×8 DVL structures. It can be seen that there are two bright oval-shaped spots above the dislocation core along the DC line, i.e. two upper Ge atoms in the neighboring dimers at the DC line appear to be two bright spots, whereas the two lower atoms to be dark. The width of the two oval shapes is found to be about 9.2 Å. In the case of the empty-state STM image, the zigzag corrugation pattern exhibits sharp contrast to the filled-state STM image: the bright spot is located not above the upper atom but the lower atom in the buckled dimer. Being different from the filled-state STM image, the dark line above the DC line appears, however, there are slightly bright spots above the DC line. This is because the height of the Ge atoms of the Ge surface above the DC line is lower compared with that at a position far from the DC line. The width of the DC line is almost the same as that of the two bright oval shapes in the filled-state image. Further discussion on the atomic structures near the DC and at the Ge surface will be given next. Accordingly, DVL and DC structures are distinguishable by looking into the bright oval shape in the filled-state STM image and the dark line in the empty-state STM image if the dimer-vacancy line and the 90° dislocation core take place, because the appearance of the bright oval shape and the width of the dark line in STM images of both structures are different from each other.

4. GROWTH OF HYDROGENATED Ge FILM ON Si (001)

Here the atomic structures and the energetics of hydrogen-covered Ge films on the Si (001) substrates in the hetero-expitaxial growth are first discussed (Fujimoto and Oshiyama 2010, Fujimoto and Oshiyama 2011). With the use of surfactants such as arsenic, antimony and hydrogen atoms, the 2D growth of Ge films on Si proceeds for more than 4 Ge ML and thus the 3D island formation is strongly suppressed. In the first subsection, we show the atomic structures of plausible H-terminated Ge surfaces and reveal their film energy diagram associated with the surfactant-mediated layer-by-layer growth mechanism the critical thickness where the 90° dislocation occurs is estimated. In the next subsection, the energy-band structure and the spatial distributions of the electronic states related to the 90° dislocation core are shown and the STM images of H-terminated Ge films on Si (001) substrates are also exhibited to observe the DC.

4.1 Structure and Energetics

We begin with studying the energetics of the hydrogenated Ge layers deposited on the Si (001) substrate and have calculated the film energy defined in Eq. (11) to discuss the stability of the Ge films on Si (001) substrate. Figure 13.7 shows the calculated film energies of the 2×1 reconstructed, the 2×24 DVL and the 90° DC structures as a function of the number of the deposited Ge layers on Si. For the 2×1 reconstructed structure, the film energy of the 2×1 structure $\gamma_{2\times1}$ decreases as the number of the deposited Ge layers increases and approaches an asymptotic value.

The film energy defined as Eq. (11) is expected to approach a sum of the top-surface energy and the interface energy as the number of Ge layers increases since any coupling between the Ge top-surface and the Ge/Si interface decreases, as already discussed in the previous section. To validate this expectation, the asymptotic value $\gamma^{\infty}_{2\times1}$ has been obtained, where $\gamma^{\infty}_{2\times1}$ is a sum of the surface energy $\Gamma^{Ge}_{2\times1}$ is the surface energy of the biaxially compressed hydrogenated Ge (001)-(2×1) surface and the Ge/Si interface energy I_{GeSi}. It is noted that the interface energy I_{GeSi} is the same value as calculated in the previous section. In Fig. 13.7b, the asymptotic value $\gamma^{\infty}_{2\times1}$ as the horizontal dotted line is plotted and it is found that the film energy value $\gamma_{2\times1}$ is quantitatively in agreement with the asymptotic value of $\gamma^{\infty}_{2\times1}$.

It is reported that the reconstruction of Ge surface varies depending on Ge layer thickness when the Ge layers are deposited on the Si substrate without surfactant, implying the competition of several interaction energies. The hydrogen-surfactant mediated hetero-epitaxitial growth of Ge/Si (001) has not been observed experimentally and accordingly, the H-terminated Ge/Si (001)-(2 × *N*) DVL reconstructed surface is also expected to be one of the possible surface structures. The film energies of 2 × *N* DVL structures including H surfactants for $N = 16$, 20, 24, 26 and 28 have been examined here and the film energy at $N = 24$ is found to be the lowest among the biaxially compressed (2 × *N*)-DVL Ge (001) surfaces. Thus, the 2 × 24 DVL structure is chosen as a representative of the 2 × *N* DVL structures. To clarify stability of the DVL structures associated with H surfactants, the film energy of the 2 × 24 DVL structure has been calculated as a function of the number of Ge layers and the film energy $\gamma_{2\times24\mathrm{DVL}}$ of the 2 × 24 DVL structure is plotted in Fig. 13.7b. It is noticed that the film energy of $\gamma_{2\times24\mathrm{DVL}}$ is larger than that of $\gamma_{2\times1}$ by about 5-10 meV/Å^2. Again the film energy approaches an asymptotic value $\gamma^{\infty}_{2\times24\mathrm{DVL}}$ in this case, which is calculated as a sum of the corresponding surface energy $\Gamma^{Ge}_{2\times24\mathrm{DVL}}$ and the interface energy I_{GeSi}.

We are now in a position to discuss the 90° DC structure which contains the five- and seven-membered rings. The 90° DC structure consists of two types of Ge layers: i.e. the compressed and the strain-relaxed Ge layers. The former is the biaxially compressed Ge layer which intervenes between the DC and the Ge/Si interface and the latter is the strain-relaxed Ge layer which is located above the DC. Therefore, the strain-released mechanism of the 90° DC structure should depend on the position of the DC in the Ge film and it is of great importance to know where the DC is located in the Ge film.

In order to determine the position of the DC, the film energy of the 90° DC structure has been calculated, in which the DC is located at the Ge top surface as shown in A1, A2 and A3 in Fig. 13.7a. As the number of the Ge layers between the DC and the interface increases, the film energy decreases. It is found that the film energy in the structure of the model A2 already reaches an asymptotic value γ^{∞}_{DC}. The asymptotic value γ^{∞}_{DC} in Fig. 13.7b plotted as the horizontal dotted line. Here, the asymptotic value γ^{∞}_{DC} is defined as a sum of the surface energy Γ^{Ge}_{DC} and the interface energy I_{GeSi}, i.e. $\gamma^{\infty}_{DC} = \Gamma^{Ge}_{DC} + I_{GeSi}$. The surface energy Γ^{Ge}_{DC} is calculated for the DC structure consisting of only Ge element with the DC being located at the top surface. For further discussion, the DC is therefore placed at the position as in the structure A2 in which three Si atoms participate in forming a five-membered ring. It is noteworthy that the film energy of the structure A2 is slightly lower in energy than that of the structure A3 where the five-membered ring resides in the Ge layers. This is presumably because existence of the five-membered ring causes additional strains in compressed Ge overlayers. Segregation of large impurities in five- and seven-membered rings in Si is discussed in terms of local strains (Kaplan et al. 2000).

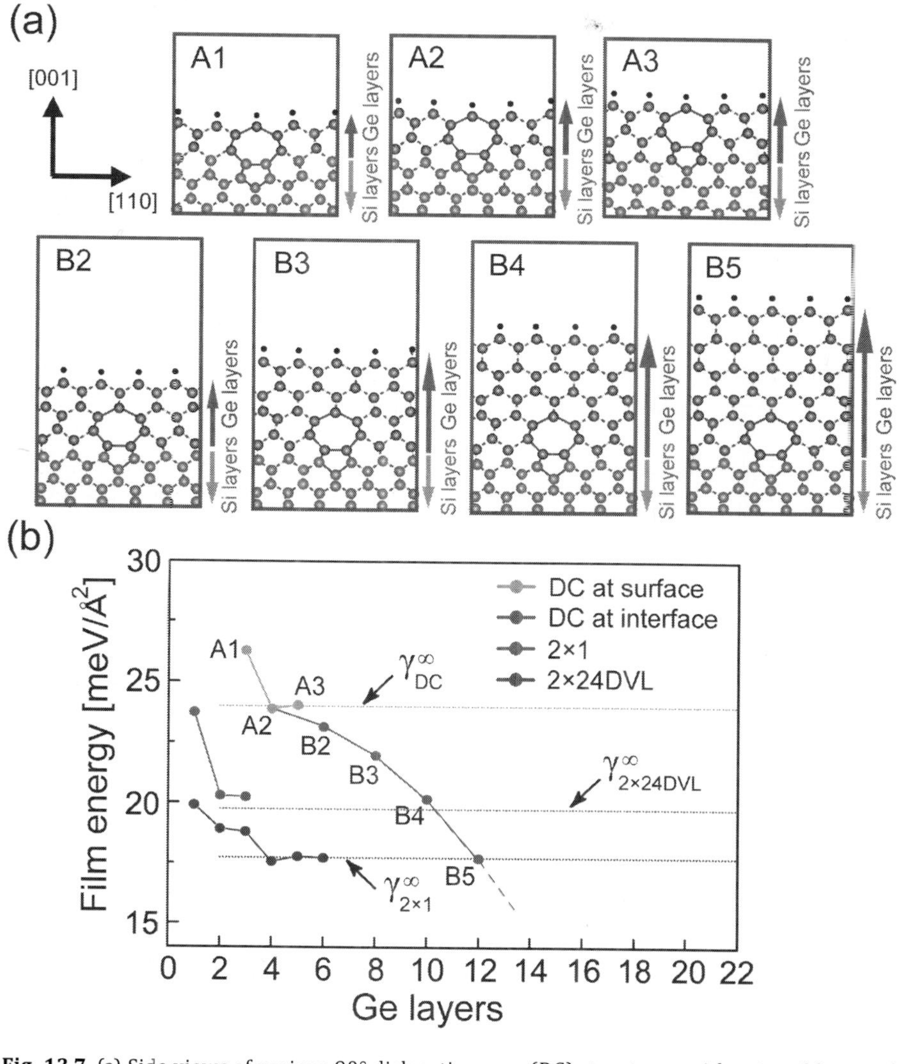

Fig. 13.7 (a) Side views of various 90° dislocation core (DC) structures with pairs of five- and seven-membered rings. In A1, A2 and A3, the 90° DC is located at the top-surface of the Ge film, whereas it is at the interface in B2-B5, hereby forming the strain-relaxed Ge overlayers above 90° DC. (b) Film energies for various Ge/Si (001) structures as a function of the number of Ge layers. Reproduced with permission from Fujimoto and Oshiyama 2010, copyright 2010 the American Physical Society.

We next consider how the film energy γ_{DC} behaves with increasing the number of the strain-relaxed Ge layers deposited above the DC. We consider four structures shown in B2, B3, B4 and B5 in Fig. 13.7a where the number n_l of the strain-relaxed Ge layers is 2, 4, 6 and 8, respectively. It is found that the film energies for the DC structures of B2-B5 decrease with increasing numbers of the deposited Ge layers as depicted in Fig. 13.7b. The structure of the Ge layers deposited above the DC becomes similar to that of the strain-relaxed Ge film with increasing the number of the Ge layers. The film energy of the 90° DC structure is therefore expected to diminish as in $n_l\Delta\gamma$ where $\Delta\gamma$ is the energy difference per layer between the compressed and the strain-relaxed Ge bulks as defined earlier and the energy difference $\Delta\gamma$ is −1.18 meV/Å^2. The calculated film energies for B2-B5 structures unequivocally exhibit this reduction as the Ge-film thickness increases. In Fig. 13.7b, $\nu_\lambda\Delta\gamma$ is again plotted as an oblique dashed line for more than 12 Ge layers.

Based on the calculated film energies for Ge overlayers on Si (001) shown in Fig. 13.7b, we now discuss the critical thickness of Ge overlayers at which the 90° dislocation covered by the relaxed Ge layers takes place. From the results of the film-energy diagram (Fig. 13.7b), it is clearly shown that the film energy of the 2 × 1 structure is always lower than that of the 2 × 24 DVL structure during the early stage of Ge layer growth. The film energy of the 2 × 1 structure becomes close to be the asymptotic value $\gamma^{\infty}_{2\times2}$ for more than four Ge overlayers, while the film energy of the 90° DC structure monotonically decreases with increasing the number of the deposited Ge overlayers. The present calculations have suggested that the 90° DC structure becomes energetically preferable rather than the 2 × 1 structure without the DC when more than 12 Ge layers are deposited on the Si (001) substrate. It is thus concluded that the critical thickness causing the transition from 2 × 1 structure to 90° DC structure is 12 Ge layers.

Here we discuss how the core structure in the 90° dislocation varies with increasing the relaxed Ge-film thickness. Table 13.1 shows the optimized structural parameters of the 90° DC structure at several numbers of the deposited relaxed Ge layers, where d_1 and d_2 are the characteristic bond lengths and θ_1 and θ_2 are the bond angles of the DC (Fig. 13.1). The height D denotes the difference between the highest and the lowest atomic positions of the hydrogen bonded at the Ge top-surface layer. It is therefore clear that the top-surface atomic layer shows indentation even after the Ge layers are deposited. The bond angles θ_1 and θ_2 are considerably different from the ideal tetrahedral bond angle 109.5°. In particular, θ_2 is larger by ~30% compared with the ideal angle of 109.5°. The bond length d_1 is slightly larger than that of Ge bulk, 2.43 Å and the bond length d_2 is larger by 8%. The large bond-angle and small bond-length variations show the feature of the structural relaxation in covalently bonded materials (Liu et al. 1995). These values become almost unchanged without depending on the number of the deposited Ge layers. Such peculiar behavior in atomic geometry would modify the electronic structures of Ge films as shown below. On the other hand, it is found that the height D value decreases with increasing the number of the Ge layers. Thus, the roughness near the Ge surface is improved, i.e. the surface becomes flat when the Ge layers are sufficiently deposited because the strain energy is relieved.

Table 13.1 Structural parameters of 90° dislocation core (DC) structures. The bond lengths d_1 and d_2 and the bond angles θ_1 and θ_2 are depicted in Fig. 13.1. See text for height *D*. Data taken from Fujimoto and Oshiyama 2010.

Ge layers	θ_1 (deg)	θ_2 (deg)	d_1 (Å)	d_2 (Å)	Height D (Å)
4	99.3	141.3	2.47	2.63	0.88
6	99.8	138.5	2.47	2.62	0.65
8	99.7	134.6	2.46	2.64	0.52
10	99.4	133.0	2.47	2.63	0.45
12	98.9	132.3	2.45	2.63	0.42

4.2 Total Charge Density

Figure 13.8 exhibits the optimized atomic structure and the isosurface of the total charge density of the 90° DC structure, where the Ge film is composed of 12 Ge overlayers deposited on Si (001) as shown in the structure B5 of Fig. 13.7a. The spatial distribution of electron density mainly resides between the neighboring two atoms and is also observed in the typical sp^3-bonded materials. On the other hand, the electron densities at two bonds above the seven-membered ring are found to be considerably lower compared with other bonds because the bond length d_2 as well as the bond angle θ_2 (Table 13.1) is considerably large, compared with other bond lengths and angles. It is interesting that the bond-length and bond-angle distortions in the dislocation core have sizable effects on the spatial distribution of the electron density.

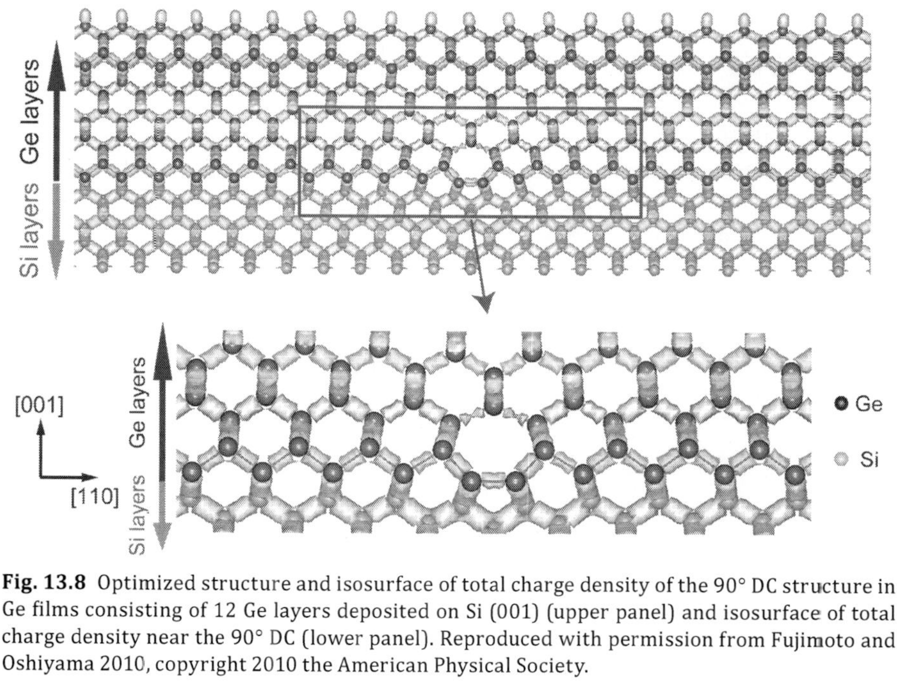

Fig. 13.8 Optimized structure and isosurface of total charge density of the 90° DC structure in Ge films consisting of 12 Ge layers deposited on Si (001) (upper panel) and isosurface of total charge density near the 90° DC (lower panel). Reproduced with permission from Fujimoto and Oshiyama 2010, copyright 2010 the American Physical Society.

4.3 Electronic Structure

Figure 13.9a shows the energy-band structure of the 90° DC structure consisting of 12 Ge overlayers. The zero energy is the center of the Valence-Band Maximum (VBM) and the Conduction-Band Minimum (CBM). The 90° DC structure is found to possess the direct bandgap with a gap value of 0.09 eV. It is interesting that the similar feature in the bandgap is also observed in a pair of the edge dislocations in Si crystal (Liu et al. 1995). The VBM and CBM reside within the calculated bandgap of Si bulk, 0.61 eV. Figures 13.9b and 13.9c show the electron density plots of the VBM and CBM states, respectively. In the case of the VBM state, the electron density plot is found to be localized along the DC. On the other hand, in the case of the CBM state, the electron density is found to be highly localized above the DC. It is expected that this remarkable feature such as the localization of the electron density near the DC and the direct-gap nature might provide possible optoelectronic-device applications. The localization of the electron density also gives rise to peculiar spatial distributions of the local density of states (LDOS), which will lead to distinctive STM images as will be shown below.

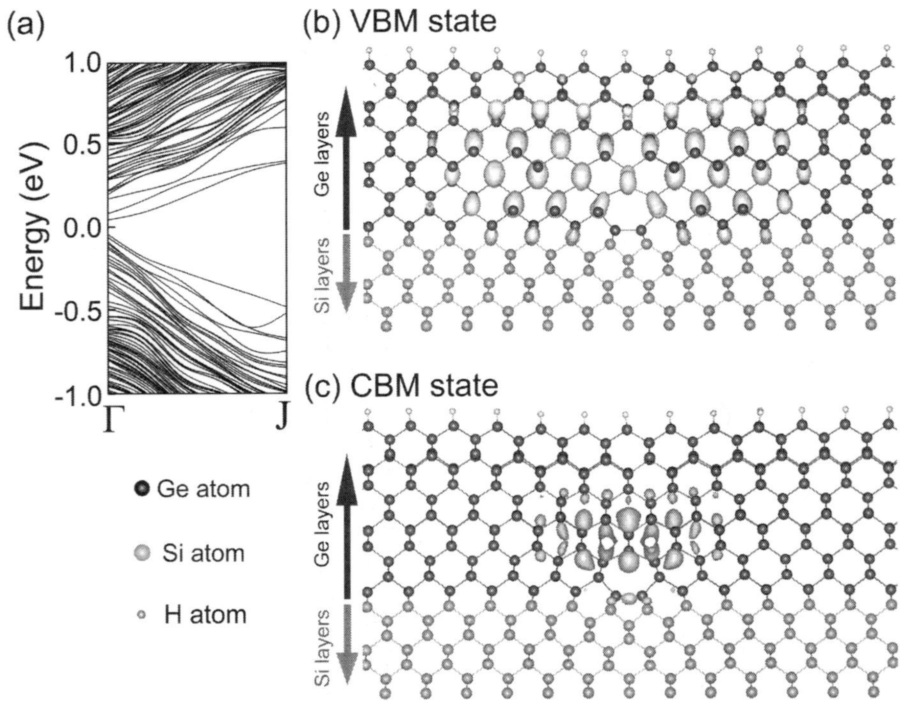

Fig. 13.9 (a) Energy bands of the 90° DC structure shown in B5 in Fig. 13.7a. The electron densities of (b) The VBM. (c) The CBM states are shown. The Fermi energy being located at the middle between VBM and CBM are set to be zero. Data taken from Fujimoto and Oshiyama 2010.

4.4 Scanning Tunneling Microscopy Image

The STM images of the 90° DC structure which consists of the 12 Ge layers are demonstrated here (see the structure B5 in Fig. 13.7a). Figures 13.10a and 13.10b show the simulated STM images at the bias voltage of −0.5 eV (filled state) and +0.5 eV (empty state), respectively. The filled-state STM image has the dark line on the DC running along the dislocation line (Fig. 13.10a) and the dark line width is about 24.7 Å. The bright oval shape forming on the dimer aligns along [110] direction (see three black arrows). The empty-state STM image exhibits sharp contrast to the filled-state STM image: The empty-state STM image possesses a bright line along the dislocation line and the bright oval shape is seen between the dimer rows. The width of the bright line along the DC is about 28.5 Å. The difference of the spatial distributions between the electronic densities for the VBM and CBM states induced by the DC leads to the distinctive filled-state and the empty-state STM images: In the filled state STM image, the electronic density is mainly localized at both sides of the DC as shown in Fig. 13.9b. As a result, the filled-state STM image has two bright protrusions not on but along the DC. On the other hand, the empty-state image has an electronic state localized largely at the upper position of the DC as shown in Fig. 13.9c. This makes a bright protrusion above the DC. It is thus expected that the 90° dislocation line is observable in atomic scale by STM experiments.

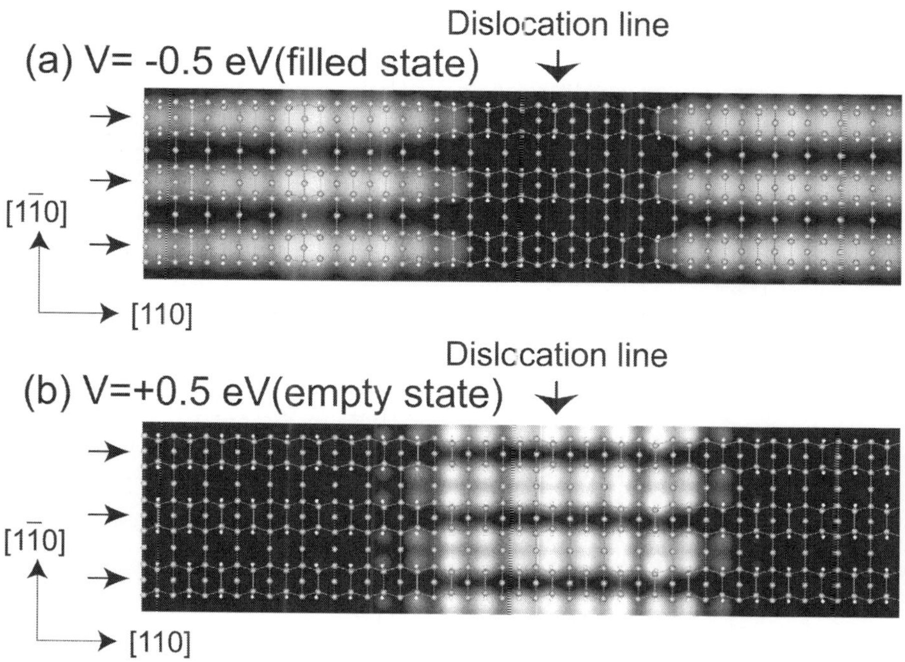

Fig. 13.10 Simulated STM images of the 90° DC structure shown in B5 in Fig. 13.7a: (a) Filled-state image. (b) Empty-state image. The arrows on the left hand represent the dimer row position. Gray and small white balls represent Ge and H atoms, respectively. Data taken from Fujimoto and Oshiyama 2010.

5. CONCLUSIONS

Based on the first-principles total-energy calculations, the atomic structures, energetic and electronic structures of various Ge films on Si (001) substrates have been reviewed. The present film-energy calculations have suggested that the 2×8 DVL structure becomes energetically favorable than the $p(2 \times 2)$ structure for more than two Ge overlayers. The film energy of the 90° DC structure consisting of the five to seven membered rings decreases with increasing the number of relaxed Ge overlayers and it is found that the 90° DC structure becomes energetically preferable compared with the 2×8 DVL structure for more than 14 Ge layers. The STM images of 2×8 DVL and 90° DC structures are demonstrated. The dimer-vacancy line appears to be dark in the STM image of the 2×8 DVL structure, whereas 90° dislocation line is seen to be bright in the filled-state STM image of the 90° DC structure. Thereby, both structures could be distinguished from each other by using STM experiments.

The epitaxial growth and electronic structures of the H-covered Ge film on Si (001) have been also examined. The 2×1 structure without DVL becomes energetically stable compared with $2 \times N$ structures with DVL without depending on the Ge film thickness. The 90° DC structure is found to be stable energetically for the deposition of more than 12 Ge overlayers. It is also found that the 90° dislocation lines are clearly observable in the simulated STM images, with being a dark line in the filled-state image and a bright line in the empty-state image.

6. ACKNOWLEDGMENTS

This work was partly supported by JSPS KAKENHI Grant No. 26390062. Computations were done at Institute for Solid State Physics, the University of Tokyo and at Cybermedia Center of Osaka University.

7. REFERENCES

Arias, T. A. and J. D. Joannopoulos. 1994. *Ab initio* theory of dislocation interactions: from close-range spontaneous annihilation to the long-range continuum limit. Phys. Rev. Lett. 73: 680-683.

Berbezier, I. and A. Ronda. 2009. SiGe nanostructures. Surf. Sci. Rep. 64: 47-98.

Bolkhovityanov, Y. B., A. S. Deryabin, A. K. Gutakovskii, M. A. Revenko and L. V. Sokolov. 2004a. Enhanced strain relaxation in a two-step process of Ge_xSi_{1-x}/Si (001) heterostructures grown by low-temperature molecular-beam epitaxy. Appl. Phys. Lett. 84: 4599-4601.

Bolkhovityanov, Y. B., A. S. Deryabin, A. K. Gutakovskii, M. A. Revenko and L. V. Sokolov. 2004b. Strain relaxation of GeSi/Si (001) heterostructures grown by low-temperature molecular-beam epitaxy. J. Appl. Phys. 96: 7665-7674.

Bolkhovityanov, Y. B., A. S. Deryabin, A. K. Gutakovskii and L. V. Sokolov. 2008. Formation of edge misfit dislocations in Ge_xSi_{1-x} (x~0.4-0.5) films grown on misoriented (0 0 1)→(1 1 1) Si substrates. J. Cryst. Growth 310: 3422-3427.

Boucaud, P., M. E. Kurdi, A. Ghrib, M. Prost, M. de Kersauson, S. Sauvage, et al. 2013. Recent advances in germanium emission. Photonics Res. 1: 102-109.

Boztug, C., J. R. Sanchez-Perez, F. Cavallo, M. Lagally and R. Paiella. 2014. Strained-germanium nanostructures for infrared photonics. ACS Nano 8: 3136-3151.

Cannon, D. D., J. Liu, D. T. Danielson, S. Jongthammanurak, U. U. Enuha, K. Wada, et al. 2007. Germanium-rich silicon-germanium films epitaxially grown by ultrahigh vacuum chemical-vapor deposition directly on silicon substrates. Appl. Phys. Lett 91: 252111_1-252111_3.

Capellini, G., M. De Seta, Y. Busby, M. Pea, F. Evangelisti, G. Nicotra, et al. 2010. Strain relaxation in high Ge content SiGe layers deposited on Si. J. Appl. Phys. 107: 063504_1-063504_8.

Chelikowsky, J. R., N. Troullier and Y. Saad. 1994a. Finite-difference-pseudopotential method: electronic structure calculations without a basis. Phys. Rev. Lett. 72: 1240-1243.

Chelikowsky, J. R., N. Troullier, K. Wu and Y. Saad. 1994b. Higher-order finite-difference pseudopotential method: an application to diatomic molecules. Phys. Rev. B 50: 11355-11364.

Chelikowsky, J. R., X. Jing, K. Wu and Y. Saad. 1996. Molecular dynamics with quantum forces: vibrational spectra of localized systems. Phys. Rev. B 53: 12071-12079.

Colace, L., G. Masini, F. Galluzzi, G. Assanto, G. Capellini, L. Di Gaspare et al. 1998. Metal–semiconductor–metal near-infrared light detector based on epitaxial Ge/Si. Appl. Phys. Lett. 72: 3175-3177.

Colace, L., G. Masini, G. Assanto, H.-C. Luan, K. Wada and L. C. Kimerling. 2000. Efficient high-speed near-infrared Ge photodetectors integrated on Si substrates. App. Phys. Lett. 76: 1231-1233.

Copel, M., M. C. Reuter, E. Kaxiras and R. M. Tromp. 1989. Surfactants in epitaxial growth. Phys. Rev. Lett. 63: 632-635.

Currie, M. T., S. B. Samavedam, T. A. Langdo, C. W. Leitz and E. A. Fitzgerald. 1998. Controlling threading dislocation densities in Ge on Si using graded SiGe layers and chemical-mechanical polishing. Appl. Phys. Lett. 72: 1718-1720.

Dentel, D., J. L. Bischoff, T. Angot and L. Kubler. 1998. The influence of hydrogen during the growth of Ge films on Si (001) by solid source molecular beam epitaxy. Surf. Sci. 402-404: 211-214.

Eaglesham, D. J. and M. Cerullo. 1990. Dislocation-free stranski-krastanow growth of Ge on Si (100). Phys. Rev. Lett. 64: 1943-1946.

Eaglesham, D. J. and M. Cerullo. 1991. Low-temperature growth of Ge on Si (100). Appl. Phys. Lett. 58: 2276-2278.

Fujimoto, Y., H. Okada, K. Endo, T. Ono, S. Tsukamoto and K. Hirose. 2001. Images of scanning tunneling microscopy on the Si (001)-$p(2 \times 2)$ reconstructed surface. Mater. Trans. 42: 2247-2252.

Fujimoto, Y., H. Okada, K. Inagaki, H. Goto, K. Endo and K. Hirose. 2003. Theoretical study of scanning tunneling microscopy on Si (001) surface. Jpn. J. Appl. Phys. 42: 5267-5268.

Fujimoto, Y., K. Hirose and T. Ohno. 2005. Calculation of surface electronic structures by overbridging boundary matching method. Surf. Sci. 586: 74-81.

Fujimoto, Y. and A. Oshiyama. 2010. Atomic structures and energetics of 90° dislocation cores in Ge films on Si (001). Phys. Rev. B 81: 205309_1-205309_5.

Fujimoto, Y. and A. Oshiyama. 2011. Formation and stability of 90 degree dislocation cores in Ge films on Si (001). AIP Conf. Proc. 1399: 185-186.

Fujimoto, Y. and A. Oshiyama. 2013. Structural stabilities and scanning tunnelling microscopy images of strained Ge films on Si (001). Phys. Rev. B 87: 075323_1-075323_5.

Goldman, R. S., K. L. Kavanagh, H. H. Wieder, S. N. Ehrlich and R. M. Feenstra. 1998. Effects of GaAs substrate misorientation on strain relaxation in $In_xGa_{1-x}As$ $Inx_{Ga1-x}As$ films and multilayers. J. Appl. Phys. 83: 5137-5149.

Goo, J.-S., Q. Xiang, Y. Takamure, F. Arasnia, E. N. Paton, P. Besser, et al. 2003. Band offset induced threshold variation in strained-Si n-MOSFETs. IEEE Electron Device Lett. 24: 568-570.

Hartmann, J. M., A. Abbadie, A. M. Papon, P. Holliger, G. Rolland, T. Billon, et al. 2004. Reduced pressure–chemical vapor deposition of Ge thick layers on Si (001) for 1.3-1.55-μm photodetection. J. Appl. Phys. 95: 5905-5913.

Hirose, K., T. Ono, Y. Fujimoto and S. Tsukamoto. 2005. First-Principles Calculations in Real-Space Formalism. Imperial College Press, London.

Hohenberg, P. and W. Kohn. 1964. Inhomogeneous electron gas. Phys. Rev. 136: B864-B871.

Horn-von Hoegen, M., B. H. Müller and A. Al-Falou. 1994. Strain relief by microroughness in surfactant-mediated growth of Ge on Si (001). Phys. Rev. B 50: 11640-11652.

Iwata, J.-I., D. Takahashi, A. Oshiyama, T. Boku, K. Shiraishi, S. Okada, et al. 2010. A massively-parallel electronic-structure calculations based on real-space density functional theory. J. Comput. Phys. 229: 2339-2363.

Kahng, S.-J., Y. H. Ha, J.-Y. Park, S. Kim, D. W. Moon and Y. Kuk. 1998. Hydrogen-surfactant mediated growth of Ge on Si (001). Phys. Rev. Lett. 80: 4931-4934.

Kahng, S.-J., Y. H. Ha, J.-Y. Park, S. Kim, D. W. Moon and Y. Kuk. 2000. Strained Ge overlayer on a Si (001)–(2 × 1) surface. Phys. Rev. B 61: 10827-10831.

Kaplan, T., F. Liu, M. Mostoller, M. F. Chisholm and V. Milman. 2000. First-principles study of impurity segregation in edge dislocations in Si. Phys. Rev. B 61: 1674-1676.

Kohn, W. and L. J. Sham. 1965. Self-consistent equations including exchange and correlation effects. Phys. Rev. 140: A1133-A1138.

Köhler, U., O. Jusko, B. Műller, M. Horn-von Hoegen and M. Pook. 1992. The interplay of surface morphology and strain relief in surfactant mediated growth of Ge on Si (111). Ultramicroscopy 42-44: 832-837.

Lee, M. L., E. A. Fitzgerald, M. T. Bulsara, M. T. Currie and A. Lochtefeld. 2005. Strained Si, SiGe and Ge channels for high-mobility metal-oxide-semiconductor field effect transistors. J. Appl. Phys. 97: 011101_1-011101_27.

Liu, F., M. Mostoller, V. Milman, M. Chisholm and T. Kaplan. 1995. Electronic and elastic properties of edge dislocations in Si. Phys. Rev. B 51: 17192-17195.

Liu, F., W. Fang and M. G. Lagally. 1997. Effect of strain on structure and morphology of ultrathin Ge films on Si (001). Chem. Rev. 97: 1045-1061.

Liu, J., H. J. Kim, O. Hul'ko, Y. H. Xie, S. Sahni, P. Bandaru, et al. 2004. Ge films grown on Si substrates by molecular-beam epitaxy below 450°C. J. Appl. Phys. 96: 916-918.

Liu, J., L. C. Kimerling and J. Michel. 2012. Monolithic Ge-on-Si lasers for large-scale electronic-photonic integration. Semicond. Sci. Technol. 27: 094006_1-094006_13.

Liu, J. L., C. D. Moore, G. D. U'Ren, Y. H. Luo, Y. Lu, G. Lin, et al. 1999. A surfactant-mediated relaxed $Si_{0.5}Ge_{0.5}$ graded layer with a very low threading dislocation density and smooth surface. Appl. Phys. Lett. 75: 1586-1588.

Liu, J. L., Z. Yang and K. L. Wang. 2006. Sb surfactant-mediated SiGe graded layers for Ge photodiodes integrated on Si. J. Appl. Phys. 99: 024504_1-024504_8.

Lu, G. H., M. Cuma and F. Liu. 2005. First-principles study of strain stabilization of Ge (105) facet on Si (001). Phys. Rev. B 72: 125415_1-125415_6.

Luan, H.-C., D. R. Lim, K. K. Lee, K. M. Chen, J. G. Sandland, K. Wada, et al. 1999. High-quality Ge epilayers on Si with low threading-dislocation densities. Appl. Phys. Lett. 75: 2909-2911.

Masini, G., L. Colace and G. Assanto. 2003. 2.5 Gbit/s polycrystalline germanium-on-silicon photodetector operating from 1.3 to 1.55 μm. Appl. Phys. Lett. 82: 2524-2526.

Michel, J., J. Liu and L. C. Kimerling. 2010. High-performance Ge-on-Si photodetectors. Nat. Photonics 4: 527-534.

Mo, Y. W. and M. G. Lagally. 1991. Scanning tunneling microscopy studies of the growth process of Ge on Si (001). J. Cryst. Growth 111: 876-881.

Mo, Y. W. and M. G. Lagally. 1992. Scanning tunneling microscopy studies of the initial stages of germanium growth on Si (001). Mater. Sci. Eng. B 14: 311-316.

Mochizuki, S., A. Sakai, O. Nakatsuka, H. Kondo, K. Yukawa, K. Izunome, et al. 2007. Strain relaxation of patterned Ge and SiGe layers on Si (001) substrates. Semicond. Sci. Technol. 22: S132-S136.

Okada, H., Y. Fujimoto, K. Endo, K. Hirose and Y. Mori. 2001. Detailed analysis of scanning tunneling microscopy images of the Si (001) reconstructed surface with buckled dimers. Phys. Rev. B 63: 195324_1-195324_7.

Olsen, S. H., A. G. O'Neill, D. J. Norris, A. G. Cullis, N. J. Woods, J. Zhang, et al. 2002. Strained Si/SiGe n-channel MOSFETs: impact of cross-hatching on device performance. Semicond. Sci. Technol. 17: 655-661.

Oshiyama, A. 2002. First-principle calculations for mechanisms of semiconductor epitaxial growth. J. Cryst. Growth 237-239: 1-7.

People, R. and J. C. Bean. 1985. Calculation of critical thickness versus lattice mismatch for Ge_xSi_{1-x}/Si strained-layer heterostructures. Appl. Phys. Lett. 47: 322-324.

Perdew, J. P. and A. Zunger. 1981. Self-interaction correction to density-functional approximations for many-electron systems. Phys. Rev. B 23: 5048-5079.

Perdew, J. P., K. Burke and M. Ernzerhof. 1996. Generalized gradient approximation mode simple. Phys. Rev. Lett. 77: 3865-3868.

Portavoce, A., I. Berbezier and A. Ronda. 2004. Sb-surfactant-mediated growth of Si and Ge nanostructures. Phys. Rev. B 69: 155416_1-155416_8.

Posthuma, N. E., J. Van der Heide, G. Flamand and J. Poortmans. 2007. Emitter formation and contact realization by diffusion for germanium photovoltaic devices. IEEE Trans. Electron Devices 54: 1210-1215.

Sakai, A. and T. Tatsumi. 1994. Ge growth on Si using atomic hydrogen as a surfactant. Appl. Phys. Lett. 64: 52-54.

Sakai, A., T. Tatsumi and K. Aoyama. 1997. Growth of strain-relaxed Ge films on Si (001) surfaces. Appl. Phys. Lett. 71: 3510-3512.

Sakai, A., N. Taoka, O. Nakatsuka and S. Zaima. 2005. Pure-edge dislocation network for strain-relaxed SiGe/Si (001) systems. Appl. Phys. Lett. 86: 221916_1-221916_3.

Schmidt, T., R. Kroger, T. Clausen, J. Falta, A. Janzen, M. Kammler, et al. 2005. Surfactant-mediated epitaxy of Ge on Si (111): beyond the surface. Appl. Phys. Lett. 86: 111910_1-111910_3.

Shang, H., M. Frank, E. P. Gusev, J. O. Chu, S. W. Bedell, K. W. Guarini, et al. 2006. Germanium channel MOSFETs: opportunities and challenges. IBM J. Res. Dev. 50: 377-386.

Sheldon, P., B. G. Yacobi, K. M. Jones and D. J. Dunlavy. 1985. Growth and characterization of GaAs/Ge epilayers grown on Si substrates by molecular beam epitaxy. J. Appl. Phys. 58: 4186-4193.

Shiraki, Y. and A. Sakai. 2005. Fabrication technology of SiGe hetero-structures and their properties. Surf. Sci. Rep. 59: 153-207.

Speck, J. S., M. A. Brewer, G. Beltz, A. E. Romanov and W. Pompe. 1996. Scaling laws for the reduction of threading dislocation densities in homogeneous buffer layers. J. Appl. Phys. 80: 3808-3816.

Takagi, Y., K. Nakatsuji, Y. Yoshimoto and F. Komori. 2007. Superstructure manipulation on a clean Ge (001) surface by carrier injection using an STM. Phys. Rev. B 75: 115304_1-115304_10.

Tersoff, J. and D. R. Hamann. 1983. Theory and application for the scanning tunneling microscope. Phys. Rev. Lett. 50: 1998-2001.

Tersoff, J. and D. R. Hamann. 1985. Theory of the scanning tunneling microscopy. Phys. Rev. B 31: 805-813.

Troullier, N. and J. L. Martins. 1991. Efficient pseudopotentials for plane-wave calculations. Phys. Rev. B 43: 1993-2006.

Varga, K., L. Wang, S. Pantelides and Z. Zhang. 2004. Critical layer thickness in Stranski–Krastanow growth of Ge on Si (0 0 1). Surf. Sci. 562: L225-L230.

Voigtländer, B. 2001. Fundamental processes in Si/Si and Ge/Si epitaxy studied by scanning tunneling microscopy during growth. Surf. Sci. Rep. 43: 127-254.

von Känel, H., M. Kummer, G. Isella, E. Müller and T. Hackbarth. 2002. Very high hole mobilities in modulation-doped Ge quantum wells grown by low-energy plasma enhanced chemical vapor deposition. Appl. Phys. Lett. 80: 2922-2924.

Wietler, T. F., E. Bugiel and K. R. Hofmann. 2005. Surfactant-mediated epitaxy of relaxed low-doped Ge films on Si (001) with low defect densities. Appl. Phys. Lett. 87: 182102-182103.

Wietler, T. F., E. Bugiel and K. R. Hofmann. 2006. Surfactant-mediated epitaxy of high-quality low-doped relaxed germanium films on silicon (001). J. Cryst. Growth 508: 6-9.

Index

L

M

N

O

P

R

S

T

V